Applied Mathematical Sciences

Founding Editors
F. John, New York University, New York, NY, USA
J. P. LaSalle, Brown University, Providence, RI, USA
L. Sirovich, Brown University, Providence, RI, USA

Volume 218

Series Editors

Anthony Bloch, Department of Mathematics, University of Michigan, Ann Arbor, MI, USA
C. L. Epstein, Department of Mathematics, University of Pennsylvania, Philadelphia, PA, USA
Alain Goriely, Department of Mathematics, University of Oxford, Oxford, UK
Leslie Greengard, New York University, New York, NY, USA

Advisory Editors

J. Bell, Center for Computational Sciences and Engineering, Lawrence Berkeley National Laboratory, Berkeley, CA, USA
P. Constantin, Department of Mathematics, Princeton University, Princeton, NJ, USA
R. Durrett, Department of Mathematics, Duke University, Durham, CA, USA
R. Kohn, Courant Institute of Mathematical Sciences, New York University, New York, NY, USA
R. Pego, Department of Mathematical Sciences, Carnegie Mellon University, Pittsburgh, PA, USA
L. Ryzhik, Department of Mathematics, Stanford University, Stanford, CA, USA
A. Singer, Department of Mathematics, Princeton University, Princeton, NJ, USA
A. Stevens, Department of Applied Mathematics, University of Münster, Münster, Germany
S. Wright, Computer Sciences Department, University of Wisconsin, Madison, WI, USA

The mathematization of all sciences, the fading of traditional scientific boundaries, the impact of computer technology, the growing importance of computer modeling and the necessity of scientific planning all create the need both in education and research for books that are introductory to and abreast of these developments. The purpose of this series is to provide such books, suitable for the user of mathematics, the mathematician interested in applications, and the student scientist. In particular, this series will provide an outlet for topics of immediate interest because of the novelty of its treatment of an application or of mathematics being applied or lying close to applications. These books should be accessible to readers versed in mathematics or science and engineering, and will feature a lively tutorial style, a focus on topics of current interest, and present clear exposition of broad appeal. A compliment to the Applied Mathematical Sciences series is the Texts in Applied Mathematics series, which publishes textbooks suitable for advanced undergraduate and beginning graduate courses.

Hildeberto E. Cabral • Lúcia Brandão Dias

Normal Forms and Stability of Hamiltonian Systems

Springer

Hildeberto E. Cabral
Department of Mathematics
Federal University of Pernambuco
Recife, Pernambuco, Brazil

Lúcia Brandão Dias
Department of Mathematics and Statistics
Federal University of Rondônia
Ji-Paraná, Rondônia, Brazil

ISSN 0066-5452 ISSN 2196-968X (electronic)
Applied Mathematical Sciences
ISBN 978-3-031-33048-3 ISBN 978-3-031-33046-9 (eBook)
https://doi.org/10.1007/978-3-031-33046-9

This Springer imprint is published by the registered company Springer Nature Switzerland AG
The registered company address is: Gewerbestrasse 11, 6330 Cham, Switzerland

O LAMENTO DAS COISAS

Augusto dos Anjos (1914)

Triste, a escutar pancada por pancada,
A sucessividade dos segundos,
Ouço, em sons subterrâneos, do Orbe oriundos,
O chôro da Energia abandonada!

É a dor da Fôrça desaproveitada
- O cantochão dos dínamos profundos,
Que, podendo mover milhões de mundos,
Jazem ainda na estática do Nada!

É o soluço da forma ainda imprecisa . . .
Da transcendência que se não realiza . . .
Da luz que não chegou a ser lampejo . . .

E é em suma, o subconsciente ai formidando
Da Natureza que parou, chorando,
No rudimentarismo do Desejo!

From Hildeberto,
To my wife Elisabete and my sons and daughters Hermano, Elane, Henrique, and Érica.

"Look deep into nature, and then you will understand everything better."

Albert Einstein

From Lúcia,
To my husband Alcides and my sons Davi and Daniel.

"For from him and through him and to him are all things."

Rm 11:36a

Foreword

One of the leading research centers in the theory of Hamiltonian dynamics is found in the city Recife on the Atlantic coast of Brazil at the Federal University of Pernambuco. Researchers from Europe, Asia, and the Americas come to spend a week, a month, or a year to discuss their current research and future plans with the faculty and graduate students. Often, the researcher presents a seminar on his/her research and then writes up the lectures for publication. Indeed, the collection of research articles found in *Classical and Celestial Mechanics, The Recife Lectures* Princeton University Press edited by Hildeberto E. Cabral and Florin Diacu is a direct result of this seminar series.

In the spring of 1995, I visited the mathematics department of the University of Pernambuco and gave a seminar to audience of faculty and advanced graduate students while also enjoying the sunny beaches of Recife. I subsequently wrote up the lectures and published them in the Springer Lecture Notes series.

The driving force for the last 40 years at this research center is Professor Hildeberto E. Cabral. He taught graduation courses and seminars in Hamiltonian dynamical systems, which attracted active graduate students who went on to receive a PhD and then continue their research. He invited researchers from other world centers of research to come to Recife to join in the research environment.

His own research includes many different aspects of the theory, including the topology of the integral manifolds of the three-body problem, various stability and instability criteria based on the Kolmogorov–Arnold–Moser theory and Lyapunov's and Chetaev's methods, normalization by Hori–Deprit procedure, and the existence and stability of periodic solutions.

This book is a development of the theory and applications of Hamiltonian systems with a special emphasis on the stability of autonomous and periodic systems. All the background needed is presented from the beginnings with the Hamiltonian formulation of the laws of motion to the application of the Krein–Gelfand–Lidskii theory of strongly stable systems.

Kenneth Meyer

Preface

In the first part of this Preface we give a short account of the historical evolution of the theory of Hamiltonian systems. We do it in a descriptive way without formulas, with technical terms defined informally; for some of them even a definition is not given, as in the case of the Poisson bracket, which is mentioned in some places. We believe that this is not a difficulty for an overall appraisal of what is being exposed. Many actors in the development of the theory are mentioned and references to them in the Bibliography at the end of the book can be easily identified by the reader.

In the second part of the Preface we describe the content of the main body of the book with some considerations about the subjects studied in each chapter.

Hamiltonian systems form an important class of ordinary differential equations. Its rich theory has been developed since the nineteenth century with the work of W. R. Hamilton, who passed from the Lagrangian formulation of Analytical Mechanics in the phase space of positions and velocities to the phase space of positions and momenta.

Through the work of L. Euler and J. L. Lagrange on variational methods in mechanics, this science, under the restriction of forces deriving from a potential, was freed from the concept of force used by Isaac Newton and passed to relying on minimal principles such as the principle of least action formulated in terms of functions of positions and velocities with these physical concepts determined by generalized coordinates. This approach gives rise to the Euler–Lagrange equations of the variational problem. These equations are of second order in configuration space, the space of positions, and when viewed as a system of first order in the phase space of positions and velocities, they lack symmetry. The introduction of momenta in place of velocities makes the new first-order equations known as the Hamiltonian equations of motion exhibit a nice symmetry. The phase space in this case is formed by the generalized coordinates of positions and momenta. The connection between the Lagrangian and the Hamiltonian formulations is provided by the Legendre transformation.

The problems formulated in the Hamiltonian context require us to search for good coordinate systems, which, simplifying the Hamiltonian function in terms of which the equations are written, make these equations more tractable. Although the Lagrangian equations are invariant under any coordinate transformation, the Hamiltonian equations are invariant only under canonical transformations, which are those for which the Jacobian matrix is symplectic. These can be generated by functions that involve the old and new variables and are determined in specific ways. Using the original Hamiltonian, the generating function, and the defining relations of the sought-after canonical transformation, one gets an equation for a most simplified new Hamiltonian. This is a partial differential equation known as the Hamilton–Jacobi equation. When a nondegenerate particular solution of this equation is known, the equations of motion can be integrated by quadratures.

An integrable system is a Hamiltonian system that can be solved by quadratures. A symmetry allows us to find a first integral of the equations of motion, that is a nonconstant function on phase space that remains constant on each trajectory of the system, a conservation law, such as energy and angular momentum. The energy in a conservative mechanical system is the Hamiltonian function of the system. The first integral of the system is characterized by the annihilation of the Poisson bracket between it and the Hamiltonian function. If it is possible to find as many independent first integrals as the dimension of the configuration space and they are pairwise in involution, that is, the Poisson bracket of any two of them is zero, then one has an integrable system. This is the Arnold–Liouville theorem.

Through his studies on periodic orbits and on the stability of equilibria of conservative systems Lyapunov also made a contribution to the development of Hamiltonian systems; his theorem on the existence of a family of periodic orbits emanating from an equilibrium when a certain nonresonance condition holds is a nice example.

By the end of the nineteenth century Poincaré developed further the Hamiltonian system theory. Very much interested in understanding the dynamics of the restricted three-body problem he was led to a broader program on periodic orbits and integrability in Hamiltonian systems. His well-known book in three volumes on the qualitative methods of celestial mechanics contains a wealth of methods on Hamiltonian mechanics. In Sect. 3 of the first volume he formulated what he called the general problem of dynamics, the problem of studying the Hamiltonian systems, which are perturbations of integrable Hamiltonian systems.

Following Poincaré on his studies of dynamics G. D. Birkhoff advanced further the theory of canonical transformations and described a technique to write the Hamiltonian function in normal form, a process that disregards the question of convergence of the formal power series involved. Interested not only in understanding the dynamics of the restricted three-body problem but also the qualitative dynamical behavior of trajectories of mechanical systems, Birkhoff poses in his book on dynamical systems the problem of describing the topology of the integral manifolds of the n-body problem, one of the main problems in Hamiltonian dynamics.

Birkhoff's normal form requires that no resonance relation exists among the frequencies of the quadratic part of the Hamiltonian. Under the presence of resonance relations a modified normal form was first considered by F. G. Gustavson, aiming at constructing formal first integrals of a Hamiltonian system in the neighborhood of an equilibrium. At about the same time G. Hori used a theorem of S. Lie on canonical transformations to describe a new way of simplifying a Hamiltonian function given as a power series in a small parameter. A. Deprit further developed this theory describing an inductive formula that is very efficient for calculations with a computer. This technique of constructing the resonant normal form of a Hamiltonian function is now known as the Deprit–Hori method. It also includes the nonresonant case of the Birkhoff normal form.

These processes of obtaining the normal form use formal series expansions and disregard the question of convergence. C. L. Siegel proved that the series involved in the normalization process of Birkhoff are in general divergent, despite V. I. Arnold, using the Birkhoff normal form up to fourth order, obtaining a stability criterion for the equilibrium of a nonlinear two-degree of freedom Hamiltonian system. Arnold's result can be proved using a theorem of J. K. Moser on the existence of closed invariant curves for twist mappings of the plane. The works of Arnold and Moser came under the trend of the communication of Kolmogorov in 1954 at the International Congress of Mathematicians regarding the general problem of dynamics defined by Poincaré. The acronym KAM, from Kolmogorov–Arnold–Moser, appearing in the expression KAM theory, refers to the techniques and results related to the study of the existence of the invariant tori described by Kolmogorov and the dynamics of the system concerning them.

Stability in time-dependent Hamiltonian systems, just as in the autonomous case of a higher degree of freedom, is a rather difficult problem, so for these systems attention is restricted to the stability of the linearized system around the equilibrium. The stability of periodic linear Hamiltonian systems received a great deal of attention from the Russian mathematicians in the 1950s, mainly in the works of Krein, Gelfand, and Lidskii. The book by Yakubovich and Starzhinskii is a good reference for this subject.

In his studies of the restricted three-body problem Poincaré also developed much work on topology and on the geometrical study of ordinary differential equations. Together with Poincaré, the contribution of Birkhoff on the geometrical studies of dynamics gave rise to the modern theory of Dynamical Systems. This geometrization approach to differential equations was extended to Analytical Mechanics and, in particular, to the theory of Hamiltonian systems with consideration of the methods of differential topology leading to what is now known as Geometrical Mechanics. In this geometrical setting the phase space is a manifold on which a nondegenerate closed two form, a symplectic form, provides the necessary structure for looking at the Hamiltonian system in an intrinsic way as the gradient of a smooth function with respect to the symplectic form. This allows a global approach to Hamiltonian systems in which much can be done in a context free of coordinates. The work of S. Smale on Topology and Mechanics, of K. R. Meyer, and also of J. Marsden with A. Weinstein on symplectic reduction using symmetries and integrals are along this trend.

The books by Abraham and Marsden and by Souriau are the first representatives of this geometrization of Classical Mechanics. The research of the geometer Élie Cartan on integral invariants, extending the work of Poincaré to the context of time-dependent Hamiltonians, was later presented in the book [15], which, although not using the terminology of smooth manifolds for the configuration space, is certainly one that deserves to be included in this geometrization process.

For local questions such as the study of the Lyapunov stability of an equilibrium, the local approach still provides the main tools.

In this book we are interested in local questions, mainly in the question of the stability of an equilibrium and the theory of Hamiltonian systems will be developed with this goal.

In the short Chap. 1 we provide a brief review of some results of Linear Algebra and Advanced Calculus that will be used in the main body of the book, at the same time as fixing the notation of the intrinsic calculus.

Besides this introductory chapter, the book has seven other chapters.

Chapter 2 deals with the basic theory of Hamiltonian systems that pervades throughout the book.

Chapter 3 is dedicated to the study of normal forms of Hamiltonian systems, including the Birkhoff normal form and the Deprit–Hori scheme for resonant normal forms.

Chapter 4 deals with normalizing Hamiltonian matrices in the nondegenerate case: distinct eigenvalues or multiple purely imaginary eigenvalues with diagonalizable matrix.

Chapter 5 treats the normalization of Hamiltonian matrices in full generality. We construct the normalizing symplectic bases and give the normal forms of Hamiltonian matrices. These normal forms will be used in Chap. 7 in the study of the regions of parametric resonance.

Chapter 6 is about the stability of equilibria of nonlinear systems. We prove two stability lemmas and use them to prove the Arnold theorem in a version that is also valid when resonance terms are present in the normal form.

Chapter 7 is on the stability of linear Hamiltonian systems. We prove the Krein–Gelfand–Lidskii theorem on strongly stable systems and make an exposition of the Gelfand–Lidskii theory on the topological structure of the space of strongly stable systems.

Chapter 8 deals with systems that depend on parameters. The goal is to study the regions of parametric resonance, determining a decomposition of the parameter space into a succession of stable and unstable regions. The Deprit-Hori formula of Chap. 2 and the normal forms seen in Chap. 4 are needed in this chapter.

This book reflects the experience of the authors with the subject of Hamiltonian systems through seminars, courses, and research work. It can be used both as a textbook for courses at the graduate level and as a reference book for researchers who need these techniques in dealing with the analysis of stability of equilibria in problems of classical mechanics.

An introductory chapter on Linear Algebra and Advanced Calculus is included, which recalls the basic facts of these subjects and fixes the notation of the intrinsic calculus of vector functions. This chapter can be skipped by those who have this basic knowledge or can be consulted every time that it is cited in the text.

Recife, Brazil Hildeberto E. Cabral
Ji-Paraná, Brazil Lúcia Brandão Dias

Acknowledgement

The authors warmly thank Kenneth Meyer for his suggestions and continued support throughout the preparation of the book. His encouragement and advice were stimuli in this process. Also, thanks are due to Dieter Schmidt whose suggestions much improved the presentation of the book. We thank Alain Albouy for his relevant considerations on Chap. 2 and discussions on historical matters in the Preface. We thank Carlos Paiva for discussions on the Krein–Gelfand–Lidskii theory. The authors are also thankful to the entire production team especially Mr. Robinson dos Santos, Mr. Vignesh Viswanathan and Ms. A. Meenahkumary for their support.

Contents

Chapter 1
Preliminaries on Advanced Calculus

1.1 Introduction

In this book we use the approach of the intrinsic differential calculus, which frequently saves time, space and gives clarity to the presentation. Therefore, we recall some facts from Advanced Calculus. For more details the reader can consult any text on Advanced Calculus, for instance the books by Goffman [25], Loomis-Sternberg [37] or Rudin [52].

1.2 Differential Calculus in n-Space

We denote by $\mathbf{R}^n$ the set of n-tuples of real numbers $\mathbf{x} = (x_1, \ldots, x_n)$. This set becomes an n-dimensional real vector space with the sum and scalar multiplication defined by

$$\mathbf{x} + \mathbf{y} = (x_1 + y_1, \ldots, x_n + y_n) \quad \text{and} \quad \lambda\mathbf{x} = (\lambda x_1, \ldots, \lambda x_n). \tag{1.1}$$

Among the inner products and norms in $\mathbf{R}^n$, we will work specially with the Euclidean inner product and the corresponding norm defined by

$$\langle \mathbf{x}, \mathbf{y} \rangle = x_1 y_1, \ldots, x_n y_n \quad \text{and} \quad \|\mathbf{x}\| = \sqrt{\langle \mathbf{x}, \mathbf{x} \rangle} = \sqrt{x_1^2 + \ldots + x_n^2}. \tag{1.2}$$

Among the bases of $\mathbf{R}^n$ we will use most frequently the canonical basis formed by the vectors

$$\mathbf{e}_1 = (1, 0, \ldots, 0), \quad \mathbf{e}_2 = (0, 1, \ldots, 0), \quad \ldots, \quad \mathbf{e}_n = (0, 0, \ldots, 1). \tag{1.3}$$

H. E. Cabral, L. Brandão Dias, *Normal Forms and Stability of Hamiltonian Systems*, Applied Mathematical Sciences 218,
https://doi.org/10.1007/978-3-031-33046-9_1

An important identity involving the inner product is the one that relates a real square matrix A to its transpose A^T, namely

$$\langle A\mathbf{x}, \mathbf{y}\rangle = \langle \mathbf{x}, A^T\mathbf{y}\rangle. \tag{1.4}$$

Let $\mathbf{f} : U \to \mathbf{R}^m$ be a mapping defined on an open subset of the Euclidean space $\mathbf{R}^n$. We say that $\mathbf{f}$ is differentiable at the point $\mathbf{x} \in U$ if there exists a linear mapping $L : \mathbf{R}^n \to \mathbf{R}^m$ such that

$$\mathbf{f}(\mathbf{x} + \boldsymbol{\xi}) = \mathbf{f}(\mathbf{x}) + L \cdot \boldsymbol{\xi} + R(\mathbf{x}, \boldsymbol{\xi}) \quad \text{with} \quad \lim_{\|\boldsymbol{\xi}\| \to 0} \frac{R(\mathbf{x}, \boldsymbol{\xi})}{\|\boldsymbol{\xi}\|} = 0. \tag{1.5}$$

If $\mathbf{f}$ is differentiable at $\mathbf{x}$ the mapping L is unique and is called the derivative of $\mathbf{f}$ at the point $\mathbf{x}$, denoted by $D\mathbf{f}(\mathbf{x})$.

Given differentiable mappings $\mathbf{f} : U \to V$ and $\mathbf{g} : V \to W$, where $U \subset \mathbf{R}^m$, $V \subset \mathbf{R}^n$, $W \subset \mathbf{R}^p$ the composite mapping $\mathbf{g} \circ \mathbf{f} : U \to \mathbf{R}^p$ defined by $(\mathbf{g} \circ \mathbf{f})(\mathbf{x}) = \mathbf{g}(\mathbf{f}(\mathbf{x}))$ is differentiable and its derivative is given by the chain rule

$$D(\mathbf{g} \circ \mathbf{f})(\mathbf{x}) = D\mathbf{g}(\mathbf{f}(\mathbf{x})) \circ D\mathbf{f}(\mathbf{x}), \tag{1.6}$$

that is,

$$D(\mathbf{g} \circ \mathbf{f})(\mathbf{x}) \cdot \boldsymbol{\xi} = D\mathbf{g}(\mathbf{f}(\mathbf{x})) \cdot (D\mathbf{f}(\mathbf{x}) \cdot \boldsymbol{\xi}), \quad \boldsymbol{\xi} \in \mathbf{R}^m.$$

For a differentiable real-valued function $f(\mathbf{x})$, $\mathbf{x} \in U \subset \mathbf{R}^n$ we use the notation $df(\mathbf{a})$ for the derivative instead of $Df(\mathbf{a})$. Sometimes we use the symbol $f_{\mathbf{x}}(\mathbf{a})$ for $df(\mathbf{a})$.

The partial derivative of f with respect to the variable x_i at the point $\mathbf{a} = (a_1, \ldots, a_n)$ is defined as the limit $\dfrac{\partial f}{\partial x_i}(\mathbf{a}) = \lim\limits_{s \to 0} \dfrac{f(\mathbf{a} + s\mathbf{e}_i) - f(\mathbf{a})}{s}$ if it exists, where $\mathbf{e}_i$ is the ith canonical basis vector of $\mathbf{R}^n$. If $f(\mathbf{x})$ is differentiable, then $\dfrac{\partial f}{\partial x_i}(\mathbf{a}) = d\mathbf{f}(\mathbf{a}) \cdot \mathbf{e}_i$.

In the definition of derivative (1.5) the target space $\mathbf{R}^n$ can be replaced by any normed finite-dimensional vector space E. We meet this case below when we talk about the derivative of the mapping $\mathbf{x} \mapsto df(\mathbf{x}) \in \mathbf{R}^{n*}$, the dual space of $\mathbf{R}^n$. By the way, we can do that also for the source space $\mathbf{R}^m$ or for both the target and source spaces. We can also take infinite-dimensional normed vector spaces, but here we must be more careful, requiring the mapping L to be a continuous linear mapping and the normed spaces to be complete (Banach spaces).

Example 1.1

(1) A linear mapping $\mathbf{f} : E \to F$ is differentiable and its derivative is itself,

$$D\mathbf{f}(\mathbf{x}) \cdot \boldsymbol{\xi} = \mathbf{f}(\boldsymbol{\xi});$$

(2) A bilinear mapping $\mathbf{f} : E \times F \to G$ is differentiable and its derivative is given by

$$D\mathbf{f}(\mathbf{x}, \mathbf{y}) \cdot (\boldsymbol{\xi}, \boldsymbol{\eta}) = \mathbf{f}(\boldsymbol{\xi}, \mathbf{y}) + \mathbf{f}(\mathbf{x}, \boldsymbol{\eta}).$$

By the symbol of partial derivative $\dfrac{\partial f}{\partial \mathbf{x}}(\mathbf{a})$ we may denote the gradient vector $\nabla f(\mathbf{a})$ and sometimes the linear mapping $df(\mathbf{x})$. We recall that the gradient and derivative are related by the identity

$$\langle \nabla f(\mathbf{a}), \boldsymbol{\xi} \rangle = df(\mathbf{a}) \cdot \boldsymbol{\xi}, \quad \text{for any} \quad \boldsymbol{\xi} \in \mathbf{R}^n. \tag{1.7}$$

This equality can be used as an intrinsic definition of the gradient vector. Using it we can express the gradient vector in any basis of $\mathbf{R}^n$. The canonical basis gives the usual expression in terms of the partial derivatives of the function

$$\nabla f(\mathbf{x}) = \Big(\frac{\partial f}{\partial x_1}(\mathbf{x}), \ldots, \frac{\partial f}{\partial x_n}(\mathbf{x})\Big).$$

If the mapping $\mathbf{x} \mapsto df(\mathbf{x}) \in \mathbf{R}^{n*}$ has a derivative, $ddf(\mathbf{x})$ we denote it by $d^2 f(\mathbf{x})$. We recall that this is a bilinear function in $\mathbf{R}^n \times \mathbf{R}^n$ and that if the function is of class C^2, that is, the mapping $\mathbf{x} \mapsto d^2 f(\mathbf{x})$ is continuous, this bilinear function is symmetric, $d^2 f(\mathbf{x})(\boldsymbol{\xi}, \boldsymbol{\eta}) = d^2 f(\mathbf{x})(\boldsymbol{\eta}, \boldsymbol{\xi})$ (Schwarz theorem) .

The Hessian matrix of a real-valued twice differentiable function at the point $\mathbf{x}$ is the matrix of the bilinear form $d^2 f(\mathbf{x})$ in the canonical basis of $\mathbf{R}^n$, that is,

$$G(\mathbf{x}) = \begin{bmatrix} \frac{\partial^2 f}{\partial x_1^2} & \cdots & \frac{\partial^2 f}{\partial x_1 \partial x_n} \\ \vdots & & \vdots \\ \frac{\partial^2 f}{\partial x_n \partial x_1} & \cdots & \frac{\partial^2 f}{\partial x_n^2} \end{bmatrix}. \tag{1.8}$$

The Hessian determinant or simply the Hessian of f at $\mathbf{x}$ is the determinant of $G(\mathbf{x})$. The function f is of class C^2 if and only if the second-order partial derivatives are continuous. A real-valued function that has continuous partial derivatives of all orders is said to be a function of class C^∞. An analytic function is a C^∞ function whose Taylor series converges.

Let $\phi : U \to V$, $\mathbf{x} = \phi(\mathbf{u})$, be a differentiable change of variables between open sets of $\mathbf{R}^m$. Given the differentiable real function $f(\mathbf{x})$, let $g(\mathbf{u}) = f(\phi(\mathbf{u}))$. Then we have the following result.

Proposition 1.1 *The relation between the gradients of f and g is given by the equation*

$$\nabla g(\mathbf{u}) = \mathbf{x}_u^T \nabla f(\mathbf{x}), \tag{1.9}$$

where $\mathbf{x}_u = D\phi(\mathbf{u})$.

Proof By the chain rule (1.6) we have for all $\boldsymbol{\xi}$, $dg(\mathbf{u}) \cdot \boldsymbol{\xi} = df(\mathbf{x}) \cdot \big(D\phi(\mathbf{u}) \cdot \boldsymbol{\xi}\big)$; hence, using (1.7) we pass to the equation

$$\langle \nabla g(\mathbf{u}), \boldsymbol{\xi} \rangle = \langle \nabla f(\mathbf{x}), D\phi(\mathbf{u}) \cdot \boldsymbol{\xi} \rangle = \langle D\phi(\mathbf{u})^T \nabla f(\mathbf{x}), \boldsymbol{\xi} \rangle.$$

Since this equality holds for all $\boldsymbol{\xi}$ this proves (1.9). □

Proposition 1.2 *Let* $\mathbf{x} = \phi(\mathbf{u})$ *be a differentiable change of variables and given the path* $\mathbf{u} = \mathbf{u}(t)$ *consider the image path* $\mathbf{x} = \phi(\mathbf{u}(t))$. *Then, we have*

$$\dot{\mathbf{x}} = \mathbf{x}_u \, \dot{\mathbf{u}}, \tag{1.10}$$

the dot over the letters denoting a derivative with respect to the "time" t.

Proof That is just the chain rule, $\dfrac{d\mathbf{x}}{dt} = D\phi(\mathbf{u}(t))\dfrac{d\mathbf{u}}{dt}$. □

For a real-valued function $f(\mathbf{x}, \mathbf{y})$ the partial derivatives $\dfrac{\partial f}{\partial \mathbf{x}}(\mathbf{x}, \mathbf{y})$, $\dfrac{\partial f}{\partial \mathbf{y}}(\mathbf{x}, \mathbf{y})$ are the derivatives of the partial mappings $\mathbf{x} \mapsto f(\mathbf{x}, \mathbf{y})$, $\mathbf{y}$ fixed and $\mathbf{y} \mapsto f(\mathbf{x}, \mathbf{y})$, $\mathbf{x}$ fixed. We will also view $f_{\mathbf{x}} = \dfrac{\partial f}{\partial \mathbf{x}}(\mathbf{x}, \mathbf{y})$ and $f_{\mathbf{y}} = \dfrac{\partial f}{\partial \mathbf{y}}(\mathbf{x}, \mathbf{y})$ as the gradients of f with respect to $\mathbf{x}$ and $\mathbf{y}$ respectively.

If $\mathbf{x} = \mathbf{x}(t)$ and $\mathbf{y} = \mathbf{y}(t)$ are differentiable, then we have the equality

$$\frac{d}{dt} f(\mathbf{x}(t), \mathbf{y}(t)) = \big\langle f_{\mathbf{x}}, \dot{\mathbf{x}}(t) \big\rangle + \big\langle f_{\mathbf{y}}, \dot{\mathbf{y}}(t) \big\rangle, \tag{1.11}$$

where $f_{\mathbf{x}} = f_{\mathbf{x}}(\mathbf{x}(t), \mathbf{y}(t))$ and $f_{\mathbf{y}} = f_{\mathbf{y}}(\mathbf{x}(t), \mathbf{y}(t))$.

For a mapping $\mathbf{f} = (f_1, \ldots, f_m)$ from an open set U of $\mathbf{R}^n$ to $\mathbf{R}^m$, the Jacobian matrix is the matrix of the linear mapping $D\mathbf{f}(\mathbf{x})$ in the canonical bases of theses spaces, namely

$$J(\mathbf{x}) = \begin{bmatrix} \frac{\partial f_1}{\partial x_1} & \cdots & \frac{\partial f_1}{\partial x_n} \\ \vdots & & \vdots \\ \frac{\partial f_m}{\partial x_1} & \cdots & \frac{\partial f_m}{\partial x_n} \end{bmatrix}. \tag{1.12}$$

1.3 Exponential of a Linear Operator

We denote by $\mathcal{M}_n(\mathbf{R})$ the space of real square matrices of order n and by $\mathcal{L}(\mathbf{R}^n, \mathbf{R}^n)$ the space of linear operators from $\mathbf{R}^n$ to $\mathbf{R}^n$.

Let E denote one of the normed spaces $\mathcal{M}_n(\mathbf{R})$ or $\mathcal{L}(\mathbf{R}^n, \mathbf{R}^n)$. An important convergent series in E is the one that defines the *exponential* of a square matrix or of a linear operator.

Proposition 1.3 *Given $T \in E$, the sequence in E defined by*

$$S_j(T) = \sum_{k=1}^{j} \frac{1}{k!} T^k \tag{1.13}$$

is convergent; its limit, denoted by e^T, depends continuously on T.

Proof For $i < j$, we have by the triangle inequality

$$||S_j(T) - S_i(T)|| \le \sum_{k=i+1}^{j} \frac{1}{k!} ||T||^k. \tag{1.14}$$

Therefore, $\big(S_j(T)\big)$ is a Cauchy sequence in E and since this space is complete, the sequence converges. Denote its limit by e^T. Now, given $T_0 \in E$, there is a constant $a > 0$ such that for all T in the closed unit ball B centered at T_0, we have $||T|| \le a$. From (1.14) we see that $\big(S_j(T)\big)$ is a uniform Cauchy sequence in B, so this sequence converges uniformly to the operator e^T, which therefore depends continually on T, since each $S_j(T)$ is continuous in T. □

Definition 1.4 The linear operator

$$\mathrm{e}^T = I + \frac{1}{1!}T + \frac{1}{2!}T^2 + \frac{1}{3!}T^3 + \ldots, \tag{1.15}$$

is called the *exponential* of T.

An important property of this operator is given in the following proposition for the proof of which we refer to [30], §3 Chap. 5.

Proposition 1.5 *If $A, B \in E$ commute, then $e^{A+B} = e^A e^B$.*

This property is used in the proof of the following result.

Proposition 1.6 *For a matrix (or operator) $A \in E$ we have*

$$\frac{d}{dt} e^{tA} = A e^{tA}. \tag{1.16}$$

Proof Since tA commutes with hA we have $\mathrm{e}^{(t+h)A} = \mathrm{e}^{hA+tA} = \mathrm{e}^{hA}\mathrm{e}^{tA}$; hence,

$$\frac{1}{h}\big[\mathrm{e}^{(t+h)A} - \mathrm{e}^{tA}\big] = \frac{1}{h}\big[\mathrm{e}^{hA} - I\big]\mathrm{e}^{tA}.$$

Using (1.15) and letting $h \to 0$ results in the formula (1.16) for the derivative of e^{tA}. □

We consider now some examples that will be used in Chap. 8.

Definition 1.7 The matrix $J = \begin{bmatrix} O & I \\ -I & O \end{bmatrix}$ where O is the zero matrix and I is the identity matrix, both of order n, is called the standard symplectic matrix of order $2n$.

Example 1.2 For the matrix $J = \begin{bmatrix} O & I \\ -I & O \end{bmatrix}$ we have

$$e^{\alpha J} = \begin{bmatrix} \cos\alpha I & \sin\alpha I \\ -\sin\alpha I & \cos\alpha I \end{bmatrix}.$$

We use the symbol I for the identity matrix, leaving its order implicit by the context. We compute

$$J^2 = -I, \quad J^3 = -J, \quad J^4 = I, \quad J^5 = J, \ldots,$$

so we get at once

$$e^{\alpha J} = \left(1 - \frac{\alpha^2}{2!} + \frac{\alpha^4}{4!} - \ldots\right) I + \left(\frac{\alpha}{1!} - \frac{\alpha^3}{3!} + \frac{\alpha^5}{5!} \ldots\right) J = \cos\alpha I + \sin\alpha J,$$

which gives the mentioned expression of $e^{\alpha J}$.

Example 1.3 For $B = \begin{bmatrix} 0 & 1 & 0 & 0 \\ -1 & 0 & 0 & 0 \\ 0 & 0 & 0 & 1 \\ 0 & 0 & -1 & 0 \end{bmatrix}$, we have $e^{\beta B} = \begin{bmatrix} R_\beta & O \\ O & R_\beta \end{bmatrix}$,

where $R_\beta = \begin{bmatrix} \cos\beta & \sin\beta \\ -\sin\beta & \cos\beta \end{bmatrix}$ is the counter-clockwise rotation by the angle β.

Let $J_2 = \begin{bmatrix} 0 & 1 \\ -1 & 0 \end{bmatrix}$. Then, Example 1.2 shows that R_β is the exponential of βJ_2. As B is block diagonal we get the above expression for $e^{\beta B}$.

Example 1.4 For the matrix $C = \begin{bmatrix} 0 & 1 & 0 \\ -2 & 0 & 2 \\ 0 & -1 & 0 \end{bmatrix}$ we have

$$e^{\gamma C} = \frac{1}{2} \begin{bmatrix} 1 + \cos 2\gamma & \sin 2\gamma & 1 - \cos 2\gamma \\ -2\sin 2\gamma & 2\cos 2\gamma & 2\sin 2\gamma \\ 1 - \cos 2\gamma & -\sin 2\gamma & 1 + \cos 2\gamma \end{bmatrix}. \tag{1.17}$$

Indeed, let $\tilde{C} = \begin{bmatrix} 1 & 0 & -1 \\ 0 & 2 & 0 \\ -1 & 0 & 1 \end{bmatrix}$. Then, $\tilde{C}C = 2C$ and since $C^2 = -2\tilde{C}$ we readily compute

$$C^3 = -2^2 C, \quad C^4 = 2^3 \tilde{C}, \quad C^5 = 2^4 C, \quad C^6 = -2^5 \tilde{C}, \ldots$$

so we get

$$e^{\gamma C} = I + \Big(\frac{\gamma}{1!} - \frac{(2\gamma)^3}{2 \cdot 3!} + \frac{(2\gamma)^5}{2 \cdot 5!} - \ldots \Big) C + \Big(- \frac{(2\gamma)^2}{2 \cdot 2!} + \frac{(2\gamma)^4}{2 \cdot 4!} - \frac{(2\gamma)^6}{2 \cdot 6!} + \ldots \Big) \tilde{C},$$

from which it follows that

$$e^{\gamma C} = \Big(I - \frac{1}{2}\tilde{C} \Big) + \frac{1}{2} \sin 2\gamma\, C + \frac{1}{2} \cos 2\gamma\, \tilde{C},$$

which gives the matrix (1.17).

Example 1.5 The symplectic matrix $J = J_4$ and the matrix B of Example 1.3 commute, so we have

$$e^{\alpha J + \beta B} = e^{\alpha J} e^{\beta B} = \begin{bmatrix} \cos \alpha R_\beta & \sin \alpha R_\beta \\ -\sin \alpha R_\beta & \cos \alpha R_\beta \end{bmatrix},$$

where R_β is the counter-clockwise rotation by the angle β.

Example 1.6 For the matrices $E_1 = \begin{bmatrix} 0 & 0 & 0 \\ a & 0 & 0 \\ 0 & b & 0 \end{bmatrix}$ and $E_2 = \begin{bmatrix} 0 & a & 0 \\ 0 & 0 & b \\ 0 & 0 & 0 \end{bmatrix}$, we have

$$e^{xE_1} = \begin{bmatrix} 1 & 0 & 0 \\ ax & 1 & 0 \\ abx^2 & bx & 1 \end{bmatrix} \quad \text{and} \quad e^{yE_2} = \begin{bmatrix} 1 & ay & aby^2 \\ 0 & 1 & by \\ 0 & 0 & 1 \end{bmatrix}.$$

We compute $E_1^2 = \begin{bmatrix} 0 & 0 & 0 \\ 0 & 0 & 0 \\ ab & 0 & 0 \end{bmatrix}$ and $E_1^3 = 0$, so $e^{xE_1} = I + xE_1 + \frac{1}{2}x^2 E_1^2$, which gives the expression for e^{xE_1}. Computations for E_2 are similar.

Example 1.7 For the matrices

$$F_1 = \begin{bmatrix} 0\,0\,0\,0 \\ 0\,0\,0\,0 \\ 1\,0\,0\,0 \\ 0\,1\,0\,0 \end{bmatrix} \quad \text{and} \quad F_2 = \begin{bmatrix} 0\,0\,0\,0 \\ 1\,0\,0\,0 \\ 0\,0\,0\,0 \\ 0\,0\,1\,0 \end{bmatrix},$$

we have

$$\mathrm{e}^{xF_1} = \begin{bmatrix} 1\;0\;0\;0 \\ 0\;1\;0\;0 \\ x\;0\;1\;0 \\ 0\;x\;0\;1 \end{bmatrix} \quad \text{and} \quad \mathrm{e}^{yF_2} = \begin{bmatrix} 1\;0\;0\;0 \\ y\;1\;0\;0 \\ 0\;0\;1\;0 \\ 0\;0\;y\;1 \end{bmatrix}. \tag{1.18}$$

Since $F_1^2 = 0$ and $F_2^2 = 0$ we have $\mathrm{e}^{xF_1} = I + xF_1$ and $\mathrm{e}^{yF_2} = I + yF_2$, which give the matrices (1.18).

We check that $F_1F_2 = F_2F_1$, so we have $\mathrm{e}^{xF_1+yF_2} = \mathrm{e}^{xF_1}\mathrm{e}^{yF_2}$. Since $\mathrm{e}^{xF_1} = \begin{bmatrix} I & O \\ xI & I \end{bmatrix}$ and $\mathrm{e}^{yF_2} = \begin{bmatrix} I_y & O \\ O & I_y \end{bmatrix}$, where I is the 2×2 identity matrix and $I_y = \begin{bmatrix} 1 & 0 \\ y & 1 \end{bmatrix}$, we have

$$\mathrm{e}^{xF_1+yF_2} = \begin{bmatrix} I_y & O \\ xI_y & I_y \end{bmatrix} = \begin{bmatrix} 1 & 0 & 0 & 0 \\ y & 1 & 0 & 0 \\ x & 0 & 1 & 0 \\ xy & x & y & 1 \end{bmatrix}. \tag{1.19}$$

1.4 Formula of Variation of Parameters

In this section we consider the solution of linear differential equations with constant coefficients. Let A be a constant real $n \times n$ matrix.

Proposition 1.8 *The solution of the homogeneous linear system*

$$\dot{\mathbf{x}} = A\mathbf{x}, \quad \mathbf{x}(0) = \boldsymbol{\xi}, \tag{1.20}$$

is given by

$$\mathbf{x}(t) = e^{tA}\boldsymbol{\xi}. \tag{1.21}$$

Proof This follows immediately from Proposition 1.6. □

Now we consider a nonhomogeneous linear system with constant coefficients,

$$\dot{\mathbf{x}} = A\mathbf{x} + \mathbf{b}, \qquad \mathbf{x}(0) = \boldsymbol{\xi} \tag{1.22}$$

where A is a constant real square matrix and $\mathbf{b} = \mathbf{b}(t)$.

To solve this equation we use the method of variation of parameters.

Proposition 1.9 *The solution of the initial value problem (1.22) is given by the formula*

$$\mathbf{x}(t) = e^{tA}\boldsymbol{\xi} + e^{tA}\int_0^t e^{-sA}\mathbf{b}(s)ds \tag{1.23}$$

known as the formula of variation of parameters.

Proof By Proposition 1.8 the general solution of the homogeneous equation is given by $\mathbf{x} = \mathrm{e}^{tA}\mathbf{u}$, where $\mathbf{u}$ is a constant vector. Taking $\mathbf{u}$ as a function of time and assuming that $\mathbf{x} = \mathrm{e}^{tA}\mathbf{u}$ is a solution of the nonhomogeneous equation (1.22) we get $\dot{\mathbf{u}} = \mathrm{e}^{-tA}\mathbf{b}$; hence,

$$\mathbf{u}(t) = \mathbf{u}(0) + \int_0^t e^{-sN}\mathbf{b}(s)ds.$$

Since $\mathbf{u}(0) = \mathbf{x}(0)$ the equality $\mathbf{x}(t) = \mathrm{e}^{tA}\mathbf{u}(t)$ gives $\mathbf{x}(t) = \mathrm{e}^{tA}\left[\mathbf{x}(0) + \int_0^t \mathrm{e}^{-sN}\mathbf{b}(s)ds\right]$, which is seen to be the formula (1.23). □

This chapter contains the material on Calculus and fixes the notation that we will use in the text. We assume that the reader is familiar with the main theorems of Calculus in the intrinsic approach such as the inverse function theorem, the implicit function theorem, etc.

Chapter 2
Hamiltonian Systems Theory

2.1 Introduction

This chapter deals with the basic theory of Hamiltonian systems. After the definition of a Hamiltonian system and two examples from Mechanics we show how the general setting of a variational problem leads to a Hamiltonian system via the Legendre transformation, which establishes the equivalence of Euler–Lagrange equations with the Hamiltonian equations. Besides the two mentioned examples, we also present in the form of examples some mechanical problems giving the corresponding Hamiltonian functions. Among them we include the Kepler problem, the harmonic oscillator, the pendulum, the rigid body problem, the satellite problem, the Lagrange top, and others. Then the coordinate transformations that preserve the structure of Hamilton's equations and the generating functions of such transformations are studied. Along the way, we meet the concepts of symplectic and Hamiltonian matrices, studying their basic properties. The symplectic group is introduced and at the end of the chapter some information is given on its geometry. A brief discussion on the stability of equilibria of an autonomous system of differential equations is presented in this chapter but a deeper study of stability is carried out only in Chap. 6.

2.2 Hamiltonian Systems

Many systems of ordinary differential equations can be expressed in the form

$$\dot{x}_k = H_{y_k}, \qquad \dot{y}_k = -H_{x_k}, \qquad (k = 1, \ldots, n,) \tag{2.1}$$

H. E. Cabral, L. Brandão Dias, *Normal Forms and Stability of Hamiltonian Systems*, Applied Mathematical Sciences 218,
https://doi.org/10.1007/978-3-031-33046-9_2

with

$$H = H(x_1, \ldots, x_n, y_1, \ldots, y_n, t), \tag{2.2}$$

where the dot over the letter denotes a derivative with respect to the time t.

Systems of differential equations of the form (2.1) are called *Hamiltonian systems* and the function H is called the *Hamiltonian function* or, simply, the *Hamiltonian* of the system. We say that the system (2.1) is *autonomous* if the Hamiltonian does not depend on the time t. The variables $y_1, \ldots, y_n$ are called the *conjugate variables* of $x_1, \ldots, x_n$. The integer n is called *the number of degrees of freedom* of the Hamiltonian system.

Setting $\mathbf{x} = (x_1, \ldots, x_n)^T$, $\mathbf{y} = (y_1, \ldots, y_n)^T$ the Hamiltonian system (2.1) can be written in the form

$$\dot{\mathbf{x}} = H_{\mathbf{y}}, \quad \dot{\mathbf{y}} = -H_{\mathbf{x}} \tag{2.3}$$

where $H_{\mathbf{x}} = \nabla_{\mathbf{x}} H$, $H_{\mathbf{y}} = \nabla_{\mathbf{y}} H$ are the gradients of H relative to the vectors $\mathbf{x}$, $\mathbf{y}$.

With $\mathbf{z} = \begin{bmatrix} \mathbf{x} \\ \mathbf{y} \end{bmatrix}$ and using the standard symplectic matrix $J = \begin{bmatrix} 0 & I \\ -I & 0 \end{bmatrix}$ the Hamiltonian system can be written in the more compact form

$$\dot{\mathbf{z}} = J \nabla H(\mathbf{z}). \tag{2.4}$$

An immediate result for autonomous Hamiltonian systems is the fact that the function H is a *first integral* of the system, that is, it is a nonconstant function that remains constant along every solution of the system.

Proposition 2.1 *For an autonomous system, the Hamiltonian H is a first integral.*

Proof This is an immediate consequence of the chain rule since if $\big(\mathbf{x}(t), \mathbf{y}(t)\big)$ is a solution of the system, then we have by (1.11)

$$\frac{d}{dt} H\big(\mathbf{x}(t), \mathbf{y}(t)\big) = \langle H_{\mathbf{x}}, \dot{\mathbf{x}} \rangle + \langle H_{\mathbf{y}}, \dot{\mathbf{y}} \rangle = \langle -\dot{\mathbf{y}}, \dot{\mathbf{x}} \rangle + \langle \dot{\mathbf{x}}, \dot{\mathbf{y}} \rangle = 0.$$

□

Notice that for a time-dependent Hamiltonian $H(\mathbf{x}, \mathbf{y}, t)$ this is no longer true since $\frac{dH}{dt} = \frac{\partial H}{\partial t}$.

We will now give two examples from mechanics, with specific cases in the field of celestial mechanics.

Example 2.1 (System of Mass Particles) Consider a system of n mass particles in space with position vectors $\mathbf{r}_1, \ldots, \mathbf{r}_n$, moving under the action of a field of forces derived from a potential $V = V(\mathbf{r}_1, \ldots, \mathbf{r}_n)$ (Fig. 2.1).

The Newtonian equations of motion are

$$m_k \ddot{\mathbf{r}}_k = -\nabla_{\mathbf{x}_r} V.$$

Fig. 2.1 System of mass particles

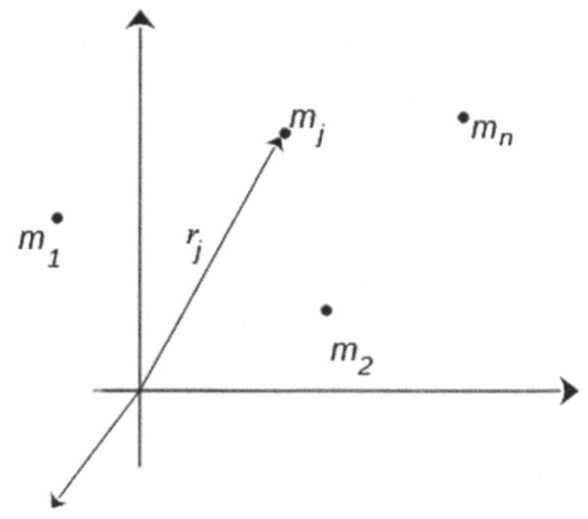

Setting $\mathbf{x}_k = \mathbf{r}_k$ and $\mathbf{y}_k = m_k\dot{\mathbf{r}}_k$, where m_k is the mass of the kth particle, the equations of motion can be written as the system of ordinary differential equations of first order

$$\dot{\mathbf{x}}_k = \frac{1}{m_k}\mathbf{y}_k, \quad \dot{\mathbf{y}}_k = -\nabla_{\mathbf{x}_k} V.$$

Considering the function H of the variables $\mathbf{x} = (\mathbf{x}_1, \ldots, \mathbf{x}_n)$, $\mathbf{y} = (\mathbf{y}_1, \ldots, \mathbf{y}_n)$ defined by

$$H(\mathbf{x}, \mathbf{y}) = \sum_{j=1}^{n} \frac{1}{2m_j}|\mathbf{y}_j|^2 + V(\mathbf{x}),$$

this system assumes the form

$$\dot{\mathbf{x}}_k = H_{\mathbf{y}_k}, \quad \dot{\mathbf{y}}_k = -H_{\mathbf{x}_k},$$

where $H_{\mathbf{x}_k}$ and $H_{\mathbf{y}_k}$ denote the gradients of H with respect to the vectors $\mathbf{x}_k$ and $\mathbf{y}_k$ respectively. A particular case is the Newtonian n-body problem of celestial mechanics where

$$V = -\sum_{j<k} \frac{Gm_j m_k}{|\mathbf{x}_j - \mathbf{x}_k|}, \tag{2.5}$$

and G is the constant of universal gravitation.

Example 2.2 (Rotating Coordinates in the Plane) Consider a fixed coordinate system in the plane and a second coordinate system rotating about the fixed origin with constant angular speed ω. Taking the vectors $\mathbf{e}_1 = (\cos\omega t, \sin\omega t)$, $\mathbf{e}_2 = (-\sin\omega t, \cos\omega t)$, we have $\dot{\mathbf{e}}_1 = \omega\mathbf{e}_2$, $\dot{\mathbf{e}}_2 = -\omega\mathbf{e}_1$, so differentiating the position vector $\mathbf{r} = \xi\mathbf{e}_1 + \eta\mathbf{e}_2$ of a variable point P twice we obtain

$$\ddot{\mathbf{r}} = (\ddot{\xi} - 2\omega\dot{\eta} - \omega^2\xi)\mathbf{e}_1 + (\ddot{\eta} + 2\omega\dot{\xi} - \omega^2\eta)\mathbf{e}_2.$$

The dynamics of a unit mass particle which moves in the plane under the action of a field of forces deriving from a potential $V = V(\mathbf{x})$ is described by the differential equation

$$\ddot{\mathbf{r}} = -\nabla V.$$

Expressing the gradient of V in the basis $\mathbf{e}_1$, $\mathbf{e}_2$, $\nabla V = \frac{\partial V}{\partial \xi}\mathbf{e}_1 + \frac{\partial V}{\partial \eta}\mathbf{e}_2$, the equation of motion can be written as the following system of equations in ξ, η,

$$\ddot{\xi} - 2\omega\dot{\eta} - \omega^2\xi = -\frac{\partial V}{\partial \xi}, \quad \ddot{\eta} + 2\omega\dot{\xi} - \omega^2\eta = -\frac{\partial V}{\partial \eta}. \tag{2.6}$$

Setting $x_1 = \xi$, $x_2 = \eta$, $y_1 = \dot{\xi} - \omega\eta$, $y_2 = \dot{\eta} + \omega\xi$, and introducing the function

$$H(x_1, x_2, y_1, y_2) = \frac{1}{2}\left(y_1^2 + y_2^2\right) + \omega(x_2 y_1 - x_1 y_2) + V(x_1, x_2), \tag{2.7}$$

the system (2.6) can be written in the form

$$\dot{x}_1 = H_{y_1}, \quad \dot{x}_2 = H_{y_2}, \quad \dot{y}_1 = -H_{x_1}, \quad \dot{y}_2 = -H_{x_2}.$$

A specific example that falls under the pattern (2.7) is the restricted three-body problem, which consists in the study of the motion of a particle of infinitesimal mass under the attraction of two primaries moving in circular orbits about the common center of mass (Fig. 2.2). The potential V of this system is given by

$$V = \frac{\mu}{\rho_1} + \frac{1-\mu}{\rho_2}, \tag{2.8}$$

Fig. 2.2 Restricted three-body problem

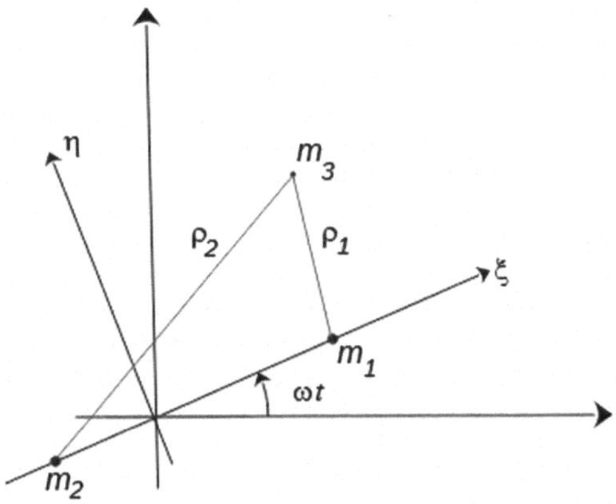

where $m_1 = \mu$ and $m_2 = 1-\mu$ are the masses of the primaries and ρ_1 and ρ_2 are the distances of the infinitesimal mass to the particles of masses m_1 and m_2 respectively. In the coordinates $x_1 = \xi$, $x_2 = \eta$ of the rotating system we have

$$\rho_1^2 = (x_1 - 1 + \mu)^2 + x_2^2, \quad \rho_2^2 = (x_1 + \mu)^2 + x_2^2. \tag{2.9}$$

For three mass particles the potential (2.5) defines the famous *three-body problem*, an unsolved problem to this day with origin in Newton's *Principia*. Oscar II, King of Sweden and Norway, offered a prize to whomever could solve this problem. Poincaré won the prize, not because he reached this goal, but because in his efforts to solve it, particularly through his studies on the restricted three-body problem, he created new methods in celestial mechanics and originated new branches in the mathematical sciences such as algebraic topology, qualitative theory of dynamical systems, and others. The three-volume set [50] continues to be a rich source of mathematical techniques in dynamical systems.

As the equations of motion of the three-body problem are very difficult to solve, one tries to get information on the problem through their analysis. Much qualitative behavior of the system of particles can be described without solving the equations. In the next example we list some facts that are known.

Example 2.3 (The Three-Body Problem)

(1) If the angular momentum $\sum m_j(\mathbf{r}_j \times \dot{\mathbf{r}}_j)$ is zero, then the motion is planar, that is, the three particles move in a fixed plane.
(2) When the angular momentum is not zero the plane through the origin and orthogonal to the angular momentum is called the invariable plane. If the motion is spatial, then at an instant where the particles cross the invariable plane they lie along a straight line, a syzygy configuration.
(3) If a solution ends in a triple collision then the angular momentum is zero.
(4) Double collisions can be regularized but not triple collisions.
(5) The homographic solutions generated by the combination of an elliptic orbit of the Kepler problem with a central configuration of Euler or Lagrange (see the Example 2.5), are the periodic solutions that have been known for a long time. Recent work by Chenciner and Montgomery [17] shows a new periodic orbit of the three-body problem in which three particles of equal masses follow one after the other along a figure-of-eight orbit.
(6) A solution is called isosceles if the triangle formed by the three particles is always isosceles with collinear or equilateral configurations happening only for isolated values of time, so generically the basis of the isosceles triangle is well defined. It is known that the masses at the basis of an isosceles solution are equal.
(7) There are three types of isosceles solutions, one with zero angular momentum and a fixed line of symmetry and two with nonzero angular momentum, one with the axis of the angular momentum as a symmetry line of the masses at the base, the other with the invariable plane as a symmetry plane for these masses.

(8) The plane that contains the three particles at every instant is called the plane of motion. For a noncollinear solution it is generically defined. This plane is fixed for a planar motion and coincides with the invariable plane if this is defined, that is, if the angular momentum is not zero, but is always moving for a nonplanar motion and never coincides with the invariable plane.
(9) If the angular momentum is not zero, the inclination angle is the angle formed by the plane of motion and the invariable plane. The inclination angle is equal to $\pi/2$ for the spatial isosceles solutions. The only solutions that have an inclination angle equal to $\pi/2$ are the spatial isosceles solutions [10]. It is unknown whether there is a solution with a constant inclination angle between zero and $\pi/2$.
(10) If the moment of inertia of a solution is constant then the solution is a relative equilibrium (Saari's conjecture), proved for $n = 3$ in certain cases but still an open problem for $n > 3$ (see [22] and [53]).
(11) There is a classification of motions in the three-body problem by Jean Chazy based on their asymptotic behavior as time goes to infinity [16].
(12) A singularity of a solution of the n-body problem is a time t_* beyond which the solution cannot be continued. In the three-body problem, any singularity is due to a collision. For $n \geq 4$ interchange of groups of particles could happen as t approaches t_* and the system could undergo wild oscillations without collisions. This was shown by Z. Xia [63] in the 5-body problem.

This is a problem with a rich history and the references [7, 38] bring much information on it. The Historical Notes in Wintner's book [62] is also a rich source of information.

The case of two mass particles, known as the *two-body problem* is reduced to the case of a central force problem under an inverse square law and is known as the *Kepler problem*. This case is completely solved and we give some information on it in the next example.

Example 2.4 (The Kepler Problem) Let $\mathbf{x}_1$ and $\mathbf{x}_2$ be the position vector of the particles with masses m_1 and m_2 respectively. Consider the vector $\mathbf{x} = \mathbf{x}_2 - \mathbf{x}_1$ that defines the relative position of m_2 with respect to m_1. Then, using (2.5) with $n = 2$,

$$\ddot{\mathbf{x}}_1 = \frac{Gm_2}{\|\mathbf{x}_2 - \mathbf{x}_1\|^\beta}(\mathbf{x}_2 - \mathbf{x}_1) \quad \text{and} \quad \ddot{\mathbf{x}}_2 = \frac{Gm_1}{\|\mathbf{x}_1 - \mathbf{x}_2\|^\beta}(\mathbf{x}_1 - \mathbf{x}_2),$$

we get

$$\ddot{\mathbf{x}} = -\frac{\kappa}{\|\mathbf{x}\|^3}\mathbf{x}, \quad \text{with} \quad \kappa = G(m_1 + m_2). \tag{2.10}$$

The problem defined by the Eq. (2.10) is known as the *Kepler problem*.

Notice that, setting $\mathbf{y} = \dot{\mathbf{x}}$, the Hamiltonian of the Kepler problem is given by

$$H(\mathbf{x}, \mathbf{y}) = \frac{1}{2}\|\mathbf{y}\|^2 - \frac{\kappa}{\|\mathbf{x}\|}. \tag{2.11}$$

All the information on the dynamics of this problem derives from the following conservation laws:

Proposition 2.2 *The scalar function* h *(energy), the vector function* $\mathbf{C}$ *(angular momentum), and the vector function* $\mathbf{e}$ *(Laplace vector), defined below are integrals of motion of the Kepler problem.*

(1) $h = \frac{1}{2}\|\dot{\mathbf{x}}\|^2 - \frac{\kappa}{\|\mathbf{x}\|}$;
(2) $\mathbf{C} = \mathbf{x} \times \dot{\mathbf{x}}$
(3) $\kappa\left(\frac{\mathbf{x}}{\|\mathbf{x}\|} + \mathbf{e}\right) = \dot{\mathbf{x}} \times \mathbf{C}$.

Proof

(1) Since $\frac{d}{dt}\|\mathbf{x}\| = \frac{\mathbf{x}\cdot\dot{\mathbf{x}}}{\|\mathbf{x}\|}$, we have

$$\dot{h} = \dot{\mathbf{x}}\cdot\ddot{\mathbf{x}} + \frac{\kappa}{\|\mathbf{x}\|^3}(\mathbf{x}\cdot\dot{\mathbf{x}}) = \dot{\mathbf{x}}\cdot\left(\ddot{\mathbf{x}} + \frac{\kappa}{\|\mathbf{x}\|^3}\mathbf{x}\right) = 0, \quad \text{by} \quad (2.10);$$

(2) $\dot{\mathbf{C}} = \dot{\mathbf{x}} \times \dot{\mathbf{x}} + \mathbf{x} \times \ddot{\mathbf{x}} = 0$, obviously.
(3) First we notice that, using the dot product and the cross product in $\mathbf{R}^3$, we have

$$\frac{d}{dt}\frac{\mathbf{x}}{\|\mathbf{x}\|} = \frac{\dot{\mathbf{x}}}{\|\mathbf{x}\|} - \frac{1}{\|\mathbf{x}\|^2}\frac{\mathbf{x}\cdot\dot{\mathbf{x}}}{\|\mathbf{x}\|}\mathbf{x} = \frac{(\mathbf{x}\cdot\mathbf{x})\dot{\mathbf{x}} - (\mathbf{x}\cdot\dot{\mathbf{x}})\mathbf{x}}{\|\mathbf{x}\|^3} = \frac{(\mathbf{x}\times\dot{\mathbf{x}})\times\mathbf{x}}{\|\mathbf{x}\|^3};$$

hence,

$$\frac{d}{dt}\left(\kappa\frac{\mathbf{x}}{\|\mathbf{x}\|}\right) = \mathbf{C} \times \left(\frac{\kappa}{\|\mathbf{x}\|^3}\mathbf{x}\right) = \mathbf{C} \times (-\ddot{\mathbf{x}}) = \frac{d}{dt}(\dot{\mathbf{x}} \times \mathbf{C}),$$

so there is a constant vector $\mathbf{e}$ such that

$$\kappa\left(\frac{\mathbf{x}}{\|\mathbf{x}\|} + \mathbf{e}\right) = \dot{\mathbf{x}} \times \mathbf{C}.$$

□

Consequences Item (3) of the Proposition shows that if $\mathbf{C} = 0$, then $\mathbf{e} \neq 0$ and also that the motion is rectilinear motion along the axis defined by $\mathbf{e}$. If $\mathbf{C} \neq 0$, by item (2) the motion $\mathbf{x}(t)$ occurs in the plane passing through the origin and orthogonal to $\mathbf{C}$, the orbital plane. Since $\mathbf{x}$ and $\dot{\mathbf{x}} \times \mathbf{C}$ are orthogonal to $\mathbf{C}$ taking the inner product of equation in (3) by $\mathbf{C}$ we get $\mathbf{e}\cdot\mathbf{C} = 0$, so the Laplace vector $\mathbf{e}$ lies inside the orbital plane.

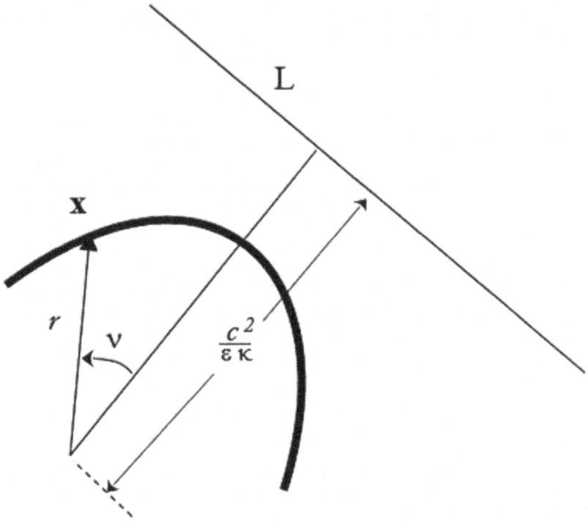

Fig. 2.3 Polar equation of a conic

Taking the inner product of the equation in item (3) with the vector $\mathbf{x}$ we get

$$\kappa(\|\mathbf{x}\| + \mathbf{e} \cdot \mathbf{x}) = c^2, \quad \text{where} \quad c = \|\mathbf{C}\|. \tag{2.12}$$

Let ν be the angle of $\mathbf{x}$ with the axis defined by $\mathbf{e}$ and let $r = \|\mathbf{x}\|$ and $\epsilon = \|\mathbf{e}\|$ (Fig. 2.3). Then, from the above equation we get $r(1 + \epsilon \cos \nu) = c^2/\kappa$; hence,

$$r = \frac{c^2/\kappa}{1 + \epsilon \cos \nu}. \tag{2.13}$$

Equation (2.13) is the polar equation of a conic of eccentricity ϵ. To look at the geometry we re-write the equation in the form

$$r = \epsilon\Big(c^2/\epsilon\kappa - r \cos \nu\Big)$$

which says that the distance r from $\mathbf{x}$ to the attracting center O is ϵ times the distance from $\mathbf{x}$ to the line L orthogonal to the Laplace vector $\mathbf{e}$ at a distance $c^2/\epsilon\kappa$ from O. This is the definition of the conic of eccentricity ϵ, directrix L and axis along the vector $\mathbf{e}$.

Equation (2.13) represents Newton's generalization of Kepler's first law "*the planets describe elliptical orbits around the Sun with the Sun at one of the foci*". But what kind of conic is the orbit?

To answer this question, square the equation in item (3) and use the energy equation (1) to get

$$\kappa^2\Big(1 + \frac{2(\mathbf{e} \cdot \mathbf{x})}{\|\mathbf{x}\|} + \epsilon^2\Big) = c^2\|\dot{\mathbf{x}}\|^2 = c^2\Big(2h + \frac{2\kappa}{\|\mathbf{x}\|}\Big).$$

Using the expression $\mathbf{e} \cdot \mathbf{x} = c^2/\kappa - \|\mathbf{x}\|$ obtained from (2.12) the above equation reduces to

$$\epsilon^2 - 1 = 2c^2 h.$$

Therefore, the orbit is elliptic, parabolic or hyperbolic according to $h < 0$, $h = 0$ or $h > 0$. Taking the orbital plane as the xy-plane of a rectangular coordinate system and writing $\mathbf{x} = r(\cos\nu, \sin\nu, 0)$ we have $\dot{\mathbf{x}} = \dot{r}(\cos\nu, \sin\nu, 0) + r\dot{\nu}(-\sin\nu, \cos\nu, 0)$, hence by equation in item (2), $\mathbf{C} = r^2\dot{\nu}(0, 0, 1)$, so the conservation of angular momentum means

$$r^2\dot{\nu} = c. \tag{2.14}$$

This is Kepler's second law "The radius vector of a planet describes equal areas in equal intervals of time." Indeed, the area in polar coordinates s computed with the integral $A = \frac{1}{2}\int r^2 d\nu = \frac{1}{2}\int r^2\dot{\nu}\,dt$, hence $\dot{A} = \frac{1}{2}c$, so $A(t) - A(t_0) = \frac{1}{2}c(t - t_0)$. In the case of an elliptic orbit, let a be the major semi-axis of the ellipse. Then, the polar equation (2.13) gives $2a = \frac{c^2/\kappa}{1-\epsilon} + \frac{c^2/\kappa}{1+\epsilon}$, hence $c^2/\kappa = a(1-\epsilon^2)$. If b is the minor semi-axis, the area of the ellipse is πab, so if P is the period of the orbit then $\pi ab = \frac{1}{2}cP$. Squaring this equation and using the relation between the two axes of an ellipse $b = a\sqrt{1-\epsilon}$ we get $\pi^2 a^4(1-\epsilon^2) = \frac{1}{4}c^2P^2$. Since $c^2/\kappa = a(1-\epsilon^2)$, after simplification we get Newton's formulation of Kepler's third law, which says that "the ratio between the cube of the major semi-axis and square of the period of any planet is constant." Newton's formulation of this law is the equation

$$\frac{a^3}{P^2} = \frac{\kappa}{4\pi^2}. \tag{2.15}$$

Because $\kappa = G(m_1 + m_2)$, this equation shows that the constant in the statement of Kepler's third law depends on the masses of the planets. The mass of the Sun, m_1, is overwhelmingly bigger than that of any planet, m_2, so Kepler could not detect this particularity of his law. Let us use (2.15) to compute the mass of a planet that has satellites. Applying it to the systems Sun–Jupiter and Jupiter–Callisto, we get the relation

$$\frac{M_S + M_J}{M_J + M_C} = \left(\frac{a_J}{a_C}\right)^3 \left(\frac{P_C}{P_J}\right)^2. \tag{2.16}$$

Using the data $a_J = 5.2 \times 1.4959787 \times 10^{11} m$, $a_C = 1883 \times 10^6 m$, $P_J = 4332.59$ days, $P_C = 16.689$ days and disregarding the mass of Callisto *vis-a-vis* the mass of Jupiter we find $M_S = 1047 M_J$. Using (2.16) with data for the systems Sun–Earth and Earth–Moon taking M_J as mass M_E of Earth and M_C as mass of Moon we find that $M_S = 329796 M_E$. Therefore, Jupiter is about 315 times as massive as the Earth. These are crude estimates but they give a good idea of the sizes of the celestial bodies of the Solar System. To find the mass of the Earth we equate the weight of a mass particle on its surface to the gravitational pull it suffers to the center of the Earth

$$mg = \frac{GmM_E}{R^2},$$

where R is the radius of the Earth. From this we get $M_E = \frac{gR^2}{G}$. Using the value of G (H. Cavendish, 1798), the value of g (C. Huygens, 1659) and the value of $R = 6378$km (Erasthotenes, third century BC), we find that $M_E = 5.97 \times 10^{24}$kg. Using this we find the estimate for the mass of the Sun, $M_S = 1.969 \times 10^{30}$kg. Computing the mass of each planet that has satellites we will see that the mass of the Sun represents more than 98.8% of the sum of their masses. Mercury and Venus do not have satellites but their masses will not alter this percentage much. Considering all the satellites of the Solar system and, despite their huge number, the comets and asteroids too, the total mass of these bodies do not add very much and probably we can say that the Sun accounts for more than 97% of the total mass of the Solar System. To conclude these considerations on the Solar System, we compute the volume of the Earth, $V_E = \frac{4}{3}\pi R^3$ and find the estimate $\rho = M_E/V_E = 5.5$ g/cm^3 for its density, which shows that the heavy part of the Earth lies in its core, since on the average the material on the surface has density below 3 g/cm^3.

Writing the equations of motion for the three-body problem (Fig. 2.4) with masses m_1, m_2, m_3 and position vectors $\mathbf{r}_1, \mathbf{r}_2, \mathbf{r}_3$ we have

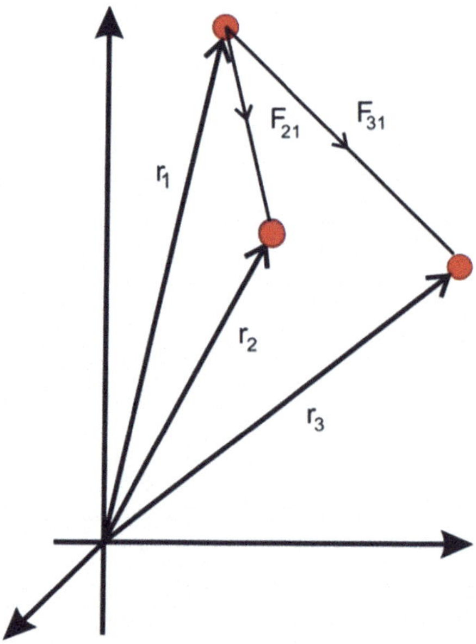

Fig. 2.4 Three-body problem

$$\ddot{\mathbf{r}}_1 = \frac{Gm_2}{\|\mathbf{r}_2-\mathbf{r}_1\|^3}(\mathbf{r}_2 - \mathbf{r}_1) + \frac{Gm_3}{\|\mathbf{r}_3-\mathbf{r}_1\|^3}(\mathbf{r}_3 - \mathbf{r}_1),$$

$$\ddot{\mathbf{r}}_2 = \frac{Gm_1}{\|\mathbf{r}_1-\mathbf{r}_2\|^3}(\mathbf{r}_1 - \mathbf{r}_2) + \frac{Gm_3}{\|\mathbf{r}_3-\mathbf{r}_2\|^3}(\mathbf{r}_3 - \mathbf{r}_2),$$

$$\ddot{\mathbf{r}}_3 = \frac{Gm_1}{\|\mathbf{r}_1-\mathbf{r}_3\|^3}(\mathbf{r}_1 - \mathbf{r}_3) + \frac{Gm_2}{\|\mathbf{r}_2-\mathbf{r}_3\|^3}(\mathbf{r}_2 - \mathbf{r}_3).$$

We see that the third equation does not contain the mass m_3. The first two equations with $m_3 = 0$ describe the two-body problem. Taking an elliptic solution $\mathbf{r}_1(t)$, $\mathbf{r}_2(t)$ of this problem and inserting it into the third equation we get a time-periodic differential equation that describes the motion of the zero mass particle. This is what we call the restricted three-body problem.

Take a length L, a time τ and $m = m_1 + m_2$, then set $\mathbf{r}_j = L\mathbf{x}_j$, $m_k = m\mu_k$ and $t = \tau s$. Then, the third equation is written in the form

$$\frac{d^2\mathbf{x}_3}{ds^2} = \frac{Gm\tau^2}{L^3}\left(\frac{\mu_1}{\|\mathbf{x}_1(t) - \mathbf{x}_3\|^3}(\mathbf{x}_1(t) - \mathbf{x}_3) + \frac{\mu_2}{\|\mathbf{x}_2(t) - \mathbf{x}_3\|^3}(\mathbf{x}_2(t) - \mathbf{x}_3)\right).$$

As the dimensionality of G is $\mathrm{m}^3 \cdot \mathrm{kg}^{-1} \cdot s^{-2}$, the quotient $\frac{Gm\tau^2}{L^3}$ is dimensionless and choosing L and τ to make it have value one, we get the equation of motion of the restricted three-body problem in the dimensionless form, with $\mathbf{x} = \mathbf{x}_3$,

$$\frac{d^2\mathbf{x}}{ds^2} = \nabla V(\mathbf{x}, t, \mu), \quad \text{with} \quad V = \frac{\mu_1}{\|\mathbf{x}_1(t) - \mathbf{x}\|} + \frac{\mu_2}{\|\mathbf{x}_2(t) - \mathbf{x}\|}. \tag{2.17}$$

If the solution $\mathbf{r}_1(t)$, $\mathbf{r}_2(t)$ is circular with angular speed ω, then in a rotating system with angular speed ω, the potential V has the expression (2.8) with ρ_1 and ρ_2 given in (2.9) and the corresponding Hamiltonian system is autonomous.

Example 2.5 (Periodic Orbits of the n-Body Problem) We consider the n-body problem in the plane. Take n complex numbers $\boldsymbol{\zeta}_1, \ldots, \boldsymbol{\zeta}_n$ and a complex vector function $\mathbf{z} = \mathbf{z}(t)$ of the time t. Imposing the condition that the vectors $\mathbf{r}_j = \mathbf{z}\boldsymbol{\zeta}_j$, $j = 1, \ldots, n$ define a solution of the n-body problem with masses $m_1, \ldots, m_n$, we have

$$\left(\ddot{\mathbf{z}}|\mathbf{z}|^3\mathbf{z}^{-1}\right)m_j\boldsymbol{\zeta}_j = \sum_{i\neq j} \frac{Gm_im_j}{|\boldsymbol{\zeta}_i - \boldsymbol{\zeta}_j|^3}(\boldsymbol{\zeta}_i - \boldsymbol{\zeta}_j), \quad j = 1, \ldots, n. \tag{2.18}$$

The right-hand side does not depend on t, so the expression inside the parenthesis on the left-hand side does not depend on t either. Setting it equal to $-\kappa$, the above equation decouples into two parts the system of algebraic equations

$$-\kappa m_j\boldsymbol{\zeta}_j = \sum_{i\neq j} \frac{Gm_im_j}{|\boldsymbol{\zeta}_i - \boldsymbol{\zeta}_j|^3}(\boldsymbol{\zeta}_i - \boldsymbol{\zeta}_j), \quad j = 1, \ldots, n \tag{2.19}$$

and the dynamical equation

$$\ddot{\mathbf{z}} = -\frac{\kappa}{|\mathbf{z}|^3}\mathbf{z}. \tag{2.20}$$

Notice that the right-hand side of Eq. (2.19) is $\nabla_{\zeta_j} V(\boldsymbol{\zeta})$, where

$$V(\boldsymbol{\zeta}) = \sum_{i<j} \frac{Gm_i m_j}{|\zeta_i - \zeta_j|} \quad \text{with} \quad \boldsymbol{\zeta} = (\zeta_1, \ldots, \zeta_n).$$

Since V is homogeneous of degree -1, by Euler's theorem we have $\sum_{j=1}^n \zeta_j \cdot \nabla_{\zeta_j} V(\boldsymbol{\zeta}) = -V(\boldsymbol{\zeta})$. On the other hand, we have

$$\sum_{j=1}^n m_j \zeta_j \cdot \zeta_j = \sum_{j=1}^n m_j |\zeta_j|^2 = \frac{1}{2} I(\boldsymbol{\zeta}),$$

the moment of inertia of the configuration $\boldsymbol{\zeta} = (\zeta_1, \ldots, \zeta_n)$. Therefore, taking the inner product by ζ_j with the equations of the system (2.18), that is

$$-\kappa m_j \zeta_j = \sum_{i \neq j} \nabla_{\zeta_j} V(\boldsymbol{\zeta}), \qquad j = 1, \ldots, n \tag{2.21}$$

and adding we get $\kappa = \dfrac{2V(\boldsymbol{\zeta})}{I(\boldsymbol{\zeta})}$, that is κ is a positive constant, so the dynamical equation (2.20) defines the Kepler problem in the plane. Taking a configuration $\boldsymbol{\zeta} = (\zeta_1, \ldots, \zeta_n)$, which is a solution of the system (2.19) and an elliptic solution $\mathbf{z}(t)$ of the Kepler problem (2.20), we get a periodic solution $\mathbf{r}_1(t) = \mathbf{z}(t)\zeta_1, \ldots, \mathbf{r}_n(t) = \mathbf{z}(t)\zeta_n$ of the n-body problem (Fig. 2.5).

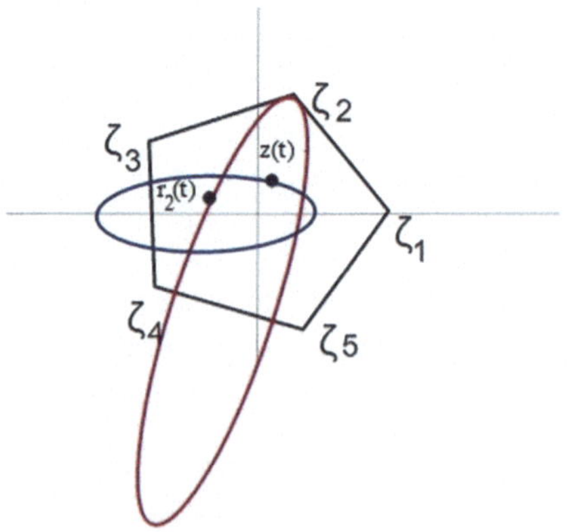

Fig. 2.5 A periodic solution of the five-body problem

The fundamental point, then, is the question of existence of configurations $\boldsymbol{\zeta}$ solving the system of equations (2.19). This is not an easy question. For $n = 3$ it was solved by Euler, who proved in 1767 that there are essentially three collinear configurations, that is, the points $\boldsymbol{\zeta}_1, \boldsymbol{\zeta}_2, \boldsymbol{\zeta}_3$ lie on a straight line passing through the center of mass, the origin. Five years later, in 1772, Lagrange proved that besides the Euler collinear configurations there are two other solutions where the three particles form the vertices of an equilateral triangle. For many years these were the only known solutions of the system (2.19). In 1906, Moulton [47] proved that for each natural number n there are exactly $\frac{n!}{2}$ collinear solutions of these equations. Given $\boldsymbol{\zeta} = (\boldsymbol{\zeta}_1, \ldots, \boldsymbol{\zeta}_n)$ consider the configuration $\lambda\boldsymbol{\zeta}$ obtained by multiplying each point $\boldsymbol{\zeta}_j$ by λ and the configuration $R\boldsymbol{\zeta}$ obtained by applying a rotation R to each $\boldsymbol{\zeta}_j$. Then, if $\boldsymbol{\zeta}$ is a solution of the system (2.19) so are the configurations $\lambda\boldsymbol{\zeta}$ and $R\boldsymbol{\zeta}$. When we refer to a solution of (2.19) we mean the whole equivalence class of solutions defined by these operations. An equivalence class is called a *central configuration*. The terminology comes from the fact that the acceleration of each particle is directed toward the center (see Eq. (2.21)). For $n = 3$, there are exactly five central configurations, the three Euler configurations and the two Lagrangian equilateral configurations. For any natural number n there are exactly $\frac{n!}{2}$ collinear central configurations, those found by Moulton. The search for noncollinear central configurations is a hard problem for $n \geq 4$. For $n = 4$ the question was settled by Alain Albouy in the case of equal masses, see [1]. For $n \geq 5$ the question is wide open. Even the question of whether or not the set of central configurations for a given set of arbitrary masses $m_1, \ldots, m_n$ is finite is a very hard open question. This question was included in Smale's list of open problems for the twenty-first century (see [56]). If one allows negative masses, it has been proved that there are infinitely many central configurations for $n = 5$ particles (see Roberts [51]). Now we observe that for very special sets of masses it may be possible to find central configurations, for instance, for any natural number n the regular n-gon with equal masses placed at the vertices of the polygon gives a central configuration. It is an amazing fact that for $n = 3$ the masses can have any values!

When the solution $\mathbf{z}(t)$ of the Kepler problem (2.20) is circular and $\boldsymbol{\zeta}$ is a central configuration, the n particles $\mathbf{r}_j = \mathbf{z}(t)\boldsymbol{\zeta}_j$ describe circular orbits about the vertical axis, all with the same angular speed ω. In a coordinate system rotating about the vertical axis with this angular speed the circular orbits correspond to equilibria of the system in the rotating coordinates. That is why such periodic solutions of the n-body problem are called *relative equilibria*.

When the periodic solution is induced by an elliptic orbit $\mathbf{z}(t)$ then each particle $\mathbf{r}_j$ describes an ellipse of the same nature as $\mathbf{z}(t)$.

If we consider the restricted problem of a zero mass particle attracted by the n mass particles $\mathbf{r}_j = \mathbf{z}(t)\boldsymbol{\zeta}_j$ the Hamiltonian is periodic in time. If $\mathbf{z}(t)$ is a circular solution then the Hamiltonian in a rotating system about the vertical axis with the same angular speed becomes autonomous. If $\mathbf{z}(t)$ is an elliptic solution the dynamical system in a rotating coordinate system about the vertical axis is non-autonomous. However, if we take the coordinate system rotating about the vertical axis through the angle given by the true anomaly ν, make a scaling by the radius vector (2.13) and use ν as the new time, then we can describe the dynamics of this restricted problem by a very convenient Hamiltonian. We will do that in the next example.

Example 2.6 (Pulsating Coordinates) The equation of motion of the zero mass particle located at the position $\mathbf{r} \in \mathbf{R}^3$ is

$$\ddot{\mathbf{r}} = \sum_{j=1}^{n} \frac{\mu_j}{|\mathbf{r}_j(t) - \mathbf{r}|^3}(\mathbf{r}_j(t) - \mathbf{r}). \tag{2.22}$$

Let $\mathbf{r} = \Omega\mathbf{x}$, where Ω is the rotation by the angle ν about the vertical axis. We have

$$\Omega = \begin{bmatrix} \cos\nu & -\sin\nu & 0 \\ \sin\nu & \cos\nu & 0 \\ 0 & 0 & 1 \end{bmatrix} \quad \text{and} \quad \dot{\Omega} = \dot{\nu}\Omega\Sigma, \quad \text{where} \quad \Sigma = \begin{bmatrix} 0 & -1 & 0 \\ 1 & 0 & 0 \\ 0 & 0 & 0 \end{bmatrix}.$$

We compute

$$\ddot{\mathbf{r}} = \Omega\big(\ddot{\mathbf{x}} + 2\dot{\nu}\Sigma\dot{\mathbf{x}} + \ddot{\nu}\Sigma\mathbf{x} + \dot{\boldsymbol{\nu}}^2\Sigma^2\mathbf{x}\big).$$

Now, let $\mathbf{x} = r\boldsymbol{\xi}$, where $r = \dfrac{p}{1 + \epsilon\cos\nu}$, with $p = c^2/\kappa = a(1 - \epsilon^2)$, see (2.13), the parameter of the elliptic orbit.

Denoting by a prime derivative with respect to ν, we compute

$$\dot{\mathbf{x}} = \dot{r}\boldsymbol{\xi} + r\dot{\nu}\boldsymbol{\xi}', \quad \ddot{\mathbf{x}} = r\dot{\nu}^2\boldsymbol{\xi}'' + 2\dot{r}\dot{\nu}\boldsymbol{\xi}' + \ddot{r}\boldsymbol{\xi} + r\ddot{\nu}\boldsymbol{\xi}.$$

Since $r^2\dot{\nu} = c$ we have $r\ddot{\nu} + 2\dot{r}\dot{\nu} = 0$ and the equation for $\ddot{x}$ becomes $\ddot{\mathbf{x}} = r\dot{\nu}^2\boldsymbol{\xi}'' + {}$ $+\ddot{r}\boldsymbol{\xi}$. Replacing the expressions of $\mathbf{x}$, $\dot{\mathbf{x}}$ and $\ddot{\mathbf{x}}$ into the expression of $\ddot{\mathbf{r}}$ we find

$$\ddot{\mathbf{r}} = r\dot{\nu}^2\Omega\Big(\boldsymbol{\xi}'' + 2\Sigma\boldsymbol{\xi}' + \Sigma^2\boldsymbol{\xi} + \frac{\ddot{r}}{r\dot{\nu}^2}\boldsymbol{\xi}\Big). \tag{2.23}$$

Differentiating the expression of r we get $\dot{r} = \dfrac{\epsilon p\sin\nu}{(1 + \epsilon\cos\nu)^2}\dot{\nu} = \dfrac{c\epsilon}{p}\sin\nu$ and differentiating a second time we find for the quotient $\dfrac{\ddot{r}}{r\dot{\nu}^2}$ the expression $\dfrac{r}{p}\epsilon\cos\nu$. Now, if $\boldsymbol{\zeta} = (\boldsymbol{\zeta}_1, \ldots, \boldsymbol{\zeta}_n)$ is the fixed central configuration that defines the planar motion $\mathbf{r}_1(t), \ldots, \mathbf{r}_n(t)$ through the Kepler solution $\mathbf{z}(t)$, then the right-hand side of the Eq. (2.22) is written in the form

$$\frac{1}{r^2}\Omega\sum_{j=1}^{n}\frac{\mu_j}{|\boldsymbol{\zeta}_j - \boldsymbol{\xi}|^3}(\boldsymbol{\zeta}_j - \boldsymbol{\xi}) = \frac{1}{r^2}\Omega\nabla V(\boldsymbol{\xi}), \quad \text{with} \quad V(\boldsymbol{\xi}) = \sum_{j=1}^{n}\frac{\mu_j}{|\boldsymbol{\zeta}_j - \boldsymbol{\xi}|}. \tag{2.24}$$

Since $r\dot{\nu}^2 = \dfrac{c^2}{r^3}$ and $\dfrac{r}{p} = \dfrac{1}{1+\epsilon\cos\nu}$, we finally get from (2.23) and (2.24) the equation

$$\boldsymbol{\xi}'' + 2\Sigma\boldsymbol{\xi}' + \Sigma^2\boldsymbol{\xi} + \frac{\epsilon\cos\nu}{1+\epsilon\cos\nu}\boldsymbol{\xi} = \frac{r}{c^2}\nabla V(\boldsymbol{\xi}). \tag{2.25}$$

Since $p = c^2/\kappa$ we have $\dfrac{r}{c^2} = \dfrac{1}{\kappa(1+\epsilon\cos\nu)}$ and setting $\boldsymbol{\eta} = \boldsymbol{\xi}' + \Sigma\boldsymbol{\xi}$ we can write the second-order equation (2.25) as the system of first-order differential equations

$$\boldsymbol{\xi}' = \boldsymbol{\eta} - \Sigma\boldsymbol{\xi}, \quad \boldsymbol{\eta}' = -\Sigma\boldsymbol{\eta} - \nabla U, \tag{2.26}$$

where

$$U(\boldsymbol{\xi}, \nu, \epsilon) = -\frac{1}{1+\epsilon\cos\nu}\left(\frac{1}{2}\epsilon\cos\nu\|\boldsymbol{\xi}\|^2 + \frac{1}{\kappa}V(\boldsymbol{\xi})\right) \tag{2.27}$$

with $V(\boldsymbol{\xi})$ given in (2.24). This is a periodic Hamiltonian system with the Hamiltonian function given by

$$H(\boldsymbol{\xi}, \boldsymbol{\eta}, \nu, \epsilon) = \frac{1}{2}\|\boldsymbol{\eta}\|^2 - \langle\Sigma\boldsymbol{\xi}, \boldsymbol{\eta}\rangle + U(\boldsymbol{\xi}, \nu, \epsilon). \tag{2.28}$$

We now give the example of a most important Hamiltonian system, the *mathematical pendulum*, and its first approximation, the simplest of all Hamiltonian systems, the *harmonic oscillator*.

The pendulum played an important role in the considerations of Isaac Newton on the theory of gravitation as he was interested in knowing how the acceleration of gravity would change in points far apart on the Earth and many observers overseas measured the period of a pendulum in several places around the Globe. Regarding this we quote from Newton's "System of the World" in the *Principia*, [48] "After this, M. Couplet, the son, in the month of July 1697, at the Royal Observatory of Paris, so fitted his pendulum clock to the mean motion of the sun, that for a considerable time together the clock agreed with the motion of the sun. In November following, upon his arrival at Lisbon, he found his clock to go slower than before at the rate of 2' 13" in 24 hours. And next March coming to Paraiba, he found his clock to go slower than at Paris, and at the rate 4' 12" in 24 hours; and he affirms, that the pendulum vibrating in seconds was shorter at Lisbon by 2 1/2 lines, and at Paraiba by 3 2/3 lines, than at Paris."

Example 2.7 (The Pendulum and the Harmonic Oscillator) The mathematical pendulum is given by a mass m attached at the end of a massless rod of length l, which can rotate without friction about a fixed hinge located at its other end, the motion of the mass being due only to the action of the gravity field (Fig. 2.6).

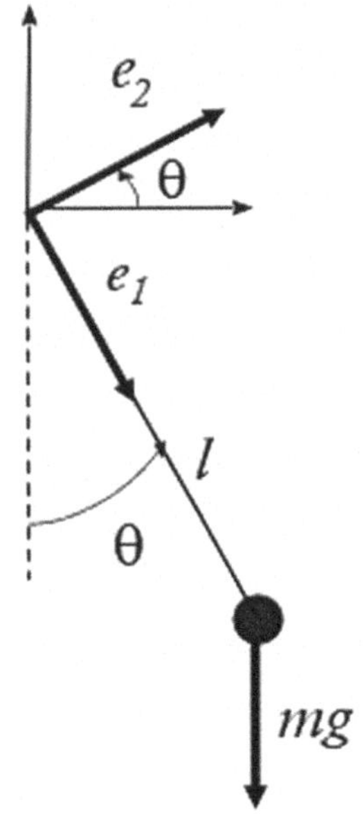

Fig. 2.6 The pendulum

Let θ be the angle that the rod makes with the vertical and T the tension on the rod. With the origin of an inertial coordinate system at the hinge, the position vector of m is $\mathbf{r} = l\mathbf{e}_1$, where $\mathbf{e}_1 = (\sin\theta, -\cos\theta)$. Let $\mathbf{e}_2 = (\cos\theta, \sin\theta)$. Then $\dot{\mathbf{e}}_1 = \dot{\theta}\mathbf{e}_2$, $\dot{\mathbf{e}}_2 = -\dot{\theta}\mathbf{e}_1$ and consequently $\ddot{\mathbf{r}} = -l\dot{\theta}^2\mathbf{e}_1 + l\ddot{\theta}\mathbf{e}_2$. The forces on the bob of the pendulum are the weight $-mg\,\mathbf{j}$ of the bob and the reaction to the tension $-T\mathbf{e}_1$; here $\mathbf{j} = (0, 1)$ is the second basis vector of the coordinate system. Since $-\mathbf{j} = \cos\theta\mathbf{e}_1 - \sin\theta\mathbf{e}_2$ comparing the coefficients of $\mathbf{e}_1$ and $\mathbf{e}_2$ in the equation given by Newton's second law of dynamics, $m\ddot{\mathbf{r}} = Force$ we get the equations

$$\ddot{\theta} + \alpha\sin\theta = 0, \quad \text{and} \quad T = mg\cos\theta + ml\dot{\theta}^2, \tag{2.29}$$

where $\alpha = \dfrac{g}{l}$. The first equation in (2.29) defines the dynamics of the pendulum and the second gives the intensity of the tension on the rod. Setting $x = \theta$ and $y = \dot{\theta}$ the second-order differential equation of the pendulum becomes the first-order system

$$\dot{x} = y, \quad \dot{y} = -\alpha\sin x,$$

which is a Hamiltonian system with Hamiltonian function

$$H(x, y) = \frac{1}{2}y^2 - \alpha\cos x. \tag{2.30}$$

With $x = \theta$ and $\alpha = \omega^2$ we write the equation of the pendulum as $\ddot{x} + \omega^2\sin x = 0$. Considering the Taylor expansion of $\sin x$ around the origin and disregarding

Fig. 2.7 Mass-spring system

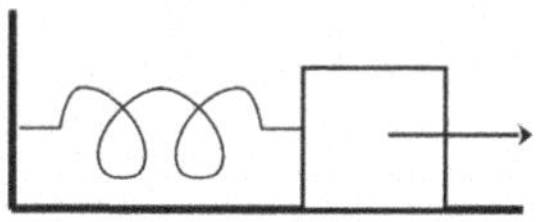

terms of order higher than 2 in the equation of the pendulum we get the differential equation

$$\ddot{x} + \omega^2 x = 0, \tag{2.31}$$

which is the equation of the harmonic oscillator.

This is the equation of a mass-spring system with the spring obeying Hooke's law (see Fig. 2.7). The general solution of this equation is given by $x = A\sin(\omega t + \phi)$; the constant A is the amplitude of the motion, ϕ is the phase, and ω is the frequency. The period of motion is $P = 2\pi/\omega$, that is, $P = 2\pi\sqrt{\frac{l}{g}}$. This formula shows that the period of a harmonic oscillation depends on neither the amplitude nor the mass of the bob, which is not the case for the period of the pendulum. The Hamiltonian of the harmonic oscillator is, with $y = \frac{\dot{x}}{\omega}$,

$$H(x, y) = \frac{1}{2}\omega(x^2 + y^2), \tag{2.32}$$

Consider a planar system of ordinary differential equation

$$\dot{x} = f(x, y), \quad \dot{y} = g(x, y)$$

defined in an open set $U \subset \mathbf{R}^2$, the *phase space* of the system. If f and g are at least of class C^1, then for each point (ξ, η) of U there is a unique solution of the system, $x = x(t, \xi, \eta)$, $y = y(t, \xi, \eta)$, such that $x(0, \xi, \eta) = \xi$ and $y(0, \xi, \eta) = \eta$. Therefore, U is a disjoint union of the trajectories defined by the solution paths. The collection of these trajectories is called the *phase portrait* of the system and it gives an overall picture of the solution paths.

Let us consider an autonomous one-degree of freedom Hamiltonian system

$$H(x, y) = \frac{1}{2}y^2 + V(x)$$

defined by a potential function $V = V(x)$, and let us describe its phase portrait based on the qualitative behavior of the function V.

Since the system is autonomous the function H is constant along any solution, say $H(x(t), y(t)) = h$, the energy of the solution. Since the level set $h = H(x, y)$ is one-dimensional the trajectory of the solution path $(x(t), y(t))$ is the curve $h =$

$H(x, y)$. Now we have $h - V(x(t)) = \frac{1}{2}y(t)^2$, so $x(t)$ stays in the set $V(x) \leq h$ and $y(t)$ is computed from the equation $|y(t)| = \sqrt{2(h - V(x(t)))}$; hence, we get two values of $y(t)$, which are symmetric with respect to the origin.

To fix the ideas, we take the function $V(x)$ defined for $x > 0$, decreasing from infinity to a local minimum, then increasing to a local maximum and finally decreasing to zero as x goes to $+\infty$.

We draw the graph G_V of V in the plane (x, y) and right below it we consider the phase space with the y axis in the same vertical as the first figure.

For a given value h we draw vertical lines from the points of intersection of G_V with the horizontal line $y = h$ and consider the intersections of these lines with the x-axis of the figure on the bottom. The intervals of this axis on which the graph G_V lies below the line $y = h$ are the regions of possible motions with energy h. For each point x in these intervals we find the two values $y = \pm\sqrt{h - V(x)}$. The phase points corresponding to the intersections of G_V with the line $y = h$ are points where the trajectory cuts the x-axis of the phase plane. With this procedure we get the curves in phase space that have energy h; they are symmetric with respect to the x-axis (Fig. 2.8).

Notice that the local minimum corresponds to a stable equilibrium, surrounded by closed paths that describe periodic motions. From the stable equilibrium departs a trajectory that comes back to it and is the limiting position of the closed trajectories. This curve defines the separatrix that separates the closed trajectories from the unbounded ones. The local maximum corresponds to an unstable equilibrium. To this equilibrium a separatrix contained in the half-plane $y < 0$ and another separatrix contained in the half-plane $y > 0$ also arrive. These two curves separate the trajectories that come closest to the origin from the trajectories that stay to the right of the unstable equilibrium. The curves obtained in this way describe the phase portrait of the system.

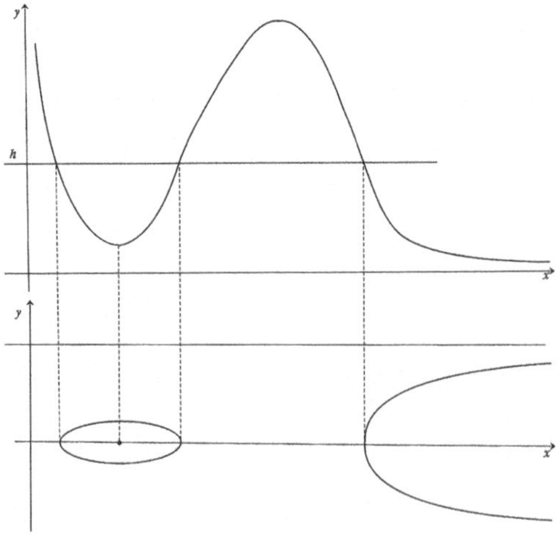

Fig. 2.8 Phase portrait of a one-degree of freedom system

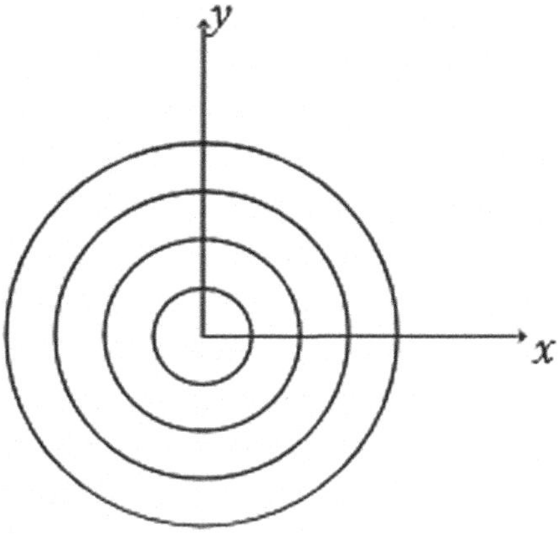

Fig. 2.9 Phase portrait of a harmonic oscillator

In the next example we describe the phase portrait of the two one-degree of freedom Hamiltonian systems seen above, namely the harmonic oscillator and the pendulum.

Example 2.8 (Phase Portraits of the Harmonic Oscillator and the Pendulum) Since the Hamiltonian of the harmonic oscillator is (2.32) its potential function is $V(x) = \omega x^2/2$. The graph of $y = V(x)$ is a parabola with a minimum.

So the phase portrait (Fig. 2.9) consists of one stable equilibrium surrounded by closed curves whose diameters increase from zero do infinity. They are given by $x = A\sin(\omega t + \phi)$, $y = A\omega\cos(\omega t + \phi)$ and represent circles depending on the initial conditions. The Hamiltonian of the pendulum is given in (2.30), so the potential function in this case is $V(x) = \alpha\cos x$.

The graph of the co-sine has a local minimum at every odd multiple of π and a local maximum at every even multiple of π (Fig. 2.10). So there is an alternating sequence of stable and unstable equilibria. There are infinitely many separatrices with any pair of them being symmetrically situated with respect to the x-axis, one going, in the upper half-plane, from one unstable equilibrium to the next one, the other going back from this unstable equilibrium to the previous one, in the lower half-plane. In the center of the region bounded by these two separatrices lies, on the x-axis the stable equilibrium, surrounded by a family of closed orbits that fill up the whole region.

The next example deals with the motion of a rigid body about a fixed point in the absence of external forces. We first make some considerations about the motion of a rigid body about a fixed point.

Consider two rectangular coordinate systems with the same origin, $OXYZ$ and $Oxyz$ (Fig. 2.11). Denote by E_X, E_Y, E_Z and $\mathbf{e}_x, \mathbf{e}_y, \mathbf{e}_z$ their oriented frames, so $E_Z = E_X \times E_Y$ and $\mathbf{e}_z = \mathbf{e}_x \times \mathbf{e}_y$. The *line of nodes* is the line given by the intersection of the planes OXY and Oxy when it is defined. Let ON be the axis defined by the line of nodes oriented by the unit vector $\mathbf{e}_N = E_Z \times \mathbf{e}_z$.

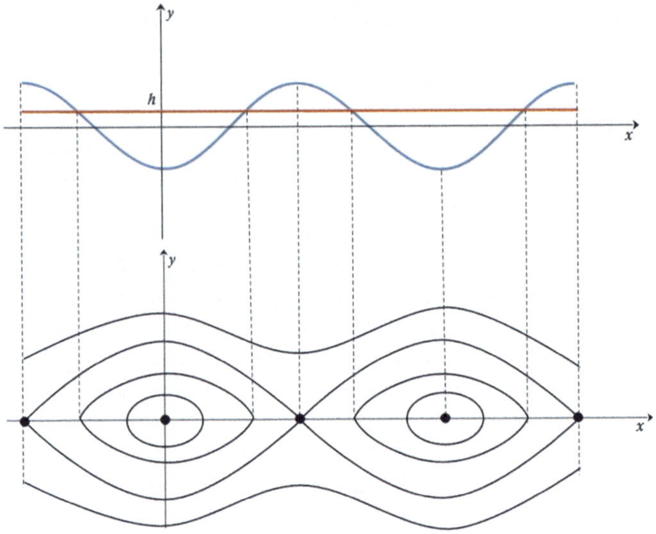

Fig. 2.10 Phase portrait of a pendulum

Fig. 2.11 Euler angles

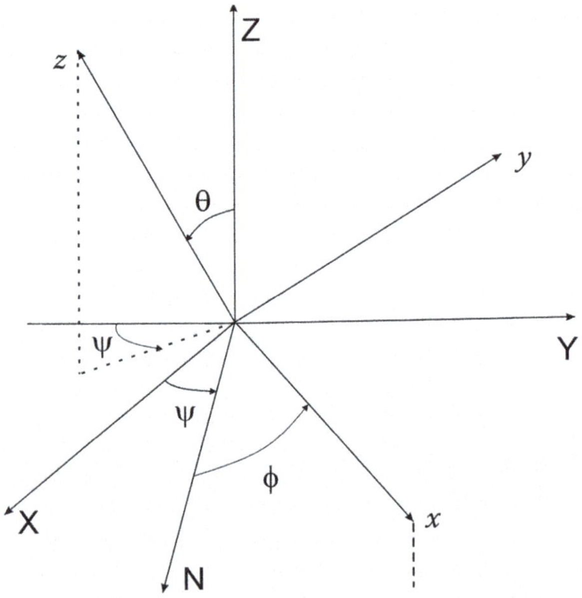

The angle ψ from $\mathbf{e}_x$ to $\mathbf{e}_N$ is called the *precession angle*, the angle θ from E_Z to $\mathbf{e}_z$ is the *nutation angle* and the angle ϕ from $\mathbf{e}_N$ to $\mathbf{e}_x$ is the *angle of proper rotation*. These are called the *Euler angles* of the two frames.

The orthogonal projection of the axis Oz onto the plane OXY makes with the axis OY an angle equal to ψ; hence, we have $\mathbf{e}_z = a_{13}E_X + a_{23}E_Y + a_{33}E_Z$, with

$$a_{13} = \sin\theta \sin\psi, \quad a_{23} = -\sin\theta\cos\psi, \quad a_{33} = \cos\theta. \tag{2.33}$$

Obviously, we have $\mathbf{e}_N = \cos\psi E_X + \sin\psi E_Y$; hence, for the vector $\mathbf{e}_N^{\perp} = \mathbf{e}_z \times \mathbf{e}_N$ we get the expression

$$\mathbf{e}_N^{\perp} = -\cos\theta\sin\psi E_X + \cos\theta\cos\psi\, E_Y + \sin\theta\, E_Z.$$

Since $\mathbf{e}_x = \cos\phi\mathbf{e}_N + \sin\phi\mathbf{e}_N^{\perp}$ and $\mathbf{e}_y = -\sin\phi\mathbf{e}_N + \cos\phi\mathbf{e}_N^{\perp}$, we immediately get the linear combinations $\mathbf{e}_x = a_{11}E_X + a_{21}E_Y + a_{31}E_Z$, $\mathbf{e}_y = a_{12}E_X + a_{22}E_Y + a_{32}E_Z$, and find that the transition matrix $\mathcal{A} = [a_{ij}]$ from the basis E_X, E_Y, E_Z to the basis $\mathbf{e}_1, \mathbf{e}_2, \mathbf{e}_3$ is the orthogonal matrix

$$\mathcal{A} = \begin{bmatrix} \cos\phi\cos\psi - \sin\phi\cos\theta\sin\psi & -\sin\phi\cos\psi - \cos\phi\cos\theta\sin\psi & \sin\theta\sin\psi \\ \cos\phi\sin\psi + \sin\phi\cos\theta\cos\psi & -\sin\phi\sin\psi + \cos\phi\cos\theta\cos\psi & -\sin\theta\cos\psi \\ \sin\theta\sin\phi & \sin\theta\cos\phi & \cos\theta \end{bmatrix}. \tag{2.34}$$

Consider now a rigid body S with a center of mass fixed at the origin of the inertial rectangular coordinate system, $OXYZ$. Denote by $\mathbf{r}_0$ the position vectors of points of S.

The linear mapping $\mathbf{I} : \mathbf{R}^3 \to \mathbf{R}^3$ defined by

$$\mathbf{I}(\mathbf{x}) = \int_S \mathbf{r}_0 \times (\mathbf{x} \times \mathbf{r}_0)\, dm \tag{2.35}$$

is symmetric because since $\mathbf{r}_0 \times (\mathbf{x} \times \mathbf{r}_0) = \langle \mathbf{r}_0, \mathbf{r}_0\rangle \mathbf{x} - \langle \mathbf{r}_0, \mathbf{x}\rangle \mathbf{r}_0$ we have

$$\langle \mathbf{I}(\mathbf{x}), \mathbf{y}\rangle = \int_S (\langle \mathbf{r}_0, \mathbf{r}_0\rangle\langle \mathbf{x}, \mathbf{y}\rangle - \langle \mathbf{r}_0, \mathbf{x}\rangle\langle \mathbf{r}_0, \mathbf{y}\rangle)\, dm$$

and the value of the integral on the right-hand side of this equality does not change if we interchange the vectors $\mathbf{x}$ and $\mathbf{y}$, so $\mathbf{I}$ is a symmetric operator.

The mapping (2.35) is called the *inertia operator* or the *tensor of inertia* of the body.

From linear algebra we know that $\mathbf{I}$ has an orthonormal basis of eigenvectors, $\mathbf{e}_1, \mathbf{e}_2, \mathbf{e}_3$. If $\mathbf{I}(\mathbf{e}_k) = I_k \mathbf{e}_k$, we see from the above expression of $\langle \mathbf{I}(\mathbf{x}), \mathbf{y}\rangle$ that

$$I_k = \langle \mathbf{I}(\mathbf{e}_k), \mathbf{e}_k\rangle = \int_S (\|\mathbf{r}_0\|^2 - \langle \mathbf{r}_0, \mathbf{e}_k\rangle^2)\, dm = \int_S d_k^2\, dm,$$

where d_k is the distance from $\mathbf{r}_0$ to the axis defined by $\mathbf{e}_k$, so I_k is the moment of inertia of the body relative to this axis. The numbers I_1, I_2, I_3 are called the *principal moments of inertia* of the body and the axes defined by $\mathbf{e}_1, \mathbf{e}_2, \mathbf{e}_3$ are called the *principal axes of inertia*.

Consider the quadratic form $Q(\mathbf{x}) = \langle \mathbf{I}(\mathbf{x}), \mathbf{x}\rangle$. If $\mathbf{x} = x_1\mathbf{e}_1 + x_2\mathbf{e}_2 + x_3\mathbf{e}_3$, then $Q(\mathbf{x}) = I_1x_1^2 + I_2x_2^2 + I_3x_3^2$. The quadric surface

$$I_1x_1^2 + I_2x_2^2 + I_3x_3^2 = 1$$

is called the *ellipsoid of inertia* of the body.

Since $\|\mathbf{r}_0\|^2 - \langle \mathbf{r}_0, \mathbf{e}_k\rangle^2 = d_k^2 \geq 0$, if none of the moments of inertia is zero this is indeed an ellipsoid with semi-axes $\frac{1}{\sqrt{I_1}}, \frac{1}{\sqrt{I_2}}, \frac{1}{\sqrt{I_3}}$. If only one of the moments of inertia is zero the quadric is a cylinder; if two are zero it is a pair of parallel planes.

Now we consider the body in motion. Take a rectangular coordinate system $Oxyz$ with the same origin as the inertial system, rigidly attached to the body, defined by the principal axes of inertia with the frame $\mathbf{e}_1, \mathbf{e}_2, \mathbf{e}_3$ of the body at rest.

At any instant of time the position of the body relative to the inertial system is determined by a rotation $R = R(t)$, so $S_t = R(t)S$. The matrix of $R(t)$ in the basis E_X, E_Y, E_Z is the above matrix $A = A(t)$ given in terms of the Euler angles of the frames E_X, E_Y, E_Z and $\mathbf{e}_1(t), \mathbf{e}_2(t), \mathbf{e}_3(t)$, with $\mathbf{e}_k(t) = R(t)\mathbf{e}_k$, $k = 1, 2, 3$.

Each point of the body in motion, $S_t = R(t)S$, comes from its position $\mathbf{r}_0 \in S$ at time $t = 0$ and at the instant t it has the position vector $\mathbf{r}(t) = R(t)\mathbf{r}_0$, relative to the inertial system, so

$$\dot{\mathbf{r}}(t) = \dot{R}(t)\mathbf{r}_0 = \Sigma(t)\mathbf{r}(t), \quad \text{where} \quad \Sigma = \dot{R}R^{-1}\mathbf{r}.$$

Since R is an orthogonal matrix Σ is a skew-symmetric matrix and if we write it as

$$\Sigma = \begin{bmatrix} 0 & -\omega_3 & \omega_2 \\ \omega_3 & 0 & -\omega_1 \\ -\omega_2 & \omega_1 & 0 \end{bmatrix}$$

and consider the vector $\boldsymbol{\omega} = (\omega_1, \omega_2, \omega_3)$ we have $\Sigma\mathbf{x} = \boldsymbol{\omega} \times \mathbf{x}$ for any vector $\mathbf{x} \in \mathbf{R}^3$. Therefore, we have

$$\dot{\mathbf{r}} = \boldsymbol{\omega} \times \mathbf{r}. \tag{2.36}$$

The vector $\boldsymbol{\omega} = \boldsymbol{\omega}(t)$ is called the instantaneous *angular velocity* of the body in motion.

The angular momentum of an infinitesimal mass dm of the body in motion, located at the position vector $\mathbf{r}$ is $dm(\mathbf{r} \times \dot{\mathbf{r}})$, so the total angular momentum of the rigid body is given by

$$\mathbf{c} = \int_{S_t} (\mathbf{r} \times \dot{\mathbf{r}})\,dm = \int_{S_t} \mathbf{r} \times (\boldsymbol{\omega} \times \mathbf{r})\,dm.$$

Let $\mathbf{r} = R(t)\mathbf{r}_0$ and $\boldsymbol{\omega} = R(t)\boldsymbol{\Omega}$. Since rotations preserve cross products, that is $R(\mathbf{x} \times \mathbf{y}) = R\mathbf{x} \times R\mathbf{y}$, we have $\mathbf{r} \times (\boldsymbol{\omega} \times \mathbf{r}) = R(t)\big(\mathbf{r}_0 \times (\boldsymbol{\Omega} \times \mathbf{r}_0)\big)$; hence, we get

$$\mathbf{c} = \int_{S_t} R(t)\big(\mathbf{r}_0 \times (\boldsymbol{\Omega} \times \mathbf{r}_0)\big)\,dm = R(t)\int_S \big(\mathbf{r}_0 \times (\boldsymbol{\Omega} \times \mathbf{r}_0)\big)\,dm = R(t)\mathbf{C},$$

where by (2.35), $\mathbf{C} = \mathbf{I}(\boldsymbol{\Omega})$.

Let $\mathbf{I}_t(\mathbf{x}) = \displaystyle\int_{S_t} \mathbf{r} \times (\mathbf{x} \times \mathbf{r})\,dm$ be the inertia operator for S_t. Then, setting $\mathbf{x} = R(t)\mathbf{x}_0$, we get

$$\mathbf{I}_t(\mathbf{x}) = R(t)\int_S \mathbf{r}_0 \times (\mathbf{x}_0 \times \mathbf{r}_0)\,dm = R(t)\mathbf{I}(\mathbf{x}_0),$$

so $\mathbf{I}_t = R(t)\mathbf{I}R(t)^{-1}$. Therefore, $\mathbf{I}_t\mathbf{e}_k(t) = R(t)\mathbf{I}\mathbf{e}_k = I_k R(t)\mathbf{e}_k$ that is

$$\mathbf{I}_t\mathbf{e}_k(t) = I_k\mathbf{e}_k(t).$$

We see that the principal moments of inertia stay constant along the motion of the body and the principal axes of inertia are rotated from those of the body at rest through the rotation $R(t)$.

Example 2.9 (Equations of Motion of a Free Rigid Body) Suppose that a rigid body moves about its center of mass, fixed in space, and in the absence of external forces. Take a fixed inertial system $OXYZ$ with origin at the center of mass.

For translational motion, Newton's second law says that in an inertial coordinate system the rate of change of the linear momentum is equal to the total external force. For rotatory motion, the rate of change of the angular momentum is equal to the moment of total external force, or torque.

Since the external force on the body is zero, the torque is also zero, so the angular momentum $\mathbf{c}$ in the inertial coordinate system is constant.

Differentiating $\mathbf{c} = R(t)\mathbf{C}$ we get, using the fact that $\dot{R} = \Sigma R$,

$$0 = \dot{\mathbf{c}} = \dot{R}\mathbf{C} + R\dot{\mathbf{C}} = R\big(R^{-1}\Sigma R\mathbf{C} + \dot{\mathbf{C}}\big), \quad \text{hence} \quad R^{-1}\Sigma R\mathbf{C} + \dot{\mathbf{C}} = 0.$$

But $\Sigma R\mathbf{C} = \boldsymbol{\omega} \times R\mathbf{C}$, so $R^{-1}\Sigma R\mathbf{C} = R^{-1}(\boldsymbol{\omega} \times R\mathbf{C}) = \boldsymbol{\Omega} \times \mathbf{C}$. Therefore, we get

$$\dot{\mathbf{C}} = \mathbf{C} \times \boldsymbol{\Omega}. \tag{2.37}$$

Let $\boldsymbol{\Omega} = p\mathbf{e}_1 + q\mathbf{e}_2 + r\mathbf{e}_3$. Then, since $\mathbf{I}\mathbf{e}_k = I_k\mathbf{e}_k$, we get $\mathbf{C} = \mathbf{I}(\boldsymbol{\Omega}) = I_1 p\mathbf{e}_1 + I_2 q\mathbf{e}_2 + I_3 r\mathbf{e}_3$ and the vector equation (2.37) gives the system of Euler equations for the motion of a free rigid body about its center of mass

$$I_1\dot{p} = (I_2 - I_3)qr, \quad I_2\dot{q} = (I_3 - I_1)pr, \quad I_3\dot{r} = (I_1 - I_2)pq. \tag{2.38}$$

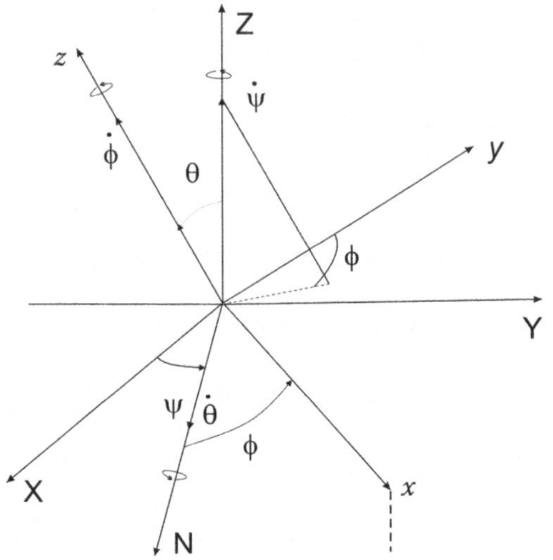

Fig. 2.12 Angular velocities of the Euler frame

We will find the Hamiltonian for the free rigid body after we discuss the Lagrangian equations of motion in the next section.

To get the Hamiltonian we need the expressions of p, q, r in terms of the Euler angles and also the expression of the kinetic energy of the moving body. Let us consider this now.

In connection with the definition of the Euler angles let R_1 be the rotation about the axis OZ by the angle $\dot{\psi}$, let R_2 be the rotation about the axis ON by the angle $\dot{\theta}$, and R_3 the rotation about the axis Oz by the angle $\dot{\phi}$; their angular velocities are therefore $\dot{\psi} E_Z, \dot{\theta}\mathbf{e}_N$ and $\dot{\phi}\mathbf{e}_3(t)$ respectively (Fig. 2.12). The composition of the rotations $R_3 \circ R_2 \circ R_1$ is a rotation with angular velocity

$$\boldsymbol{\omega} = \dot{\psi} E_Z + \dot{\theta}\mathbf{e}_N + \dot{\phi}\mathbf{e}_3(t),$$

so

$$\boldsymbol{\Omega} = R(t)^{-1}\boldsymbol{\omega} = \dot{\psi} R(t)^{-1} E_Z + \dot{\theta} R(t)^{-1}\mathbf{e}_N + \dot{\phi} R(t)^{-1}\mathbf{e}_3(t). \tag{2.39}$$

The transition matrix from $\mathbf{e}_1(t), \mathbf{e}_2(t), \mathbf{e}_3(t)$ to E_X, E_Y, E_Z is the transpose of the matrix A given in (2.34) in terms of the Euler angles of these two frames. Therefore, we have

$$\begin{aligned} E_X &= a_{11}\mathbf{e}_1(t) + a_{12}\mathbf{e}_2(t) + a_{13}\mathbf{e}_3(t), \\ E_Y &= a_{21}\mathbf{e}_1(t) + a_{22}\mathbf{e}_2(t) + a_{23}\mathbf{e}_3(t), \\ E_Z &= a_{31}\mathbf{e}_1(t) + a_{32}\mathbf{e}_2(t) + a_{33}\mathbf{e}_3(t); \end{aligned} \tag{2.40}$$

hence, for the vector $\mathbf{e}_N = \cos\psi E_X + \sin\psi E_Y$ we have $\mathbf{e}_N = (\cos\psi a_{11} + \sin\psi a_{21})\mathbf{e}_1(t) + (\cos\psi a_{12} + \sin\psi a_{22})\mathbf{e}_2(t) + (\cos\psi a_{13} + \sin\psi a_{23})\mathbf{e}_3(t)$ and using the expressions of the entries a_{ij} of A we find, for $\mathbf{e}_N$ and E_Z:

$$\mathbf{e}_N = \cos\phi\,\mathbf{e}_1(t) - \sin\phi\,\mathbf{e}_2(t), \quad E_Z = \sin\theta\sin\psi\,\mathbf{e}_1(t) - \sin\theta\cos\psi\,\mathbf{e}_2(t) + \cos\theta\,\mathbf{e}_3(t).$$

Inserting these into (2.39) and recalling that $\mathbf{e}_k(t) = R(t)\mathbf{e}_k$, $k = 1, 2, 3$ we get

$$\boldsymbol{\Omega} = (\dot\theta\cos\phi + \dot\psi\sin\theta\sin\phi)\mathbf{e}_1 + (-\dot\theta\sin\phi + \dot\psi\sin\theta\cos\phi)\mathbf{e}_2 + (\dot\phi + \dot\psi\cos\theta)\mathbf{e}_3.$$

Since $\boldsymbol{\Omega} = p\mathbf{e}_1 + q\mathbf{e}_2 + r\mathbf{e}_3$, we obtain Euler's kinematic equations

$$p = \dot\psi\sin\theta\sin\phi + \dot\theta\cos\phi, \quad q = \dot\psi\sin\theta\cos\phi - \dot\theta\sin\phi, \quad r = \dot\psi\cos\theta + \dot\phi. \tag{2.41}$$

Now we consider the question of the kinetic energy of the moving body. For an infinitesimal mass dm located at the point $\mathbf{r}$ of the moving body S_t, the kinetic energy is $\frac{1}{2}dm\|\dot{\mathbf{r}}\|^2 = \frac{1}{2}\|\boldsymbol{\omega}\times\mathbf{r}\|^2 dm$, so the kinetic energy of the body S_t is given by

$$T = \frac{1}{2}\int_{S_t}\|\boldsymbol{\omega}\times\mathbf{r}\|^2 dm = \frac{1}{2}\int_S \|\boldsymbol{\Omega}\times\mathbf{r}_0\|^2 dm.$$

Using the identity $\langle\mathbf{r}_0\times(\mathbf{x}\times\mathbf{r}_0), \mathbf{y}\rangle = \langle\mathbf{x}\times\mathbf{r}_0, \mathbf{y}\times\mathbf{r}_0,\rangle$, we get from (2.35), $\langle\mathbf{I}(\mathbf{x}), \mathbf{x}\rangle = \int_S \|\mathbf{x}\times\mathbf{r}_0\|^2 dm$ and therefore we get the expression of the kinetic energy of the body S_t

$$T = \frac{1}{2}\langle\mathbf{I}(\boldsymbol{\Omega}), \boldsymbol{\Omega}\rangle = \frac{1}{2}(Ap^2 + Bq^2 + Cr^2), \tag{2.42}$$

where A, B, C are the principal moments of inertia of the rigid body. We will use (2.41) and (2.42) in the examples of the next section.

Example 2.10 (Lagrange's Top) We consider a symmetrical rigid body moving about one of its points O, fixed in space, under the action of the gravity field of the Earth, *Lagrange's top* (Fig. 2.13).

We take a fixed inertial coordinate system $OXYZ$ with origin at O and axis OZ directed upward. We denote by $\mathbf{r}_0$ the position vector of the center of mass of the top and assume that the symmetry axis of the top is the axis defined by the vector $\mathbf{r}_0$, which is given by

$$\mathbf{r}_0 = l\big(\sin\theta\sin\psi E_X - \sin\theta\cos\psi E_Y + \cos\theta E_Z\big), \tag{2.43}$$

where l is the distance from O to the center of mass of the top and ψ, θ, ϕ are the Euler angles defined using the frame E_X, E_Y, E_Z and the frame $\mathbf{e}_1, \mathbf{e}_2, \mathbf{e}_3$ given by the principal axes of inertia of the top.

Fig. 2.13 Lagrange's top

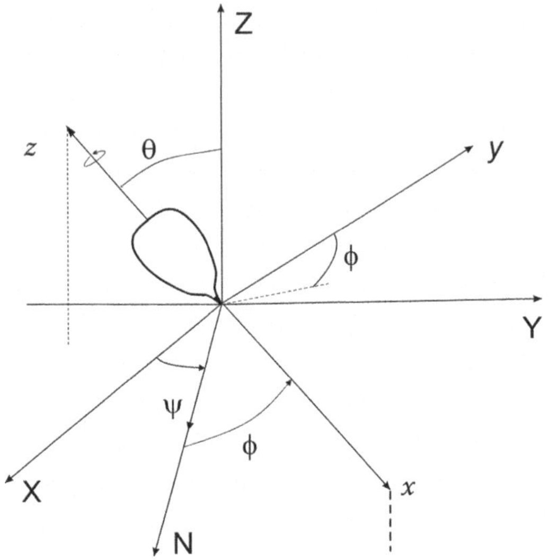

The force is $\mathbf{F} = -\left(\int_{S_t} g\,dm\right)\mathbf{k} = -Mg\mathbf{k}$, where M is the mass of the top and g is the acceleration of gravity.

The torque about the vector $\mathbf{r}_0$ is $\mathbf{M}_0 = \mathbf{r}_0 \times \mathbf{F} = Mg\sin\theta(\cos\psi E_X + \sin\psi E_Y)$,, that is,

$$\mathbf{M}_0 = Mgl\sin\theta\,\mathbf{e}_N,$$

where $\mathbf{e}_N = \cos\psi E_X + \sin\psi E_Y$ is a unit vector along the node. The dynamical equation defined by Newton's second law, $\dot{\mathbf{c}} = \mathbf{M}_0$, corresponding to (2.37) is now

$$\dot{\mathbf{C}} = \mathbf{C} \times \boldsymbol{\Omega} + Mgl\sin\theta R(t)^{-1}\mathbf{e}_N. \tag{2.44}$$

Since $R(t)^{-1}\mathbf{e}_N = \cos\phi\,\mathbf{e}_1 - \sin\phi\,\mathbf{e}_2$, then with $\Omega = p\mathbf{e}_1 + q\mathbf{e}_2 + r\mathbf{e}_3$, instead of (2.38) we now have the following differential equations:

$$\begin{aligned} I_1\dot{p} &= (I_2 - I_3)qr + Mgl\sin\theta\cos\phi, \\ I_2\dot{q} &= (I_3 - I_1)\,pr - Mgl\sin\theta\sin\phi, \\ I_3\dot{r} &= (I_1 - I_2)\,pq. \end{aligned} \tag{2.45}$$

Since the symmetry axis of the top is along the vector $\mathbf{r}_0$ we have the relation $I_1 = I_2$ among the moments of inertia, so the Eqs. (2.45) become

$$\dot{p} = \frac{I_2 - I_3}{I_1}qr + \frac{Mgl}{I_1}\sin\theta\cos\phi, \quad \dot{q} = \frac{I_3 - I_1}{I_2}pr - \frac{Mgl}{I_2}\sin\theta\sin\phi, \quad \dot{r} = 0.$$

We see that the variable

$$r = \dot{\psi}\cos\theta + \dot{\phi}$$

remains constant along the motion, that is, the function $r = r(\psi, \theta, \phi)$ is a first integral.

2.3 Variational Systems

A general context in which Hamiltonian systems appear naturally is that of variational problems. This is the problem of finding the paths in the set $\Omega = \{\mathbf{x} : [a, b] \to U \subset \mathbf{R}^n\}$, which minimize the *action*, that is, the functional defined by

$$\mathcal{A}_L = \int_a^b L(\mathbf{x}, \dot{\mathbf{x}})dt, \tag{2.46}$$

corresponding to a function $L : U \times \mathbf{R}^n \to \mathbf{R}$ called the *Lagrangian* of the problem. We assume that all the data are at least of class C^2 and that the Hessian matrix $L_{\dot{\mathbf{x}}\dot{\mathbf{x}}}$ of L with respect to the variable $\dot{\mathbf{x}}$ has nonzero determinant; in this case we say that the Lagrangian is regular.

A *proper variation* of the path $\mathbf{x} : [a, b] \to U$ is a one-parameter family of paths $\mathbf{x}(t, \alpha)$ of class C^2 in $[a, b] \times (-\epsilon, \epsilon)$ such that (Fig. 2.14)

$$\begin{aligned} &\mathbf{x}(t, 0) = \mathbf{x}(t) \quad \text{and} \quad \mathbf{x}(a, \alpha) = \mathbf{x}(a), \\ &\mathbf{x}(b, \alpha) = \mathbf{x}(b), \text{ for all } \alpha. \end{aligned}$$

Let $\mathcal{A}(\alpha)$ be the value of the action on the path $\mathbf{x}_\alpha(t)$. Then,

$$\mathcal{A}_L(\alpha) = \int_a^b L\left(\mathbf{x}(t, \alpha), \frac{\partial \mathbf{x}}{\partial t}(t, \alpha)\right) dt.$$

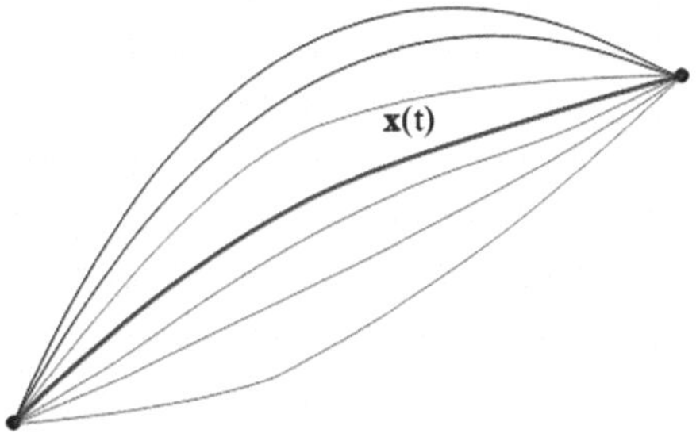

Fig. 2.14 Proper variation

The derivative $\mathcal{A}_L'(\alpha)$ is found by differentiating with respect to α under the integral sign. Using the notation $\mathbf{u}(t,\alpha) = \left(\mathbf{x}(t,\alpha), \frac{\partial \mathbf{x}}{\partial t}(t,\alpha)\right)$ we obtain by (1.11)

$$\frac{\partial L}{\partial \alpha} = \left\langle L_{\mathbf{x}}\big(\mathbf{u}(t,\alpha)\big), \frac{\partial \mathbf{x}}{\partial \alpha}(t,\alpha)\right\rangle + \left\langle L_{\dot{\mathbf{x}}}\big(\mathbf{u}(t,\alpha)\big), \frac{\partial}{\partial \alpha}\frac{\partial \mathbf{x}}{\partial t}(t,\alpha)\right\rangle dt.$$

Since $\frac{\partial}{\partial \alpha}\frac{\partial \mathbf{x}}{\partial t} = \frac{\partial}{\partial t}\frac{\partial \mathbf{x}}{\partial \alpha}$, we have that

$$\left\langle L_{\dot{\mathbf{x}}}\big(\mathbf{u}(t,\alpha)\big), \frac{\partial}{\partial \alpha}\frac{\partial \mathbf{x}}{\partial t}(t,\alpha)\right\rangle = \frac{d}{dt}\left\langle L_{\dot{\mathbf{x}}}\big(\mathbf{u}(t,\alpha)\big), \frac{\partial \mathbf{x}}{\partial \alpha}\right\rangle - \left\langle \frac{d}{dt}L_{\dot{\mathbf{x}}}\big(\mathbf{u}(t,\alpha)\big), \frac{\partial \mathbf{x}}{\partial \alpha}\right\rangle;$$

hence,

$$\frac{\partial L}{\partial \alpha} = \frac{d}{dt}\left\langle L_{\dot{\mathbf{x}}}\big(\mathbf{u}(t,\alpha)\big), \frac{\partial \mathbf{x}}{\partial \alpha}(t,\alpha)\right\rangle + \left\langle L_{\mathbf{x}}\big(\mathbf{u}(t,\alpha)\big) - \frac{d}{dt}L_{\dot{\mathbf{x}}}\big(\mathbf{u}(t,\alpha)\big), \frac{\partial \mathbf{x}}{\partial \alpha}(t,\alpha)\right\rangle.$$

Since the variation is proper, we have $\left\langle L_{\dot{\mathbf{x}}}\big(\mathbf{u}(t,\alpha)\big), \frac{\partial \mathbf{x}}{\partial \alpha}(t,\alpha)\right\rangle\Big|_a^b = 0$; hence, upon integration we get for the derivative of the action the expression

$$\mathcal{A}_L'(\alpha) = \int_a^b \left\langle \mathcal{L}(\mathbf{x}, \dot{\mathbf{x}}), \frac{\partial \mathbf{x}}{\partial \alpha}(t,\alpha)\right\rangle dt, \tag{2.47}$$

where $\mathcal{L}(\mathbf{x}, \dot{\mathbf{x}}) = L_{\mathbf{x}}\big(\mathbf{u}(t,\alpha)\big) - \frac{d}{dt}L_{\dot{\mathbf{x}}}\big(\mathbf{u}(t,\alpha)\big)$ is a vector field along $\mathbf{x}(t)$ (Fig. 2.15).

Now, take the proper variation of $\mathbf{x}(t)$ given by

$$\mathbf{x}(t,\alpha) = \mathbf{x}(t) + \phi(t,\alpha)\mathcal{L}(\mathbf{x}, \dot{\mathbf{x}})$$

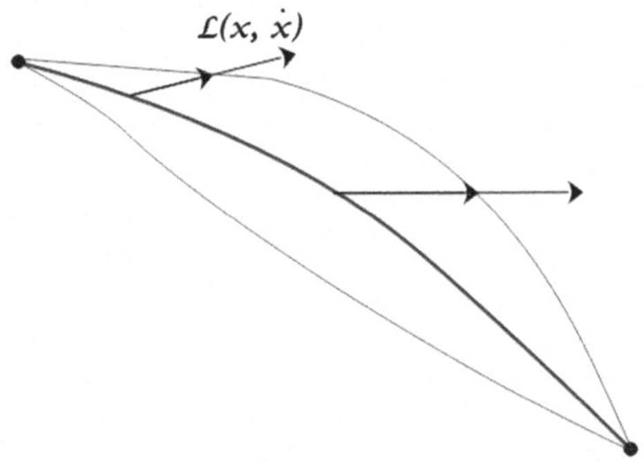

Fig. 2.15 Variation induced by $\mathcal{L}$

where $\phi(t,\alpha) = \alpha(t-a)(b-t)$, considering ϵ so small that $\mathbf{x}(t,\alpha) \in U$, for all $(t,\alpha) \in [a,b] \times (-\epsilon,\epsilon)$. As $\dfrac{\partial \phi}{\partial \alpha}(t,0) > 0$ in (a,b), it follows from (2.47) that

$$\mathcal{A}_L'(0) = 0 \quad \text{if, and only if,} \quad \frac{d}{dt} L_{\dot{\mathbf{x}}}\big(\mathbf{x}(t), \dot{\mathbf{x}}(t)\big) - L_{\mathbf{x}}\big(\mathbf{x}(t), \dot{\mathbf{x}}(t)\big) = 0. \tag{2.48}$$

The conclusion of the above discussion is that:

The path $\mathbf{x}(t)$ is a stationary point of the action (2.46) if, and only if, it is a solution of the differential equation defined in (2.48), known as the Euler–Lagrange equation of the variational problem.

Given a real-valued function $f(\mathbf{x}, \dot{\mathbf{x}})$ the *Lagrangian derivative* of f along the path $\mathbf{x}(t)$ is defined as in (2.48)

$$\mathcal{L}_{\mathbf{x}} f = \frac{d}{dt} \frac{\partial f}{\partial \dot{\mathbf{x}}}(\mathbf{x}(t), \dot{\mathbf{x}}(t)) - \frac{\partial f}{\partial \mathbf{x}}(\mathbf{x}(t), \dot{\mathbf{x}}(t)), \tag{2.49}$$

so the path $\mathbf{x}(t)$ is a solution of the variational problem if, and only if, $\mathcal{L}_{\mathbf{x}} L = 0$.

Change variables by means of $\mathbf{x} = \Phi(\mathbf{y})$ and set $\dot{\mathbf{x}} = D\Phi(\mathbf{y}) \cdot \dot{\mathbf{y}}$ and $g(\mathbf{y}, \dot{\mathbf{y}}) = f(\mathbf{x}, \dot{\mathbf{x}})$. Then we have

Proposition 2.3 *The Lagrangian derivative changes according to the formula*

$$\mathcal{L}_{\mathbf{y}} g = D\Phi(\mathbf{y})^T \cdot \mathcal{L}_{\mathbf{x}} f,$$

so the Euler–Lagrange equation is invariant under an arbitrary change of coordinates.

Proof By the chain rule we have $\dfrac{\partial g}{\partial \dot{\mathbf{y}}} \cdot \eta = \left\langle \dfrac{\partial f}{\partial \dot{\mathbf{x}}}, D\Phi(\mathbf{y}) \cdot \eta \right\rangle$, so

$$\frac{d}{dt}\left(\frac{\partial g}{\partial \dot{\mathbf{y}}} \cdot \eta\right) = \left\langle \frac{d}{dt}\left(\frac{\partial f}{\partial \dot{\mathbf{x}}}\right), D\Phi(\mathbf{y}) \cdot \eta \right\rangle + \left\langle \frac{\partial f}{\partial \dot{\mathbf{x}}}, \left(D^2\Phi(\mathbf{y}) \cdot \dot{\mathbf{y}}\right) \cdot \eta \right\rangle.$$

On the other hand,

$$\frac{\partial g}{\partial \mathbf{y}} \cdot \eta = \left\langle \frac{\partial f}{\partial \mathbf{x}}, D\Phi(\mathbf{y}) \cdot \eta \right\rangle + \left\langle \frac{\partial f}{\partial \dot{\mathbf{x}}}, \left(D^2\Phi(\mathbf{y}) \cdot \eta\right) \cdot \dot{\mathbf{y}} \right\rangle.$$

Because $D^2\Phi(\mathbf{y})$ is symmetric, subtraction yields

$$\left(\frac{d}{dt}\left(\frac{\partial g}{\partial \dot{\mathbf{y}}}\right) - \frac{\partial g}{\partial \mathbf{y}}\right) \cdot \eta = \left\langle \frac{d}{dt}\left(\frac{\partial f}{\partial \dot{\mathbf{x}}}\right) - \frac{\partial f}{\partial \mathbf{x}}, D\Phi(\mathbf{y}) \cdot \eta \right\rangle,$$

for all vectors $\eta \in \mathbf{R}^N$. From this it follows that $\mathcal{L}_{\mathbf{y}} g = D\Phi(\mathbf{y})^T \cdot \mathcal{L}_{\mathbf{x}} f$. □

The Euler–Lagrange equation (2.48) is an ordinary differential equation of second order

$$\frac{\partial^2 L}{\partial \dot{\mathbf{x}}^2}\ddot{\mathbf{x}} + \frac{\partial^2 L}{\partial \mathbf{x}\partial \dot{\mathbf{x}}}\dot{\mathbf{x}} - \frac{\partial L}{\partial \mathbf{x}} = 0.$$

As the matrix $\frac{\partial^2 L}{\partial \dot{\mathbf{x}}^2}$ is nonsingular we get $\ddot{\mathbf{x}} = F(\mathbf{x}, \dot{\mathbf{x}}, t)$ and write the above equation as a system of first-order differential equations

$$\dot{\mathbf{x}} = \mathbf{y}, \quad \dot{\mathbf{y}} = F(\mathbf{x}, \dot{\mathbf{x}}, t).$$

We can write this system in a symmetric form as follows. Let the Lagrangian $L(\mathbf{x}, \dot{\mathbf{x}})$ be regular, that is, $\det \frac{\partial^2 L}{\partial \dot{\mathbf{x}}^2}(\mathbf{x}, \dot{\mathbf{x}}) \neq 0$ and introduce the *conjugate moment* vector $\mathbf{y} = \frac{\partial L}{\partial \dot{\mathbf{x}}}(\mathbf{x}, \dot{\mathbf{x}})$. Since L is regular the implicit function theorem says that we can locally solve the defining equation of $\mathbf{y}$ to get $\dot{\mathbf{x}} = \phi(\mathbf{x}, \mathbf{y})$. Now, define the function $H(\mathbf{x}, \mathbf{y})$ by the equality

$$H(\mathbf{x}, \mathbf{y}) = \langle \dot{\mathbf{x}}, \mathbf{y} \rangle - L(\mathbf{x}, \dot{\mathbf{x}}). \tag{2.50}$$

Proposition 2.4 *The partial derivatives of H are given by*

$$\frac{\partial H}{\partial \mathbf{x}} = -\frac{\partial L}{\partial \mathbf{x}}, \quad \frac{\partial H}{\partial \mathbf{y}} = \dot{\mathbf{x}}. \tag{2.51}$$

Proof Using the chain rule we compute

$$\frac{\partial H}{\partial \mathbf{x}} \cdot \boldsymbol{\xi} = \langle \frac{\partial \phi}{\partial \mathbf{x}} \cdot \boldsymbol{\xi}, \mathbf{y} \rangle - \langle \frac{\partial L}{\partial \mathbf{x}}, \boldsymbol{\xi} \rangle - \langle \frac{\partial L}{\partial \dot{\mathbf{x}}}, \frac{\partial \phi}{\partial \mathbf{x}} \cdot \boldsymbol{\xi} \rangle = \langle -\frac{\partial L}{\partial \mathbf{x}}, \boldsymbol{\xi} \rangle + \langle \mathbf{y} - \frac{\partial L}{\partial \dot{\mathbf{x}}}, \frac{\partial \phi}{\partial \mathbf{x}} \cdot \boldsymbol{\xi} \rangle,$$

$$\frac{\partial H}{\partial \mathbf{y}} \cdot \eta = \langle \dot{\mathbf{x}}, \eta \rangle + \langle \frac{\partial \phi}{\partial \mathbf{y}} \cdot \eta, \mathbf{y} \rangle - \langle \frac{\partial L}{\partial \dot{\mathbf{x}}}, \frac{\partial \phi}{\partial \mathbf{y}} \cdot \eta \rangle = \langle \dot{\mathbf{x}}, \eta \rangle + \langle \mathbf{y} - \frac{\partial L}{\partial \dot{\mathbf{x}}}, \frac{\partial \phi}{\partial \mathbf{y}} \cdot \eta \rangle.$$

By definition of the moment $\mathbf{y}$ and the arbitrariness of $\boldsymbol{\xi}$, η we get equalities (2.51). □

Corollary 2.4.1 *Given $\mathbf{x}(t)$ let $\mathbf{y}(t) = \frac{\partial L}{\partial \dot{\mathbf{x}}}(\mathbf{x}(t), \dot{\mathbf{x}}(t))$ Then, $\mathbf{x}(t)$ is a solution of the Euler–Lagrange equation (2.48) if and only if $(\mathbf{x}(t), \mathbf{y}(t))$ is a solution of the Hamiltonian equations*

$$\dot{\mathbf{x}} = \frac{\partial H}{\partial \mathbf{y}}, \quad \dot{\mathbf{y}} = -\frac{\partial H}{\partial \mathbf{x}}. \tag{2.52}$$

Proof Since $\dot{\mathbf{y}} = \mathcal{L}_{\mathbf{x}} L + \dfrac{\partial L}{\partial \mathbf{x}}$, Eqs. (2.51) show that Eqs. (2.52) are equivalent to the Euler–Lagrange equation $\mathcal{L}_{\mathbf{x}} L = 0$. □

We have to agree that Eqs. (2.52) exhibit a beautiful symmetry.

The position $\mathbf{x}$ and the moment $\mathbf{y}$ are said to be *conjugate variables*. Variables $x_1, \ldots, x_n, y_1, \ldots, y_n$ for which a system of ordinary differential equations is expressed in the form (2.52) are called *canonical coordinates*. The integer n is called the *number of degrees of freedom* of the system.

Equality (2.50) defines the *Legendre transformation*. It transforms the Lagrangian L to the Hamiltonian H and conversely, the Hamiltonian H to the Lagrangian L.

An important variational context is that of Analytical Mechanics in which the Lagrangian L is the following combination of kinetic and potential energies,

$$L(\mathbf{x}, \dot{\mathbf{x}}) = \frac{1}{2}\|\dot{\mathbf{x}}\|^2 - V(\mathbf{x}),$$

where the norm is that associated with the inner product defined by

$$\langle \mathbf{u}, \mathbf{v} \rangle = \sum_j m_j(\mathbf{u}_j \cdot \mathbf{v}_j),$$

with $m_1, \ldots, m_n$ positive real numbers, $\mathbf{u} = (\mathbf{u}_1, \ldots, \mathbf{u}_n)$, $\mathbf{v} = (\mathbf{v}_1, \ldots, \mathbf{v}_n)$, $\mathbf{u}_j, \mathbf{v}_j \in \mathbf{R}^3$ and, for the function $V : A \to \mathbf{R}$ the argument is the vector of positions $\mathbf{x} = (\mathbf{x}_1, \ldots, \mathbf{x}_n)$, $\mathbf{x}_k \in \mathbf{R}^3$.

Example 2.11 (The Newtonian n-Body Problem) This problem is the mechanical system where $\mathbf{r} = (\mathbf{r}_1, \ldots, \mathbf{r}_n)$, $\mathbf{r}_i \in \mathbf{R}^3$ and the potential is given by $V = -\sum_{j<k} \dfrac{Gm_jm_k}{\|\mathbf{r}_j - \mathbf{r}_k\|}$.

The kinetic and potential energies of the system are given by

$$T = \frac{1}{2}\sum_{j=1}^{n} m_j\|\dot{\mathbf{r}}_j\|^2, \quad V = -\sum_{j<k} \frac{Gm_jm_k}{\|\mathbf{r}_j - \mathbf{r}_k\|},$$

and the corresponding Lagrangian $L = T - V$ gives rise, via the Legendre transformation, to the Hamiltonian function

$$H(\mathbf{x}, \mathbf{y}) = \sum_{j=1}^{n} \frac{1}{2m_j}\|\mathbf{y}_j\|^2 + V(\mathbf{x}),$$

where $\mathbf{x}_j = \mathbf{r}_j$ and $\mathbf{y}_j = m_j\dot{\mathbf{x}}_j$. (See Example 2.1).

Example 2.12 (Hamiltonian of the Free Rigid Body) We consider a rigid body moving around its center of mass fixed at a point of the space, in the absence of external forces. Since the total torque is zero the potential for the rotatory dynamics is $V = 0$ and the Lagrangian function is $L = T - V = T$ where the expression of the kinetic energy is given in (2.42).

The variables are the Euler angles $\mathbf{x} = (\psi, \theta, \phi)$ and their conjugate momenta $\mathbf{y} = (P_\psi, P_\theta, P_\phi)$, where $P_\psi = \frac{\partial L}{\partial \dot\psi}$, $P_\theta = \frac{\partial L}{\partial \dot\theta}$, $P_\phi = \frac{\partial L}{\partial \dot\phi}$. Using (2.42) we find

$$\mathbf{y} = (Ap \sin\theta \sin\phi + Bq \sin\theta \cos\phi + Cr \cos\theta, Ap \cos\phi - Bq \sin\phi, Cr). \tag{2.53}$$

Now, from (2.41) we find

$$\dot{\mathbf{x}} = \left(\frac{\sin\phi}{\sin\theta} p + \frac{\cos\phi}{\sin\theta} q,\ \cos\phi\, p - \sin\phi\, q,\ -\frac{\cos\theta \sin\phi}{\sin\theta} p - \frac{\cos\theta\cos\phi}{\sin\theta} q + r\right). \tag{2.54}$$

Computing$\langle \dot{\mathbf{x}}, \mathbf{y}\rangle$ we get $\langle \dot{\mathbf{x}}, \mathbf{y}\rangle = Ap^2 + Bq^2 + Cr^2 = 2T$; hence, Legendre's transformation (2.50) gives

$$H(\mathbf{x}, \mathbf{y}) = \frac{1}{2}\left(Ap^2 + Bq^2 + Cr^2\right).$$

Replacing the expressions of p, q, r obtained from (2.53) into $H(\mathbf{x}, \mathbf{y})$ and setting $\theta_A = \frac{A}{B}$, $\theta_C = \frac{C}{B}$ we get for H the expression

$$\begin{aligned} 2AH =& \frac{1}{\sin^2\theta}(\sin^2\phi + \theta_A \cos^2\phi)P_\psi^2 + (\cos^2\phi + \theta_A \sin^2\phi)P_\theta^2 + \\ &+\left[(\sin^2\phi + \theta_A\cos^2\phi)\cot^2\theta + \frac{\theta_A}{\theta_C}\right]P_\phi^2 \\ &+\frac{\sin 2\phi}{\sin\theta}(1-\theta_A)P_\psi P_\theta - \frac{2\cos\theta}{\sin^2\theta}(\sin^2\phi + \theta_A\cos^2\phi)P_\psi P_\phi + \\ &-\cot\theta \sin 2\phi(1-\theta_A)P_\theta P_\phi. \end{aligned} \tag{2.55}$$

Example 2.13 (Hamiltonian of Lagrange's Top) We consider the symmetrical rigid body of Example 2.10. We can write

$$\mathbf{F} = -\left(\int_{S_t} \nabla(gz)dm\right) = -\nabla\left(g\int_{S_t} z dm\right) = -\nabla(Mgl\cos\theta).$$

Therefore, for the potential $V = V(\psi, \theta, \phi)$ we have the expression

$$V = Mgl\cos\theta. \tag{2.56}$$

Notice that $l\cos\theta$ is the height of the center of mass of the top.

The kinetic energy T is given by (2.42) with p, q, r given in (2.41). Since the symmetry axis of the top is along the vector $\mathbf{r}_0$ we have the relation $I_1 = I_2 = A$ and $I_3 = C$, so we get the following expression of T

$$T = \frac{1}{2}A\left(\dot\psi^2\sin^2\theta + \dot\theta^2\right) + \frac{1}{2}C(\dot\psi\cos\theta + \dot\phi)^2. \tag{2.57}$$

Using (2.56) and (2.57) the Lagrangian $L = T - V$ is given by[1]

$$L = \frac{1}{2}A\left(\dot{\psi}^2 \sin^2\theta + \dot{\theta}^2\right) + \frac{1}{2}C(\dot{\psi}\cos\theta + \dot{\phi})^2 - Mgl\cos\theta. \tag{2.58}$$

Setting $\mathbf{x} = (\psi, \theta, \phi)$ and $\mathbf{y} = (P_\psi, P_\theta, P_\phi)$, where $P_\psi = \frac{\partial L}{\partial \dot{\psi}}$, $P_\theta = \frac{\partial L}{\partial \dot{\theta}}$, $P_\phi = \frac{\partial L}{\partial \dot{\phi}}$ we compute the inner product $\langle \dot{\mathbf{x}}, \mathbf{y} \rangle$ and use Legendre's transformation (2.50) to get the expression of the Hamiltonian.

Writing down the Lagrangian equations $\dfrac{d}{dt}\dfrac{\partial L}{\partial \dot{\mathbf{x}}} - \dfrac{\partial L}{\partial \mathbf{x}} = 0$, we get

$$\begin{aligned}
\frac{d}{dt}\left[A\dot{\psi}\sin^2\theta + C(\dot{\psi}\cos\theta + \dot{\phi})\cos\theta\right] &= 0,\\
A\ddot{\theta} + A\dot{\psi}^2\sin\theta\cos\theta - C(\dot{\psi}\cos\theta + \dot{\phi})\dot{\psi}\sin\theta + Mgl\sin\theta &= 0,\\
C\frac{d}{dt}(\dot{\psi}\cos\theta + \dot{\phi}) &= 0.
\end{aligned}$$

From these equations we see that not only the function $\dot{\psi}\cos\theta + \dot{\phi}$ is a first integral, which we already knew from Example 2.10, but also the function

$$A\dot{\psi}\sin^2\theta + C(\dot{\psi}\cos\theta + \dot{\phi})\cos\theta.$$

is a first integral of the dynamics of this problem.

In the next example we consider the *satellite problem* , that is the case of a rigid body moving in space with its center of mass describing an elliptic orbit of the Kepler problem defined by a fixed large attracting mass M. This case differs from that of the free rigid body in two respects. One is that the Lagrangian now has a nonzero potential V, which is due to the torque of the gravitational pull of the central body M on the satellite. The other is that the center of mass of the satellite is not fixed but is moving along the elliptic orbit and we have to consider its angular velocity about the normal to the orbital plane. Let us first obtain the potential V.

We assume that the center of mass O of the body moves along an elliptic orbit $\mathbf{r}_0 = \mathbf{r}_0(t)$ of the Kepler problem defined by the mass M. We consider an orbital coordinate system $OXYZ$ with the axis OX along the vector $\mathbf{r}_0$ and the axis OZ orthogonal to the plane of motion of the center of mass of the body. This coordinate system corresponds to the fixed system $OXYZ$ considered when we defined the Euler angles. The corresponding frame of the coordinate system $\mathbf{e}_x, \mathbf{e}_y, \mathbf{e}_z$ we denote here by $\mathbf{e}_1(t), \mathbf{e}_2(t), \mathbf{e}_3(t)$ and therefore the Euler angles ψ, θ, ϕ are functions of t.

[1] Our expression of the Lagrangian differs from those given in [6] and [26] because in these references the authors interchange the variables ψ and ϕ.

The force of the attracting center M on an infinitesimal mass dm of the body located at the position $\mathbf{r} \in S_t$ is $-GM\dfrac{\mathbf{r}_0+\mathbf{x}}{\|\mathbf{r}_0+\mathbf{x}\|^3}dm = GM\nabla\Big(\dfrac{1}{\|\mathbf{r}_0+\mathbf{x}\|}\Big)dm$, where $\mathbf{x} = \mathbf{r} - \mathbf{r}_0$ is the vector from the center of mass $\mathbf{r}_0$ to the position of dm. Therefore, the total force on the body S_t is

$$\mathbf{F} = \nabla\Big(GM\int_{S_t}\frac{1}{\|\mathbf{r}_0+\mathbf{x}\|}dm\Big), \tag{2.59}$$

Now, $\|\mathbf{r}_0+\mathbf{x}\| = \|\mathbf{r}_0\|(1+u)^{-\frac{1}{2}}$, where $u = \dfrac{2\langle\mathbf{r}_0,\mathbf{x}\rangle}{\|\mathbf{r}_0\|^2} + \dfrac{\|\mathbf{r}_0\|^2}{\|\mathbf{x}\|^2}$ and so we get

$$\frac{1}{\|\mathbf{r}_0+\mathbf{x}\|} = \frac{1}{\|\mathbf{r}_0\|}\Big(1 - \frac{\langle\mathbf{r}_0,\mathbf{x}\rangle}{\|\mathbf{r}_0\|^2} - \frac{\|\mathbf{x}_0\|^2}{2\|\mathbf{r}_0\|^2} + \frac{3}{2}\frac{\langle\mathbf{r}_0,\mathbf{x}\rangle^2}{\|\mathbf{r}_0\|^4}\Big) + O_3\Big(\frac{\|\mathbf{x}\|}{\|\mathbf{r}_0\|}\Big).$$

Therefore, disregarding the integration of the higher order terms $O_3\Big(\frac{\|\mathbf{x}\|}{\|\mathbf{r}_0\|}\Big)$, we have

$$\mathbf{F} = \nabla\left(\frac{GM}{\|\mathbf{r}_0\|}\int_{S_t}\Big(1 - \frac{\langle\mathbf{r}_0,\mathbf{x}\rangle}{\|\mathbf{r}_0\|^2} - \frac{\|\mathbf{x}_0\|^2}{2\|\mathbf{r}_0\|^2} + \frac{3}{2}\frac{\langle\mathbf{r}_0,\mathbf{x}\rangle^2}{\|\mathbf{r}_0\|^4}\Big)dm\right) \tag{2.60}$$

If M_S is the mass of the satellite, then $\mathbf{r}_0 = \frac{1}{M_{S_t}}\int_{S_t}\mathbf{r}dm = \mathbf{r}_0 + \frac{1}{M_{S_t}}\int_{S_t}\mathbf{x}dm$, so we have $\int_{S_t}\mathbf{x}dm = 0$. Therefore, disregarding the constant term we are left with

$$\mathbf{F} = \nabla\Big(\frac{GM}{2\|\mathbf{r}_0\|^5}\int_{S_t}(3\langle\mathbf{r}_0,\mathbf{x}\rangle^2 - \|\mathbf{r}_0\|^2\|\mathbf{x}\|^2)dm\Big). \tag{2.61}$$

Decompose $\mathbf{x}$ along the principal axes of inertia at the instant t,

$$\mathbf{x} = x\mathbf{e}_1(t) + y\mathbf{e}_2(t) + z\mathbf{e}_3(t).$$

From formulas (2.40), we have $E_X = a_{11}\mathbf{e}_1(t) + a_{12}\mathbf{e}_2(t) + a_{13}\mathbf{e}_3(t)$. Since the axis OX is in the direction of $\mathbf{r}_0$ we have $\mathbf{r}_0 = \|\mathbf{r}_0\|E_X$; hence,

$$\langle\mathbf{r}_0,\mathbf{x}\rangle^2 = \|\mathbf{r}_0\|^2\big(a_{11}x + a_{12}y + a_{13}z\big)^2. \tag{2.62}$$

Denote by A, B, C the moments of inertia about the axes Ox, Oy, and Oz, that is,

$$A = \int_{S_t}(y^2+z^2)dm, \quad B = \int_{S_t}(z^2+x^2)dm, \quad C = \int_{S_t}(x^2+y^2)dm.$$

Then we have $\displaystyle\int_{S_t}\|\mathbf{x}\|^2dm = \int_{S_t}(x^2+y^2+z^2)dm = \frac{A+B+C}{2}$. This term is constant, so it can be disregarded in the integration of (2.61). Since the coordinate system $Oxyz$ is defined by the principal axes of inertia, the integral of any mixed term from (2.62) is zero, and for the integral in (2.61) we are left with

$$\mathbf{F} = \nabla\Big(\frac{3GM}{2\|\mathbf{r}_0\|^3}\int_{S_t}(a_{11}^2x^2 + a_{12}^2y^2 + a_{13}^2z^2)\,dm\Big). \tag{2.63}$$

But $\int_{S_t} x^2 dm = \int_{S_t}((x^2+y^2+z^2)-(y^2+z^2))\,dm = \frac{A+B+C}{2} - A$, that is

$$\int_{S_t} x^2 dm = \frac{A+B+C}{2} - A, \quad \int_{S_t} y^2 dm = \frac{A+B+C}{2} - B, \quad \int_{S_t} z^2 dm = \frac{A+B+C}{2} - C.$$

Therefore, noticing that $a_{11}^2 + a_{12}^2 + a_{13}^2 = 1$ there appears a constant term involving $\frac{A+B+C}{2}$ in the integral (2.63), which we can disregard so that finally we get

$$\mathbf{F} = -\nabla\Big(\frac{3GM}{2\|\mathbf{r}_0\|^3}(Aa_{11}^2 + Ba_{12}^2 + Ca_{13}^2)\Big)$$

and the potential energy V for the rotational motion of the satellite about its center of mass, which is moving along the elliptic orbit $\mathbf{r}_0$, is given in terms of Euler's angles ψ, θ, ϕ by the expression

$$V = \frac{3GM}{2\|\mathbf{r}_0\|^3}\Big(Aa_{11}^2 + Ba_{12}^2 + Ca_{13}^2\Big). \tag{2.64}$$

Let us now consider the motion of the center of mass $\mathbf{r}_0(t)$ (Fig. 2.16). As it moves about the axis OZ with angular speed $\dot{\nu}$, that is, with angular velocity $\dot{\nu}E_Z$, the angular velocity of the satellite is

$$\boldsymbol{\omega} = (\dot{\nu} + \dot{\psi})E_Z + \dot{\theta}\mathbf{e}_N + \dot{\phi}\mathbf{e}_3(t),$$

Fig. 2.16 $\dot{\nu}E_Z$

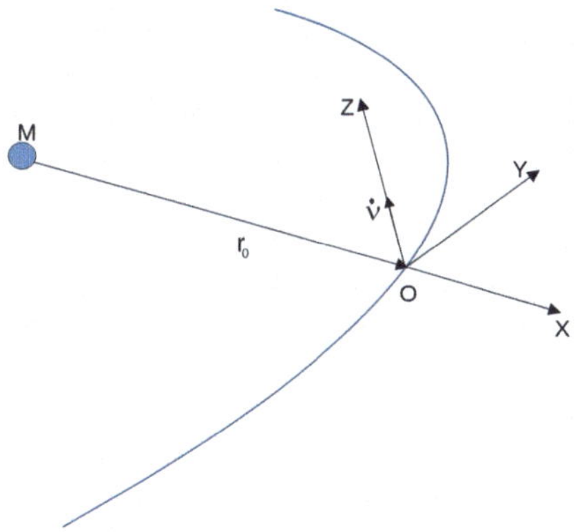

so, instead of (2.39) we get

$$\boldsymbol{\Omega} = R(t)^{-1}\boldsymbol{\omega} = (\dot{\nu} + \dot{\psi})R(t)^{-1}E_Z + \dot{\theta}R(t)^{-1}\mathbf{e}_N + \dot{\phi}R(t)^{-1}\mathbf{e}_3(t). \tag{2.65}$$

Therefore, in Euler's kinematic equations (2.41) we replace $\dot{\psi}$ by $\dot{\psi} + \dot{\nu}$. The consequence of this for the velocity $\dot{\mathbf{x}}$ is just the addition of $-\dot{\nu}$ to the first component in the expression of $\dot{\mathbf{x}}$ in (2.54). It follows that

$$\langle \dot{\mathbf{x}}, \mathbf{y}\rangle = 2T - \langle(\dot{\nu}, 0, 0), (P_\psi, P_\theta, P_\phi)\rangle = 2T - \dot{\nu}P_\psi. \tag{2.66}$$

Example 2.14 (Hamiltonian of the Satellite Problem, I) The function V in (2.64) and the kinetic energy (2.42) define the Lagrangian function

$$L = T - V$$

for the dynamics of the motion of the solid body S around its center of mass, which is moving along the elliptic orbit $\mathbf{r}_0$ of the Kepler problem defined by the central mass M.

In view of (2.66), for the part of the Hamiltonian coming from the kinetic energy we just have to add the term $-\dot{\nu}P_\psi$ to the right-hand side of the expression (2.55) divided by $2A$.

Now, let us consider the part of the Hamiltonian coming from the potential energy (2.64). Since the matrix $\mathcal{A}$ in (2.34) is orthogonal, we have $a_{11}^2 + a_{12}^2 + a_{13}^2 = 1$; hence, the expression (2.64) takes the form

$$V = \frac{3GM}{2\|\mathbf{r}_0\|^3}\big(a_{11}^2(A - B) + Ca_{13}^2(C - B) + B\big).$$

The term $\frac{3GM}{\|\mathbf{r}_0\|^3}B$ does not depend on the variables $\mathbf{x} = (\psi, \theta, \phi)$, so it is irrelevant for the dynamics and can be ignored. With $\theta_A = A/B$ and $\theta_C = C/B$, we have $A - B = A\dfrac{\theta_A - 1}{\theta_A}$ and $C - B = A\dfrac{\theta_C - 1}{\theta_A}$; hence, the potential V can be written in the form

$$V = A\frac{3GM}{2\|\mathbf{r}_0\|^3}\left(\frac{\theta_A - 1}{\theta_A}a_{11}^2 + \frac{\theta_C - 1}{\theta_A}a_{13}^2\right). \tag{2.67}$$

By hypothesis the center of mass $\mathbf{r}_0$ describes an elliptic orbit of the Kepler problem

$$\mathbf{r} = -\frac{\kappa}{\|\mathbf{r}\|^3}\mathbf{r}, \quad \text{with} \quad \kappa = GM.$$

We recall the formulas (2.13), (2.14) and (2.15) for an elliptic motion,

$$r_0 = \frac{c^2/\kappa}{1+\epsilon\cos\nu}, \qquad r_0^2\dot{\nu} = c, \qquad \frac{a^3}{P^2} = \frac{\kappa}{4\pi^2}$$

with $r_0 = \|\mathbf{r}_0\|$ and $c^2/\kappa = a(1-\epsilon^2)$. Denoting $\omega_0 = \frac{2\pi}{P}$ the mean motion of the elliptic orbit, from these formulas we get the following expressions of $\dfrac{GM}{\|\mathbf{r}_0\|^3}$ and $\dot{\nu}$:

$$\frac{GM}{\|\mathbf{r}_0\|^3} = \frac{\omega_0^2}{(1-\epsilon^2)^3}(1+\epsilon\cos\nu)^3 \quad \text{and} \quad \dot{\nu} = \frac{\omega_0}{(1-\epsilon^2)^{3/2}}(1+\epsilon\cos\nu)^2. \qquad (2.68)$$

The Hamiltonian is $\mathcal{H} = T + V$ with $T = \frac{1}{2A}E - \dot{\nu}P_\psi$ where E is the expression in the right-hand side of the equality (2.55) and V is given by (2.67). If we take

$$P_\psi = \frac{A\omega_0}{(1-\epsilon^2)^{3/2}}p_\psi, \quad P_\theta = \frac{A\omega_0}{(1-\epsilon^2)^{3/2}}p_\theta, \quad P_\phi = \frac{A\omega_0}{(1-\epsilon^2)^{3/2}}p_\phi, \qquad (2.69)$$

and use (2.68) we get

$$\mathcal{H} = \frac{A\omega_0^2}{(1-\epsilon^2)^3}(1+\epsilon\cos\nu)^2 H,$$

where

$$\begin{aligned}
H =& \frac{1}{2(1+\epsilon\cos\nu)^2\sin^2\theta}(\sin^2\phi + \theta_A\cos^2\phi)p_\psi^2 + \\
&+ \frac{1}{2(1+\epsilon\cos\nu)^2}(\cos^2\phi + \theta_A\sin^2\phi)p_\theta^2 + \\
&+ \frac{1}{2(1+\epsilon\cos\nu)^2}\left[(\sin^2\phi + \theta_A\cos^2\phi)\cot^2\theta + \frac{\theta_A}{\theta_C}\right]p_\phi^2 \\
&+ \frac{1-\theta_A}{2(1+\epsilon\cos\nu)^2}\frac{\sin 2\phi}{\sin\theta}p_\psi p_\theta - \qquad (2.70)\\
&- \frac{1}{(1+\epsilon\cos\nu)^2}\frac{\cos\theta}{\sin^2\theta}(\sin^2\phi + \theta_A\cos^2\phi)p_\psi p_\phi - \\
&- \frac{1-\theta_A}{2(1+\epsilon\cos\nu)^2}\cot\theta\sin 2\phi p_\theta p_\phi - p_\psi + \\
&+ \frac{3}{2}(1+\epsilon\cos\nu)\left(\frac{\theta_A - 1}{\theta_A}a_{11}^2 + \frac{\theta_C - 1}{\theta_A}a_{13}^2\right).
\end{aligned}$$

This function H, 2π-periodic in ν, is the Hamiltonian of the satellite problem if we take the true anomaly ν as new time.

Indeed, let $F = \dfrac{A\omega_0^2}{(1-\epsilon^2)^3}(1+\epsilon\cos\nu)^2$. Then, since $\dot{\psi} = \mathcal{H}_{P_\psi}$ and $\mathcal{H} = FH$, we get denoting by a prime derivative with respect to ν,

$$\psi'\dot{\nu} = FH_{p_\psi}\frac{dp_\psi}{dP_\psi},$$

and computing $F\frac{dp_\psi}{dP_\psi}$ we find the expression for $\dot{\nu}$ given in (2.68). Therefore, we have $\psi' = H_{p_\psi}$. From $\dot{P}_\psi = -\mathcal{H}_\psi$ we get $\dfrac{A\omega_0}{(1-\epsilon^2)^{3/2}}p'_\psi\dot{\nu} = -FH_\psi$ and we also conclude that $p'_\psi = -H_\psi$. The other Hamiltonian equations also hold for the other variables $\theta, \phi, p_\theta, p_\phi$.

In Chap. 8 we will use another expression for the Hamiltonian of the satellite, which we present in the next example.

Example 2.15 (Hamiltonian of the Satellite Problem, II) In this case we take the orbital coordinate system with the axis OZ along the radius vector $\mathbf{r}_0$ of the center of mass of the satellite and the axis OY orthogonal to the orbital plane. As before, we set $\mathbf{x} = (\psi, \theta, \phi)$ and $\mathbf{y} = (P_\psi, P_\theta, P_\phi)$. From the expression (2.53) of $\mathbf{y}$ we get

$$\begin{aligned}
Ap &= \frac{\sin\phi}{\sin\theta}P_\psi + \cos\phi P_\theta - \frac{\sin\phi\cos\theta}{\sin\theta}P_\phi,\\
Bq &= \frac{\cos\phi}{\sin\theta}P_\psi - \sin\phi P_\theta - \frac{\cos\phi\cos\theta}{\sin\theta}P_\phi,\\
Cr &= P_\phi.
\end{aligned} \tag{2.71}$$

Since $2T = \frac{1}{A}(Ap)^2 + \frac{1}{B}(Bq)^2 + \frac{1}{C}(Cr)^2$ we readily see that

$$\begin{aligned}
T =& \frac{1}{2A}\left[\frac{\sin\phi}{\sin\theta}(P_\psi - \cos\theta P_\phi) + \cos\phi P_\theta\right]^2 + \\
&+\frac{1}{2B}\left[\frac{\cos\phi}{\sin\theta}(P_\psi - \cos\theta P_\phi) - \sin\phi P_\theta\right]^2 + \frac{1}{2C}P_\phi^2.
\end{aligned} \tag{2.72}$$

One difference from the previous case is that now we have the vector E_Z instead of E_X in the direction of the center of mass. Since $E_Z = a_{31}\mathbf{e}_1(t) + a_{32}\mathbf{e}_2(t) + a_{33}\mathbf{e}_3(t)$ we now have

$$\langle \mathbf{r}_0, \mathbf{x}\rangle^2 = \|\mathbf{r}_0\|^2\left(a_{31}x + a_{32}y + a_{33}z\right)^2$$

and so instead of (2.64) we now have for the potential V the expression

$$V = \frac{3GM}{2\|\mathbf{r}_0\|^3}\left(Aa_{31}^2 + Ba_{32}^2 + Ca_{33}^2\right). \tag{2.73}$$

The second difference is that now the angular velocity of the center of mass is about the axis OY, that is, $\dot{\nu} E_Y$. Hence, $\boldsymbol{\omega} = \dot{\psi} E_Z + \dot{\theta}\mathbf{e}_N + \dot{\phi}\mathbf{e}_3(t) + \dot{\nu} E_Y$ and for the components of $\boldsymbol{\Omega} = p\mathbf{e}_1 + q\mathbf{e}_2 + r\mathbf{e}_3$, where $\mathbf{e}_k = R(t)^{-1}\mathbf{e}_k(t)$, we have

$$\begin{aligned} &p = \dot{\psi}\sin\theta\sin\phi + \dot{\theta}\cos\phi + \dot{\nu}a_{21}, \\ &q = \dot{\psi}\sin\theta\cos\phi - \dot{\theta}\sin\phi + \dot{\nu}a_{22}, \ \ r = \dot{\psi}\cos\theta + \dot{\phi} + \dot{\nu}a_{23}. \end{aligned} \tag{2.74}$$

Then, for $\dot{\mathbf{x}} = (\dot{\psi}, \dot{\theta}, \dot{\phi})$ in this case we add to the right-hand side of (2.54) the vector

$$-\dot{\nu}\Big(\frac{\sin\phi}{\sin\theta}a_{21} + \frac{\cos\phi}{\sin\theta}a_{22}, \cos\phi a_{21} - \sin\phi a_{22}, \big(\frac{\sin\phi}{\sin\theta}a_{21} + \frac{\cos\phi}{\sin\theta}a_{22}\big)\cos\theta - a_{23}\Big).$$

Since $\sin\phi a_{21} + \cos\phi a_{22} = \cos\theta\cos\psi$, the third component is equal to $-\dfrac{\cos\psi}{\sin\theta}$ and for $\langle \dot{\mathbf{x}}, \mathbf{y}\rangle$ we get the expression

$$\langle \dot{\mathbf{x}}, \mathbf{y}\rangle = 2T - \dot{\nu}\Big(\cos\psi\cot\theta\, P_\psi + \sin\psi\, P_\theta - \frac{\cos\psi}{\sin\theta}P_\phi\Big) \tag{2.75}$$

Since $a_{31}^2 + a_{32}^2 + a_{33}^2 = 1$, we re-write the potential (2.73) in the form, observing that the irrelevant term for the dynamics, $\dfrac{3GM}{2\|\mathbf{r}_0\|^3}A$, has been deleted,

$$V = \frac{3GM}{2\|\mathbf{r}_0\|^3}\big[(B-A)a_{32}^2 + (C-A)a_{33}^2\big]. \tag{2.76}$$

Using (2.72), (2.75) and (2.76), we get by Legendre's transformation (2.50) the expression of the Hamiltonian $\mathcal{H} = \langle \dot{\mathbf{x}}, \mathbf{y}\rangle - L$. Considering the equalities (2.68) and (2.69) and using the true anomaly of the elliptical orbit of the center of mass as the new time we obtain the final expression of the Hamiltonian, 2π-periodic in ν:

$$\begin{aligned} H =& \frac{\sin^2\phi + \frac{A}{B}\cos^2\phi}{2(1+\epsilon\cos\nu)^2\sin^2\theta}(p_\psi - \cos\theta p_\phi)^2 + \frac{\cos^2\phi + \frac{A}{B}\sin^2\phi}{2(1+\epsilon\cos\nu)^2}p_\theta^2 + \\ &+ \frac{A}{2C(1+\epsilon\cos\nu)^2}p_\phi^2 + \frac{\big(1-\frac{A}{B}\big)\sin 2\phi}{2(1+\epsilon\cos\nu)^2\sin\theta}p_\theta(p_\psi - \cos\theta p_\phi) - \\ &- \cos\psi\cot\theta\, p_\psi - \sin\psi\, p_\theta + \frac{\cos\psi}{\sin\theta}p_\phi + \\ &+ \frac{3}{2}(1+\epsilon\cos\nu)\Big[\Big(\frac{B}{A}-1\Big)\sin^2\theta\cos^2\phi + \Big(\frac{C}{A}-1\Big)\cos^2\theta\Big]. \end{aligned} \tag{2.77}$$

2.4 Canonical Transformations and Generating Functions

We have seen in Proposition 2.3 that the Euler–Lagrange equation $\mathcal{L}_\mathbf{x} L = 0$ is invariant under an arbitrary change of coordinates. This is not the case for the Hamiltonian equations (2.52) and we investigate this point in the first subsection.

2.4.1 Symplectic Transformations

Given the Hamiltonian function $H(\mathbf{x}, \mathbf{y})$ consider the change of coordinates

$$\mathbf{x} = \phi(\mathbf{u}, \mathbf{v}) \qquad \mathbf{y} = \psi(\mathbf{u}, \mathbf{v}), \tag{2.78}$$

and let $\mathcal{H}(\mathbf{u}, \mathbf{v}) = H(\mathbf{x}, \mathbf{y})$. For smooth functions $\mathbf{u}(t), \mathbf{v}(t)$ we have by (1.10)

$$\dot{\mathbf{x}} = \mathbf{x_u}\dot{\mathbf{u}} + \mathbf{x_v}\dot{\mathbf{v}}, \quad \dot{\mathbf{y}} = \mathbf{y_u}\dot{\mathbf{u}} + \mathbf{y_v}\dot{\mathbf{v}}, \quad \text{that is} \quad \begin{bmatrix} \dot{\mathbf{x}} \\ \dot{\mathbf{y}} \end{bmatrix} = M \begin{bmatrix} \dot{\mathbf{u}} \\ \dot{\mathbf{v}} \end{bmatrix}, \tag{2.79}$$

where $\mathbf{x_u}$ is the Jacobian matrix of $\frac{\partial \phi}{\partial \mathbf{u}} = D_\mathbf{u}\phi$ and

$$M = \begin{bmatrix} \mathbf{x_u} & \mathbf{x_v} \\ \mathbf{y_u} & \mathbf{y_v} \end{bmatrix} \tag{2.80}$$

is the Jacobian matrix of the transformation (2.78). With $H_\mathbf{x}$, $H_\mathbf{y}$ the gradients of H with respect to $\mathbf{x}$, $\mathbf{y}$ and $\mathcal{H}_\mathbf{u}$, $\mathcal{H}_\mathbf{v}$ the gradients of $\mathcal{H}$ with respect to $\mathbf{u}$, $\mathbf{u}$ we have[2]

$$\mathcal{H}_\mathbf{u} = \mathbf{x}_\mathbf{u}^T H_\mathbf{x} + \mathbf{y}_\mathbf{u}^T H_\mathbf{y}, \quad \mathcal{H}_\mathbf{v} = \mathbf{x}_\mathbf{v}^T H_\mathbf{x} + \mathbf{y}_\mathbf{v}^T H_\mathbf{y}, \quad \text{that is} \quad \begin{bmatrix} \mathcal{H}_\mathbf{u} \\ \mathcal{H}_\mathbf{v} \end{bmatrix} = M^T \begin{bmatrix} H_\mathbf{x} \\ H_\mathbf{y} \end{bmatrix}. \tag{2.81}$$

The requirement that the transformation (2.78) takes the Hamiltonian system $\dot{\mathbf{x}} = H_\mathbf{y}$, $\dot{\mathbf{y}} = -H_\mathbf{x}$ to the Hamiltonian system

$$\dot{\mathbf{u}} = \mathcal{H}_\mathbf{v}, \quad \dot{\mathbf{v}} = -\mathcal{H}_\mathbf{u} \tag{2.82}$$

leads, in view of (2.79), to the equations $H_\mathbf{y} = \mathbf{x_u}\mathcal{H}_\mathbf{v} - \mathbf{x_v}\mathcal{H}_\mathbf{u}$, $-H_\mathbf{x} = \mathbf{y_u}\mathcal{H}_\mathbf{v} - \mathbf{y_v}\mathcal{H}_\mathbf{u}$, so

$$J \begin{bmatrix} H_\mathbf{x} \\ H_\mathbf{y} \end{bmatrix} = \begin{bmatrix} -\mathbf{x_v} & \mathbf{x_u} \\ -\mathbf{y_v} & \mathbf{y_u} \end{bmatrix} \begin{bmatrix} \mathcal{H}_\mathbf{u} \\ \mathcal{H}_\mathbf{v} \end{bmatrix}. \tag{2.83}$$

---

[2] Applying the derivative of $\mathcal{H}(\mathbf{u}, \mathbf{v})$ with respect to $\mathbf{u}$ to a vector X, we have by the chain rule $D_\mathbf{u}\mathcal{H} \cdot X = D_\mathbf{x} H \cdot (D_\mathbf{u}\phi \cdot X) + D_\mathbf{y} H \cdot (D_\mathbf{u}\psi \cdot X)$, so passing to an equality in terms of gradients we get $\langle \mathcal{H}_\mathbf{u}, X\rangle = \langle H_\mathbf{x}, \mathbf{x_u} X\rangle + \langle H_\mathbf{y}, \mathbf{y_u} X\rangle = \langle \mathbf{x}_\mathbf{u}^T H_\mathbf{x}, X\rangle + \langle \mathbf{y}_\mathbf{u}^T H_\mathbf{y}, X\rangle$; hence, $\mathcal{H}_\mathbf{u} = \mathbf{x}_\mathbf{u}^T H_\mathbf{x} + \mathbf{y}_\mathbf{u}^T H_\mathbf{y}$.

Since $\begin{bmatrix} -\mathbf{x_v} & \mathbf{x_u} \\ -\mathbf{y_v} & \mathbf{y_u} \end{bmatrix} = \begin{bmatrix} \mathbf{x_u} & \mathbf{x_v} \\ \mathbf{y_u} & \mathbf{y_v} \end{bmatrix} \begin{bmatrix} O & I \\ -I & O \end{bmatrix}$, we get from (2.80), (2.81) and (2.83) the following equality

$$J\nabla H = MJM^T\nabla H.$$

Therefore, if the change of variables (2.78) preserves the Hamiltonian form of the equations for any Hamiltonian H, then the matrix M satisfies the equation

$$MJM^T = J. \tag{2.84}$$

Conversely, if the Jacobian matrix M satisfies Eq. (2.84), then the change of variables (2.78) takes the system (2.52) to the system (2.82), where $\mathcal{H}(\mathbf{u}, \mathbf{v}) = H(\mathbf{x}, \mathbf{y})$.

Indeed, from (2.84) we have $M^{-1} = JM^TJ^{-1}$, so by (2.79) and (2.81)

$$\begin{bmatrix} \dot{\mathbf{u}} \\ \dot{\mathbf{v}} \end{bmatrix} = M^{-1} \begin{bmatrix} \dot{\mathbf{x}} \\ \dot{\mathbf{y}} \end{bmatrix} = JM^TJ^{-1}J \begin{bmatrix} H_\mathbf{x} \\ H_\mathbf{y} \end{bmatrix} = JM^T \begin{bmatrix} H_\mathbf{x} \\ H_\mathbf{y} \end{bmatrix} = J \begin{bmatrix} \mathcal{H}_\mathbf{u} \\ \mathcal{H}_\mathbf{v} \end{bmatrix} \tag{2.85}$$

which gives (2.82).

The following identities involving the standard symplectic matrix J appear so frequently

$$J^2 = -I, \quad J^T = -J, \quad J^{-1} = -J, \quad J^TJ = I \tag{2.86}$$

that sometimes we use them even without mentioning it. An example of this is offered at once. The matrix equation (2.84) is equivalent to the following

$$M^TJM = J. \tag{2.87}$$

Indeed, from (2.84) we have $M = JM^{-T}J^{-1}$; hence,

$$M^TJM = M^TJJM^{-T}J^{-1} = -M^TM^{-T}J^{-1} = J.$$

Conversely, from (2.87), we have $M^T = JM^{-1}J^{-1}$; hence,

$$MJM^T = MJJM^{-1}J^{-1} = -MM^{-1}J^{-1} = J.$$

Definition 2.5 A real or complex square matrix M of order $2n$ satisfying the Eq. (2.87) is called a symplectic matrix.

Proposition 2.6 *The following properties of symplectic matrices hold:*

(a) The identity matrix I_{2n} *and the matrix* $J = \begin{bmatrix} O & I_n \\ -I_n & O \end{bmatrix}$ *are symplectic;*
(b) If M_1, M_2 *are symplectic, then so is the product* $M_1 M_2$*;*
(c) If M *is symplectic, then so are the transpose,* M^T*, and* $-M$*;*
(d) If M *is symplectic, then its inverse is given by* $M^{-1} = JM^T J^{-1}$*, so it is also symplectic;*
(e) If M *of order* $2n$ *has the block form* $M = \begin{bmatrix} A & B \\ C & D \end{bmatrix}$ *with square matrices* A, B, C, D *then* M *is symplectic if, and only if, the following hold*

$$A^T C,\ B^T D \text{ are symmetric} \quad \text{and} \quad A^T D - C^T B = I. \tag{2.88}$$

Proof The verification of these properties is straightforward and is left as an exercise. □

Definition 2.7 A $2n \times 2n$ real or complex matrix M is μ-symplectic or *symplectic with multiplier* $\mu \neq 0$ if

$$M^T JM = \mu J. \tag{2.89}$$

Proposition 2.8 *The matrix* M *is* μ*-symplectic if, and only if,* $M^{-1} = \frac{1}{\mu} J^{-1} M^T J$.

If M *and* N *are symplectic matrices with multipliers* μ *and* ν*, then* αM*,* α *a nonzero scalar,* M^T*,* M^{-1} *and* MN *are symplectic with multipliers* $\alpha^2\mu$*,* μ*,* μ^{-1} *and* $\mu\nu$ *respectively.*

Writing M *of order* $2n$ *in block form, then* M *is* μ*-symplectic if, and only if, the conditions stated in item (e) of Proposition 2.6 hold with* I *replaced by* μI*.*

Proof The first statement is clear. From the expression for M^{-1} we get $JM^T = \mu M^{-1} J$, so $(M^T)^T JM^T = MJM^T = \mu J$, hence M^T is μ-symplectic. The other statements are left as exercises. □

Proposition 2.9 *If the Jacobian matrix* M *of (2.78) is* μ*-symplectic, the transformation still carries the Hamiltonian system (2.52) to the Hamiltonian system (2.82) but now the new Hamiltonian is defined by*

$$\mathcal{H}(\mathbf{u}, \mathbf{v}) = \frac{1}{\mu} H(\mathbf{x}, \mathbf{y}). \tag{2.90}$$

Proof With $\mathcal{H}(\mathbf{u}, \mathbf{v}) = \frac{1}{\mu} H(\mathbf{x}, \mathbf{y})$ the right-hand sides of (2.81) are multiplied by $\frac{1}{\mu}$. Since $MJM^T = \mu J$ we have $M^{-1} = \frac{1}{\mu} JM^T J^{-1}$ and the chain of computations (2.85) shows that the statement is correct. □

Definition 2.10 The change of variables (2.78) is said to be a *canonical change of variables* or a *symplectic transformation* if its Jacobian matrix is symplectic. If $\mu \neq 1$, we call it a canonical transformation with multiplier μ or a μ-symplectic transformation.

Example 2.16 (Action-Angles Variables) Using (2.119) one readily checks that the following change of variables is symplectic

$$x_j = \sqrt{2I_j}\cos\phi_j, \quad y_j = \sqrt{2I_j}\sin\phi_j, \qquad (j = 1, \dots, n)$$

The variables $I_1, \dots, I_n, \phi_1, \dots, \phi_n$ are called *action-angles variables.*

2.4.2 Constructing Symplectic Transformations

One way of getting symplectic transformations is to construct them using a generating function. We start with a function $W(\mathbf{x}, \mathbf{v})$ having nonsingular Hessian, $\det[W_{x_i v_j}] \neq 0$, and define $\mathbf{y}$ and $\mathbf{u}$ by the equations

$$\mathbf{y} = W_{\mathbf{x}}(\mathbf{x}, \mathbf{v}), \quad \mathbf{u} = W_{\mathbf{v}}(\mathbf{x}, \mathbf{v}). \tag{2.91}$$

Because the Hessian is nonzero by the inverse function theorem we can solve $\mathbf{u} = W_{\mathbf{v}}(\mathbf{x}, \mathbf{v})$ locally to get $\mathbf{x} = \phi(\mathbf{u}, \mathbf{v})$. Inserting this into the first equation, we get $\mathbf{y} = W_{\mathbf{x}} = \psi(\mathbf{u}, \mathbf{v})$. It turns out that this is a symplectic transformation.

Indeed, differentiating Eqs. (2.91) with respect to $\mathbf{u}$ and $\mathbf{v}$ we have

$$\mathbf{y}_{\mathbf{u}} = W_{\mathbf{xx}}\mathbf{x}_{\mathbf{u}}, \quad \mathbf{y}_{\mathbf{v}} = W_{\mathbf{xx}}\mathbf{x}_{\mathbf{v}} + W_{\mathbf{xv}}, \quad I = W_{\mathbf{vx}}\mathbf{x}_{\mathbf{u}}, \quad 0 = W_{\mathbf{vx}}\mathbf{x}_{\mathbf{v}} + W_{\mathbf{vv}}, \tag{2.92}$$

and we see that the Jacobian matrix $M = \begin{bmatrix} \mathbf{x}_{\mathbf{u}} & \mathbf{x}_{\mathbf{v}} \\ \mathbf{y}_{\mathbf{u}} & \mathbf{y}_{\mathbf{v}} \end{bmatrix}$ has the form

$$M = \begin{bmatrix} W_{\mathbf{vx}}^{-1} & -W_{\mathbf{vx}}^{-1} W_{\mathbf{vv}} \\ W_{\mathbf{xx}} W_{\mathbf{vx}}^{-1} & W_{\mathbf{xv}} - W_{\mathbf{xx}} W_{\mathbf{vx}}^{-1} W_{\mathbf{vv}} \end{bmatrix}.$$

Taking into account that the matrices $W_{\mathbf{xx}}$, $W_{\mathbf{vx}}$, $W_{\mathbf{xv}}$ and $W_{\mathbf{vv}}$ are symmetric we easily check that the conditions (2.119) are satisfied, so $\mathcal{M}$ is symplectic. The function $W(\mathbf{x}, \mathbf{v})$ is called the *generating function* of this symplectic transformation.

In a similar way we can generate symplectic transformations by means of a function of old moment $\mathbf{y}$ and new position $\mathbf{u}$, $W(\mathbf{y}, \mathbf{u})$, through the equations

$$\mathbf{x} = W_{\mathbf{y}}, \quad \mathbf{v} = W_{\mathbf{u}}.$$

If the function W is given as a function of old and new positions, $W(\mathbf{x}, \mathbf{u})$, or old and new moments, $W(\mathbf{y}, \mathbf{v})$ the transformations are defined by setting

$$\mathbf{y} = W_{\mathbf{x}}, \quad \mathbf{v} = -W_{\mathbf{u}} \qquad \text{or} \qquad \mathbf{x} = W_{\mathbf{y}}, \quad \mathbf{u} = -W_{\mathbf{v}}.$$

In all cases we assume that the Hessian of W is nonzero, so we can apply the implicit function theorem to solve the equations locally.

We leave as an exercise the proof that these conditions generate symplectic transformations and also the task of writing the conditions to generate μ-symplectic transformations.

Example 2.17 (Mathieu Transformation) A diffeomorphism $\mathbf{x} = \mathbf{f}(\boldsymbol{\xi})$ defined in an open set $U \subset \mathbf{R}^n$ can be completed to a symplectic mapping

$$\mathbf{x} = \mathbf{f}(\boldsymbol{\xi}), \quad \mathbf{y} = D\mathbf{f}(\boldsymbol{\xi})^{-T}\boldsymbol{\eta} \tag{2.93}$$

in $U \times \mathbf{R}^n$.

Let $W(\boldsymbol{\xi}, \mathbf{y}) = \langle \mathbf{f}(\boldsymbol{\xi}), \mathbf{y}\rangle$. The function W has Hessian $D\mathbf{f}(\boldsymbol{\xi})$, which is nonsingular because $\mathbf{f}$ is a diffeomorphism. Hence, W generates a symplectic mapping if we set $\mathbf{x} = W_{\mathbf{y}} = \mathbf{f}(\boldsymbol{\xi})$ and $\boldsymbol{\eta} = W_{\boldsymbol{\xi}} = D\mathbf{f}(\boldsymbol{\xi})^T\mathbf{y}$, which gives (2.93). This symplectic mapping is called a Mathieu transformation.

We give an illustration of the use of this transformation in the following example.

Example 2.18 The Hamiltonian of the Kepler problem (2.11) is given in polar coordinates by

$$H(r, \theta, R, \Theta) = \frac{1}{2}\left(R^2 + \frac{\Theta^2}{r^2}\right) - \frac{\kappa}{r}. \tag{2.94}$$

To see this complete the mapping in $\mathbf{R}^2$ given by polar coordinates $x = r\cos\theta$, $y = r\sin\theta$ to a Mathieu transformation in $\mathbf{R}^4$ as in the previous example. Denoting by X, Y the conjugate moments of x, y and by R, Θ those of r, θ, we get $X = R\cos\theta - \frac{\Theta}{r}\sin\theta$, $Y = R\sin\theta + \frac{\Theta}{r}\cos\theta$. Replacing these coordinates into the Kepler Hamiltonian (2.11) we get (2.94).

2.4.3 Time-Dependent Symplectic Transformations

Now assume that we make a time-dependent symplectic transformation, that is, the functions ϕ and ψ in (2.78) depend also on the time t,

$$\mathbf{x} = \phi(\mathbf{u}, \mathbf{v}, t) \qquad \mathbf{y} = \psi(\mathbf{u}, \mathbf{v}, t). \tag{2.95}$$

We take the Hamiltonian to be time-dependent and modify the definition of $\mathcal{H}$ by adding an extra term, so let

$$\mathcal{H}(\mathbf{u}, \mathbf{v}, t) = H(\mathbf{x}, \mathbf{y}, t) + R(\mathbf{u}, \mathbf{v}, t). \tag{2.96}$$

Since

$$\dot{\mathbf{x}} = \mathbf{x}_{\mathbf{u}}\dot{\mathbf{u}} + \mathbf{x}_{\mathbf{v}}\dot{\mathbf{v}} + \mathbf{x}_t, \quad \dot{\mathbf{y}} = \mathbf{y}_{\mathbf{u}}\dot{\mathbf{u}} + \mathbf{y}_{\mathbf{v}}\dot{\mathbf{v}} + \mathbf{y}_t,$$

the equations corresponding to (2.79) and (2.81) are now

$$\begin{bmatrix} \dot{\mathbf{x}} \\ \dot{\mathbf{y}} \end{bmatrix} = M \begin{bmatrix} \dot{\mathbf{u}} \\ \dot{\mathbf{v}} \end{bmatrix} + \begin{bmatrix} \mathbf{x}_t \\ \mathbf{y}_t \end{bmatrix} \quad \text{and} \quad \begin{bmatrix} \mathcal{H}_{\mathbf{u}} \\ \mathcal{H}_{\mathbf{v}} \end{bmatrix} = M^T \begin{bmatrix} H_{\mathbf{x}} \\ H_{\mathbf{y}} \end{bmatrix} + \begin{bmatrix} R_{\mathbf{u}} \\ R_{\mathbf{v}} \end{bmatrix}, \tag{2.97}$$

whereas the equation corresponding to (2.83) is

$$J \begin{bmatrix} H_{\mathbf{x}} \\ H_{\mathbf{y}} \end{bmatrix} = \begin{bmatrix} -\mathbf{x}_{\mathbf{v}} & \mathbf{x}_{\mathbf{u}} \\ -\mathbf{y}_{\mathbf{v}} & \mathbf{y}_{\mathbf{u}} \end{bmatrix} \begin{bmatrix} \mathcal{H}_{\mathbf{u}} \\ \mathcal{H}_{\mathbf{v}} \end{bmatrix} + \begin{bmatrix} \mathbf{x}_t \\ \mathbf{y}_t \end{bmatrix}.$$

Since $\begin{bmatrix} -\mathbf{x}_{\mathbf{v}} & \mathbf{x}_{\mathbf{u}} \\ -\mathbf{y}_{\mathbf{v}} & \mathbf{y}_{\mathbf{u}} \end{bmatrix} = \begin{bmatrix} \mathbf{x}_{\mathbf{u}} & \mathbf{x}_{\mathbf{v}} \\ \mathbf{y}_{\mathbf{u}} & \mathbf{y}_{\mathbf{v}} \end{bmatrix} \begin{bmatrix} O & I \\ -I & O \end{bmatrix}$, from these equations we have

$$J \nabla H = M J M^T \nabla H + M J \begin{bmatrix} R_{\mathbf{u}} \\ R_{\mathbf{v}} \end{bmatrix} + \begin{bmatrix} \mathbf{x}_t \\ \mathbf{y}_t \end{bmatrix}.$$

But M is symplectic, so $MJM^T = J$; hence, this equation reduces to

$$\nabla R = M^T J \mathbf{z}_t, \quad \text{where} \quad \mathbf{z}_t = \begin{bmatrix} \mathbf{x}_t \\ \mathbf{y}_t \end{bmatrix}. \tag{2.98}$$

Lemma 2.11 *The mapping $X(\mathbf{u}, \mathbf{v}, t) = M^T J \mathbf{z}_t(\mathbf{u}, \mathbf{v})$ has a symmetric Jacobian matrix.*

Proof Denote by $F = (\phi, \psi)$ the transformation (2.78). The Jacobian matrix $DF = M$ is that in (2.80). Since $\mathbf{z}_t = F_t$, interchanging the order of differentiation in DF_t we have $D\mathbf{z}_t = \frac{\partial}{\partial t} DF = M_t$. Notice that $M\mathbf{e}_i = \frac{\partial F}{\partial u_i}$ and $M\mathbf{e}_{n+i} = \frac{\partial F}{\partial v_i}$. Now,

$$\langle X(\mathbf{u}, \mathbf{v}, t), \mathbf{e}_j \rangle = \langle J\mathbf{z}_t, M\mathbf{e}_j \rangle = \left\langle J\mathbf{z}_t, \frac{\partial F}{\partial u_j} \right\rangle.$$

As J is constant, we have $D(J\mathbf{z}_t) = JD\mathbf{z}_t = JM_t$ so we get, for $i, j \le n$

$$\langle DX \cdot \mathbf{e}_i, \mathbf{e}_j \rangle = \left\langle (DJ\mathbf{z}_t) \cdot \mathbf{e}_i, \frac{\partial F}{\partial u_j} \right\rangle + \left\langle J\mathbf{z}_t, \frac{\partial^2 F}{\partial u_i \partial u_j} \right\rangle = \left\langle JM_t \cdot \mathbf{e}_i, \frac{\partial F}{\partial u_j} \right\rangle + \left\langle J\mathbf{z}_t, \frac{\partial^2 F}{\partial u_i \partial u_j} \right\rangle.$$

Similarly, we get

$$\langle DX \cdot \mathbf{e}_j, \mathbf{e}_i \rangle = \left\langle JM_t \cdot \mathbf{e}_j, \frac{\partial F}{\partial u_i} \right\rangle + \left\langle J\mathbf{z}_t, \frac{\partial^2 F}{\partial u_j \partial u_i} \right\rangle.$$

Therefore, we have

$$\langle DX \cdot \mathbf{e}_i, \mathbf{e}_j\rangle - \langle DX \cdot \mathbf{e}_j, \mathbf{e}_i\rangle = \left\langle JM_t \cdot \mathbf{e}_i, \frac{\partial F}{\partial u_j}\right\rangle - \left\langle JM_t \cdot \mathbf{e}_j, \frac{\partial F}{\partial u_i}\right\rangle.$$

But $\frac{\partial F}{\partial u_i} = DF(\mathbf{u}, \mathbf{v}) \cdot \mathbf{e}_i = M \cdot \mathbf{e}_i$, so the right-hand side of this equality is equal to

$$\langle M^T JM_t \cdot \mathbf{e}_i, \mathbf{e}_j\rangle - \langle M^T JM_t \cdot \mathbf{e}_j, \mathbf{e}_i\rangle. \tag{2.99}$$

Since M is symplectic, $M^T JM = J$; hence, differentiating with respect to t we get $M_t^T JM + M^T JM_t = 0$ and taking transpose we see that $(M_t^T JM)^T - M_t^T JM = 0$, that is, $M_t^T JM$ is symmetric. Therefore, $M^T JM_t = -(M_t^T JM)^T$ is symmetric. This says that the difference (2.99) is zero; hence, $\langle DX \cdot \mathbf{e}_i, \mathbf{e}_j\rangle = \langle DX \cdot \mathbf{e}_j, \mathbf{e}_i\rangle$. Therefore, the Jacobian matrix DX is symmetric. □

By the lemma the vector field $X = M^T J\mathbf{z}_t(\mathbf{u}, \mathbf{v})$ has a symmetric Jacobian matrix, so locally, or globally in a star-shaped domain, the vector function X is the gradient of some function R.

With this function R, called the *remainder function*, the function $\mathcal{H}(\mathbf{u}, \mathbf{v}, t)$ in (2.96) defines the Hamiltonian equations in the variables $\mathbf{u}$, $\mathbf{v}$.

Proposition 2.12 *If the symplectic transformation (2.95) is generated by $W(\mathbf{x}, \mathbf{v}, t)$, then the remainder function is the time derivative of the generating function, that is,*

$$R(\mathbf{u}, \mathbf{v}, t) = W_t(\phi(\mathbf{u}, \mathbf{v}, t), \mathbf{v}, t). \tag{2.100}$$

Proof Differentiating the equations $\mathbf{y} = W_{\mathbf{x}}(\mathbf{x}, \mathbf{v}, t)$ and $\mathbf{u} = W_{\mathbf{v}}(\mathbf{x}, \mathbf{v}, t)$ with respect to t, we get

$$\dot{\mathbf{y}} = W_{\mathbf{xx}}\dot{\mathbf{x}} + W_{\mathbf{xv}}\dot{\mathbf{v}} + W_{\mathbf{xt}}, \quad \dot{\mathbf{u}} = W_{\mathbf{vx}}\dot{\mathbf{x}} + W_{\mathbf{vv}}\dot{\mathbf{v}} + W_{\mathbf{vt}}.$$

These equations together with (2.97) lead to the following equations

$$\begin{aligned}(\mathbf{y_u} - W_{\mathbf{xx}}\mathbf{x_u})\dot{\mathbf{u}} + (\mathbf{y_v} - W_{\mathbf{xx}}\mathbf{x_v} - W_{\mathbf{xv}})\dot{\mathbf{v}} + \mathbf{y}_t - W_{\mathbf{xx}}\mathbf{x}_t &= W_{\mathbf{x}t}, \\ (I - W_{\mathbf{vx}}\mathbf{x_u})\dot{\mathbf{u}} - (W_{\mathbf{vx}}\mathbf{x_v} + W_{\mathbf{vv}})\dot{\mathbf{v}} - W_{\mathbf{vx}}\mathbf{x}_t &= W_{\mathbf{v}t}.\end{aligned}$$

From (2.92) we see that the coefficients of $\dot{\mathbf{u}}$ and $\dot{\mathbf{v}}$ vanish, so the above equations reduce to

$$\mathbf{y}_t - W_{\mathbf{xx}}\mathbf{x}_t = W_{\mathbf{x}t}, \quad -W_{\mathbf{vx}}\mathbf{x}_t = W_{\mathbf{v}t}. \tag{2.101}$$

The partial derivatives of the function $\beta(\mathbf{u}, \mathbf{v}, t) = W_t(\phi(\mathbf{u}, \mathbf{v}, t), \mathbf{v}, t)$ with respect to $\mathbf{u}$ and $\mathbf{v}$ are given by[3] $\beta_{\mathbf{u}} = \mathbf{x}_{\mathbf{u}}^T W_{t\mathbf{x}}$, $\beta_{\mathbf{v}} = \mathbf{x}_{\mathbf{v}}^T W_{t\mathbf{x}} + W_{t\mathbf{v}}$; hence, in view of (2.101) we have

$$\beta_{\mathbf{u}} = \mathbf{x}_{\mathbf{u}}^T \mathbf{y}_t - \mathbf{x}_{\mathbf{u}}^T W_{\mathbf{xx}}\mathbf{x}_t, \quad \beta_{\mathbf{v}} = \mathbf{x}_{\mathbf{v}}^T \mathbf{y}_t - \mathbf{x}_{\mathbf{v}}^T W_{\mathbf{xx}}\mathbf{x}_t - W_{\mathbf{vx}}\mathbf{x}_t. \tag{2.102}$$

But differentiating $\mathbf{y} = W_{\mathbf{x}}(\mathbf{x}, \mathbf{v}, t)$ with respect to $\mathbf{u}$ and $\mathbf{v}$, we get

$$\mathbf{y}_{\mathbf{u}} = W_{\mathbf{xx}}\mathbf{x}_{\mathbf{u}} \quad \text{and} \quad \mathbf{y}_{\mathbf{v}} = W_{\mathbf{xx}}\mathbf{x}_{\mathbf{v}} + W_{\mathbf{xv}}.$$

Therefore, since the matrices $W_{\mathbf{xx}}$ and $W_{\mathbf{xv}}$ are symmetric equations (2.102) lead to

$$\beta_{\mathbf{u}} = \mathbf{x}_{\mathbf{u}}^T \mathbf{y}_t - \mathbf{y}_{\mathbf{u}}^T \mathbf{x}_t, \quad \beta_{\mathbf{v}} = \mathbf{x}_{\mathbf{v}}^T \mathbf{y}_t - \mathbf{y}_{\mathbf{v}}^T \mathbf{x}_t.$$

Consequently, we get

$$\nabla\beta = M^T J \mathbf{z}_t.$$

Comparing this with (2.98) we see that $R = \beta + const$. Since the constant is irrelevant for the dynamics we are done. □

Therefore, we can state the following proposition.

Proposition 2.13 *If the symplectic transformation (2.95) is generated by the function $W(\mathbf{x}, \mathbf{v}, t)$ then the new Hamiltonian is given by*

$$\mathcal{H}(\mathbf{u}, \mathbf{v}, t) = H(\mathbf{x}, \mathbf{y}, t) + W_t(\phi(\mathbf{u}, \mathbf{v}, t), \mathbf{v}, t). \tag{2.103}$$

2.4.4 Hamiltonian After a Time-Dependent μ-Symplectic Transformation

Under a time-independent μ-symplectic transformation the new Hamiltonian is given by (2.90). Under a time-dependent symplectic transformation generated by the function $W(\mathbf{x}, \mathbf{v}, t)$ the new Hamiltonian is given by (2.103).

The next proposition treats the case of a time-dependent μ-symplectic transformation.

Proposition 2.14 *Let the time-dependent mapping (2.95) be a μ-symplectic transformation generated by the function $W(\mathbf{x}, \mathbf{v}, t)$. Then, the new Hamiltonian is given by*

[3] By the chain rule (1.6) we have $D_{\mathbf{u}}\beta \cdot \xi = D_{\mathbf{x}} W_t \cdot (D_{\mathbf{u}}\phi \cdot \xi)$, and passing to gradients, see (1.7), we get $\langle \nabla_{\mathbf{u}}\beta, \xi\rangle = \langle \nabla_{\mathbf{x}} W_t, \mathbf{x}_{\mathbf{u}} \cdot \xi\rangle = \langle \mathbf{x}_{\mathbf{u}}^T \nabla_{\mathbf{x}} W_t, \xi\rangle$, or $\langle \beta_{\mathbf{u}}, \xi\rangle = \langle \mathbf{x}_{\mathbf{u}}^T W_{t\mathbf{x}}; \xi\rangle$ hence, $\beta_{\mathbf{u}} = \mathbf{x}_{\mathbf{u}}^T W_{t\mathbf{x}}$.

$$\mathcal{H}(\mathbf{u}, \mathbf{v}, t) = \frac{1}{\mu} H(\mathbf{x}, \mathbf{y}, t) + \frac{1}{\mu} W_t(\phi(\mathbf{u}, \mathbf{v}, t), \sqrt{\mu}\mathbf{v}, t). \tag{2.104}$$

Proof The mapping $\mathbf{u} = \frac{1}{\sqrt{\mu}}\boldsymbol{\xi}, \mathbf{v} = \frac{1}{\sqrt{\mu}}\boldsymbol{\eta}$ is $\frac{1}{\mu}$-symplectic; hence, the composite mapping $\mathbf{x} = \phi(\mathbf{u}, \mathbf{v}, t) = \tilde{\phi}(\boldsymbol{\xi}, \boldsymbol{\eta}, t)$, $\mathbf{y} = \psi(\mathbf{u}, \mathbf{v}, t) = \tilde{\psi}(\boldsymbol{\xi}, \boldsymbol{\eta}, t)$ is a time-dependent symplectic mapping. Therefore, by (2.103) the Hamiltonian $H(\mathbf{x}, \mathbf{y}, t)$ is transformed to

$$\tilde{H}(\boldsymbol{\xi}, \boldsymbol{\eta}, t) = H(\mathbf{x}, \mathbf{y}, t) + W_t(\mathbf{x}, \boldsymbol{\eta}, t).$$

Now, the mapping $\boldsymbol{\xi} = \sqrt{\mu}\mathbf{u}$, $\boldsymbol{\eta} = \sqrt{\mu}\mathbf{v}$ is a time-independent μ-symplectic mapping, so by (2.90) we get

$$H(\mathbf{u}, \mathbf{v}, t) = \frac{1}{\mu}\tilde{H}(\boldsymbol{\xi}, \boldsymbol{\eta}, t) = \frac{1}{\mu} H(\mathbf{x}, \mathbf{y}, t) + \frac{1}{\mu} W_t(\phi(\mathbf{u}, \mathbf{v}, t), \sqrt{\mu}\mathbf{v}, t).$$

□

2.4.5 The Generating Function of a Rotation

On several occasions it is important to know the generating function of the symplectic mapping given by a rotation

$$\begin{aligned} x &= \xi \cos \omega t - \eta \sin \omega t, \\ y &= \xi \sin \omega t + \eta \cos \omega t. \end{aligned} \tag{2.105}$$

This is a counter-clockwise rotation by the angle ωt (see Fig. 2.17).

Proposition 2.15 *The generating function of the rotation (2.105) is given by*

$$W = \frac{1}{2}(x^2 + \eta^2)\frac{\sin \omega t}{\cos \omega t} + \frac{x\eta}{\cos \omega t}$$

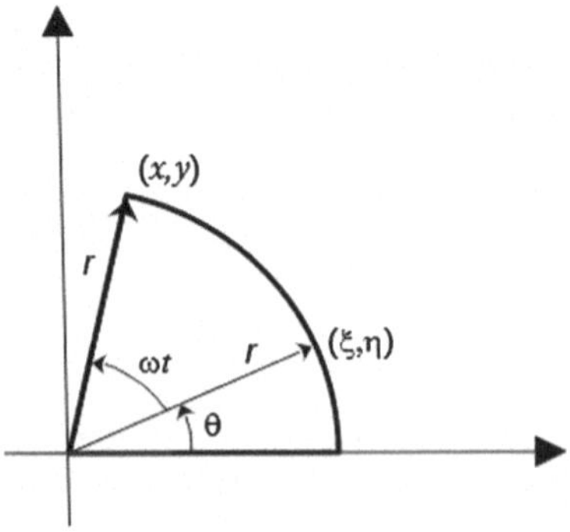

Fig. 2.17 Counter-clockwise rotation

and its time derivative is

$$W_t = \frac{\omega}{2}(\xi^2 + \eta^2). \tag{2.106}$$

Proof Let $W(x, \eta, t)$ be the generating function of the symplectic transformation (2.105). Then,

$$y = W_x \quad \text{and} \quad \xi = W_\eta.$$

From the first equation in (2.105) we have $\xi = \dfrac{x + \eta \sin \omega t}{\cos \omega t}$; hence,

$$W_x = \xi \sin \omega t + \eta \cos \omega t = \frac{x + \eta \sin \omega t}{\cos \omega t} \sin \omega t + \eta \cos \omega t = \frac{x \sin \omega t}{\cos \omega t} + \frac{\eta}{\cos \omega t},$$

and integrating with respect to x we get

$$W(x, \eta, t) = \frac{1}{2}x^2 \frac{\sin \omega t}{\cos \omega t} + \frac{\eta x}{\cos \omega t} + g(\eta);$$

hence,

$$W_\eta = \frac{x}{\cos \omega t} + g'(\eta) = \xi = \frac{x + \eta \sin \omega t}{\cos \omega t},$$

so

$$g'(\eta) = \frac{\eta \sin \omega t}{\cos \omega t} \quad \text{hence} \quad g(\eta) = \frac{1}{2}\eta^2 \frac{\sin \omega t}{\cos \omega t}.$$

Therefore, the generating function is that stated in the proposition. We now compute

$$W_t = \frac{\omega}{2}(x^2 + \eta^2)\frac{1}{\cos^2 \omega t} + \frac{wx\eta \sin \omega t}{\cos^2 \omega t}$$

and using the expression of x in (2.105) we find (2.106). □

Example 2.19 As an illustration consider the one degree of freedom Hamiltonian

$$H(q, p, t; \epsilon) = H_0(q, p) + \epsilon H_1(q, p, t) + \epsilon^2 H_2(q, p, t) + \dots,$$

in which $H_0 = \dfrac{\omega}{2}(q^2 + p^2)$.

Through a rotation by the angle $-\omega t$, a clockwise rotation, we can get rid of the term independent of ϵ since to get the new Hamiltonian we must add W_t to the old one.

2.5 The Symplectic Group

The set of real symplectic matrices of order $2n$ is a subgroup of the group of invertible real square matrices of order $2n$ under matrix multiplication. This is so because it contains the identity matrix, the product of any two of its matrices and the inverse of each one of its matrices (see items (a), (b) and (d) of Proposition 2.6).

This group is called the *real symplectic group* of order $2n$ and is denoted by $Sp(2n, \mathbf{R})$.

The symplectic group permeates the whole of Hamiltonian systems theory. Besides the aforementioned properties of symplectic matrices many properties will be seen in the future, as the polar decomposition of a symplectic matrix (Proposition 4.7), the J-orthogonality of generalized eigenspaces of a symplectic matrix (Lemma 4.37) and others.

At the end of this chapter we consider some aspects of the geometry of the real symplectic group $Sp(2n, \mathbf{R})$ and in Sect. 7.7.2 we study its topology, which is essential in the study of the Gelfand–Lidskii Theory made in Chap. 7.

2.6 The Symplectic Product and the Poisson Bracket

The standard sympletic matrix J of order $2n$ induces in the space $\mathbf{R}^{2n}$ a nondegenerate skew-symmetric bilinear form defined by

$$\{\mathbf{z}, \boldsymbol{\zeta}\} = \mathbf{z}^T J \boldsymbol{\zeta} = \langle \mathbf{z}, J\boldsymbol{\zeta} \rangle, \tag{2.107}$$

where $\langle\ ,\ \rangle$ denotes the Euclidean inner product defined in (1.2).

Definition 2.16 The bilinear form defined by (2.107) is called the *symplectic product* or the *standard symplectic form* in $\mathbf{R}^{2n}$.

If $\mathbf{z} = (\mathbf{x}, \mathbf{y})$, $\boldsymbol{\zeta} = (\boldsymbol{\xi}, \boldsymbol{\eta}) \in \mathbf{R}^n \times \mathbf{R}^n = \mathbf{R}^{2n}$ the symplectic product (2.107) is given by

$$\{\mathbf{z}, \boldsymbol{\zeta}\} = \langle \mathbf{x}, \boldsymbol{\eta} \rangle - \langle \mathbf{y}, \boldsymbol{\xi} \rangle.$$

Symplectic forms in general are studied in Chap. 4 but the standard symplectic form (2.107) plays a dominant role in the study of normal forms of Hamiltonian matrices, which are made in Chaps. 4 and 5. It is also used extensively in Chaps. 7 and 8 on the stability of periodic linear Hamiltonian systems.

An important concept related to the symplectic product is that of the Poisson bracket.

Definition 2.17 The Poisson bracket of two smooth functions $f(\mathbf{z})$, $g(\mathbf{z})$ on an open subset U of $\mathbf{R}^{2n}$ is defined by

$$\{f, g\}(\mathbf{z}) = \nabla f(\mathbf{z})^T J \nabla g(\mathbf{z}) = \{\nabla f(\mathbf{z}), \nabla g(\mathbf{z})\}. \tag{2.108}$$

This is a skew-symmetric bilinear form in the space of smooth functions, that is, functions at least continuously differentiable in an open domain U of $\mathbf{R}^{2n}$.

The Poisson bracket is invariant under symplectic transformations. In fact, we have

Proposition 2.18 *The mapping* $\mathbf{z} = \Phi(\boldsymbol{\zeta})$ *is symplectic if, and only if,*

$$\{f \circ \Phi, g \circ \Phi\}(\boldsymbol{\zeta}) = \{f, g\}(\mathbf{z}), \tag{2.109}$$

for any functions f, g.

Proof We have $\{f \circ \Phi, g \circ \Phi\}(\boldsymbol{\zeta}) = \nabla(f \circ \Phi)(\boldsymbol{\zeta})^T J \nabla(g \circ \Phi)(\boldsymbol{\zeta})$ and $\{f, g\}(\mathbf{z}) = f(\mathbf{z})^T J \nabla g(\mathbf{z})$.

But $\nabla(f \circ \phi)(\boldsymbol{\zeta}) = D\phi(\boldsymbol{\zeta})^T \nabla f(\mathbf{z})$ and $\nabla(g \circ \phi)(\boldsymbol{\zeta}) = D\phi(\boldsymbol{\zeta})^T \nabla g(\mathbf{z})$, so

$$\{f \circ \Phi, g \circ \Phi\}(\boldsymbol{\zeta}) = \nabla f(\mathbf{z})^T \left[D\phi(\boldsymbol{\zeta}) J D\phi(\boldsymbol{\zeta})^T\right] \nabla g(\mathbf{z}).$$

Therefore, (2.109) holds for any functions f, g if, and only if, $D\phi(\boldsymbol{\zeta}) J D\phi(\boldsymbol{\zeta})^T = J$. □

If $\mathbf{z} = (\mathbf{x}, \mathbf{y}) \in \mathbf{R}^n \times \mathbf{R}^n = \mathbf{R}^{2n}$, the Poisson bracket is given by

$$\{f, g\}(\mathbf{z}) = \langle \nabla_{\mathbf{x}} f, \nabla_{\mathbf{y}} g\rangle - \langle \nabla_{\mathbf{y}} f, \nabla_{\mathbf{x}} g\rangle = \sum_{i=1}^{n} \left(\frac{\partial f}{\partial x_i} \frac{\partial g}{\partial y_i} - \frac{\partial f}{\partial y_i} \frac{\partial g}{\partial x_i} \right). \tag{2.110}$$

Poisson brackets can be used to test for the symplecticity of a transformation $\mathbf{z} = \Phi(\boldsymbol{\zeta})$. Let $\mathbf{z} = (\mathbf{x}, \mathbf{y})$ and $\boldsymbol{\zeta} = (\boldsymbol{\xi}, \boldsymbol{\eta})$. Then, $x_j = \Phi_j(\boldsymbol{\xi}, \boldsymbol{\eta})$ and $y_j = \Phi_{n+j}(\boldsymbol{\xi}, \boldsymbol{\eta})$.

Proposition 2.19 *The mapping* $\mathbf{z} = \Phi(\boldsymbol{\zeta})$ *is symplectic if, and only if,*

$$\{x_i, x_j\} = 0, \quad \{y_i, y_j\} = 0 \quad \text{and} \quad \{x_i, y_j\} = \delta_{ij}, \tag{2.111}$$

where $\delta_{ij} = 0$ *for* $i \neq j$ *and* $\delta_{ij} = 1$ *for* $i = j$ *(Kronecker delta).*

Proof First, notice that a matrix M is symplectic if and only if its rows form a symplectic basis. Indeed,

$$(MJM^T)_{kl} = \langle \mathbf{e}_k, MJM^T \mathbf{e}_l\rangle = \langle M^T \mathbf{e}_k, JM^T \mathbf{e}_l\rangle = \{M_k, M_l\},$$

since $M^T \mathbf{e}_k$ is the kth column of M^T, that is, the kth row M_k of M. Therefore, $MJM^T = J$ if and only if $\{M_k, M_l\} = J_{kl}$, which proves the statement.

Now, the equalities (2.111) say that the rows of the Jacobian matrix of Φ form a symplectic basis. Therefore, Φ is symplectic if, and only if, the equalities (2.111) hold. □

First integrals of Hamiltonian systems can be formulated in terms of the Poisson bracket since the total derivative of a function f of $\mathbf{z} = (\mathbf{x}, \mathbf{y})$ along a solution $(\mathbf{x}(t), \mathbf{y}(t))$ of the Hamiltonian system $\dot{\mathbf{x}} = H_{\mathbf{y}}, \dot{\mathbf{y}} = -H_{\mathbf{x}}$, is given by (see (1.11))

$$\frac{df}{dt} = \langle \nabla_{\mathbf{x}} f, \dot{\mathbf{x}} \rangle + \langle \nabla_{\mathbf{y}} f, \dot{\mathbf{y}} \rangle = \langle \nabla_{\mathbf{x}} f, \nabla_{\mathbf{y}} H \rangle - \langle \nabla_{\mathbf{y}} f, \nabla_{\mathbf{x}} H \rangle = \{f, H\}(\mathbf{z}).$$

Therefore, $f(\mathbf{x}, \mathbf{y})$ is a first integral of the system $\dot{\mathbf{x}} = H_{\mathbf{y}}, \dot{\mathbf{y}} = -H_{\mathbf{x}}$ if, and only if, $\{f, H\} \equiv 0$.

An important property of the Poisson bracket is the Jacobi identity

$$\{\{f, g\}, h\} + \{\{h, f\}, g\} + \{\{g, h\}, f\} = 0,$$

valid for any three smooth functions f, g, h. The proof is simple and we omit it. From this identity we see that if two first integrals f, g of the Hamiltonian system $\dot{\mathbf{x}} = H_{\mathbf{y}}, \dot{\mathbf{y}} = -H_{\mathbf{x}}$ are known, a third one can be found, namely $\{f, g\}$.

The Poisson bracket will play an important role in the study of normal forms of nonlinear Hamiltonian systems (see Deprit formula, Theorem 3.10 and the associated homological equation (3.61). See also its presence in the formulation of the normal form by Kamel, Theorem 3.22).

2.7 Stability of Equilibria

Consider an ordinary differential equation

$$\dot{\mathbf{x}} = \mathbf{f}(\mathbf{x}) \tag{2.112}$$

with $\mathbf{f} : \Omega \to \mathbf{R}^n$ of class at least C^1 in the open set $\Omega \subset \mathbf{R}^n$, so local existence and uniqueness of solutions are guaranteed.

An *equilibrium* point of (2.112) is a zero of the vector field defining the equation, that is a point $\mathbf{x}_0$ such that $\mathbf{f}(\mathbf{x}_0) = 0$.

Definition 2.20 The equilibrium $\mathbf{x}_0$ is *Lyapunov stable* (in the future) if given an arbitrary $\epsilon > 0$ there is a $\delta > 0$ such that for any $\mathbf{x}$ in the open ball $B_\delta(\mathbf{x}_0)$ the solution $\mathbf{x}(t)$ with $\mathbf{x}(0) = \mathbf{x}$ is defined for all $t > 0$ and remains inside the open ball $B_\epsilon(\mathbf{x}_0)$.[4] The equilibrium $\mathbf{x}_0$ is *unstable* if there is an $\epsilon > 0$, such that for any $\delta > 0$, some solution $\mathbf{x}(t)$ with $\mathbf{x}(0) = \mathbf{x}$ in $B_\delta(\mathbf{x}_0)$ either is not defined for all $t > 0$ or it does not remain inside $B_\epsilon(\mathbf{x}_0)$.

Theorem 2.21 (Lyapunov) *Let $\lambda_1, \dots, \lambda_n$ be the eigenvalues of the $n \times n$ matrix $A = D\mathbf{f}(\mathbf{x}_0)$. If the equilibrium $\mathbf{x}_0$ is stable, then $Re\lambda_j \leq 0$, for all j; if $Re\lambda_j > 0$ for some j, the equilibrium is unstable.*

---

[4] We can omit the requirement *defined for all $t > 0$* because since the solution remains inside the compact set $\overline{B_\delta(\mathbf{x}_0)}$ for $t > 0$ in its interval of definition this interval is the half-line $t \geq 0$.

Proof See the proof in Chapter 9 of [30]. □

Theorem 2.22 (Dirichlet) *If in a neighborhood of an equilibrium* $\mathbf{x}^*$ *of the Eq. (2.112) there is a first integral that is positive (or negative) definite, then the equilibrium is stable.*

Proof Let Ψ be a first integral in the neighborhood $B_r(\mathbf{x}^*)$ of the equilibrium point $\mathbf{x}^*$. We can assume that it is positive definite. Then, for any positive $\epsilon < r$ the minimum m of Ψ on the sphere $\|\mathbf{x} - \mathbf{x}^*\| = \epsilon$ is positive. Consider the open set $\Omega = \Psi^{-1}[0, m) \subset B_\epsilon(\mathbf{x}^*)$. Let $\delta > 0$ be such that the ball $B_\delta(\mathbf{x}^*)$ is contained in Ω. Then, for any $\boldsymbol{\xi} \in B_\delta(\mathbf{x}^*)$, the solution $\phi(t, \boldsymbol{\xi})$ remains inside the ball $B_\epsilon(\mathbf{x}^*)$ otherwise it would cross the sphere $\|\mathbf{x} - \mathbf{x}^*\| = \epsilon$, where the minimum value of Ψ is m, which is not possible because as Ψ is a first integral $\Psi(\phi(t, \boldsymbol{\xi})) \equiv \Psi(\boldsymbol{\xi}) < m$. As the solution $\phi(t, \boldsymbol{\xi})$ remains inside the compact set $\overline{B_\epsilon(\mathbf{x}^*)}$ it is defined for all $t \geq 0$; hence, the equilibrium $\mathbf{x}^*$ is stable. □

For an instability result we will state and prove Chetaev's theorem in Sect. 6.2.

Example 2.20 (Stability of Harmonic Oscillator) Since for the linear system

$$\dot{x} = \omega y, \qquad \dot{y} = -\omega x,$$

the function $I = x^2 + y^2$ has zero derivative, it is a first integral and as it is positive definite, Dirichlet's theorem says that the equilibrium (0, 0) is stable.

Example 2.21 (Stability of Stationary Rotations of a Free Rigid Body) Recall from (2.38) the equations of motion for the free rigid body about its fixed center of mass, namely

$$I_1\dot{p} = (I_2 - I_3)qr, \quad I_2\dot{q} = (I_3 - I_1)pr, \quad I_3\dot{r} = (I_1 - I_2)pq, \tag{2.113}$$

where $\boldsymbol{\Omega} = p\mathbf{e}_1 + q\mathbf{e}_2 + r\mathbf{e}_3$.

Besides the trivial equilibrium $\Omega = 0$ (body at rest), these equations have three more equilibrium points, namely the rotations corresponding to the cases when two components of the angular momentum are zero and the third is not zero.[5] This means that in the absence of external forces the rigid body admits stationary rotations, that is, rotations with Ω constant, about each one of the principal axes of inertia.

Let us prove that the equilibria corresponding to stationary rotations about the minor and the major semi-axes of the ellipsoid of inertia are stable.

Suppose that the principal moments of inertia are ordered so that $I_1 \leq I_2 \leq I_3$. If $I_1 = I_3$, then the three moments of inertia are equal and the Eqs. (2.113) show that all the rotations of the rigid body are stationary; hence, all the points are stable equilibria.

---

[5] These are the only possibilities, if the three moments of inertia are pairwise distinct; otherwise, there are infinitely many equilibria.

Now suppose that $I_1 < I_3$. If $I_1 = I_2$, the third equation of the system (2.113)) shows that r is constant and the first two equations reduce to

$$\dot{p} = -\alpha q, \qquad \dot{q} = \alpha p,$$

with $\alpha = \frac{(I_3 - I_1)r}{I_1} > 0$, so the equilibrium is stable by the previous example.

Let us examine the generic case $I_1 < I_2 < I_3$. Assume that $p = a \neq 0$, $q = r = 0$, which corresponds to a stationary rotation about the major semi-axis. Making the translation $x = p - a$, $y = q$, $z = r$ the system (2.113) becomes

$$I_1\dot{x} = (I_2 - I_3)yz, \quad I_2\dot{y} = (I_3 - I_1)(x + a)z, \quad I_3\dot{z} = (I_1 - I_2)(x + a)y,$$

with the equilibrium now at the origin (0, 0, 0). Now, consider the function

$$v(x, y, z) = \left(I_1x^2 + 2aI_1x + I_2y^2 + I_3z^2\right)^2 + \frac{I_2 - I_1}{I_3}y^2 + \frac{I_3 - I_1}{I_2}z^2$$

Clearly, $v(x, y, z) \geq 0$ and if $v(x, y, z) = 0$ then $y = z = 0$; hence, $(I_1x^2 + 2aI_1x)^2 = 0$. This implies that if we take x small then $x = 0$ and therefore $v(x, y, z)$ is positive definite in a neighborhood of the origin.

Now, computing the time derivative of $v(x, y, z)$ along solutions of the above system we find it to vanish identically; hence, by Theorem 2.22 the origin is a stable equilibrium, which means that the stationary rotation about the major axis is stable. Analogously, the stability of the stationary rotation about the minor axis is proved.

In Chap. 6 we see, as an application of Chetaev's theorem, that the equilibrium corresponding to rotation about the intermediary axis is unstable.

Setting $\mathbf{z} = \begin{bmatrix} \mathbf{x} \\ \mathbf{y} \end{bmatrix}$ we can write the Hamiltonian system (2.52) in the form

$$\dot{\mathbf{z}} = J\nabla H(\mathbf{z}), \tag{2.114}$$

where J is the standard symplectic matrix (see Definition 1.7).

If $Z = \mathbf{z} - \mathbf{z}_0$, then the linearized system in a neighborhood of the equilibrium $\mathbf{z}_0$ is

$$\dot{Z} = AZ, \tag{2.115}$$

where $A = JG$, with $G = D_{zz}H(\mathbf{z}_0)$ the Hessian matrix of H at $\mathbf{z}_0$.

Theorem 2.21 implies that if an equilibrium of a Hamiltonian system is stable then all the eigenvalues of the linearized system are purely imaginary numbers. This is because if there is an eigenvalue with a nonzero real part, then there is one with a positive real part because of the following fact:

Proposition 2.23 *The characteristic polynomial of $A = JG$ is even.*

Proof Indeed, we have

$$\begin{aligned}p(\lambda) &= \det(JG - \lambda I) = \det(JG - \lambda I)^T = \det(-GJ - \lambda I)\\ &= \det[J(JG + \lambda I)J] = \det(JG + \lambda I) = p(-\lambda).\end{aligned}$$

□

Now, the question of stability of an equilibrium of a system for which all the eigenvalues have zero real part is quite delicate as is shown by the example below.

Example 2.22 The following system in the plane

$$\dot{x} = -y + \epsilon x\left(x^2 + y^2\right)^n, \quad \dot{y} = x + \epsilon y\left(x^2 + y^2\right)^n, \tag{2.116}$$

has an equilibrium at the origin with eigenvalues $\pm i$. Passing to polar coordinates $x = r\cos\theta$, $y = r\sin\theta$, the system becomes

$$\dot{r} = \epsilon r^{n+1}, \quad \dot{\theta} = 1$$

which can be easily integrated to give $r(t) = r_0/\sqrt[n]{1 - \epsilon n r_0^n t}$, $\theta(t) = t + \theta_0$. Therefore, when $\epsilon < 0$ the solutions spiral in toward the origin as $t \to \infty$, and when $\epsilon > 0$ they are not defined for all $t > 0$ and in fact spiral out, becoming unbounded as t approaches the value $1/\epsilon n r_0^n$. In the former case the origin $(0, 0)$ is a stable equilibrium; in the latter case, an unstable equilibrium. The perturbation of the linear system is of the order $O\left((x^2 + y^2)^n\right)$ so in the neighborhood of the origin it is small if n is large and we see that the origin $(0, 0)$ for the perturbed system may become stable or unstable, depending on the sign of ϵ.

If the Hamiltonian $H = H(\mathbf{z})$ is analytic we can write its Taylor expansion in a neighborhood of the equilibrium $\mathbf{z}_0 = 0$ in the form

$$H(\mathbf{z}) = H_2(\mathbf{z}) + H_3(\mathbf{z}) + H_4(\mathbf{z}) + \dots, \tag{2.117}$$

since $\nabla H(\mathbf{z}_0) = 0$ and the constant term can be omitted as it is irrelevant for the dynamics; here, $H_j(\mathbf{z})$ is a real homogeneous polynomial of degree j in the variables $x_1, \dots, x_n, y_1, \dots, y_n$.

If the quadratic polynomial

$$H_2(\mathbf{z}) = \frac{1}{2}\mathbf{z}^T G\mathbf{z}, \quad G = D_{\mathbf{zz}}H(0), \tag{2.118}$$

is positive or negative definite, then the same is true of the Hamiltonian $H(\mathbf{z})$ in a neighborhood of $\mathbf{z}_0$. Since H is a nondegenerate first integral of the system, Theorem 2.22 implies that the equilibrium is stable.

So, the question of the stability of an equilibrium of a Hamiltonian system is trivial if some eigenvalue of JG has a nonzero real part or if the Hessian matrix G

is definite. However, if all the eigenvalues are purely imaginary and the matrix G is indefinite, then this question is very difficult as the Example 2.22 suggests.

We notice that if $\epsilon \neq 0$ the system (2.116) is not Hamiltonian because the image of a region of nonzero volume by the flow has a variable volume and this is not possible for a Hamiltonian system as we will see in Corollary 2.28.1.

In the next proposition we characterize the matrix of a linear Hamiltonian system.

Proposition 2.24 *Let A be a $2n \times 2n$ real matrix. Then, the linear system $\dot{\mathbf{x}} = A\mathbf{x}$ is Hamiltonian if and only if $A = JS$, where $S^T = S$.*

Proof If the system $\dot{\mathbf{x}} = A\mathbf{x}$ is Hamiltonian, then $A\mathbf{x} = J\nabla H(\mathbf{x})$, for some function of the $2n$-vector $\mathbf{x}$. Differentiating this equality we get for any $2n$-vector $\boldsymbol{\xi}$, $A \cdot \boldsymbol{\xi} = JS \cdot \boldsymbol{\xi}$ with $S = D^2 H(\mathbf{x})$. Therefore, $A = JS$ with $S^T = S$. For the converse take $H(\mathbf{x}) = \frac{1}{2}\mathbf{x}^T S\mathbf{x}$. Since $\nabla H(\mathbf{x}) = S\mathbf{x}$, we have $\dot{\mathbf{x}} = A\mathbf{x} = JS\mathbf{x} = J\nabla H(\mathbf{x})$, so the system is Hamiltonian. □

Definition 2.25 We say that a $2n \times 2n$ real matrix A is Hamiltonian if the linear system $\dot{\mathbf{x}} = A\mathbf{x}$ is Hamiltonian.

Proposition 2.26 *The following conditions are equivalent for a $2n \times 2n$ real matrix A:*

(1) A is Hamiltonian;
(2) $A = JS$, where $S^T = S$;
(3) $A^T J + JA = O$;
(4) JA is a symmetric matrix;
(5) If A of order $2n$ has the block form $A = \begin{bmatrix} a & b \\ c & d \end{bmatrix}$ with squares matrices a, b, c, d then A is Hamiltonian if, and only if, the following hold

$$b \text{ and } c \text{ are symmetric} \quad \text{and} \quad a^T + d = 0. \tag{2.119}$$

Moreover, if A and B are Hamiltonian matrices, then $A \pm B$, αA, α a scalar, A^T, and $[A, B] = AB - BA$ are Hamiltonian matrices.

Proof Condition (1) implies (2), by Proposition 2.24. Now, if (2) holds, then using the identities (2.86) we see that $A^T J + JA = S^T J^T J + JJS = S^T - S = O$, so (2) implies (3). Now, if (3) holds then $JA = -A^T J$. But $(A^T J)^T = J^T A = -JA = A^T J$, so $A^T J$; hence, JA is symmetric, that is, (3) implies (4). Finally, (4) implies (1) by Proposition 2.24. The remaining statements are straightforward; we just consider the case for the bracket $A = JR$, $B = JS$ with R and S symmetric, so $[A, B] = JV$ with $V = RJS - SJR$ symmetric; hence, $[A, B]$ is Hamiltonian. □

Remark 2.27 We can use any one of the equivalent conditions (2)–(4) of Proposition 2.26 to define real Hamiltonian matrices. We extend the definition of a Hamiltonian matrix to any complex $2n \times 2n$ matrix by requiring it to satisfy one of these conditions.

Example 2.23 The conjugate of a Hamiltonian matrix by a symplectic matrix is also a Hamiltonian matrix.

Indeed, let $A = JS$ with S symmetric and let P be a symplectic matrix. Since $P^{-1} = -JP^TJ$, we have using (1.24)

$$P^{-1}AP = -JP^TJ(JS)P = JP^TSP = J\tilde{S},$$

with $\tilde{S} = P^TSP$ a symmetric matrix; hence, $P^{-1}AP$ is a Hamiltonian matrix.

Now we prove the following result on Hamiltonian flows.

Proposition 2.28 *The flow* $\mathbf{z} = \phi(t, \boldsymbol{\zeta})$ *of a real Hamiltonian system* $\dot{\mathbf{z}} = J\nabla_{\mathbf{z}}H(\mathbf{z}, t)$ *is a canonical transformation. Conversely, if the flow of an ordinary differential equation* $\dot{\mathbf{z}} = \mathbf{f}(\mathbf{z}, t)$ *is a canonical transformation, then the equation is locally a Hamiltonian system.*

Proof The solution $\phi(t, \boldsymbol{\zeta})$ with initial condition $\boldsymbol{\zeta}$ is defined by

$$\frac{\partial}{\partial t}\phi(t, \boldsymbol{\zeta}) = J\nabla_{\mathbf{z}}H(\phi(t, \boldsymbol{\zeta}), t), \qquad \phi(0, \boldsymbol{\zeta}) = \boldsymbol{\zeta}.$$

Differentiating the equation with respect to $\boldsymbol{\zeta}$ and interchanging the order of differentiation we get the equation

$$\frac{\partial}{\partial t}D_{\boldsymbol{\zeta}}\phi(t, \boldsymbol{\zeta}) = JH_{\mathbf{zz}}D_{\boldsymbol{\zeta}}\phi(t; \boldsymbol{\zeta})$$

hence, taking into account the identities (2.86) we have

$$\frac{\partial}{\partial t}\left[D_{\boldsymbol{\zeta}}\phi(t, \boldsymbol{\zeta})^T J D_{\boldsymbol{\zeta}}\phi(t, \boldsymbol{\zeta})\right] = \left(D_{\boldsymbol{\zeta}}\phi\right)^T H_{\mathbf{zz}}J^T J D_{\boldsymbol{\zeta}}\phi + \left(D_{\boldsymbol{\zeta}}\phi\right)^T JJH_{\mathbf{zz}}D_{\boldsymbol{\zeta}}\phi = 0.$$

Now, the equality $D_{\boldsymbol{\zeta}}\phi(t, \boldsymbol{\zeta})^T J D_{\boldsymbol{\zeta}}\phi(t, \boldsymbol{\zeta}) = J$ holds when $t = 0$, since $D_{\boldsymbol{\zeta}}\phi(0, \boldsymbol{\zeta}) = I$, so it holds identically, which shows that $D_{\boldsymbol{\zeta}}\phi(t, \boldsymbol{\zeta})$ is symplectic, that is $\mathbf{z} = \phi(t, \boldsymbol{\zeta})$ is a canonical transformation.

To prove the converse, let $\phi(t, \boldsymbol{\zeta})$ be the flow of $\dot{\mathbf{z}} = \mathbf{f}(\mathbf{z})$ and set $X(t, \boldsymbol{\zeta}) = D_{\boldsymbol{\zeta}}\phi(t, \boldsymbol{\zeta})$. By hypothesis, X is a symplectic matrix, for all $(t, \boldsymbol{\zeta})$; hence, $X^TJX = J$. Differentiating with respect to t, we have $\dot{X}^TJX + X^TJ\dot{X} = 0$; hence, $(\dot{X}X^{-1})^TJ + J(\dot{X}X^{-1}) = 0$, so by item (3) of Theorem 2.26 the matrix $\dot{X}X^{-1}$ is Hamiltonian. Therefore, by item (2) of the same proposition $\dot{X}X^{-1} = JS$, where $S = S(t, \boldsymbol{\zeta})$ is a symmetric matrix.

Since $\frac{\partial}{\partial t}\phi(t, \boldsymbol{\zeta}) = \mathbf{f}(\phi(t, \boldsymbol{\zeta}), t)$, differentiating this identity with respect to $\boldsymbol{\zeta}$ and then interchanging the order of differentiation we get the linear differential equation for $X(t, \boldsymbol{\zeta})$,

$$\dot{X} = D_{\mathbf{z}}\mathbf{f}(\phi(t, \boldsymbol{\zeta}), t)X \quad \text{or} \quad \dot{X}X^{-1} = D_{\mathbf{z}}\mathbf{f}(\phi(t, \boldsymbol{\zeta}), t).$$

Consequently, $JS = D_{\mathbf{z}}\mathbf{f}(\phi(t, \zeta), t)$, which means that the vector field $J^{-1}\mathbf{f}(\mathbf{z}, t)$ defined on an open set $U \times I \subset \mathbf{R}^{2n} \times \mathbf{R}$ has a symmetric Jacobian matrix. Therefore, for a star-shaped domain $V \subset U$, $J^{-1}\mathbf{f}(\mathbf{z}, t) = \nabla_{\mathbf{z}} H(\mathbf{z}, t)$ for some function H defined in $V \times I$, so $\mathbf{f}(\mathbf{z}, t) = J\nabla_{\mathbf{z}} H(\mathbf{z}, t)$ proving that the equation $\dot{\mathbf{z}} = \mathbf{f}(\mathbf{z}, t)$ defined in $V \times I$ is Hamiltonian. □

Corollary 2.28.1 *The flow of a Hamiltonian system preserves volume.*

Proof Since the determinant of a symplectic matrix has absolute value equal to one, in fact it is equal to one; as we will see in Corollary 2.31.1, we have from the change of variables formula

$$\mathrm{Vol}(\phi_t(R)) = \int_{\phi_t(R)} d\mathbf{z} = \int_R |\mathrm{det} D_\zeta \phi_t(\zeta)| d\zeta = \int_R d\zeta = \mathrm{Vol}(R).$$

□

For linear systems Proposition 2.28 has a much simpler proof.

Definition 2.29 The *matrizant* of the linear system $\dot{\mathbf{z}} = A(t)\mathbf{z}$ is the fundamental matrix solution, which is the identity at $t = 0$, that is, $\dot{Z} = A(t)Z$ and $Z(0) = I$.

The flow of $\dot{\mathbf{z}} = A(t)\mathbf{z}$ is $\phi(t, \zeta) = Z(t)\zeta$.

Proposition 2.30 *If $A(t)$ is Hamiltonian, then $Z(t)$ is symplectic, for each t. Conversely, if $Z(t)$ is symplectic, for each t, then A is Hamiltonian.*

Proof We will give the proof to show how in the linear case the proof is so much simpler. Let $U = Z^T JZ$. Then, $\dot{U} = \dot{Z}^T JZ + Z^T J\dot{Z} = Z^T(A^T J + JA)Z$. From this we see that if A is Hamiltonian then $U(t)$ is constant, hence $Z^T JZ = U(0) = J$, so $Z(t)$ is symplectic. Conversely, if $Z(t)$ is symplectic for each t, then $U = J$, so $\dot{U} = 0$; hence, from the above expression of $\dot{U}$ we see that A is Hamiltonian. □

Clearly, the determinant of a real symplectic matrix M is ± 1, since $|\mathrm{det} M|^2 \mathrm{det} J = \mathrm{det} J$. In fact, it is equal to $+1$ as we will prove below using the polar decomposition of a nonsingular matrix.

Proposition 2.31

(a) Any nonsingular real matrix A has two polar decompositions

$$A = PO \quad \text{and} \quad A = O'P',$$

where matrices P, P' are positive definite and O, O' are orthogonal. Each decomposition is unique.

(b) If the real matrix A is symplectic, then P, O, P', O' are symplectic.

Proof The symmetric matrices AA^T and A^TA are positive definite, because A is nonsingular; therefore, they have square roots.[6] Let P be the positive definite square root of the first and P' that of the second. Now, set $O = P^{-1}A$ and $O' = AP'^{-1}$. Then, O and O' are orthogonal matrices and we have the equalities $A = PO$ and $A = O'P'$. It remains to show that these decompositions are unique. Assume, for instance, that we have a second decomposition $A = \tilde{P}\tilde{O}$. Then, $\tilde{P}^{-1}P = \tilde{O}O^{-1} = O^*$. The matrix O^* is orthogonal, so its eigenvalues are complex numbers with absolute value equal to one. If $O^*\mathbf{v} = \lambda\mathbf{v}$, then we get $P\mathbf{v} = \lambda\tilde{P}\mathbf{v}$; hence, $\langle P\mathbf{v}, \mathbf{v}\rangle = \lambda\langle\tilde{P}\mathbf{v}, \mathbf{v}\rangle$, so λ is a real positive number; therefore, $\lambda = 1$. We conclude that $O^*\mathbf{v} = \mathbf{v}$ for all the eigenvectors of O^*, so $O^* = I$. Consequently, $\tilde{P} = P$ and $\tilde{O} = O$. Analogously, we prove the uniqueness of the second polar decomposition. This concludes the proof of part (a).

As to part (b), recall that a matrix B is symplectic if, and only if, $B^{-1} = J^{-1}B^TJ$. Now, as A is symplectic $A^{-1} = J^{-1}A^TJ$ and since $A = PO$ we get

$$O^{-1}P^{-1} = A^{-1} = J^{-1}O^TP^TJ = (J^{-1}O^TJ)(J^{-1}P^TJ).$$

The first matrix inside parenthesis on the right-hand side is orthogonal and the second matrix inside the parenthesis is positive definite. Since O^{-1} is orthogonal and P^{-1} is positive definite we get by uniqueness, $O^{-1} = J^{-1}O^TJ$ and $P^{-1} = J^{-1}P^TJ$, so O and P are symplectic. □

Corollary 2.31.1 *The determinant of a real symplectic matrix is equal to* $+1$.

Proof Let $A = PO$ be a polar decomposition of the symplectic matrix A. Since the determinant of P is positive, it suffices to prove that the determinant of O is positive also, because then the determinant of A will be $+1$. Because O is orthogonal and symplectic, then in block form we have $O = \begin{bmatrix} a & b \\ -b & a \end{bmatrix}$. Considering the matrix $Q = \frac{1}{2}\begin{bmatrix} I & iI \\ I & -iI \end{bmatrix}$ whose inverse is $Q^{-1} = \frac{1}{2}\begin{bmatrix} I & I \\ -iI & iI \end{bmatrix}$, we compute

$$QOQ^{-1} = \begin{bmatrix} a - ib & 0 \\ 0 & a + ib \end{bmatrix};$$

hence,

$$\det O = \det(a - ib)\det(a + ib) = \overline{\det(a + ib)}\det(a + ib) > 0.$$

□

[6] $B = AA^T$ is a real symmetric matrix, so it is diagonalizable through a real matrix P, that is $P^{-1}BP = \mathrm{diag}[\lambda_1, \ldots, \lambda_n]$, with λ_j real. Since B is positive definite each $\lambda_j > 0$. We see that $\sqrt{B} = P\mathrm{diag}[\sqrt{\lambda_1}, \ldots, \sqrt{\lambda_n}]P^{-1}$ is one of the square roots of B.

Remark 2.32 For a complex sympletic matrix we still have $|\det A|^2 = 1$, so $\det A$ is a point in the unit circle. But, in fact, it is also equal to 1. The proof can be seen in [23].

Consider a time-independent one degree of freedom Hamiltonian function $H = H(x, y)$, with an equilibrium at the origin (0, 0) and quadratic part $H_2 = \frac{1}{2}(ax^2 + 2bxy + cy^2)$, $a > 0$. If H_2 is definite by the Dirichlet theorem the equilibrium is stable. If H_2 is indefinite, then $b^2 - ac > 0$ and the matrix JG of the linear system has real eigenvalues, so the equilibrium is unstable. The stability of the equilibrium (0, 0) in the time-dependent case $H = H(x, y, t)$ is not trivial.

For a two degree of freedom autonomous Hamiltonian system the stability of an equilibrium can be analyzed via the Arnold stability theorem, which is formulated in terms of the normal form of the Hamiltonian at the equilibrium, the theme of the next chapter. The Arnold theorem is seen in Chap. 6. There is no such result as the Arnold theorem if the autonomous Hamiltonian system has more than two degrees of freedom or if it is time dependent.

2.8 On the Geometry of the Real Symplectic Group

The set $E = \mathcal{M}_{2n}(\mathbf{R})$ of real square matrices is a normed vector space with the norm

$$||A|| = \sup\{\|A\mathbf{x}\|;\ \mathbf{x} \in \mathbf{R}^{2n},\ \|\mathbf{x}\| = 1\}.$$

The set $\mathcal{A}_{2n}(\mathbf{R})$ of skew-symmetric matrices is a subspace of $\mathcal{M}_{2n}(\mathbf{R})$. The mapping

$$\mathbf{f}: M_{2n}(\mathbf{R}) \to \mathcal{A}_{2n}(\mathbf{R}), \quad \mathbf{f}(M) = M^T J M$$

is differentiable and its derivative, obtained from that of the bilinear mapping $(M, N) \mapsto M^T J N$ (see Example 1.1) is given by

$$D\mathbf{f}(M) \cdot X = X^T J M + M^T J X. \tag{2.120}$$

Lemma 2.33 *If the matrix M is symplectic, the kernel of $D\mathbf{f}(M)$ is given by*

$$ker\, D\mathbf{f}(M) = \{X = -MJS, \ \ where\ \ S^T = S\}.$$

Proof By (2.120), $X \in \ker D\mathbf{f}(M)$ if, and only if, $X^T J M + M^T J X = 0$. Since M is symplectic we have $M^T J M = J$, hence $JM = M^{-T} J$ and $M^T J = J M^{-1}$. Inserting these into the above equation for X it can be written as $(JM^{-1}X)^T = JM^{-1}X$, so $JM^{-1}X = S$ is a symmetric matrix. Therefore, $X = -MJS$ with $S^T = S$. □

For an invertible matrix B the mapping $Y \mapsto Z = BY$ is an isomorphism of $\mathcal{M}_{2n}(\mathbf{R})$, since it has the inverse $Z \mapsto Y = B^{-1}Z$. Taking $B = -MJ$ we see from the Lemma that if M is symplectic then $\ker D\mathbf{f}(M)$ is isomorphic to the subspace of symmetric matrices, $\mathcal{S}_{2n}(\mathbf{R})$, hence its dimension is the dimension $n(2n+1)$ of $\mathcal{S}_{2n}(\mathbf{R})$. Also notice that the image of $D\mathbf{f}(M)$ is contained in $\mathcal{A}_{2n}(\mathbf{R})$.

Since $X = \frac{1}{2}(X + X^T) + \frac{1}{2}(X - X^T)$ we have $\mathcal{M}_{2n}(\mathbf{R}) = \mathcal{S}_{2n}(\mathbf{R}) \oplus \mathcal{A}_{2n}(\mathbf{R})$; hence, for all $M \in \mathbf{f}^{-1}(J) = Sp(2n, \mathbf{R})$, the mapping (2.120) is surjective, which means that J is a regular value of the mapping $\mathbf{f}$; hence, $\mathbf{f}^{-1}(J) = Sp(2n, \mathbf{R})$ is a smooth manifold of dimension $4n^2 - \dim\mathcal{A}_{2n}(\mathbf{R})$, the dimension of $\mathcal{S}_{2n}(\mathbf{R})$. Moreover, the tangent space at each point M is the subspace $\ker D\mathbf{f}(M)$.

The mappings $(M, N) \mapsto MN$ and $M \mapsto M^{-1}$ are smooth[7] so $Sp(2n, \mathbf{R})$ is an algebraic group that has a differentiable structure in which the operations of multiplication and inversion are differentiable, that is, $Sp(2n, \mathbf{R})$ is a Lie group.

Taking $M = I$ in (2.120) we see that the tangent space of $Sp(2n, \mathbf{R})$ at the identity is the space of Hamiltonian matrices. This is the Lie algebra of the Lie group $Sp(2n, \mathbf{R})$, denoted by $sp(2n, \mathbf{R})$.

Because the one-point set $\{J\}$ is closed in $\mathcal{A}_{2n}(\mathbf{R})$ and the mapping $\mathbf{f}$ is continuous $Sp(2n, \mathbf{R})$ is a closed subset of $\mathcal{M}_{2n}(\mathbf{R})$ but it is not compact. Indeed, the sequence of symplectic matrices $M_n = \begin{bmatrix} \epsilon_n I & O \\ O & \frac{1}{\epsilon_n} I \end{bmatrix}$ is unbounded since $||M_n|| = \frac{1}{\epsilon_n} \to \infty$ as $\epsilon_n \to 0$; hence, by the Heine–Borel theorem $Sp(2n, \mathbf{R})$ is not compact. In Chap. 7 we give more information on the topology of $Sp(2n, \mathbf{R})$.

[7] The derivative of $g(M) = M^{-1}$ is $Dg(M) \cdot X = -X^{-1}MX^{-1}$.

Chapter 3
Normal Forms of Hamiltonian Systems

3.1 Introduction

Given an analytic Hamiltonian system with an equilibrium at the origin

$$\dot{\mathbf{z}} = J\nabla H(\mathbf{z}) = JG\,\mathbf{z} + \cdots ,$$

we have

$$H(\mathbf{z}) = H_2(\mathbf{z}) + H_3(\mathbf{z}) + \ldots, \qquad \mathbf{z} = (x_1, \ldots, x_n; y_1, \ldots, y_n) \tag{3.1}$$

where $H_k(\mathbf{z})$ is a homogeneous polynomial of degree k in the variables $x_1, \ldots, x_n$, $y_1, \ldots, y_n$. We will look for a canonical change of coordinates in a neighborhood of the origin, $\mathbf{z} = \phi(\boldsymbol{\zeta})$, which makes simpler the structure of the new Hamiltonian

$$\Gamma(\boldsymbol{\zeta}) = \Gamma_2(\boldsymbol{\zeta}) + \Gamma_3(\boldsymbol{\zeta}) + \cdots ,$$

where $\Gamma_k(\boldsymbol{\zeta})$ is a homogeneous polynomial of degree k in the new variables $\xi_1, \ldots, \xi_n, \eta_1, \ldots, \eta_n$. This is important to simplify the study of the dynamics. For instance, if we could get $\Gamma(\boldsymbol{\zeta})$ depending on $\boldsymbol{\zeta}$ in terms of the actions only

$$I_k = \frac{1}{2}\left(\xi_k^2 + \eta_k^2\right), \tag{3.2}$$

with conjugate variables $\theta_k = \arctan(\eta_k/\xi_k)$, then the new system of equations defined by the Hamiltonian $\Gamma = \Gamma(\mathbf{I}, \boldsymbol{\theta})$, $\dot{\boldsymbol{\zeta}} = J\Gamma_{\boldsymbol{\zeta}}$, would take the form

$$\dot{I}_k = 0, \qquad \dot{\theta}_k = -\Gamma_{I_k}, \qquad (k = 1, \cdots, n)$$

H. E. Cabral, L. Brandão Dias, *Normal Forms and Stability of Hamiltonian Systems*, Applied Mathematical Sciences 218,
https://doi.org/10.1007/978-3-031-33046-9_3

and we could immediately find its solutions

$$I_k = const., \quad \theta_k = \omega_k t + \omega_k^{(0)}, \quad (k = 1, \cdots, n); \tag{3.3}$$

the numbers $\omega_k = -\Gamma_{I_k}(\mathbf{I})$ are called the *frequencies* of the motion.

It turns out (see Corollary 32.1) that such a symplectic transformation is possible if the pure imaginary eigenvalues $\lambda_1, \ldots, \lambda_n$ are linearly independent over the rational numbers. This was proved by G. D. Birkhoff in the 1920s and the new Hamiltonian became known as the *Birkhoff normal form*. However, the process for obtaining this new Hamiltonian is formal, resulting from a comparison of coefficients to solve, formally, the equations. This process is generically divergent, as was shown by C. L. Siegel [54] in 1954. This does not mean that the normal form itself is divergent. What is known (see [49]) is that if there exists a single analytic Hamiltonian whose normal form is divergent, then generically the Birkhoff normal form of an analytic Hamiltonian is divergent. Now, even being a divergent process, important information can be drawn using the normal form up to a few terms. Thus, for instance, an important result according to V. I. Arnold on the stability of equilibria of Hamiltonian systems with two degrees of freedom requires only the knowledge of a finite part of the normal form; in general, up to fourth-order terms is sufficient. We remark that if the Birkhoff normal form is convergent, the Hamiltonian system is integrable as consequence of a theorem of Liouville. The Arnold theorem will be proved in Chap. 5.

3.2 Two Structural Theorems on Normal Forms

Consider the quadratic Hamiltonian

$$H_0(\mathbf{z}) = \frac{1}{2}\mathbf{z}^T S\mathbf{z}; \tag{3.4}$$

where S is a real constant symmetric matrix of order $2n$.

Let a Hamiltonian H be given by

$$H(\mathbf{z}) = H_0(\mathbf{z}) + H_1(\mathbf{z}) + H_2(\mathbf{z}) + \ldots, \tag{3.5}$$

where $H_j(\mathbf{z})$, $j = 0, 1, 2, \ldots$ is a homogeneous polynomial of degree $j + 2$ in the coordinates of $\mathbf{z} = (x_1, \ldots, x_n, y_1, \ldots, y_n)$.

We want to transform H into a simpler Hamiltonian

$$\mathcal{H}(\boldsymbol{\zeta}) = \mathcal{H}_0(\boldsymbol{\zeta}) + \mathcal{H}_1(\boldsymbol{\zeta}) + \mathcal{H}_2(\boldsymbol{\zeta}) + \ldots, \tag{3.6}$$

where $\mathcal{H}_j(\boldsymbol{\zeta})$, $j = 0, 1, 2, \ldots$ is a homogeneous polynomial of degree $j + 2$ in the coordinates of $\boldsymbol{\zeta} = (\xi_1, \ldots, \xi_n, \eta_1, \ldots, \eta_n)$.

This will be done by means of a canonical transformation $\mathbf{z} = \Phi(\boldsymbol{\zeta})$, which begins with the identity

$$\mathbf{z} = \boldsymbol{\zeta} + \Phi_2(\boldsymbol{\zeta}) + \dots . \tag{3.7}$$

Since $\mathcal{H}(\boldsymbol{\zeta}) = H(\mathbf{z})$ and $\mathbf{z} = \Phi(\boldsymbol{\zeta})$ begins with the identity we have $\mathcal{H}_0(\boldsymbol{\zeta}) = H_0(\boldsymbol{\zeta})$.

Theorem 3.1 *Suppose that the matrix $A = JS$ is diagonalizable. Then there exists a formal canonical transformation of the type (3.7) such that for each $j = 1, 2, 3, \dots$,*

$$\mathcal{H}_j(e^{tA}\boldsymbol{\zeta}) \equiv \mathcal{H}_j(\boldsymbol{\zeta}),$$

that is, each $\mathcal{H}_j(\boldsymbol{\zeta})$ is a first integral of the linear system defined by $\mathcal{H}_0(\boldsymbol{\zeta})$.

Theorem 3.2 *Suppose that the matrix $A = JS$ is not diagonalizable. Then there exists a formal canonical transformation of the type (3.7) such that for each $j = 1, 2, 3, \dots$,*

$$\mathcal{H}_j(e^{tA^T}\boldsymbol{\zeta}) \equiv \mathcal{H}_j(\boldsymbol{\zeta}),$$

that is each $\mathcal{H}_j(\boldsymbol{\zeta})$ is a first integral of the linear system defined by $\mathcal{H}_0^T(\boldsymbol{\zeta}) = \frac{1}{2}\boldsymbol{\zeta}^T S\boldsymbol{\zeta}$.

The condition that the polynomial $\mathcal{H}_j(\boldsymbol{\zeta})$ is invariant under the linear flow is an indication that it has a simpler structure than $H_j(\mathbf{z})$, see Corollary 32.1.

When all the polynomials $\mathcal{H}_j(\boldsymbol{\zeta})$ satisfy the invariance expressed in these theorems we say that the Hamiltonian $\mathcal{H}(\boldsymbol{\zeta})$ is in *normal form.*

Remark 3.3 We call these structural theorems because the invariance of the polynomials $\mathcal{H}_j$ expressed in their statements shows the structure of the new Hamiltonian telling us how the homogeneous polynomials are formed by certain expressions but they neither give their coefficients nor the powers with regard to how these expressions enter into the polynomials. In the proof of these theorems we use a recurrence relation (see Theorem 3.10), which is an effective way of finding the homogeneous polynomials $\mathcal{H}_j$, so that the powers of these expressions in the polynomials of the new Hamiltonian together with their coefficients are found explicitly.

For example, the structure of a Birkhoff normal form is a corollary of the first theorem.

Corollary 32.1 (Birkhoff Normal Form) *Suppose that the eigenvalues of the matrix $A = JS$ are all purely imaginary numbers,*

$$\lambda_1 = i\omega_1, \dots, \lambda_n = i\omega_n; \ -\lambda_1, \dots, -\lambda_n$$

and that $\omega_1, \dots, \omega_n$ are linearly independent over the field of rational numbers.

Then, there exists a formal canonical transformation of the type (3.7) such that $\mathcal{H}_j$ *depends only on* $\xi_1^2 + \eta_1^2, \xi_2^2 + \eta_2^2, \dots, \xi_n^2 + \eta_n^2$.

Proof We will see in Proposition 3.4 that because the eigenvalues $i\omega_1, \dots, i\omega_n$ of the matrix A are all distinct there exists a linear symplectic transformation $\mathbf{z} = P\boldsymbol{\zeta}$ that carries $H_0(\mathbf{z})$ given in (3.4) to the quadratic Hamiltonian

$$\mathcal{H}_0(\boldsymbol{\zeta}) = \frac{1}{2}\sum_{j=1}^{n} \delta_j \omega_j \left(\xi_j^2 + \eta_j^2\right). \tag{3.8}$$

Let $\boldsymbol{\zeta} = (\zeta_1, \dots, \zeta_n)$ with $\zeta_j = (\xi_j, \eta_j)$ and write $\mathcal{H}(\boldsymbol{\zeta}) = \mathcal{H}(\zeta_1, \dots, \zeta_n)$.

The flow of $\mathcal{H}_0$ is given by $\Phi_0(t, \boldsymbol{\zeta}) = (R(\omega_1 t)\zeta_1, \dots, R(\omega_n t)\zeta_n)$, where $R(\omega_j t)$ is the rotation by the angle $\omega_j t$.

By Theorem 3.1 each $\mathcal{H}_j$ is a first integral of the linear system defined by $\mathcal{H}_0(\boldsymbol{\zeta})$ so we have the identity

$$\mathcal{H}_j(R(\omega_1 t)\zeta_1, R(\omega_2 t)\zeta_2, R(\omega_3 t)\zeta_3, \dots, R(\omega_n t)\zeta_n) \equiv \mathcal{H}_j(\zeta_1, \dots, \zeta_{n-1}, \zeta_n). \tag{3.9}$$

To simplify the notation let us identify the point $(a, 0)$ on the real axis with the real number a. Taking a time t_1 such that $R(\omega_1 t_1)\zeta_1 = |\zeta_1|$ we get the equality

$$\mathcal{H}_j(|\zeta_1|, R(\omega_2 t_1)\zeta_2, R(\omega_3 t_1)\zeta_3, \dots) = \mathcal{H}_j(\zeta_1, \dots, \zeta_{n-1}, \zeta_n).$$

Since $R(\omega_j t_1)R(\omega_j t) = R(\omega_j(t_1 + t))$ from this equality and the identity (3.9) we get the identity

$$\mathcal{H}_j(R(\omega_1 t)|\zeta_1|,\ R(\omega_2(t_1 + t))\zeta_2,\ R(\omega_3(t_1 + t))\zeta_3,\ \dots) \equiv \mathcal{H}_j(\zeta_1, \dots, \zeta_{n-1}, \zeta_n) \tag{3.10}$$

Now, take a time t_2 such that $R(\omega_2 t_2)\zeta_2 = |\zeta_2|$. Then the identity (3.10) gives the equality

$$\mathcal{H}_j(R(\omega_1 t_2)|\zeta_1|, R(\omega_2 t_1)|\zeta_2|, R(\omega_3(t_1 + t_2)\zeta_3, \dots)) \equiv \mathcal{H}_j(\zeta_1, \dots, \zeta_{n-1}, \zeta_n).$$

Again, we get an identity

$$\mathcal{H}_j(R(\omega_1(t_2 + t))|\zeta_1|, R(\omega_2(t_1 + t))|\zeta_2|, R(\omega_3(t_1 + t_2 + t)\zeta_3, \dots)) \equiv \mathcal{H}_j(\zeta_1, \dots, \zeta_{n-1}, \zeta_n).$$

Take a time t_3 such that $R(\omega_3(t_3)\zeta_3 = |\zeta_3|$. Then this identity (3.10) gives the equality

$$\mathcal{H}_j(R(\omega_1(t_2 + t_3))|\zeta_1|, R(\omega_2(t_1 + t_3)|\zeta_2|, R(\omega_3(t_1 + t_2)|\zeta_3|, \dots) \equiv \mathcal{H}_j(\zeta_1, \dots, \zeta_{n-1}, \zeta_n).$$

Proceeding in this way and considering the times $t^{(j)} = t_1 + \dots + \hat{t}_j + \dots + t_n$ where the hat means that t_j is omitted we get

$$\mathcal{H}_j(R(\omega_1 t^{(1)})|\zeta_1|, R(\omega_2 t^{(2)})|\zeta_2|, \dots, R(\omega_n t^{(n)})|\zeta_n|) = \mathcal{H}_j(\zeta_1, \dots, \zeta_{n-1}, \zeta_n).$$

Since $|\zeta_k| = \sqrt{\xi_k^2 + \eta_k^2}$, this proves that $\mathcal{H}_j(\boldsymbol{\zeta}) = f_j(\xi_1^2+\eta_1^2, \xi_2^2+\eta_2^2, \dots, \xi_n^2+\eta_n^2)$.
□

To illustrate the use of the second structural theorem we consider the *Sokol'ski normal form* considered in the next example.

Example 3.1 (Sokol'ski Normal Form) A Hamiltonian function is in Sokol'ski normal form if it has the following structure:

$$\mathcal{H} = \mathcal{H}_0(\boldsymbol{\zeta}) + \mathcal{H}_1(\boldsymbol{\zeta}) + \mathcal{H}_2(\boldsymbol{\zeta}) + \dots,$$

where

$$\mathcal{H}_0(\boldsymbol{\zeta}) = \omega(\xi_2\eta_1 - \xi_1\eta_2) + \frac{\delta}{2}(\xi_1^2 + \xi_2^2), \tag{3.11}$$

with $\omega > 0$ and $\delta = \pm 1$ and all the other $\mathcal{H}_j(\boldsymbol{\zeta})$ depending only on the two quantities $\xi_2\eta_1 - \xi_1\eta_2$ and $\eta_1^2 + \eta_2^2$.

Let us show that such a normal form arises when a two-degree of freedom Hamiltonian $H(\mathbf{z})$ has an equilibrium around which its quadratic part H_0 is a linear system whose matrix has double pure imaginary eigenvalues $\pm i\omega$ and is nondiagonalizable. We show in Example 5.16 that there is a canonical linear transformation $\mathbf{z} = L\boldsymbol{\zeta}$ that leads $H_0(\mathbf{z})$ to the normal form $\mathcal{H}_0(\boldsymbol{\zeta})$.

The matrix of the linear system with Hamiltonian (3.11) is given by

$$A = \begin{bmatrix} 0 & \omega & 0 & 0 \\ -\omega & 0 & 0 & 0 \\ -\delta & 0 & 0 & \omega \\ 0 & -\delta & -\omega & 0 \end{bmatrix}.$$

Then, the Hamiltonian of the adjoint system, that is the linear system with Hamiltonian $\mathcal{H}_0^T(\boldsymbol{\zeta}) = -\frac{1}{2}\boldsymbol{\zeta}^T J A^T \boldsymbol{\zeta}$ is given by

$$\mathcal{H}_0^T(\boldsymbol{\zeta}) = \omega(\xi_1\eta_2 - \xi_2\eta_1) - \frac{\delta}{2}(\eta_1^2 + \eta_2^2).$$

To write the system in a simpler form we pass to polar coordinates, so we take $\eta_1 = r\cos\theta$, $\eta_2 = r\sin\theta$ and complete this planar transformation to a symplectic transformation in $\mathbf{R}^4$ using the generating function $W(\xi_1, \xi_2, r, \theta) = \xi_1 r\cos\theta + \xi_2 r\sin\theta$. We get $R = -W_r = -\xi_1\cos\theta - \xi_2\sin\theta$, $\Theta = -W_\theta = \xi_1 r\sin\theta - \xi_2 r\cos\theta$. We have $\eta_1^2 + \eta_2^2 = r^2$ and $\Theta = \xi_1\eta_2 - \xi_2\eta_1$ so the Hamiltonian becomes

$$\mathcal{H}_0^T(r, \theta, R, \Theta) = \omega\Theta - \frac{\delta}{2}r^2.$$

The differential equations are then $\dot{r} = 0$, $\dot{\theta} = \omega$, $\dot{R} = \delta r$, $\dot{\Theta} = 0$,; hence, the flow is $\phi(t, r, \theta, R, \Theta) = (r, \omega t + \theta, \delta r t + R, \Theta)$. By Theorem 3.2 $\mathcal{H}_j(r, \theta, R, \Theta)$ is constant along this flow so it must depend only on the variables r and Θ and therefore only on the expressions $\eta_1^2 + \eta_2^2$ and $\xi_1\eta_2 - \xi_2\eta_1$. So this is Sokol'ski's normal form.

3.3 Linear Normalization, Case of Simple Eigenvalues

The starting point of the process of normalizing a Hamiltonian around an equilibrium is the normalization of the quadratic Hamiltonian H_0. In this section we see the case when the eigenvalues of the corresponding linear system are distinct. In Chap. 4 we will see the general case. We restrict our attention to the context relevant to stability questions, namely when all the eigenvalues are purely imaginary, $\lambda_j = i\omega_j$. Recall that in Hamiltonian systems, if some eigenvalue has a nonzero real part the equilibrium is unstable.

So our problem is to simplify the quadratic real Hamiltonian H_0 given in (3.4) to get a simpler Hamiltonian

$$\mathcal{H}_0(\boldsymbol{\zeta}) = \frac{1}{2}\boldsymbol{\zeta}^T \Omega \boldsymbol{\zeta}. \tag{3.12}$$

In the proposition below we prove that if the eigenvalues are purely imaginary, $\lambda_j = \pm i\omega_j$ and distinct, then we get the Hamiltonian $\mathcal{H}_0(\boldsymbol{\zeta})$ as an algebraic sum of harmonic oscillators. We will present here the normalization process as described in [39].

Let $\mathbf{r}_j + i\mathbf{s}_j$ be an eigenvector of $A = JS$ belonging to the eigenvalue $i\omega_j$, $\omega_j > 0$. The normalization is described in terms of the symplectic product defined in Sect. 2.6, $\{\mathbf{r}_j, \mathbf{s}_j\} = \langle \mathbf{r}_j, J\mathbf{s}_j \rangle$. In the proof of the proposition it will be seen that $\{\mathbf{r}_j, \mathbf{s}_j\} \neq 0$.

Proposition 3.4 *Let $\delta_j = sign\{\mathbf{r}_j, \mathbf{s}_j\}$ and $\kappa_j = 1/\sqrt{|\{\mathbf{r}_j, \mathbf{s}_j\}|}$. Then, the real matrix*

$$P = Col\big[-\kappa_1\mathbf{s}_1, \ldots, -\kappa_n\mathbf{s}_n, \delta_1\kappa_1\mathbf{r}_1, \ldots, \delta_n\kappa_n\mathbf{r}_n\big] \tag{3.13}$$

is symplectic and the canonical linear transformation $\mathbf{z} = P\boldsymbol{\zeta}$ takes the Hamiltonian (3.4) to the Hamiltonian

$$\mathcal{H}_0(\boldsymbol{\zeta}) = \frac{1}{2}\sum_j^n \delta_j\omega_j(\xi_j^2 + \eta_j^2), \tag{3.14}$$

where $\boldsymbol{\zeta} = (\xi_1, \ldots, \xi_n, \eta_1, \ldots, \eta_n)$.

Proof The linear systems given by the Hamiltonians (3.4) and (3.12) are $\dot{\mathbf{z}} = JS\mathbf{z}$ and $\dot{\boldsymbol{\zeta}} = J\Omega\boldsymbol{\zeta}$. The requirement that the transformation $\mathbf{z} = P\boldsymbol{\zeta}$ takes the first system to the second is translated into the matrix equation

$$PJ\Omega = JSP, \tag{3.15}$$

whereas the condition that L is symplectic is

$$P^T JP = J. \tag{3.16}$$

We take P in the form of a product $P = BC$, where

$$C = \begin{bmatrix} iI & I \\ -iI & I \end{bmatrix},$$

and aim at determining B so as to get P with the required properties. The condition (3.15) gives us

$$B^{-1}JSB = CJ\Omega C^{-1}. \tag{3.17}$$

We want the Hamiltonian (3.12) to be given by the expression (3.14). Therefore, $J\Omega = \begin{bmatrix} O & \Omega_0 \\ -\Omega_0 & O \end{bmatrix}$, where $\Omega_0 = \text{diag}\,[\delta_1\omega_1, \ldots, \delta_n\omega_n]$. Computing the right-hand side of (3.17) we find that it has the diagonal form $\begin{bmatrix} i\Omega_0 & O \\ O & -i\Omega_0 \end{bmatrix}$. Therefore, the matrix B in (3.17) is the diagonalizing matrix of JS; hence, the columns of B are eigenvectors of JS, that is,

$$B = \text{Col}\,[\mathbf{v}_1, \ldots, \mathbf{v}_n, \mathbf{v}_{n+1}, \ldots, \mathbf{v}_{2n}],$$

where

$$JS\mathbf{v}_k = \lambda_k\mathbf{v}_k, \quad \text{and} \quad \mathbf{v}_{n+k} = \overline{\mathbf{v}}_k, \qquad (k = 1, \ldots, n).$$

Notice that because JS is real its eigenvalues come in pairs λ_k and $\lambda_{n+k} = \overline{\lambda}_k$.

The condition (3.16) relative to the symplecticity of L can be written in the form

$$C^T FC = J \qquad \text{with} \qquad F = B^T JB. \tag{3.18}$$

The matrix F is invertible because J and C are invertible matrices and F is skew-symmetric because J is skew-symmetric.

Now, from the second Eq. (3.18), we see that the elements of F are given by

$$f_{kl} = (\mathbf{v}_k, J\mathbf{v}_l),$$

where (,) denotes the scalar product in $\mathcal{C}^{2n}$, $(\mathbf{z}, \mathbf{w}) = z_1w_1 + \cdots + z_{2n}w_{2n}$.

Consequently, as $S^T = S$ and $J^T = -J$, from the second equation in (3.18) we have

$$\begin{aligned}\lambda_l f_{kl} &= \lambda_l(\mathbf{v}_k, J\mathbf{v}_l) = (\mathbf{v}_k, J(\lambda_l \mathbf{v}_l)) = (\mathbf{v}_k, J(JS\mathbf{v}_l)) = (-S\mathbf{v}_k, \mathbf{v}_l)\\ &= (J(JS\mathbf{v}_k), \mathbf{v}_l) = \lambda_k(J\mathbf{v}_k, \mathbf{v}_l) = -\lambda_k(\mathbf{v}_k, J\mathbf{v}_l) = -\lambda_k f_{kl},\end{aligned}$$

from which we conclude that $f_{kl} = 0$ for $|k - l| \neq n$. So,

$$F = \begin{pmatrix} O & D \\ -D & O \end{pmatrix},$$

where $D = \text{diag}\,[d_{11}, \ldots, d_{nn}]$ with $d_{kk} = (\mathbf{v}_k, J\mathbf{v}_{n+k}) = (\mathbf{v}_k, J\overline{\mathbf{v}}_k)$ and $d_{kk} \neq 0$, since F is invertible.

The first equation in (3.18) shows that $D = \dfrac{1}{2i} I$; hence,

$$2i(\mathbf{v}_k, J\overline{\mathbf{v}}_k) = 1 \qquad (k = 1, \ldots, n). \tag{3.19}$$

Since $\mathbf{r}_k + i\mathbf{s}_k$ is an eigenvector of $i\omega_k$, we have that for any $c_k > 0$ and $\delta_k = \pm 1$, the vector $\mathbf{v}_k = c_k(\delta_k \mathbf{r}_k + i\mathbf{s}_k)$ is also an eigenvector belonging to $i\omega_k$. With this expression of $\mathbf{v}_k$, the symplecticity condition (3.19) for P becomes[1]

$$4c_k^2 \delta_k \langle \mathbf{r}_k, J\mathbf{s}_k \rangle = 1, \qquad (k = 1, \ldots, n) \tag{3.20}$$

and it is satisfied with the choice $\delta_k = \text{sign}\{\mathbf{r}_k, \mathbf{s}_k\}$ and $c_k = 1/2\sqrt{|\{\mathbf{r}_k, \mathbf{s}_k\}|}$.

This proves that the matrix P is symplectic and that the linear canonical transformation $\mathbf{z} = P\boldsymbol{\zeta}$ takes the system (3.4) to the system (3.14).

It remains to show that the matrix P is given by the expression (3.13).

Denoting by $X^{(j)}$ the j-th column of a matrix X, we have for $j = 1, \ldots, n$

$$P^{(j)} = BC^{(j)} = iB\mathbf{e}_j - iB\mathbf{e}_{n+j} = i(\mathbf{v}_j - \overline{\mathbf{v}}_j) = -2\text{Im}\mathbf{v}_j = -2c_j\mathbf{s}_j, \qquad \square$$

3.4 Birkhoff Normal Form

Although the Birkhoff normal form can be obtained by the general method described in the next section, it is worth becoming acquainted with its classical presentation. We therefore include it here. We start with a Hamiltonian function having the quadratic terms already normalized

[1] Notice that for real vectors $\mathbf{x}, \mathbf{y}$, we have $(\mathbf{x}, \mathbf{y}) = \langle \mathbf{x}, \mathbf{y} \rangle$, the Euclidean inner product.

$$H(\mathbf{x},\mathbf{y}) = \frac{1}{2}\sum_{j=1}^{n} \delta_j\omega_j(x_j^2 + y_j^2) + H_3(\mathbf{x},\mathbf{y}) + H_4(\mathbf{x},\mathbf{y}) + \cdots, \tag{3.21}$$

where H_j is a homogeneous polynomial of degree j in $x_1 \dots, x_n, y_1 \dots, y_n$.

We pass to complex variables through the following symplectic change of variables with multiplier $i/2$,

$$x_j = \frac{1}{2}(z_j + \bar{z}_j), \quad y_j = \frac{1}{2i}(z_j - \bar{z}_j), \tag{3.22}$$

thereby getting the Hamiltonian function $\mathbf{H}(\mathbf{z},\bar{\mathbf{z}}) = \frac{2}{i}H(\mathbf{x},\mathbf{y})$ in the form

$$\mathbf{H}(\mathbf{z},\bar{\mathbf{z}}) = -i\sum_{j=1}^{n} \delta_j\omega_j z_j\bar{z}_j + \mathbf{H}_3(\mathbf{z},\bar{\mathbf{z}}) + \mathbf{H}_4(\mathbf{z},\bar{\mathbf{z}}) + \dots. \tag{3.23}$$

We search for a symplectic change of variables $\mathbf{z} = \phi(\boldsymbol{\zeta},\overline{\boldsymbol{\zeta}})$, $\bar{\mathbf{z}} = \psi(\boldsymbol{\zeta},\overline{\boldsymbol{\zeta}})$ in the form

$$\mathbf{z} = \boldsymbol{\zeta} + \phi_2(\boldsymbol{\zeta},\overline{\boldsymbol{\zeta}}) + \phi_3(\boldsymbol{\zeta},\overline{\boldsymbol{\zeta}}) + \dots, \quad \bar{\mathbf{z}} = \overline{\boldsymbol{\zeta}} + \psi_2(\boldsymbol{\zeta},\overline{\boldsymbol{\zeta}}) + \psi_3(\boldsymbol{\zeta},\overline{\boldsymbol{\zeta}}) + \dots, \tag{3.24}$$

so as to take the Hamiltonian (3.23) to a simpler form, $\Gamma(\boldsymbol{\zeta},\overline{\boldsymbol{\zeta}})$.

Definition 3.5 A *resonant relation* of order m among the frequencies of (3.21) is a trivial linear combination of $\omega_1, \dots, \omega_n$ with integer coefficients whose absolute values add up to m,

$$m_1\omega_1 + \dots + m_n\omega_n = 0, \qquad |m_1| + \dots |m_n| = m. \tag{3.25}$$

The set of resonant relations gives rise to a $\mathbf{Z}$-module

$$M_{\delta\omega} = \{(m_1, \dots, m_n) \in \mathbf{Z}^n \,;\, \langle \mathbf{m}, \boldsymbol{\delta\omega}\rangle = m_1\delta_1\omega_1 + \dots + m_n\delta_n\omega_n = 0\}. \tag{3.26}$$

The Hamiltonian $\Gamma(\boldsymbol{\zeta},\overline{\boldsymbol{\zeta}})$ is a formal power series in $\boldsymbol{\zeta}^k\overline{\boldsymbol{\zeta}}^l = \zeta_1^{k_1}\bar{\zeta}_1^{l_1}\cdots\zeta_n^{k_n}\bar{\zeta}_n^{l_n}$.

Definition 3.6 We say that the Hamiltonian is in *normal form* if the power series $\Gamma(\boldsymbol{\zeta},\overline{\boldsymbol{\zeta}})$ contains only terms of the form $\gamma_{k_1l_1\dots k_nl_n}\zeta_1^{k_1}\bar{\zeta}_1^{l_1}\cdots\zeta_n^{k_n}\bar{\zeta}_n^{l_n}$ with $\mathbf{k}-\mathbf{l} \in M_{\delta\omega}$, that is $\gamma_{k_1l_1\dots k_nl_n} = 0$ for $\mathbf{k}-\mathbf{l} \notin M_{\delta\omega}$.

Theorem 3.7 *There exists a formal transformation (3.24) that takes (3.23) to the normal form.*

Proof By induction. Because the transformation (3.24) begins with the identity, the quadratic term of the transformed Hamiltonian $\Gamma(\zeta, \overline{\zeta})$ is already determined

$$\Gamma(\zeta, \bar{\zeta}) = -i \sum_{j=1}^{n} \delta_j \omega_j \zeta_j \overline{\zeta}_j + \Gamma_3(\zeta, \bar{\zeta}) + \Gamma_4(\zeta, \bar{\zeta}) + \dots . \tag{3.27}$$

We take the transformation (3.24) generated by $W(\mathbf{z}, \overline{\zeta}) = \mathbf{z}^T \overline{\zeta} + W_3(\mathbf{z}, \overline{\zeta}) + W_4(\mathbf{z}, \overline{\zeta}) + \dots$.

Substituting $\overline{\mathbf{z}} = W_{\mathbf{z}} = \overline{\zeta} + W_{3,\mathbf{z}}(\mathbf{z}, \overline{\zeta}) + W_{4,\mathbf{z}}(\mathbf{z}, \overline{\zeta}) + \dots$, $\zeta = W_{\overline{\zeta}} = \mathbf{z} + W_{3,\overline{\zeta}}(\mathbf{z}, \overline{\zeta}) + W_{4,\overline{\zeta}}(\mathbf{z}, \overline{\zeta}) + \dots$, into the equation $\Gamma(\zeta, \overline{\zeta}) = \mathbf{H}(\mathbf{z}, \bar{\mathbf{z}})$ we obtain the identity in $\mathbf{z}, \overline{\zeta}$:

$$\begin{aligned}
&-i \sum_{j=1}^{n} \delta_j \omega_j \big(z_j + W_{3,\overline{\zeta}_j} + \dots\big)\overline{\zeta}_j + \Gamma_3\big(\mathbf{z} + W_{3,\overline{\zeta}} + \dots, \overline{\zeta}\big) \\
&\quad + \Gamma_4(\mathbf{z} + W_{3,\overline{\zeta}} + \dots, \overline{\zeta}) + \dots = \\
&-i \sum_{j=1}^{n} \delta_j \omega_j z_j \big(\overline{\zeta}_j + W_{3,z_j} + \dots\big) + \mathbf{H}_3\big(\mathbf{z}, \overline{\zeta} + W_{3,\mathbf{z}} + \dots\big) \\
&\quad + \mathbf{H}_4\big(\mathbf{z}, \overline{\zeta} + W_{3,\mathbf{z}} + \dots\big) + \dots ,
\end{aligned}$$

which we write in the form

$$\begin{aligned}
&\textstyle\sum_{j=1}^{n} i\delta_j \omega_j \Big[\big(z_j W_{3,z_j} - \overline{\zeta}_j W_{3,\overline{\zeta}_j}\big) + \big(z_j W_{4,z_j} - \overline{\zeta}_j W_{4,\overline{\zeta}_j}\big) + \dots\Big] + \Gamma_3 + \Gamma_4 + \dots \\
&\quad = \mathbf{H}_3 + \mathbf{H}_4 + \dots .
\end{aligned}$$

Comparing in this identity the terms of degree s in $\mathbf{z}, \overline{\zeta}$, we obtain the equation

$$DW_s(\mathbf{z}, \overline{\zeta}) + \Gamma_s(\mathbf{z}, \overline{\zeta}) = \mathbf{H}_s(\mathbf{z}, \overline{\zeta}) + C_s(\mathbf{z}, \overline{\zeta}), \tag{3.28}$$

where D is the differential operator

$$D = \sum_{j=1}^{n} i\delta_j \omega_j \left(z_j \frac{\partial}{\partial z_j} - \overline{\zeta}_j \frac{\partial}{\partial \overline{\zeta}_j}\right) \tag{3.29}$$

and $C_s(\mathbf{z}, \overline{\zeta})$ consists of the terms of degree s coming from $-\Gamma_3, \dots, -\Gamma_{s-1}$, computed at $(\zeta, \overline{\zeta})$, and from $\mathbf{H}_3, \dots, \mathbf{H}_{s-1}$, computed at $(\mathbf{z}, \overline{\mathbf{z}})$, where $\zeta = \mathbf{z} + W_{3,\overline{\zeta}}(\mathbf{z}, \overline{\zeta}) + \dots + W_{s-1,\overline{\zeta}}(\mathbf{z}, \overline{\zeta})$ and $\overline{\mathbf{z}} = \overline{\zeta} + W_{3,\mathbf{z}}(\mathbf{z}, \overline{\zeta}) + \dots + W_{s-1,\mathbf{z}}(\mathbf{z}, \overline{\zeta})$. Observe that $C_3(\mathbf{z}, \overline{\zeta}) = 0$.

Now, $\Gamma(\zeta, \overline{\zeta}) = \sum_s \Gamma_s(\zeta, \overline{\zeta})$ and we want to determine the coefficients of the homogeneous polynomial $\Gamma_s(\zeta, \overline{\zeta}) = \sum_{k,l} \Gamma_{s,k_1 \dots k_n l_1 \dots l_n} \zeta_1^{k_1} \dots \zeta_n^{k_n} \overline{\zeta}_1^{l_1} \dots \overline{\zeta}_n^{l_n}$.

Since $\Gamma_s(\mathbf{z}, \overline{\boldsymbol{\zeta}}) = \sum_{k,l} \Gamma_{s,k_1...k_nl_1...l_n} z_1^{k_1} \dots z_n^{k_n} \overline{\zeta}_1^{l_1} \dots \overline{\zeta}_n^{l_n}$, then noticing that

$$D\big(z_1^{k_1} \dots z_n^{k_n} \overline{\zeta}_1^{l_1} \dots \overline{\zeta}_n^{l_n}\big) = i \langle \mathbf{k} - \mathbf{l}, \boldsymbol{\delta\omega} \rangle z_1^{k_1} \dots z_n^{k_n} \overline{\zeta}_1^{l_1} \dots \overline{\zeta}_n^{l_n}$$

where

$$\langle \mathbf{k} - \mathbf{l}, \boldsymbol{\delta\omega} \rangle = (k_1 - l_1)\delta_1\omega_1 + (k_2 - l_2)\delta_2\omega_2 + (k_n - l_n)\delta_n\omega_n, \tag{3.30}$$

we get from (3.28) the following equality among the coefficients of the polynomials W_s, Γ_s, $\mathbf{H}_s$ and C_s:

$$i \langle \mathbf{k} - \mathbf{l}, \boldsymbol{\delta\omega} \rangle W_{s,kl} + \Gamma_{s,kl} = \mathbf{H}_{s,kl} + C_{s,kl}. \tag{3.31}$$

The right-hand side of this equation is known by induction. So, if $\mathbf{k} - \mathbf{l} \in M_{\delta\omega}$, we get $\Gamma_{s,kl} = \mathbf{H}_{s,kl} + C_{s,kl}$ and for $\mathbf{k} - \mathbf{l} \notin M_{\delta\omega}$ we set $\Gamma_{s,kl} = 0$ and determine the value of $W_{s,kl}$ dividing the right-hand side of (3.31) by $i \langle \mathbf{k} - \mathbf{l}, \boldsymbol{\delta\omega} \rangle$. The coefficients $W_{s,kl}$ is taken arbitrarily when $\mathbf{k} - \mathbf{l} \in M_{\delta\omega}$, so we can take them zero. Therefore, we get by induction all the polynomials $W_s(\mathbf{z}, \overline{\boldsymbol{\zeta}})$ of $W(\mathbf{z}, \overline{\boldsymbol{\zeta}})$ and $\Gamma_s(\mathbf{z}, \overline{\boldsymbol{\zeta}})$ of $\Gamma(\mathbf{z}, \overline{\boldsymbol{\zeta}})$. Having now all the coefficients of Γ_s, for all s, we have found $\Gamma(\boldsymbol{\zeta}, \overline{\boldsymbol{\zeta}})$. □

We can now use (3.22) together with the similar transformation

$$\xi_j = \frac{1}{2}(\zeta_j + \overline{\zeta}_j), \quad \eta_j = \frac{1}{2i}(\zeta_j - \overline{\zeta}_j)$$

in order to go back to real variables and get the real normal form $\mathcal{H}(\boldsymbol{\xi}, \boldsymbol{\eta})$ of $H(\mathbf{x}, \mathbf{y})$.

If $\omega_1, \dots, \omega_n$ are linearly independent over the rational numbers, then $M_{\delta\omega} = \{0\}$ and in the normal form appears only the terms of the form

$$\zeta_1^{k_1} \dots \zeta_n^{k_n} \overline{\zeta}_1^{k_1} \dots \overline{\zeta}_n^{k_n} = (\xi_1^2 + \eta_1^2)^{k_1} \cdots (\xi_n^2 + \eta_n^2)^{k_n}.$$

The normal form is then called the *Birkhoff normal form*. In terms of the action-angles variables, the normal form depends only on the action variables (3.2), $\mathcal{H}(\mathbf{I}, \boldsymbol{\phi}) = \mathcal{H}(I_1, \dots, I_n)$.

When resonances are present among the frequencies $\omega_1, \dots, \omega_n$ the normal form is sometimes referred to as *Gustavson normal form*, considered in [27].

3.5 Fourth-Order Birkhoff Normal Form

For a two-degree of freedom autonomous Hamiltonian system assume that there is no resonance relation among the frequencies ω_1 and ω_2 up to order four. Let $\mathcal{H}_4(I_1, I_2)$ be the fourth-degree terms in Birkhoff normal form. The following theorem is the main tool on the stability of an equilibrium.

Theorem 3.8 *If $\mathcal{H}_4(\omega_2, \omega_1) \neq 0$, then the equilibrium is stable.*

This is Arnold's stability theorem, which will be proved in Chap. 5.

So to find the fourth-order Birkhoff normal form of a two-degree of freedom autonomous Hamiltonian system is an important step in the analysis of stability. This is a tiresome task for which the use of a computer may be essential. However, in simple cases we can do it by hand computation, as will be illustrated by the example below.

Let us first make some preliminary considerations regarding the computation of $C_4(\mathbf{z}, \overline{\zeta})$.

In the absence of resonance relation up to order four we have $\Gamma_3 = 0$ and since $C_3 = 0$ the Eq. (3.28) reduces to the equation

$$DW_3(\mathbf{z}, \overline{\zeta}) = \mathbf{H}_3(\mathbf{z}, \overline{\zeta}), \tag{3.32}$$

where D is the operator (3.29). From the proof of Theorem 3.7 we know that $C_4(\mathbf{z}, \overline{\zeta})$ is formed by the terms of the fourth degree in $\mathbf{H}_3(\mathbf{z}, \overline{\zeta} + W_3(\mathbf{z}, \overline{\zeta}) + \dots)$.

Expanding around $(\mathbf{z}, \overline{\zeta})$ and keeping only the terms of the fourth degree we get $C_4(\mathbf{z}, \overline{\zeta}) = \mathbf{H}_{3,\overline{\mathbf{z}}}(\mathbf{z}, \overline{\zeta})^T W_{3,\mathbf{z}}(\mathbf{z}, \overline{\zeta})$, that is,

$$C_4(\mathbf{z}, \overline{\zeta}) = \mathbf{H}_{3,\overline{\mathbf{z}}_1}(\mathbf{z}, \overline{\zeta})^T W_{3,\mathbf{z}_1}(\mathbf{z}, \overline{\zeta}) + \mathbf{H}_{3,\overline{\mathbf{z}}_2}(\mathbf{z}, \overline{\zeta})^T W_{3,\mathbf{z}_2}(\mathbf{z}, \overline{\zeta}), \tag{3.33}$$

a sum of products of homogeneous polynomials of second degree.

Now we fix an order for writing the sequence of coefficients $[P]$ of a homogeneous polynomial of degree three, $P = \sum_{\nu+\mu=3} P_{\nu_1\nu_2\mu_1\mu_2} X_1^{\nu_1} X_2^{\nu_2} Y^{\mu_1} Y_2^{\mu_2}$. We will choose the order

$$[P]=(P_{3000}, P_{2010}, P_{1020}, P_{0030}; P_{2100}, P_{1110}, P_{0120}, P_{2001}, P_{1011}, P_{0021};$$
$$P_{1200}, P_{1101}, P_{1002}, P_{0210}, P_{0111}, P_{0012}; P_{0300}, P_{0201}, P_{0102}, P_{0003}),$$

which corresponds to the power expansions of the expressions

$$(X_1 + Y_1)^3, \quad (X_1 + Y_1)^2(X_2 + Y_2), \quad (X_1 + Y_1)(X_2 + Y_2)^2, \quad (X_2 + Y_2)^3.$$

This ordering leads to a diagonal matrix D_ω whose diagonal elements are computed from (3.30) with $n = 2$.

For the case $\delta_1 = +1$ and $\delta_2 = -1$ this 20×20 diagonal matrix D_ω has the following diagonal entries

$$3\omega_1, \ \omega_1, \ -\omega_1, \ -3\omega_1; \ 2\omega_1 - \omega_2, \ -\omega_2, \ -2\omega_1 - \omega_2, \ 2\omega_1 + \omega_2, \ \omega_2, \ -2\omega_1 + \omega_2;$$
$$\omega_1 - 2\omega_2, \ \omega_1, \ \omega_1 + 2\omega_2, \ 2\omega_1 + \omega_2, \ -\omega_1, \ -\omega_1 + 2\omega_2, \ -3\omega_2, \ -\omega_2, \ \omega_2, \ 3\omega_2. \tag{3.34}$$

Then we see from (3.31) that the sequences of the coefficients $[\mathbf{H}_3]$ of $\mathbf{H}_3$ and the coefficients $[W_3]$ of W_3 are related by the equation

$$D_\omega[W_3] = \frac{1}{i}[\mathbf{H}_3]. \tag{3.35}$$

Therefore, we get at once the function W_3. Then, we can compute all the quadratic polynomials in the expression (3.33) and we can readily select the resonant terms in C_4, that is, the terms defined by the monomials $(z_1\overline{\zeta}_1)^2$, $z_1z_2\overline{\zeta}_1\overline{\zeta}_2$ and $(z_2\overline{\zeta}_2)^2$. Together with the resonant terms in $\mathbf{H}_4$ we finally obtain the fourth-degree terms of the Birkhoff normal form.

Example 3.2 (A Simple Case of Order Four) Under the hypothesis of no resonances up to order four, find the fourth-order Birkhoff normal form of the Hamiltonian

$$H = \frac{1}{2}\omega_1(x_1^2 + y_1^2) - \frac{1}{2}\omega_2(x_2^2 + y_2^2) + x_1^2 y_1 + x_2^2 y_2 + (x_1 + y_1)^2(x_2 - y_2)^2.$$

Making the symplectic transformation with multiplier $i/2$,

$$x_j = \frac{1}{2}(z_j + \overline{z}_j), \quad x_j = \frac{1}{2i}(z_j - \overline{z}_j), \qquad j = 1, 2$$

the new Hamiltonian $\mathbf{H} = \frac{2}{i}H$ is given by

$$\mathbf{H}(\mathbf{z}, \overline{\mathbf{z}}) = -i(z_1\overline{z}_1 - z_2\overline{z}_2) + \mathbf{H}_3(\mathbf{z}, \overline{\mathbf{z}}) + \mathbf{H}_4(\mathbf{z}, \overline{\mathbf{z}}))$$

where

$$\mathbf{H}_3(\mathbf{z}, \overline{\mathbf{z}}) = \frac{i}{4}\big[i(z_1^3 + z_1^2\overline{z}_1 - z_1\overline{z}_1^2 - \overline{z}_1^3) + (z_2^3 - z_2^2\overline{z}_2 - z_2\overline{z}_2^2 + \overline{z}_2^3\big])$$

and

$$\mathbf{H}_4(\mathbf{z}, \overline{\mathbf{z}}) = \frac{1}{2i}\big[2z_1\overline{z}_1 - i(z_1^2 - \overline{z}_1^2)\big]\big[2z_2\overline{z}_2 + i(z_2^2 - \overline{z}_2^2)\big].$$

Therefore,

$$[\mathbf{H}_3] = \frac{i}{4}(i, i, -i, -i; 0, 0, 0, 0, 0, 0; 0, 0, 0, 0, 0, 0; 1, -1, -1, 1),$$

so from (3.34) and (3.35) we get

$$[W_3] = \frac{1}{4}\Big(\frac{i}{3\omega_1}, \frac{i}{\omega_1}, \frac{i}{\omega_1}, \frac{i}{3\omega_1}; 0, 0, 0, 0, 0, 0; 0, 0, 0, 0, 0, 0; -\frac{1}{3\omega_2}, \frac{1}{\omega_2}, -\frac{1}{\omega_2}, \frac{1}{3\omega_2}\Big).$$

Then, for the expression (3.33) we obtain

$$C_4(\mathbf{z},\overline{\boldsymbol{\zeta}}) = -\frac{1}{16}\big(z_1^2 - 2z_1\overline{\zeta}_1 - 3\overline{\zeta}_1^2\big)\Big(\frac{1}{\omega_1}z_1^2 + \frac{2}{\omega_1}z_1\overline{\zeta}_1 + \frac{1}{\omega_1}\overline{\zeta}_1^2\Big)$$
$$= \frac{i}{16}\big(-z_2^2 - 2z_2\overline{\zeta}_2 + 3\overline{\zeta}_2^2\big)\Big(-\frac{1}{\omega_2}z_2^2 + \frac{2}{\omega_1 2}z_2\overline{\zeta}_2 - \frac{1}{\omega_2}\overline{\zeta}_2^2\Big).$$

We readily find the resonant part of C_4, namely

$$\operatorname{Res} C_4(\mathbf{z},\overline{\boldsymbol{\zeta}}) = \frac{3i}{8\omega_1}z_1^2\overline{\zeta}_1^2 - \frac{3i}{8\omega_1}z_2^2\overline{\zeta}_2^2.$$

From the expression of $\mathbf{H}_4$ we find for the resonant terms

$$\operatorname{Res}\mathbf{H}_4(\mathbf{z},\overline{\boldsymbol{\zeta}}) = -2i\, z_1\overline{\zeta}_1 z_2\overline{\zeta}_2.$$

Therefore, the fourth-degree terms of the normal form, in the complex variables $\boldsymbol{\zeta}, \overline{\boldsymbol{\zeta}}$ are

$$\Gamma_4(\boldsymbol{\zeta},\overline{\boldsymbol{\zeta}}) = \frac{3i}{8\omega_1}z_1^2\overline{\zeta}_1^2 - 2i\, z_1\overline{\zeta}_1 z_2\overline{\zeta}_2 - \frac{3i}{8\omega_2}z_2^2\overline{\zeta}_2^2.$$

Making now the symplectic transformation with multiplier $2/i$ $\zeta_j = \xi_j + i\eta_j$, $\overline{\zeta}_j = \xi_j - i\eta_j$ we get the real normal form up to fourth-order terms

$$\Gamma = \frac{\omega_1}{2}(\xi_1^2+\eta_1^2) - \frac{\omega_2}{2}(\xi_2^2+\eta_2^2) + (\xi_1^2+\eta_1^2)(\xi_2^2+\eta_2^2) - \frac{3}{16\omega_1}(\xi_1^2+\eta_1^2)^2 + \frac{3}{16\omega_2}(\xi_2^2+\eta_2^2)^2.$$

Finally, passing to action-angles variables via the symplectic transformation

$$\xi_j = \sqrt{2I_l}\cos\theta_j, \quad \eta_j = \sqrt{2I_l}\sin\theta_j, \qquad j = 1, 2$$

we get the following expression for the fourth-order real normal form

$$\mathcal{H}(I_1, I_2) = \omega_1 I_1 - \omega_2 I_2 - \frac{3}{4\omega_1}I_1^2 + \frac{3}{4\omega_2}I_2^2.$$

Notice that since

$$\mathcal{H}(\omega_2,\omega_1) = \frac{3}{4\omega_1\omega_2}(\omega_1^3 - \omega_2^3) \neq 0,$$

we conclude from the Arnold theorem that the origin is a stable equilibrium for this Hamiltonian system.

Given the Hamiltonian function (3.1) of a specific problem the main computational difficulty in finding the fourth-order Birkhoff normal form is the step of computing the third- and fourth-degree terms after the normalizing linear symplectic transformation with matrix (3.13) is substituted into $H_3(\mathbf{z})$ and $H_4(\mathbf{z})$ in order to get the Hamiltonian (3.21) with the normalized quadratic term. In the above example this step was not necessary.

Example 3.3 (The Restricted Three-Body Problem) Let us consider the Lagrangian equilibrium point L_4 of the restricted three-body problem. The potential V is given by (2.8) and it is not difficult to conclude that at the noncollinear relative equilibria we have $\rho_1 = \rho_2 = 1$. For L_4 the coordinates in the rotating frame are $\xi = \frac{1}{2} - \mu$, $\eta = \frac{3}{2}$ and we can easily compute the partial derivatives of the potential (2.8) at L_4 getting without much difficulty the expressions of H_2, H_3 and H_4. The Hessian matrix of V at L_4 is the matrix $\begin{bmatrix} a & b \\ b & c \end{bmatrix}$ with $a = -\frac{1}{4}$, $c = \frac{5}{4}$ and $b = \frac{3\sqrt{3}}{4}(1 - 2\mu)$. The characteristic equation of the linearized system at L_4 is easily found to be

$$\lambda^4 + \lambda^2 + \frac{27}{4}\mu(1 - \mu) = 0$$

so there are two distinct and pure imaginary eigenvalues if $27\mu(1 - \mu) < 1$. This means that L_4 is linearly stable. Let the eigenvalues be $\lambda_1 = i\omega_1$ and $\lambda_2 = i\omega_2$ with $\omega_1 > \omega_2$. It is not difficulty to find the eigenvector $\mathbf{v} = \mathbf{r} + i\mathbf{s}$ corresponding to one of the eigenvalues $i\omega$ and one gets

$$\mathbf{r} = \Big(-b, \omega^2 + \frac{3}{4}, \omega^2 - \frac{3}{4}, -b\Big) \quad \text{and} \quad \mathbf{s} = \Big(-2\omega, 0, -\omega b, \omega\big(\omega^2 - \frac{5}{4}\big)\Big). \tag{3.36}$$

We find the expression

$$\{\mathbf{r}, J\mathbf{s}\} = \omega\big(b^2 + \omega^4 + \frac{3}{2}\omega^2 - \frac{39}{16}\big).$$

To decide the sign of this expression notice that from the characteristic equation we have $\omega^4 - \omega^2 + \frac{27}{4}\mu(1-\mu) = 0$; hence, $\omega^2 = 1 \pm \sqrt{1 - \Delta}$, where $\Delta = 27\mu(1-\mu)$. From the expression of b we have $b^2 = \frac{27}{16} - \frac{27}{4}\mu(1 - \mu) = \frac{27}{16} - \frac{1}{4}\Delta$. Therefore, we get

$$\frac{1}{\omega}\{\mathbf{r}, J\mathbf{s}\} = -\frac{6}{8} - \frac{1}{2}\Delta + \frac{5}{4}\big(1 \pm \sqrt{1 - \Delta}\big) = \frac{1}{2}\sqrt{1 - \Delta}\Big(\sqrt{1 - \Delta} \pm \frac{5}{2}\Big). \tag{3.37}$$

From this we see that $\{\mathbf{r}, J\mathbf{s}\}$ is positive for the plus sign, that is, for the larger root ω_1 and it is negative for the minus sign, that is, for the smaller root ω_2.

With (3.36) for ω_1 and ω_2 we then have $\delta_1 = 1$, $\delta_2 = -1$, κ_1 and κ_2 of Proposition 3.4 and we readily find the matrix P in (3.13).

As the quadratic part of the Hamiltonian is indefinite to decide about the nonlinear stability of L_4 we have to compute the fourth-order Birkhoff normal. All the above computations are done without much difficulty, but now comes the insane task of computing the new third- and fourth-degree terms of the Hamiltonian after using the transformation $\mathbf{z} = P\zeta$ to arrive at (3.21). Of course with patience and time this task can be done by hand calculation, but here the help of a computer is a blessing.

After (3.21) is found, the subsequent calculations are not that difficult, we just follow the steps done for the above example, computing W_3 and proceeding as we did there.

3.6 A Resonant Case

Consider an autonomous two-degree of freedom Hamiltonian system

$$H(x_1, x_2, y_1, y_2) = H_2 + H_3 + H_4 + \dots \tag{3.38}$$

with an indefinite quadratic part

$$H_2 = \frac{1}{2}\omega_1(x_1^2 + y_1^2) - \frac{1}{2}\omega_2(x_2^2 + y_2^2), \qquad \omega_1, \omega_2 > 0.$$

Assume that the frequencies $\omega_1, \omega_2 > 0$ satisfy the resonance relation of order $m = p + q$

$$p\omega_1 - q\omega_2 = 0, \quad p, q \quad \text{relatively prime positive integers.}$$

If there are no resonances of order less than m, the normal form is given in action-angle variables $(I_1, I_2, \phi_1, \phi_2)$ (see Example 2.16), by

$$H = H_2 + H_4 + \dots + H_{2l-2} + H_m + \dots. \tag{3.39}$$

where

$$H_2 = \omega_1 I_1 - \omega_2 I_2, \tag{3.40}$$

H_{2j} is a homogeneous polynomial of degree j in the actions I_1, I_2 and $H_m(I_1, I_2, \phi_1, \phi_2)$ is a polynomial of degree m with respect to $\sqrt{I_1}, \sqrt{I_2}$; of course, H_{2j} is a polynomial of degree $2j$ in $\sqrt{I_1}, \sqrt{I_2}$.

To see why we use the square roots of the actions let us look at the structure of H_m. A typical monomial of H_m in $z_1, z_2, \overline{z}_1, \overline{z}_2$ where $z_j = x_j + iy_j$ is written as

$$\mathbf{z}^k\overline{\mathbf{z}}^l = z_1^{k_1} z_2^{k_2} \overline{z}_1^{l_1} \overline{z}_2^{l_2} \quad \text{with} \quad k_1 + k_2 + l_1 + l_2 = m$$

and if H_m is in normal form we have that $\mathbf{k} - \mathbf{l} = (k_1 - l_1, k_2 - l_2)$ belongs to the module

$$M_{\delta\omega} = \{(m_1, m_2);\ m_1\omega_1 - m_2\omega_2 = 0\}.$$

Since $p\omega_1 - q\omega_2 = 0$ and p, q are relatively prime numbers, we have $m_1 = sp$ and $m_2 = sq$ for some $s \in \mathbf{Z}$. Therefore, $k_1 = l_1 + sp$, $k_2 = l_2 + sq$ and consequently

$$\mathbf{z}^k \bar{\mathbf{z}}^l = \left(z_1\bar{z}_1\right)^{l_1}\left(z_2\bar{z}_2\right)^{l_2} z_1^{sp} z_2^{sq}.$$

Passing to action-angle variables, $z_j = \sqrt{2I_j}e^{i\phi_j}$, we have

$$z^k \bar{z}^l = (2I_1)^{l_1}(2I_2)^{l_2}(\sqrt{2I_1})^{sp}(\sqrt{2I_2})^{sq}e^{i(sp\phi_1 + sq\phi_2)} = A\sqrt{I_1}^{2l_1+sp}\sqrt{I_2}^{2l_2+sq}e^{is(p\phi_1+q\phi_2)},$$

where $A = \sqrt{2}^{2l_1+sp+2l_2+sq}$. Therefore, the normalized resonant term H_m is given by

$$H_m = H_m(I_1, I_2, p\phi_1 + q\phi_2) \tag{3.41}$$

a polynomial of degree m in $\sqrt{I_1}$, $\sqrt{I_2}$ with the angles ϕ_1, ϕ_2 appearing only in the combination $p\phi_1 + q\phi_2$; the degree is m because $2l_1 + sp + 2l_2 + sq = k_1 + l_1 + k_2 + l_2 = m$.

If we normalize the Hamiltonian H up to H_m the remainder is an analytic function given by a series in $\sqrt{I_1}$, $\sqrt{I_2}$ where the coefficients are finite sums of sines and co-sines of the angles ϕ_1, ϕ_2.

3.7 The Method of Deprit-Hori

In this section we describe a method of simplifying a Hamiltonian given as a series expansion in a small parameter ϵ.

3.7.1 Historical Considerations

In [31] Gen-ichiro Hori considered a theorem of Solphus Lie [36] on canonical transformations obtained from a function $S(\mathbf{x}, \mathbf{y})$ to introduce a new way of looking at the simplification of a Hamiltonian that is given as a power series in a small parameter.

Let D_S be the linear operator defined by the Poisson bracket $D_S f = \{f, S\}$, for any function $f(\mathbf{x}, \mathbf{y})$ and consider the sequence of operators

$$D_S^0 f = f, \quad D_S^1 f = D_S f, \quad D_S^n f = D_S(D_S^{n-1} f).$$

Assume that for ϵ small the following power series

$$\sum_{n=0}^{\infty} \frac{\epsilon^n}{n!} D_S^n f(\mathbf{x}, \mathbf{y})$$

converges for any function $f(\mathbf{x}, \mathbf{y})$.

Theorem 3.9 (S. Lie) *If the variables* $(\boldsymbol{\xi}, \boldsymbol{\eta})$ *and* $(\mathbf{x}, \mathbf{y})$ *are related by the equality*

$$f(\mathbf{x}, \mathbf{y}) = \sum_{n=0}^{\infty} \frac{\epsilon^n}{n!} D_S^n f(\boldsymbol{\xi}, \boldsymbol{\eta}) \tag{3.42}$$

for every function f *then this defines a canonical transformation* $(\boldsymbol{\xi}, \boldsymbol{\eta}) \mapsto (\mathbf{x}, \mathbf{y})$.

Proof Let $\boldsymbol{\phi}(\boldsymbol{\xi}, \boldsymbol{\eta}, \epsilon) = (\mathbf{x}(\boldsymbol{\xi}, \boldsymbol{\eta}, \epsilon), \mathbf{y}(\boldsymbol{\xi}, \boldsymbol{\eta}, \epsilon)$ be the flow of the Hamiltonian system defined by the function S with ϵ as time

$$\frac{dx_j}{d\epsilon} = \frac{\partial S}{\partial y_j}, \quad \frac{dy_j}{d\epsilon} = -\frac{\partial S}{\partial x_j}, \qquad j = 1, \ldots, n. \tag{3.43}$$

Since $D_S^n f = D_S(D_S^{n-1} f) = (D_S^{n-1}, f)$ we have along the solution $\boldsymbol{\phi}(\boldsymbol{\xi}, \boldsymbol{\eta}, \epsilon)$,

$$\begin{aligned} D_S^n f(\boldsymbol{\phi}(\boldsymbol{\xi}, \boldsymbol{\eta}, \epsilon)) &= \sum_{j=1}^{n} \left(\frac{\partial(D_S^{n-1} f)}{\partial x_j} \frac{\partial S}{\partial y_j} - \frac{\partial(D_S^{n-1} f)}{\partial y_j} \frac{\partial S}{\partial x_j} \right) \\ &= \sum_{j=1}^{n} \left(\frac{\partial(D_S^{n-1} f)}{\partial x_j} \frac{dx_j}{d\epsilon} + \frac{\partial(D_S^{n-1} f)}{\partial y_j} \frac{dy_j}{d\epsilon} \right) = \frac{d}{d\epsilon} D_S^{n-1} f, \end{aligned}$$

so, omitting the argument after the first equality sign, we get

$$D_S^n f(\boldsymbol{\phi}(\boldsymbol{\xi}, \boldsymbol{\eta}, \epsilon)) = \frac{d}{d\epsilon}(D_S^{n-1} f) = \frac{d^2}{d\epsilon^2}(D_S^{n-2} f) = \ldots = \frac{d^n}{d\epsilon^n}(D_S^0 f),$$

that is, for all n,

$$D_S^n f(\mathbf{x}(\boldsymbol{\phi}(\boldsymbol{\xi}, \boldsymbol{\eta}, \epsilon)) = \frac{d^n}{d\epsilon^n} f\big(\mathbf{x}(\boldsymbol{\xi}, \boldsymbol{\eta}, \epsilon), \mathbf{y}(\boldsymbol{\xi}, \boldsymbol{\eta}, \epsilon)\big). \tag{3.44}$$

Therefore, the right-hand side of (3.42) is the Taylor series of $f\big(\mathbf{x}(\boldsymbol{\xi}, \boldsymbol{\eta}, \epsilon), \mathbf{y}(\boldsymbol{\xi}, \boldsymbol{\eta}, \epsilon)\big)$, so the Eq. (3.42) says that for all functions f we have

$$f(\mathbf{x}, \mathbf{y}) = f\big(\mathbf{x}(\boldsymbol{\xi}, \boldsymbol{\eta}, \epsilon), \mathbf{y}(\boldsymbol{\xi}, \boldsymbol{\eta}, \epsilon)\big);$$

hence,

$$\mathbf{x} = \mathbf{x}(\boldsymbol{\xi}, \boldsymbol{\eta}, \epsilon), \quad \mathbf{y} = \mathbf{y}(\boldsymbol{\xi}, \boldsymbol{\eta}, \epsilon). \tag{3.45}$$

As the flow of (3.43) is a canonical transformation, for each fixed ϵ, taking $\epsilon = 0$ we get the symplectic mapping $(\boldsymbol{\xi}, \boldsymbol{\eta}) \mapsto (\mathbf{x}, \mathbf{y})$. □

Since $(\boldsymbol{\xi}, \boldsymbol{\eta})$ is the initial condition of the solution $\boldsymbol{\phi}(\boldsymbol{\xi}, \boldsymbol{\eta}, \epsilon)$ Eqs. (3.45) show that

$$\mathbf{x} = \boldsymbol{\xi} + O(\epsilon), \quad \mathbf{y} = \boldsymbol{\eta} + O(\epsilon)$$

is a near-identity canonical transformation.

The function $S(\mathbf{x}, \mathbf{y})$ is called the *determining function* of this transformation.

In [31] G. Hori considers an autonomous Hamiltonian dynamical system

$$\frac{dx_j}{dt} = \frac{\partial F}{\partial y_j}, \quad \frac{dy_j}{dt} = -\frac{\partial F}{\partial x_j}, \tag{3.46}$$

where

$$F(\mathbf{x}, \mathbf{y}) = F_0(\mathbf{x}, \mathbf{y}) + \sum_{k=1}^{\infty} \epsilon^k F_k(\mathbf{x}, \mathbf{y})$$

and the solution $\mathbf{x}(\boldsymbol{\xi}, \boldsymbol{\eta}, \epsilon), \mathbf{y}(\boldsymbol{\xi}, \boldsymbol{\eta}, \epsilon)$ of the system (3.43) with initial condition $(\boldsymbol{\xi}, \boldsymbol{\eta})$ when $\epsilon = 0$, S being taken as a series in ϵ of the form

$$S(\mathbf{x}, \mathbf{y}) = \sum_{k=0}^{\infty} \epsilon^k S_{k+1}(\mathbf{x}, \mathbf{y}).$$

As the Hamiltonian system (3.46) does not depend on time, F is a first integral, so we have the energy equation

$$\sum_{k=0}^{\infty} \epsilon^k F_k(\mathbf{x}, \mathbf{y}) = \sum_{k=0}^{\infty} \epsilon^k F_k^*(\boldsymbol{\xi}, \boldsymbol{\eta}).$$

As $D_S^n f(\boldsymbol{\xi}, \boldsymbol{\eta}) = \sum_{k=0}^{\infty} \epsilon^k D_{S_{k+1}}^n f(\boldsymbol{\xi}, \boldsymbol{\eta})$ comparing terms of like powers of ϵ in the energy equation, Hori expresses the function $F_k^*(\boldsymbol{\xi}, \boldsymbol{\eta})$ as $F_k(\boldsymbol{\xi}, \boldsymbol{\eta})$ plus a sum of terms involving Poisson brackets of $F_0(\boldsymbol{\xi}, \boldsymbol{\eta}), \dots, F_{k-1}(\boldsymbol{\xi}, \boldsymbol{\eta})$ with the functions $S_1(\boldsymbol{\xi}, \boldsymbol{\eta}), \dots, S_k(\boldsymbol{\xi}, \boldsymbol{\eta})$, see [31]. Rather than listing these relations we go directly to a general formula given by Deprit (see (3.50), in his paper [20]).

In 1966, André Deprit published his theory of perturbations along these lines. He considers a function $W(\mathbf{y}, Y)$ and defines the *Lie operator*

$$\exp(\epsilon L_W) = \sum_{n\geq 0} \frac{1}{n!}\epsilon^n L_W^n, \tag{3.47}$$

where L_W is the Lie derivative $L_W f = \{f, W\}$, making a detailed study of this operator. Compare (3.47) with (3.42) and notice that D_S is the same as L_S, when $\mathbf{y} = \boldsymbol{\xi}$, $Y = \boldsymbol{\eta}$.

In his theory Deprit considers the Hamiltonian system with time ϵ (see (3.43)),

$$\frac{d\mathbf{x}}{d\epsilon} = \frac{\partial W}{\partial X}, \quad \frac{dX}{d\epsilon} = -\frac{\partial W}{\partial \mathbf{x}}, \tag{3.48}$$

where

$$W(\mathbf{x}, X; \epsilon) = \sum_{k\geq 0} \frac{1}{n!}\epsilon^n W_{n+1}(\mathbf{x}, X)$$

and inserts its solution $\big(\mathbf{x}(\mathbf{y}, Y; \epsilon), X(\mathbf{y}, Y; \epsilon)\big)$ with the initial condition $\mathbf{x} = \mathbf{y}$, $X = Y$ when $\epsilon = 0$ into the series

$$f(\mathbf{x}, X; \epsilon) = \sum_{n\geq 0} \frac{1}{n!}\epsilon^n f_n(\mathbf{x}, X).$$

In terms of the differential operator

$$\Delta_W f = L_W f + \frac{\partial f}{\partial \epsilon}, \tag{3.49}$$

this gives

$$f\big(\mathbf{x}(\mathbf{y}, Y; \epsilon^n), X(\mathbf{y}, Y; \epsilon^n)\big) = \sum_{n\geq 0} \frac{1}{n!}\epsilon^n \left(\Delta_W^n f\right)_{\epsilon=0}.$$

Then, writing the kth iterate of $\Delta_W f$ in the form

$$\Delta_W^k f = \sum_{k\geq 0} \epsilon^n f_n^{(k)}(\mathbf{y}, Y),$$

he arrives at the important formula

$$f_n^{(k)} = f_{n+1}^{(k-1)} + \sum_{0\leq m\leq n} \binom{n}{m} L_{m+1} f_{n-m}^{(k-1)}, \tag{3.50}$$

where $\binom{n}{m} = \dfrac{n!}{m!(n-m)!}$ is the Newton binomial coefficient.

Formula (3.50), or rather formula (3.55) of Theorem 3.10, will be fundamental to proving the existence of the normal forms considered in Theorems 3.1 and 3.2.

In the sequel we will consider the Deprit approach following the exposition in Meyer's book [41]. Compare what we will be doing with the above historical considerations.

3.7.2 Deprit's Formula

We start with a Hamiltonian $H(\epsilon, \mathbf{x}) = H_*(\epsilon, \mathbf{x})$ given as a power series in a small parameter ϵ,

$$H_*(\epsilon, \mathbf{x}) = \sum_{i=0}^{\infty} \frac{\epsilon^i}{i!} H_i^0(\mathbf{x}), \tag{3.51}$$

where $H_i^0(\mathbf{x})$ is a homogeneous polynomial of degree $i+2$ in the components of $\mathbf{x}$ and we want to find a canonical mapping that transforms it to the Hamiltonian

$$H^*(\epsilon, \mathbf{y}) = \sum_{i=0}^{\infty} \frac{\epsilon^i}{i!} H_0^i(\mathbf{y}), \tag{3.52}$$

where $H_0^i(\mathbf{y})$ is a homogeneous polynomial of degree $i+2$ in (the components of) $\mathbf{y}$.

For $i = 0$, we take the two polynomials in $\mathbf{x}$ and $\mathbf{y}$ with the same coefficients; hence, we must do this by means of a formal symplectic transformation that begins with the identity

$$\mathbf{x} = X(\epsilon, \mathbf{y}) = \mathbf{y} + \Phi(\epsilon, \mathbf{y}).$$

The method consists in taking this symplectic transformation as the transformation defined by the set of solutions of a Hamiltonian system with Hamiltonian function $W(\epsilon, \mathbf{x})$ in which the time is the variable ϵ, that is, $\mathbf{x} = X(\epsilon, \mathbf{y})$ is the solution of the Cauchy problem

$$\dot{\mathbf{x}} = J \frac{\partial W}{\partial \mathbf{x}}(\mathbf{x}, \epsilon), \qquad X(0, \mathbf{y}) = \mathbf{y}. \tag{3.53}$$

As $\mathbf{x} = X(\epsilon, \mathbf{y})$ is a symplectic transformation that does not depend on time t, the relation between the original Hamiltonian and the transformed Hamiltonian is $H^*(\epsilon, \mathbf{y}) = H_*(\epsilon, X(\epsilon, \mathbf{y}))$.

We write $W(\epsilon, \mathbf{x})$ as a formal series

$$W(\epsilon, \mathbf{x}) = \sum_{i=0}^{\infty} \frac{\epsilon^i}{i!} W_{i+1}(\mathbf{x}), \tag{3.54}$$

where $W_i(\mathbf{x})$ is a homogeneous polynomial of degree $i + 2$ in the components of $\mathbf{x}$. Note that the first term of W is W_1, which is convenient for in the formula (3.55) below the sum of the indices, is always the same, equal to $i + j$.

The homogeneous polynomials of H_*, H^* will be related through a doubly indexed auxiliary sequence of homogeneous polynomials, H_j^i, whose terms coincide with those of the polynomials in H_* and H^* when either i or j is equal to zero.

Theorem 3.10 (Deprit Formula) *The polynomials W_k are related to the double index polynomials H_j^i by the formula*

$$H_j^i = H_{j+1}^{i-1} + \sum_{k=0}^{j} \binom{j}{k} \{H_{j-k}^{i-1}, W_{k+1}\}, \tag{3.55}$$

where $\{\ ,\ \}$ denotes the Poisson bracket of two functions.

Proof Consider the differential operator $\mathcal{D}F = \Delta_W F$ defined in (3.49) and notice that if $\mathbf{x} = X(\epsilon, \mathbf{y})$, then $\mathcal{D}F(\epsilon, \mathbf{x})$ is the total derivative of $F(\epsilon, \mathbf{x})$ along the solutions of the equation $\dot{\mathbf{x}} = J\dfrac{\partial W}{\partial \mathbf{x}}(\epsilon, \mathbf{x})$. Consider the sequence of functions defined inductively by

$$H^0 = H_*, \quad H^i = \mathcal{D}H^{i-1}, \qquad i \geq 1.$$

Let the series expansions of these functions be

$$H^i(\epsilon, \mathbf{x}) = \sum_{j=0}^{\infty} \frac{\epsilon^j}{j!} H_j^i(\mathbf{x}). \tag{3.56}$$

Since $H^i(\epsilon, \mathbf{x}) = \mathcal{D}H^{i-1}(\epsilon, \mathbf{x}) = \mathcal{D}\sum_{k=0}^{\infty} \dfrac{\epsilon^k}{k!} H_k^{i-1}(\mathbf{x})$, we have by (3.49) and (3.54)

$$H^i(\epsilon, \mathbf{x}) = \sum_{k=0}^{\infty} \frac{\epsilon^{k-1}}{(k-1)!} H_k^{i-1}(\mathbf{x}) + \left\{ \sum_{k=0}^{\infty} \frac{\epsilon^k}{k!} H_k^{i-1}(\mathbf{x}), \sum_{s=0}^{\infty} \frac{\epsilon^s}{s!} W_{s+1}(\mathbf{x}) \right\}.$$

Now, using the identity $\sum_{k=0}^{\infty} \sum_{s=0}^{\infty} a_k b_s = \sum_{j=0}^{\infty} \sum_{k=0}^{j} a_{j-k} b_k$, we obtain

$$H^i(\epsilon, \mathbf{x}) = \sum_{j=0}^{\infty} \frac{\epsilon^j}{j!}\Big[H^{i-1}_{j+1} + \sum_{k=0}^{j} \binom{j}{k}\{H^{i-1}_{j-k}, W_{k+1}\}\Big].$$

Comparing this with (3.56), we get the formula (3.55) stated in the theorem. Since $H^0 = H_*$, the polynomials H^0_k in (3.56) coincide with the polynomials of H_* in (3.51).

It remains to show that $H^*(\epsilon, \mathbf{y}) = H_*(\epsilon, X(\epsilon, \mathbf{y}))$ is given by the series (3.52). To prove this, we use the following three observations:

$$H^k = \mathcal{D}^k H_*, \quad H^i(0, \mathbf{x}) = H^i_0(\mathbf{x}), \quad \frac{d^k}{d\epsilon^k}\Big(F(\epsilon, \mathbf{x})\Big|_{\mathbf{x}=X(\epsilon,\mathbf{y})}\Big) = \mathcal{D}^k F(\epsilon, \mathbf{x})\Big|_{\mathbf{x}=X(\epsilon,\mathbf{y})}. \tag{3.57}$$

The first is the very definition of the function H^k; for the second just take $\epsilon = 0$ in (3.56). We now prove the third by induction; compare it with (3.44). It holds for $k = 1$ since $\mathcal{D}F(\epsilon, X(\epsilon, \mathbf{y}))$ is the total derivative of $F(\epsilon, \mathbf{x})$ along the curve $\mathbf{x} = X(\epsilon, \mathbf{y})$, solution of the equation $\dot{\mathbf{x}} = J\dfrac{\partial W}{\partial \mathbf{x}}(\epsilon, \mathbf{x})$. Assuming its validity for k, it is valid for $k + 1$, since

$$\begin{aligned}
\frac{d^{k+1}}{d\epsilon^{k+1}}\Big(F(\epsilon, \mathbf{x})\Big|_{\mathbf{x}=X(\epsilon,\mathbf{y})}\Big) &= \frac{d}{d\epsilon}\Big(\frac{d^k}{d\epsilon^k}\Big(F(\epsilon, \mathbf{x})\Big|_{\mathbf{x}=X(\epsilon,\mathbf{y})}\Big)\Big) = \frac{d}{d\epsilon}\Big(\mathcal{D}^k F(\epsilon, \mathbf{x})\Big|_{\mathbf{x}=X(\epsilon,\mathbf{y})}\Big) \\
&= \mathcal{D}\Big(\mathcal{D}^k F(\epsilon, \mathbf{x})\Big|_{\mathbf{x}=X(\epsilon,\mathbf{y})}\Big) = \mathcal{D}^{k+1} F(\epsilon, \mathbf{x})\Big|_{\mathbf{x}=X(\epsilon,\mathbf{y})}.
\end{aligned}$$

Now, considering the Taylor series $H^*(\epsilon, \mathbf{y}) = \displaystyle\sum_{k=0}^{\infty} \frac{\epsilon^k}{k!}\frac{d^k}{d\epsilon^k} H^*(\epsilon, \mathbf{y})\Big|_{\epsilon=0}$ and using the observations (3.57), we have

$$\begin{aligned}
H^*(\epsilon, \mathbf{y}) &= \sum_{k=0}^{\infty} \frac{\epsilon^k}{k!}\frac{d^k}{d\epsilon^k}\Big(H_*(\epsilon, \mathbf{x})\Big|_{\mathbf{x}=X(\epsilon,\mathbf{y})}\Big)_{\epsilon=0} \\
&= \sum_{k=0}^{\infty} \frac{\epsilon^k}{k!}\Big(\mathcal{D}^k H_*(\epsilon, \mathbf{x})\Big|_{\mathbf{x}=X(\epsilon,\mathbf{y})}\Big)_{\epsilon=0} \\
&= \sum_{k=0}^{\infty} \frac{\epsilon^k}{k!}\Big(H^k_*(\epsilon, \mathbf{x})\Big|_{\mathbf{x}=X(\epsilon,\mathbf{y})}\Big)_{\epsilon=0} = \sum_{k=0}^{\infty} \frac{\epsilon^k}{k!} H^k_0(\mathbf{y}),
\end{aligned}$$

which concludes the proof of the theorem. □

3.7.3 The Lie Triangle

Writing the Deprit formula (3.55) in extension, with $C_k^j = \binom{j}{k}$, we have

$$H_j^i = H_{j+1}^{i-1} + C_0^j\{H_j^{i-1}, W_1\} + C_1^j\{H_{j-1}^{i-1}, W_2\} + \ldots + C_{j-1}^j\{H_1^{i-1}, W_j\} + C_j^j\{H_0^{i-1}, W_{j+1}\}$$

and we see that a helpful scheme to write down every equation coming from it is the following schematic triangle, known as the *Lie triangle*, or the Lie H-triangle

$$\begin{array}{ccccccc}
H_0^0 & & & & & & \\
\uparrow & & & & & & \\
H_1^0 & \leftarrow & H_0^1 & & & & \\
\uparrow & & \uparrow & & & & \\
H_2^0 & \leftarrow & H_1^1 & \leftarrow & H_0^2 & & \\
\uparrow & & \uparrow & & \uparrow & & \\
H_3^0 & \leftarrow & H_2^1 & \leftarrow & H_1^2 & \leftarrow & H_0^3 \\
\vdots & & \vdots & & \vdots & & \vdots
\end{array}$$

in the use of which it is convenient to have in mind the following considerations:

- the polynomial H_j^i is obtained from those in the column at its left, up to the top;
- the polynomial to the left of H_j^i appears as the first term in the sum giving H_j^i;
- the polynomials H_{j-s}^{i-1} in the previous lines are present in this sum through the Poisson bracket with a polynomial W_k in such a way that the sum of the indices are equal to $i + j$, that is $k = s + 1$;
- the term involving H_{j-s}^{i-1} is to be multiplied by the binomial coefficient $\binom{j}{s} = \frac{j!}{s!(j-s)!}$.

  For instance, with these rules we can write at once

$$H_3^2 = H_4^1 + \binom{3}{0}\{H_3^1, W_1\} + \binom{3}{1}\{H_2^1, W_2\} + \binom{3}{2}\{H_1^1, W_3\} + \binom{3}{3}\{H_0^1, W_4\}.$$

In Sect. 3.11 we present a formula that gives directly the polynomials of the Hamiltonian $H^*(\epsilon, \mathbf{y})$ from the polynomials of the Hamiltonian $H_*(\epsilon, \mathbf{x})$. This formula, obtained by Kamel [32] from Deprit formula (3.55) by elimination of the intermediate polynomials H_j^i is that used by Markeev in his books [39] and [40].

Notice that in order to know the polynomials $H_0^1, H_0^2, \ldots$ it is necessary to know $W_1, W_2, \ldots$. Also, notice that the formula (3.55) says nothing about the structure of the target Hamiltonian, $H^*(\epsilon, \mathbf{y})$. We simply have a process to construct $H^*(\epsilon, \mathbf{y})$ from $H_*(\epsilon, \mathbf{x})$ by means of the sequence of polynomials $W_1, W_2, \ldots$.

3.7.4 A Fundamental Lemma

The following lemma is a key point in the proof of the existence of normal forms.

Lemma 3.11 *Let $\mathcal{P}_j$, $j = 0, 1, 2, \ldots$ be the space of homogeneous polynomials of degree $j + 2$ with real or complex coefficients. Let $\mathcal{Q}_j$ and $\mathcal{R}_j$, $j = 1, 2, \ldots$ be subspaces of $\mathcal{P}_j$ such that for each $P \in \mathcal{P}_j$, $j = 1, 2, \ldots$ there exist $Q \in \mathcal{Q}_j$ and $R \in \mathcal{R}_j$ satisfying the equation*

$$Q = P + \{H_0^0, R\}. \tag{3.58}$$

Then, there exists a formal series W as in (3.54) with $W_i \in \mathcal{R}_i$, $i = 1, 2, \ldots$ which generates a symplectic transformation through the solutions of the Hamiltonian system (3.53) and this transformation takes the original Hamiltonian (3.51) to the final Hamiltonian (3.52) having $H_0^i \in \mathcal{Q}_i$, $i = 1, 2 \ldots$.

Proof Consider the polynomials H_j^i, W_i and H_0^i in formula (3.55). We will prove that the following statement is true for all i and j

$$H_j^i \in \mathcal{P}_{i+j},\ 0 \leq i + j \leq n \quad \text{and} \quad W_i \in \mathcal{R}_i,\ H_0^i \in \mathcal{Q}_i,\ 1 \leq i \leq n. \tag{3.59}$$

This statement is valid for $n = 0$ because $H_0^0 \in \mathcal{P}_0$ and $W_0 \in \mathcal{R}_0$; $H_0^0 \in \mathcal{Q}_0$, by vacuity. Assume that (3.59) is valid for $n - 1$, and let us prove its validity for n. By (3.55), we have

$$H_{n-1}^1 = H_n^0 + \sum_{k=0}^{n-2} \binom{n-1}{k} \{H_{n-1-k}^0, W_{k+1}\} + \{H_0^0, W_n\}.$$

We write the formula (3.55) detaching the term $\{H_0^0, W_n\}$ because it is the unique term in the right-hand side of the equality that we cannot guarantee to lie in $\mathcal{P}_n$; as to the others, this is guaranteed by the statement (3.59) for $n - 1$. So, we have

$$H_{n-1}^1 = E_{n-1}^1 + \{H_0^0, W_n\},$$

where $E_{n-1}^1 \in \mathcal{P}_n$ is known. An induction on the columns of the Lie triangle using (3.55) shows that

$$H_{n-s}^s = E_{n-s}^s + \{H_0^0, W_n\}.$$

where $E^s_{n-s} \in \mathcal{P}_n$, for $s = 1, 2, \ldots, n$. Consequently, we have

$$H_0^n = E_0^n + \{H_0^0, W_n\}. \tag{3.60}$$

By the hypothesis (3.58) we can solve this equation for $W_n \in \mathcal{R}_n$ and $H_0^n \in \mathcal{Q}_n$. This proves that the statement (3.59) is valid for n and concludes the proof of the lemma. □

Remark 3.12 What makes the new Hamiltonian $H^*(\epsilon, \mathbf{y})$ simpler than the original $H_*(\epsilon, \mathbf{x})$ is the possibility of choosing the subspaces $\mathcal{Q}_j$ with suitable restrictions.

Remark 3.13 As the Deprit process is purely formal to get a convergent transformation and a convergent Hamiltonian we have to stop it at a finite order N. Starting with an analytic Hamiltonian $H^*(\epsilon, \mathbf{y})$ we get an analytic canonical transformation generated by an analytic function W_N and a transformed analytic Hamiltonian $H_N^*(\epsilon, \mathbf{y})$ with an analytic remainder $R_N(\epsilon, \mathbf{y})$.

Remark 3.14 Let us make the following consideration regarding the computational process of the explicit construction of the polynomials in H^* and W. In Eq. (3.60) the term E_0^n is equal to H_n^0 plus a sum of terms that do not involve W_n, so it is the same as if we computed it from the Eqs. (3.55) by setting $W_n = 0$. Let it then be denoted by $G_0^n = H_n^0 + \tilde{H}_n^0$, where $\tilde{H}_n^0$ is already known by recurrence. Then, to get H_0^n and W_n we just have to solve the homological equation

$$H_0^n = H_n^0 + \tilde{H}_n^0 + \{H_0^0, W_n\}. \tag{3.61}$$

3.8 Proof of the Existence of Normal Forms

In this section we will prove the structural theorems of Sect. 3.2.

3.8.1 Proof of Theorem 3.1

We begin with some preliminary considerations for the proof of the first Theorem.

Suppose that the real matrix A is diagonalizable and let $\mathbf{s}_1, \ldots, \mathbf{s}_{2n}$ be a basis of $\mathbf{C}^{2n}$ formed by eigenvectors of A^T corresponding to the eigenvalues $\lambda_1, \ldots, \lambda_{2n}$, that is, $A^T\mathbf{s}_j = \lambda_j \mathbf{s}_j$, $j = 1, \ldots, 2n$. As A^T is real the complex eigenvalues come in pairs $\lambda, \overline{\lambda}$ with the same multiplicity and corresponding eigenvectors $\mathbf{s}$ and $\overline{s}$. If $2r$ is the total number of complex eigenvalues arrange them so that the complex pairs are λ_j and $\lambda_{n+j} = \overline{\lambda}_j$, for $j = 1, \ldots, r$.

Consider the inner product in $\mathbf{C}^{2n}$, $(\mathbf{z}, \mathbf{w}) = z_1\overline{w}_1 + \ldots + z_{2n}\overline{w}_{2n}$. We have the following fact.

Lemma 3.15 *Every homogeneous polynomial with real coefficients can be written in the form*

$$P(\mathbf{x}) = \sum_{m_1,\ldots,m_{2n}} p_{m_1,\ldots,m_{2n}} (\mathbf{x}, \mathbf{s}_1)^{m_1} (\mathbf{x}, \mathbf{s}_2)^{m_2} \ldots, (\mathbf{x}, \mathbf{s}_{2n})^{m_{2n}},$$

with $p_{m_1,\ldots,m_{2n}}$ *real numbers.*

Proof Every vector $\mathbf{x} \in \mathbf{R}^{2n}$ is a linear combination of the basis vectors $\mathbf{s}_1, \ldots, \mathbf{s}_{2n}$;

$$\mathbf{x} = \xi_1 \mathbf{s}_1 + \ldots + \xi_{2n} \mathbf{s}_{2n}$$

hence, each coordinate x_i is a linear combination of $\xi_1, \ldots, \xi_{2n}$.

Now, taking the inner product of the above linear combination with the vector $\mathbf{s}_i$, we obtain the system of linear equations

$$(\mathbf{s}_1, \mathbf{s}_j)\xi_1 + \ldots + (\mathbf{s}_{2n}, \mathbf{s}_j)\xi_{2n} = (\mathbf{x}, \mathbf{s}_j), \qquad (j = 1, \ldots, 2n)$$

and since the matrix with entries $(\mathbf{s}_i, \mathbf{s}_j)$ is invertible, each ξ_i is a linear combination of the variables $(\mathbf{x}, \mathbf{s}_1), \ldots, (\mathbf{x}, \mathbf{s}_{2n})$, the same, consequently occurring with each x_i. Therefore, every homogeneous polynomial in the variables $x_1, \ldots, x_{2n}$ is also a homogeneous polynomial with the same degree in the variables $(\mathbf{x}, \mathbf{s}_1), \ldots, (\mathbf{x}, \mathbf{s}_{2n})$.

Since $\mathbf{x}$ is a real vector we have for $i = 1, \ldots, r$,

$$(\mathbf{x}, \mathbf{s}_i)^{m_i} (\mathbf{x}, \mathbf{s}_{n+i})^{m_{n+i}} = (\mathbf{x}, \mathbf{s}_i)^{m_i} \overline{(\mathbf{x}, \mathbf{s}_i)}^{m_i},$$

so $(\mathbf{x}, \mathbf{s}_1)^{m_1} (\mathbf{x}, \mathbf{s}_2)^{m_2} \ldots, (\mathbf{x}, \mathbf{s}_{2n})^{m_{2n}}$ is real and therefore the coefficients $p_{m_1,\ldots,m_{2n}}$ are real numbers. □

This lemma shows that the set of monomials

$$\mathcal{B} = \left\{ (\mathbf{x}, \mathbf{s}_1)^{m_1} (\mathbf{x}, \mathbf{s}_2)^{m_2} \cdots (\mathbf{x}, \mathbf{s}_{2n})^{m_{2n}};\ m_1 + m_2 + \ldots, m_{2n} = j + 2 \right\} \tag{3.62}$$

forms a basis for the space $\mathcal{P}_j$ of homogeneous real polynomials of degree $j + 2$.

We will use Lemma 3.11 in the proof of Theorem 3.1. This requires us to obtain from the Hamiltonian (3.5) another one involving a small parameter ϵ. We attain this goal by means of a scale change, taking $\mathbf{z} = \epsilon\mathbf{x}$. This is a symplectic change of coordinates with multiplier ϵ^2 so we obtain the new Hamiltonian $H(\epsilon, \mathbf{x}) = \frac{1}{\epsilon^2} H(\epsilon\mathbf{x})$ which has the following expansion in ϵ

$$H(\epsilon, \mathbf{x}) = H_0(\mathbf{x}) + \epsilon H_1(\mathbf{x}) + \epsilon^2 H_2(\mathbf{x}) + \ldots.$$

Taking $H_i^0 = i! H_i$ we are with the notation (3.51) used in the method of Deprit.

Proof of Theorem 3.1 Let $A = JS$, where $H_0(\mathbf{x}) = \frac{1}{2}\mathbf{x}^T S\mathbf{x}$.

Let $\mathcal{P}_j$ be the space of homogeneous polynomials of degree $j+2$ with real coefficients and consider the linear operator $\mathcal{F}(A) : \mathcal{P}_j \to \mathcal{P}_j$ defined by $\mathcal{F}(A)P = \{P, H_0\}$.

For the monomial $G = (\mathbf{x}, \mathbf{s}_1)^{m_1}(\mathbf{x}, \mathbf{s}_2)^{m_2} \cdots (\mathbf{x}, \mathbf{s}_{2n})^{m_{2n}} \in \mathcal{B}$, we have

$$(\nabla G(\mathbf{x}), \mathbf{y}) = DG(\mathbf{x}) \cdot \mathbf{y} = \Big[m_1 \frac{(\mathbf{y}, \mathbf{s}_1)}{(\mathbf{x}, \mathbf{s}_1)} + \ldots + m_{2n} \frac{(\mathbf{y}, \mathbf{s}_{2n})}{(\mathbf{x}, \mathbf{s}_{2n})}\Big] G.$$

Now $\mathcal{F}(A)P = \nabla P(\mathbf{x})^T J \nabla H_0(\mathbf{x}) = (\nabla P(\mathbf{x}), JS\mathbf{x}) = (\nabla P(\mathbf{x}), A\mathbf{x})$. Since $(A\mathbf{x}, \mathbf{s}_i) = (\mathbf{x}, A^T \mathbf{s}_i) = \lambda(\mathbf{x}, \mathbf{s}_i)$, taking $\mathbf{y} = A\mathbf{x}$ in the above expression of $(\nabla G(\mathbf{x}), \mathbf{y})$ we obtain

$$\mathcal{F}(A)G = \langle \mathbf{m}, \boldsymbol{\lambda} \rangle G,$$

where $\langle \mathbf{m}, \boldsymbol{\lambda} \rangle = m_1\lambda_1 + m_2\lambda_2 + \ldots + m_{2n}\lambda_{2n}$. Therefore, the vectors of the basis $\mathcal{B}$ are eigenvectors of the operator $\mathcal{F}(A)$ with eigenvalues $\langle \mathbf{m}, \boldsymbol{\lambda} \rangle$. It follows that the elements G of $\mathcal{B}$ with $\langle \mathbf{m}, \boldsymbol{\lambda} \rangle = 0$ generate the kernel of $\mathcal{F}(A)$ and since for $\langle \mathbf{m}, \boldsymbol{\lambda} \rangle \neq 0$, we have $\mathcal{F}(A)\Big(\frac{1}{\langle \mathbf{m}, \boldsymbol{\lambda} \rangle} G\Big) = G$, it follows that the elements of G with $\langle \mathbf{m}, \boldsymbol{\lambda} \rangle \neq 0$ generate the image of $\mathcal{F}(A)$. So we have the direct sum decomposition

$$\mathcal{P}_j = \mathcal{R}_j \oplus \mathcal{Q}_j, \tag{3.63}$$

where $\mathcal{R}_j = \mathcal{IF}(A)$ and $\mathcal{Q}_j = \mathcal{NF}(A)$ denote the image and the kernel of $\mathcal{F}(A)$. Therefore, given $D_j \in \mathcal{P}_j$, $j = 1, 2, \ldots$ there exist $B_j \in \mathcal{Q}_j$ and $C_j \in \mathcal{R}_j$ such that $B_j = D_j + \{H_0^0, C_j\}$. So, the subspaces $\mathcal{Q}_j$ and $\mathcal{R}_j$ of $\mathcal{P}_j$ satisfy the hypotheses of Lemma 3.11.

By this lemma there exists a formal series W as in (3.54) with $W_i \in \mathcal{R}_i$, $i = 1, 2, \ldots$ which generates a symplectic transformation $\mathbf{x} = X(\epsilon, \mathbf{y})$ through the solutions of the Hamiltonian system (3.53) and this transformation takes the original Hamiltonian (3.51) to the final Hamiltonian (3.52) having $H_0^i \in \mathcal{Q}_i$, $i = 1, 2 \ldots$.

As $\mathcal{Q}_i$ is the kernel of $\mathcal{F}(A)$ we have
$\{H_0^i(\mathbf{x}), H_0(\mathbf{x})\} = 0$. Since for each ϵ, $X(\epsilon, \mathbf{y})$ is a symplectic mapping, by Proposition 2.18 we have,

$$\{H_0^i(X(\epsilon, \mathbf{y})), H_0(X(\epsilon, \mathbf{y}))\} = \{H_0^i(\mathbf{y}), H_0(\mathbf{y})\}.$$

Consequently $\{H_0^i(\mathbf{y}), H_0(\mathbf{y})\} = 0$, so $H_0^i(\mathbf{y})$ is a first integral of the flow of $H_0(\mathbf{y})$.

The change of coordinates $\mathbf{y} = \dfrac{1}{\epsilon}\boldsymbol{\zeta}$ is symplectic with multiplier ϵ^{-2} so the Hamiltonian in the $\boldsymbol{\zeta}$ coordinates is

$$\mathcal{H}(\boldsymbol{\zeta}) = \epsilon^2 \sum_{i=0}^{\infty} \frac{\epsilon^i}{i!} \epsilon^{-(i+2)} H_0^i(\boldsymbol{\zeta}) = \sum_{i=0}^{\infty} \frac{1}{i!} H_0^i(\boldsymbol{\zeta}).$$

Therefore, the homogeneous polynomial $\mathcal{H}_j(\zeta)$ of the Hamiltonian $\mathcal{H}(\zeta)$ in (3.6) is given by $\mathcal{H}_j(\zeta) = \dfrac{1}{i!}H_0^i(\zeta)$; hence, it is a first integral of the quadratic polynomial $\mathcal{H}_0(\zeta)$. □

3.8.2 Proof of Theorem 3.2

In the proof of Theorem 3.1 it was important to decompose the space $\mathcal{P}_j$ as the direct sum of the kernel and image of the operator $\mathcal{F}(A)$. This is not possible if the matrix A is not diagonalizable. In the case of Theorem 3.2 the proof is more elaborate. It is based on the following Fredholm alternative lemma and the introduction of a special inner product in the space of linear operators on V.

Lemma 3.16 *Let V be a finite dimensional vector space with an inner product $(\ ,\ )$. Let A^* be the adjoint operator of the linear operator $\mathcal{T} : V \to V$, that is, $\langle \mathcal{T}\mathbf{x},\ \mathbf{y}\rangle = \langle\ \mathbf{x},\ \mathcal{T}^*\mathbf{y}\rangle$. Let $\mathcal{R}(\mathcal{T})$ be the image of $\mathcal{T}$ and $\mathcal{Q}(\mathcal{T}^*)$ the kernel of $\mathcal{T}^*$. Then:*

(1) $\mathcal{R}(\mathcal{T})$ is orthogonal to $\mathcal{Q}(\mathcal{T}^*)$; *(2) $V = \mathcal{R}(\mathcal{T}) \oplus \mathcal{Q}(\mathcal{T}^*)$.*

Proof Let $\mathbf{x} \in \mathcal{R}(A)$ and $\mathbf{y} \in \mathcal{Q}(A^*)$. Then, $\mathbf{x} = \mathcal{T}\mathbf{u}$ and $\mathcal{T}^*\mathbf{y} = 0$; hence,

$$\big(\mathbf{x}, \mathbf{y}\big) = \big(\mathcal{T}\mathbf{u},\ \mathbf{y}\big) = \big(\mathbf{u},\ \mathcal{T}^*\mathbf{y}\big) = 0,$$

which proves item (1).

Now, $\mathcal{R}(\mathcal{T}) + \mathcal{Q}(\mathcal{T}^*) \subset V$ and by (1), this is a direct sum. As V has finite dimension, $\dim\mathcal{R}(\mathcal{T}) + \dim\mathcal{Q}(\mathcal{T}) = \dim V$ and $\dim\mathcal{Q}(\mathcal{T}) = \dim\mathcal{Q}(\mathcal{T}^*)$. It follows that the above inclusion is an equality, which proves (2). □

To use this theorem in our context we will introduce an inner product in the space $\mathcal{P}$ of homogeneous real polynomials of a given degree s. Given a polynomial in $\mathcal{P}$,

$$P(\mathbf{x}) = \sum_{m_1+m_2+\ldots+m_k=s} p_{m_1m_2\cdots m_k} x_1^{m_1}x_2^{m_2}\cdots x_k^{m_k},$$

we consider the differential operator $P(\partial)$ defined by

$$P(\partial)Q = \sum_{m_1+m_2+\ldots+m_k=s} p_{m_1m_2\cdots m_k}\frac{\partial^{m_1+m_2+\ldots+m_k}Q}{\partial x_1^{m_1}\partial x_2^{m_2}\cdots\partial x_k^{m_k}}.$$

This defines an inner product $\big(P,\ Q\big) = P(\partial)Q$ in $\mathcal{P}$, as one immediately sees taking a polynomial

$$Q(\mathbf{x}) = \sum_{m_1+m_2+\ldots+m_k=s} q_{m_1m_2\cdots m_k} x_1^{m_1}x_2^{m_2}\cdots x_k^{m_k},$$

and observing that

$$\big(P,\ Q\big) = \sum_{m_1+m_2+\ldots+m_k=s} m_1!m_2!\cdots m_k!p_{m_1m_2\cdots m_k}q_{m_1m_2\cdots m_k}. \tag{3.64}$$

It is clear from this expression that $\big(P,\ Q\big)$ is bilinear, symmetric, and positive definite, so it defines an inner product in $\mathcal{P}$.

Let A be a real matrix of order k and consider the linear operator $\mathcal{F}(A) : \mathcal{P} \to \mathcal{P}$ defined by

$$\big(\mathcal{F}(A)P\big)(\mathbf{x}) = \nabla P(\mathbf{x})^T A\mathbf{x} = \frac{d}{dt}P(\mathrm{e}^{tA}\mathbf{x})\Big|_{t=0}. \tag{3.65}$$

Proposition 3.17 *For every matrix A we have:*

(1) $\big(P(\mathbf{x}),\ Q(A\mathbf{x})\big) = \big(P(A^T\mathbf{x}),\ Q(\mathbf{x})\big)$
(2) $\big(P,\ \mathcal{F}(A)Q\big) = \big(\mathcal{F}(A^T)P,\ Q\big)$.

Item (2) says that the adjoint operator of $\mathcal{F}(A)$ *is* $\mathcal{F}(A^T)$.

Proof Let $\mathbf{y} = A\mathbf{x}$. Then, $\dfrac{\partial y_i}{\partial x_j} = A_{ij}$; hence, $\dfrac{\partial}{\partial \mathbf{x}} = A^T\dfrac{\partial}{\partial \mathbf{y}}$ since for every function F, we have

$$\frac{\partial F}{\partial x_j} = \sum_i \frac{\partial F}{\partial y_i}\frac{\partial y_i}{\partial x_j} = \sum_i \big(A^T\big)_{ji}\frac{\partial F}{\partial y_i}.$$

Therefore, with the notation $\partial_{\mathbf{x}} = \dfrac{\partial}{\partial \mathbf{x}}$, $\partial_{\mathbf{y}} = \dfrac{\partial}{\partial \mathbf{y}}$, we have by the definition $\big(P,\ Q\big) = P(\delta)Q$,

$$\big(P(\mathbf{x}),\ Q(A\mathbf{x})\big) = P(\partial_{\mathbf{x}})Q(A\mathbf{x}) = P\big(A^T\partial_{\mathbf{y}}\big)Q(\mathbf{y}) = \big(P(A^T\mathbf{y}),\ Q(\mathbf{y})\big);$$

the inner product in the right-hand side is the same whether we use the variable $\mathbf{x}$ or a variable $\mathbf{y}$ (see expression (3.64)). This proves (1).

Since item (1) is valid for any $k \times k$ matrix A, we have for e^{tA},

$$\big(P(\mathbf{x}),\ Q(\mathrm{e}^{tA}\mathbf{x})\big) = \big(P(\mathrm{e}^{tA^T}\mathbf{x}),\ Q(\mathbf{x})\big).$$

Differentiating this with respect to t and setting $t = 0$, we obtain by (3.65) the equality of item (2). □

We now have the tools needed for the proof of the second theorem.

Proof of Theorem 3.2 Let $\mathcal{P}_j$ be the space of real homogeneous polynomials of degree $j + 2$. Consider the linear operator $\mathcal{F}(A)(P) = \{P,\ H_0\} = \nabla P(\mathbf{x})^T A\mathbf{x}$, where $H_0(\mathbf{x})$ is the quadratic polynomial after the scale change $\mathbf{z} = \epsilon\mathbf{x}$ in the original Hamiltonian (3.5) and $A = J\nabla H_0$.

By Proposition 3.17 we have $\mathcal{F}(A)^* = \mathcal{F}(A)^T = \mathcal{F}(A^T)$. So by Lemma 3.16 with $\mathcal{T} = \mathcal{F}(A)$, for each n we have $\mathcal{P}_n = \mathcal{R}(\mathcal{F}(A)) \oplus \mathcal{Q}(\mathcal{F}(A^T))$. Therefore, given $D \in \mathcal{P}_n$ we can find $Q \in \mathcal{Q}(\mathcal{F}(A^T))$ and $R \in \mathcal{R}(\mathcal{F}(A))$ such that $Q = D + \{H_0^0, R\}$.

The hypotheses of Lemma 3.11 being satisfied, for each $E^n \in \mathcal{P}_n$ and we can solve the equation

$$H_0^n = E^n + \{H_0^0, W_n\},$$

with $H_0^n \in \mathcal{Q}(\mathcal{F}(A^T))$ and $W_n \in \mathcal{Q}(\mathcal{F}(A^T))$.

So by this lemma there exists a formal series W as in (3.54) with $W_i \in \mathcal{R}_i$, $i = 1, 2, \ldots$ which generates a symplectic transformation $\mathbf{x} = X(\epsilon, \mathbf{y})$ through the solutions of the Hamiltonian system (3.53) and this transformation takes the original Hamiltonian (3.51) to the final Hamiltonian (3.52) having $H_0^i \in \mathcal{Q}_i$, $i = 1, 2 \ldots$.

As $\mathcal{Q}_i$ is the kernel of $\mathcal{F}(A^T)$ we have $\{H_0^i(\mathbf{x}), H_0^T(\mathbf{x})\} = 0$, where $H_0^T(\mathbf{x})$ is the quadratic Hamiltonian defined by the Hamiltonian matrix A^T, that is, $H_0^T(\mathbf{x}) = \frac{1}{2}\mathbf{x}^T S_1 \mathbf{x}$ where $S_1 = -JA^T$.

Therefore, each $H_0^i(\mathbf{x})$ is a first integral of the adjoint linear system $\dot{\mathbf{x}} = A^T\mathbf{x}$. Since for each ϵ, $X(\epsilon, \mathbf{y})$ is a symplectic mapping, by Proposition 2.18 we have,

$$\{H_0^i(X(\epsilon, \mathbf{y})), H_0^T(X(\epsilon, \mathbf{y}))\} = \{H_0^i(\mathbf{y}), H_0^T(\mathbf{y})\}.$$

Consequently $\{H_0^i(\mathbf{y}), H_0^T(\mathbf{y})\} = 0$, so $H_0^i(\mathbf{y})$ is a first integral of the flow of $H_0^T(\mathbf{y})$.

The change of coordinates $\mathbf{y} = \dfrac{1}{\epsilon}\boldsymbol{\zeta}$ is symplectic with multiplier ϵ^{-2} so the Hamiltonian in the $\boldsymbol{\zeta}$ coordinates is

$$\mathcal{H}(\boldsymbol{\zeta}) = \epsilon^2 \sum_{i=0}^{\infty} \frac{\epsilon^i}{i!}\epsilon^{-(i+2)} H_0^i(\boldsymbol{\zeta}) = \sum_{i=0}^{\infty} \frac{1}{i!} H_0^i(\boldsymbol{\zeta}).$$

Therefore, the homogeneous polynomial $\mathcal{H}_j(\boldsymbol{\zeta})$ of the Hamiltonian $\mathcal{H}(\boldsymbol{\zeta})$ in (3.6) is given by $\mathcal{H}_j(\boldsymbol{\zeta}) = \dfrac{1}{i!}H_0^i(\boldsymbol{\zeta})$; hence, it is a first integral of the quadratic polynomial $\mathcal{H}_0^T(\boldsymbol{\zeta})$. □

We have proved the existence of the normal form of a Hamiltonian function under the assumption that it has a nonzero quadratic part. We have done that in both cases when the linearized system has a diagonalizable or nondiagonalizable matrix. Now the process is formal and nothing is guaranteed about its convergence. In this regard, we make the following remark.

Remark 3.18 If $W(\mathbf{x}, \epsilon)$ is analytic the flow $\mathbf{x} = X(\epsilon, \mathbf{y})$ of (3.53) is analytic, so from an analytic function $H_*(\mathbf{x}, \epsilon)$ we get an analytic function $H_*(\mathbf{y}, \epsilon) = H_*(X(\epsilon, \mathbf{y}), \epsilon)$. The method gives a formal series (3.54) and we cannot guarantee that the sum of this series is the original $W(\mathbf{x}, \epsilon)$. In fact, in the case of the Birkhoff normal form, which can also be obtained by the Deprit method, Siegel [54] proved that the divergence of the normalizing transformation is the rule, that is, with a certain topology in the space of coefficients of the series the divergent normalizing transformations form a dense set. This does not mean that the series of the normalized Hamiltonian is divergent; concerning this point, it is worth looking at the reference [49].

3.9 Time-Dependent Hamiltonian Systems

Let us consider a Hamiltonian system that depends on the time t,

$$\dot{\mathbf{x}} = J\nabla_{\mathbf{x}} H(\mathbf{x}, t) \tag{3.66}$$

and let us make a time-dependent symplectic change of variables, $\mathbf{x} = X(\mathbf{y}, t)$. Let $\mathbf{y} = Y(\mathbf{x}, t)$ be its inverse and let $\hat{H}(\mathbf{y}, t) = H(X(\mathbf{y}, t), t)$. Then,

$$\frac{\partial Y}{\partial \mathbf{x}} J \Big(\frac{\partial Y}{\partial \mathbf{x}}\Big)^T = J \quad \text{and} \quad \nabla_{\mathbf{x}} H(\mathbf{x}, t) = \Big(\frac{\partial Y}{\partial \mathbf{x}}(\mathbf{x}, t)\Big)^T \nabla_{\mathbf{y}} \hat{H}(\mathbf{y}, t).$$

We then have

$$\begin{aligned} \dot{\mathbf{y}} &= \frac{\partial Y}{\partial \mathbf{x}} \dot{\mathbf{x}} + \frac{\partial Y}{\partial t} = \frac{\partial Y}{\partial \mathbf{x}} J\nabla_{\mathbf{x}} H(\mathbf{x}, t) + \frac{\partial Y}{\partial t} \\ &= \frac{\partial Y}{\partial \mathbf{x}} J \Big(\frac{\partial Y}{\partial \mathbf{x}}(\mathbf{x}, t)\Big)^T \nabla_{\mathbf{y}} \hat{H}(\mathbf{y}, t) + \frac{\partial Y}{\partial t}; \end{aligned}$$

hence,

$$\dot{\mathbf{y}} = J\nabla_{\mathbf{y}} \hat{H}(\mathbf{y}, t) + \frac{\partial Y}{\partial t}(\mathbf{x}, t)\Big|_{\mathbf{x}=Y(\mathbf{y},t)}. \tag{3.67}$$

If the transformation $\mathbf{x} = X(\mathbf{y})$ does not depend on time, then $\frac{\partial Y}{\partial t} = 0$ and the obtained system is Hamiltonian with the new Hamiltonian obtained from the first simply by the substitution of variables, $\mathbf{x} = X(\mathbf{y})$. In the time-dependent case, we have to add a remainder term as in (2.96), which is given by the time derivative of the generating function of the transformation (see Proposition 2.12).

In the present context it is important to get an integral expression for the remainder term coming from (3.67), which we will do in Theorem 3.20 below. First, we have to prove that there is a function $R(\mathbf{y}, t)$ such that for each t we have

$$\frac{\partial Y}{\partial t}(\mathbf{x}, t)\Big|_{\mathbf{x}=Y(\mathbf{y},t)} = J\nabla_{\mathbf{y}} R(\mathbf{y}, t).$$

This can be done if the domain $\mathcal{O}$ of the variables $(\mathbf{x}, t)$ is such that for each t, the subset

$$\mathcal{O}_t = \{\mathbf{y} \in \mathbf{R}^{2n},\ (\mathbf{y}, t) \in \mathcal{O}\}$$

is an open ball of $\mathbf{R}^{2n}$, because then we can apply Poincaré's lemma to guarantee the existence of this function. In fact, we use only the special case of this lemma, which guarantees that in a simply connected domain a vector field is a gradient field if, and only if, its Jacobian matrix is symmetric.

Therefore, in such special domains to prove the existence of the function $R(\mathbf{y}, t)$ it suffices to show that for each t, the Jacobian matrix of the vector field $J\frac{\partial Y}{\partial t}(\mathbf{x}, t)\Big|_{\mathbf{x}=Y(\mathbf{y},t)}$ is symmetric, that is, the matrix

$$\Gamma = J\frac{\partial^2 Y}{\partial \mathbf{x}\partial t}\Big|_{\mathbf{x}=X(\mathbf{y},t)}\frac{\partial X}{\partial \mathbf{y}}(\mathbf{y}, t)$$

is symmetric.

Differentiating the identity $\left(\frac{\partial Y}{\partial \mathbf{x}}(\mathbf{x}, t)\right)^T J\frac{\partial Y}{\partial \mathbf{x}}(\mathbf{x}, t) = J$ with respect to t, we have

$$\left(\frac{\partial^2 Y}{\partial \mathbf{x}\partial t}(\mathbf{x}, t)\right)^T J\left(\frac{\partial Y}{\partial \mathbf{x}}(\mathbf{x}, t)\right) + \left(\frac{\partial Y}{\partial \mathbf{x}}(\mathbf{x}, t)\right)^T J\left(\frac{\partial^2 Y}{\partial \mathbf{x}\partial t}(\mathbf{x}, t)\right) = 0$$

and since $\left(\frac{\partial Y}{\partial \mathbf{x}}(\mathbf{x}, t)\right)^{-1} = \frac{\partial X}{\partial \mathbf{y}}(\mathbf{y}, t)$ we obtain from the last equation the following one

$$\left(\frac{\partial X}{\partial \mathbf{y}}(\mathbf{y}, t)\right)^T\left(\frac{\partial^2 Y}{\partial \mathbf{x}\partial t}(\mathbf{x}, t)\right)^T J + J\left(\frac{\partial^2 Y}{\partial \mathbf{x}\partial t}(\mathbf{x}, t)\right)\left(\frac{\partial X}{\partial \mathbf{y}}(\mathbf{y}, t)\right) = 0.$$

Substituting $\mathbf{x} = X(\mathbf{y}, t)$ and taking into account that $J^T = -J$, we see that $-\Gamma^T + \Gamma = 0$.

We have, therefore, the following theorem.

Theorem 3.19 *The symplectic change of variables* $\mathbf{x} = X(\mathbf{y}, t)$ *in such a special domain* $\mathcal{O}$ *takes the Hamiltonian system* $\dot{\mathbf{x}} = J\nabla_{\mathbf{x}}H(\mathbf{x}, t)$ *to the Hamiltonian system* $\dot{\mathbf{y}} = J\nabla_{\mathbf{y}}\mathcal{H}\mathbf{y}, t)$, *where the new Hamiltonian is given by*

$$\mathcal{H}(\mathbf{y}, t) = H(X(\mathbf{y}, t), t) + R(\mathbf{y}, t).$$

$R(\mathbf{y}, t)$ *is called the* remainder function.

For an arbitrary open set $\mathcal{O}$ this theorem holds locally since any point $(\mathbf{x}, \tau) \in \mathcal{O}$ has a cylindrical neighborhood $B_r(\mathbf{x}) \times (\tau - \delta, \tau + \delta) \subset \mathcal{O}$.

3.10 The Lie Process in the Case of Time Dependency

Suppose now that the series (3.51) depends also on the time t,

$$H_*(\epsilon, \mathbf{x}, t) = \sum_{i=0}^{\infty} \frac{\epsilon^i}{i!} H_i^0(\mathbf{x}, t), \tag{3.68}$$

where $H_i^0(\mathbf{x}, t)$ is a homogeneous polynomial of degree $i+2$ in $\mathbf{x}$. Then we take the function $W(\epsilon, \mathbf{x}, t)$ also depending on t and write it as the formal series

$$W(\epsilon, \mathbf{x}, t) = \sum_{i=0}^{\infty} \frac{\epsilon^i}{i!} W_{i+1}(\mathbf{x}, t), \tag{3.69}$$

where $W_i(\mathbf{x}, t)$ is a homogeneous polynomial of degree $i+2$ in $\mathbf{x}$. Now we write the transformed Hamiltonian in the form

$$H^*(\epsilon, \mathbf{y}, t) = \sum_{i=0}^{\infty} \frac{\epsilon^i}{i!} H_0^i(\mathbf{y}, t), \tag{3.70}$$

where $H_0^i(\mathbf{y}, t)$ is a homogeneous polynomial of degree $i+2$ in $\mathbf{y}$.

For each fixed t the polynomials $H_0^i(\mathbf{y}, t)$ are obtained from $H_i^0(\mathbf{x}, t)$ through the auxiliary polynomials H_j^i in formula (3.55) of Theorem 3.10. The question now is how do we obtain the homogeneous polynomials that comprise the remainder function, $R(\epsilon, \mathbf{y}, t)$, in the new Hamiltonian

$$\mathcal{H}^*(\epsilon, \mathbf{y}, t) = H_*(\epsilon, X(\epsilon, \mathbf{y}, t), t) + R(\epsilon, \mathbf{y}, t). \tag{3.71}$$

Let us denote by $\mathcal{L}_W H_* = H^*$ the transformation that to each H_* assigns H^* by the process described in Theorem 3.10, so $\mathcal{L}_W H_*(\epsilon, \mathbf{y}) = H_*(\epsilon, X(\epsilon, \mathbf{y}))$.

For any function $f(\epsilon, \mathbf{x}) = \sum_{i=0}^{\infty} \frac{\epsilon^i}{i!} f_i^0(\mathbf{x})$ the series $\mathcal{L}_W f$ given by the construction

$$\mathcal{L}_W f(\epsilon, \mathbf{y}) = f(\epsilon, X(\epsilon, \mathbf{y})),$$

is called the *Lie transform* of f generated by W.

Considering t as a parameter, the remainder function $R(\epsilon, \mathbf{y}, t)$ is obtained by means of the Lie transform of $\frac{\partial W}{\partial t}$. Precisely, we have the following theorem:

Theorem 3.20 *The remainder function is given by*

$$R(\epsilon, \mathbf{y}, t) = -\int_0^{\epsilon} \mathcal{L}_W\Big(\frac{\partial W}{\partial t}\Big)(s, \mathbf{y}, t)\, ds.$$

Proof Let $\mathbf{x} = X(\epsilon, \mathbf{y}, t)$ be the symplectic transformation defined by the Cauchy problem

$$\dot{\mathbf{x}} = J\frac{\partial W}{\partial \epsilon}(\epsilon, \mathbf{x}, t), \quad \mathbf{x}(0) = \mathbf{y}. \tag{3.72}$$

Since $\left(\frac{\partial X}{\partial \mathbf{y}}\right)^{-1} = \frac{\partial Y}{\partial \mathbf{x}}$, we have $\nabla_{\mathbf{x}} H_*(\epsilon, \mathbf{x}, t) = \left(\frac{\partial Y}{\partial \mathbf{x}}\right)^T \nabla_{\mathbf{y}} H^*(\epsilon, \mathbf{y}, t)$, so $\frac{\partial X}{\partial \mathbf{y}} \cdot \dot{y} + \frac{\partial X}{\partial t} = \dot{\mathbf{x}} = J\nabla_{\mathbf{x}} H_*(\epsilon, \mathbf{x}, t)$, and since $\frac{\partial Y}{\partial \mathbf{x}}$ is symplectic we get

$$\dot{\mathbf{y}} = J\nabla_{\mathbf{y}} H^*(\epsilon, \mathbf{y}, t) - \frac{\partial Y}{\partial \mathbf{x}}\frac{\partial X}{\partial t}.$$

Therefore, we have the following expression for $\nabla_{\mathbf{y}} R(\epsilon, \mathbf{y}, t)$,

$$J\nabla_{\mathbf{y}} R(\epsilon, \mathbf{y}, t) = -\frac{\partial Y}{\partial \mathbf{x}}(\epsilon, X(\epsilon, \mathbf{y}, t), t)\frac{\partial X}{\partial t}(\epsilon, \mathbf{y}, t). \tag{3.73}$$

Substituting $\mathbf{x} = X(\epsilon, \mathbf{y}, t)$ into (3.72), then differentiating with respect to $\mathbf{x}$ and interchanging the order of differentiation, we see that

$$A(\epsilon) = \frac{\partial X}{\partial \mathbf{y}}(\epsilon, \mathbf{y}, t)$$

is a solution of the equation of variation

$$\frac{dA}{d\epsilon} = \left(J\frac{\partial^2 W}{\partial \mathbf{x}^2}(\epsilon, X(\epsilon, \mathbf{y}, t), t)\right) \cdot A(\epsilon), \quad A(0) = I. \tag{3.74}$$

On the other hand, differentiating (3.72) with respect to t we get for $B(\epsilon) = \dfrac{\partial X}{\partial t}(\epsilon, \mathbf{y}, t)$ the linear equation

$$\frac{dB}{d\epsilon} = \left(J\frac{\partial^2 W}{\partial \mathbf{x}^2}(\epsilon, X(\epsilon, \mathbf{y}, t), t)\right) \cdot B(\epsilon) + J\frac{\partial^2 W}{\partial t \partial \mathbf{x}}(\epsilon, X(\epsilon, \mathbf{y}, t), t).$$

Since by (3.74) $A(\epsilon)$ is the fundamental matrix of the homogeneous linear system and $A(0) = I$, it follows by the variation of parameter formula that

$$B(\epsilon) = \int_0^\epsilon A(\epsilon)A(s)^{-1}J\frac{\partial^2 W}{\partial t \partial \mathbf{x}}(s, X(s, \mathbf{y}, t), t)ds.$$

By the definitions of $A(\epsilon)$ and $B(\epsilon)$ it follows from (3.73) that

$$J\nabla_{\mathbf{y}} R(\epsilon, \mathbf{y}, t) = A(\epsilon)^{-1}B(\epsilon),$$

so that

$$\begin{aligned} J\nabla_{\mathbf{y}}R(\epsilon, \mathbf{y}, t) &= \int_0^{\epsilon} A(s)^{-1} J \frac{\partial^2 W}{\partial t \partial \mathbf{x}}(s, X(s, \mathbf{y}, t), t) ds \\ &= \int_0^{\epsilon} J A(s)^T \frac{\partial^2 W}{\partial t \partial \mathbf{x}}(s, X(s, \mathbf{y}, t), t) ds. \end{aligned}$$

In the last equality we have used the fact that $A(s)$ is symplectic. Now the equality

$$\frac{\partial}{\partial \mathbf{y}}\Big(\frac{\partial W}{\partial t}\Big) \cdot \eta = \frac{\partial}{\partial \mathbf{x}}\Big(\frac{\partial W}{\partial t}\Big) \cdot \Big(\frac{\partial X}{\partial \mathbf{y}} \cdot \eta\Big)$$

shows that

$$\frac{\partial}{\partial \mathbf{y}}\Big(\frac{\partial W}{\partial t}\Big) = \Big(\frac{\partial X}{\partial \mathbf{y}}\Big)^T \cdot \frac{\partial}{\partial \mathbf{x}}\Big(\frac{\partial W}{\partial t}\Big).$$

Therefore, we have

$$A(s)^T \frac{\partial^2 W}{\partial t \partial \mathbf{x}}(s, X(s, \mathbf{y}, t), t) = \frac{\partial}{\partial \mathbf{y}}\Big(\frac{\partial W}{\partial t}\Big)(s, X(s, \mathbf{y}, t), t);$$

hence,

$$J\nabla_{\mathbf{y}}R(\epsilon, \mathbf{y}, t) = J\Big[-\frac{\partial}{\partial \mathbf{y}} \int_0^{\epsilon} \frac{\partial W}{\partial t}(s, X(s, \mathbf{y}, t), t) ds\Big].$$

Since

$$\mathcal{L}_W\Big(\frac{\partial W}{\partial t}\Big)(\epsilon, \mathbf{y}, t) = \frac{\partial W}{\partial t}(\epsilon, X(\epsilon, \mathbf{y}, t), t),$$

the last expression obtained for $J\nabla_{\mathbf{y}}R$ gives immediately the expression of $R(\epsilon, \mathbf{y}, t)$ stated in the theorem. □

Now that we have an expression of the remainder function we can obtain the new Hamiltonian (3.71).

The steps to finding the Hamiltonian $\mathcal{H}^*(\epsilon, \mathbf{y}, t)$ in (3.71) are the following:

(1) Viewing t as a parameter, we find the Lie transform $H^* = \mathcal{L}_W H_*$,

$$\mathcal{L}_W H_*(\epsilon, \mathbf{y}, t) = H_*(\epsilon, X(\epsilon, \mathbf{y}, t), t)$$

by the process described in Theorem 3.10, using the Lie H-triangle as a guide.

(2) Next, by the same process, we find the Lie transform, $S^* = \mathcal{L}_W S_*$ of the series

$$S_*(\epsilon, \mathbf{x}, t) = -\frac{\partial W}{\partial t}(\epsilon, \mathbf{x}, t) \tag{3.75}$$

using the Lie S-triangle as a guide.

(3) Finally, by integrating $S^*(\epsilon, \mathbf{y}, t)$, we obtain the series $R(\epsilon, \mathbf{y}, t)$, according to Theorem 3.20.

In the Lie S-triangle

$$\begin{array}{ccccccc}
S_0^0 & & & & & & \\
\uparrow & & & & & & \\
S_1^0 & \leftarrow & S_0^1 & & & & \\
\uparrow & & \uparrow & & & & \\
S_2^0 & \leftarrow & S_1^1 & \leftarrow & S_0^2 & & \\
\uparrow & & \uparrow & & \uparrow & & \\
S_3^0 & \leftarrow & S_2^1 & \leftarrow & S_1^2 & \leftarrow & S_0^3 \\
\vdots & & \vdots & & \vdots & & \vdots
\end{array}$$

as $S_*(\mathbf{x}, t, \epsilon) = \sum_{i=0}^{\infty} \frac{\epsilon^i}{i!} S_i^0(\mathbf{x}, t)$, we see from (3.75) that the first column is formed by $-\frac{\partial W_1}{\partial t}, -\frac{\partial W_2}{\partial t}, -\frac{\partial W_3}{\partial t}, \ldots$.

If $R(\mathbf{y}, t, \epsilon) = \sum_{i=0}^{\infty} \frac{\epsilon^i}{i!} R_0^i(\mathbf{y}, t)$ and $S^*(\mathbf{y}, t, \epsilon) = \sum_{i=0}^{\infty} \frac{\epsilon^i}{i!} S_0^i(\mathbf{y}, t)$, then by Theorem 3.20 we have

$$R_0^0 = 0 \quad \text{and} \quad R_0^i(\mathbf{y}, t) = S_0^{i-1}(\mathbf{y}, t), \qquad i = 1, 2, 3, \ldots. \tag{3.76}$$

Remark 3.21 In Remark 3.14 we have described, in the time-independent case, how to compute the polynomials for H^* and W by first setting $W_n = 0$.

We apply this observation to the time-dependent case, treating t as a parameter in

$$\mathcal{H}^*(\epsilon, \mathbf{y}, t) = H^*(\epsilon, \mathbf{y}, t) + R(\epsilon, \mathbf{y}, t).$$

For the polynomials at the level n in this equation, that is, for $\mathcal{H}_0^n = H_0^n + R_0^n$, using (3.76) we get, see (3.61),

$$H_0^n = H_n^0 + \tilde{H}_n^0 + \{H_0^0, W_n\} \quad \text{and} \quad R_0^n = S_0^{n-1} = S_{n-1}^0 + \tilde{S}_{n-1}^0 + \{S_0^0, W_{n-1}\}.$$

Since $S^0_{n-1} = -\frac{\partial W_n}{\partial t}$, we get the homological equation

$$\mathcal{H}^n_0 = E^n_0 + \{H^0_0, W_n\} - \frac{\partial W_n}{\partial t}, \tag{3.77}$$

where

$$E^n_0 = H^0_n + \tilde{H}^0_n + \tilde{S}^0_{n-1} + \{S^0_0, W_{n-1}\} \tag{3.78}$$

is known by recurrence.

For instance, to find the new Hamiltonian $\mathcal{H}^*(\epsilon, \mathbf{y}, t)$ to the order of ϵ^2 we need to find E^n_0 in (3.78) for $n = 0, 1, 2$.

The Deprit formula (3.55) for $n = 1, 2$ gives for H,

$$\begin{aligned} H^1_0 &= H^0_1 + \{H^0_0, W_1\}, \\ H^2_0 &= H^0_2 + 2\{H^0_0, W_1\} + \{\{H^1_0, W_1\}, W_1\} + \{H^0_0, W_2\}. \end{aligned}$$

For S and $n = 1$ it gives $S^1_0 = S^0_1 + \{S^0_0, W_1\}$. Comparing these formulas with (3.61) we see that

$$\tilde{H}^1_0 = 0, \; \tilde{H}^2_0 = 2\{H^0_0, W_1\} + \{\{H^1_0, W_1\}, W_1\} \quad \text{and} \quad \tilde{S}^1_0 = 0.$$

Since $S^0_0 = -\frac{\partial W_1}{\partial t}$, $S^0_1 = -\frac{\partial W_2}{\partial t}$, $R^2_0 = S^1_0$ and $W_0 = 0$, we get from (3.78)

$$E^1_0 = H^0_1 \quad \text{and} \quad E^2_0 = H^0_2 + 2\{H^0_0, W_1\} + \{\{H^1_0, W_1\}, W_1\}$$

and from (3.77) we have

$$\begin{aligned} \mathcal{H}^0_0 &= H^0_0, \\ \mathcal{H}^1_0 &= H^0_1 + \{H^0_0, W_1\} - \frac{\partial W_1}{\partial t}, \\ \mathcal{H}^2_0 &= H^0_2 + 2\{H^0_0, W_1\} + \{\{H^1_0, W_1\}, W_1\} + \{H^0_0, W_2\} - \frac{\partial W_2}{\partial t} \end{aligned} \tag{3.79}$$

for the homogeneous polynomials in the Lie transform

$$\mathcal{H} = \mathcal{H}^0_0 + \frac{\epsilon}{1!}\mathcal{H}^1_0 + \frac{\epsilon^2}{2!}\mathcal{H}_2 + O(\epsilon^3).$$

Example 3.4 (Lie Transform in the Absence of the Third-Degree Term) Let the third degree term H_1 be zero in the following Hamiltonian

$$H = H_0 + \frac{\epsilon}{1!}H_1 + \frac{\epsilon^2}{2!}H_2 + O(\epsilon^3).$$

Let us find the Lie transform $K = \mathcal{L}_W H$ to the order of ϵ^2.

Since $H_1^0 = H_1 = 0$ we can take $\mathcal{H}_0^1 = 0$ and $W_1 = 0$ as the solution of the second equation in (3.79). Then, from the third equation we get $K_2 = \mathcal{H}_0^2 = H_2^0 + \{H_0^0, W_2\} - \frac{\partial W_2}{\partial t}$, and so for the Lie transform $K = \mathcal{L}_W H$ we find

$$K = K_0 + \frac{\epsilon^2}{2!}K_2 + O(\epsilon^3), \quad \text{with} \quad K_0 = H_0 \text{ and } K_2 = H_2 + \{H_0^0, W_2\} - \frac{\partial W_2}{\partial t}. \tag{3.80}$$

3.11 The Normal Form in the Formulation of Kamel

In [39] and [40], Markeev always uses a recurrence formula proved by A. Kamel in [32], which allows the polynomials of the final Hamiltonian

$$K(\epsilon, \mathbf{y}, t) = \sum_{i=0}^{\infty} \frac{\epsilon^i}{i!} K_i,$$

to be obtained directly from the polynomials of the original Hamiltonian

$$H(\epsilon, \mathbf{x}, t) = \sum_{i=0}^{\infty} \frac{\epsilon^i}{i!} H_i,$$

without the necessity of computing the intermediate polynomials H_j^i. Let W be the generating function

$$W(\epsilon, \mathbf{x}, t) = \sum_{i=0}^{\infty} \frac{\epsilon^i}{i!} W_{i+1}.$$

Then, the mentioned formula is the one described in the following theorem.

Theorem 3.22 *Given the Hamiltonian H, the polynomials of K are computed by the formula*

$$K_m = H_m + \sum_{j=1}^{m-1} \left[C_{j-1}^{m-1} L_j H_{m-j} + C_j^{m-1} K_{j,m-j} \right] + L_m H_0 - \frac{\partial W_m}{\partial t}, \tag{3.81}$$

where C_j^m is the binomial coefficient $\binom{m}{j} = \frac{m!}{j!(m-j)!}$, L_m is the operator on the smooth functions about the origin given in terms of the Poisson bracket by

$$L_m g = \{g, W_m\}$$

and $K_{j,i} = G_j K_i$, $G_j$being the operator defined recursively by the formula

$$G_j = L_j - \sum_{m=1}^{j-1} C_{m-1}^{j-1} L_m G_{j-m} \qquad (1 \leq j \leq n). \tag{3.82}$$

Proof Given $f(\epsilon, \mathbf{x}, t) = \sum_{i=1}^{\infty} \frac{\epsilon^i}{i!} f_i^0(\mathbf{x}, t)$, let

$$f(\epsilon, X(\epsilon, \mathbf{y}, t), t) = \sum_{n=0}^{\infty} \frac{\epsilon^i}{i!} f_0^i(\mathbf{y}, t).$$

Then, seeing t as a parameter, the coefficients $f_i^0(\mathbf{x}, t)$ and $f_0^i(\mathbf{y}, t)$ are computed by the formula of Theorem 3.10 using the auxiliary polynomials f_j^i. We write that recurrence formula in the form

$$f_n^k = f_{n-1}^{k+1} - \sum_{m=0}^{n-1} C_{n-1}^m L_{m+1} f_{n-m-1}^k \qquad (n \geq 1,\ k \geq 0). \tag{3.83}$$

Fixing k and writing this equality for $n = 1, 2, \ldots,$ we obtain

$$\begin{aligned}
f_1^k &= f_0^{k+1} - C_0^0 L_1 f_0^k, \\
f_2^k &= f_0^{k+2} - (C_0^0 + C_0^1) L_1 f_0^{k+1} + (C_0^0 C_0^1 L_1 L_1 - C_1^1 L_2) f_0^k, \\
&\vdots
\end{aligned}$$

and, in general, we have the expression

$$f_n^k = f_0^{k+n} - \sum_{j=0}^{n} C_j^n G_j f_0^{k+n-j} \qquad (n \geq 1,\ k \geq 0) \tag{3.84}$$

where G_j is a linear operator that involves $L_1, L_2, \ldots, L_j$. The convenience of writing the numeric coefficient C_j^n in (3.84) will become apparent on getting the expression of G_j in (3.82), obtained when we insert (3.84) in (3.83). To obtain that expression, we substitute in (3.83) the expression of f_n^k in (3.84) together with those corresponding to f_{n-1}^{k+1} and f_{n-m-1}^k, which leads to the equality

$$\sum_{j=1}^{n} C_j^n G_j f_0^{k+n-j} = \sum_{j=1}^{n-1} C_j^{n-1} G_j f_0^{k+n-j} + \sum_{s=0}^{n-1} C_s^{n-1} L_{s+1} f_0^{k+n-s-1}$$

$$- \sum_{s=0}^{n-1} \sum_{j=1}^{n-1} C_s^{n-1} C_j^{n-s-1} L_{s+1} G_j f_0^{k+n-s-1-j}. \tag{3.85}$$

Using the identities among binomial coefficients

$$C_{n-p}^n - C_{n-p}^{n-1} = C_p^{n-1}, \quad C_{n-p-1}^{n-1} = C_p^{n-1}, \quad C_s^{n-1} C_{n-s-p-1}^{n-s-1} = C_p^{n-1} C_s^{n-1-p},$$

and comparing the terms involving f_0^{k+p} in (3.85), we obtain

$$G_{n-p} f_0^{k+p} = L_{n-p} f_0^{k+p} - \sum_{s=0}^{n-1} C_s^{n-1-p} L_{s+1} G_{n-p-s-1} f_0^{k+p}.$$

Setting $s + 1 = m$, $n - p = j$ and observing that as G_k begins with $k = 1$ the sum runs only up to $m = j - 1$, we obtain from the last equality the expression for G_j in (3.84).

Taking now $k = 1$ and $f = K = H + R$, where R is the remainder function, we obtain using the equality (3.84) for K_n^1,

$$H_n^1 + R_n^1 = K_{n+1} - \sum_{j=1}^{n} C_j^n G_j K_{1+n-j}. \tag{3.86}$$

By Theorem 3.20 we have $R = \dfrac{\partial S^*}{\partial \epsilon}$, where $S^* = \mathcal{L}_W\Big(-\dfrac{\partial W}{\partial t}\Big)$, hence $R_n^1 = -\dfrac{\partial W_{n+1}}{\partial t}$. Now, by Theorem 3.10 we have for H_n^1, the expression

$$H_n^1 = H_{n+1} + \sum_{j=0}^{n} C_j^n L_{j+1} H_{n-j}.$$

Substituting these expressions in (3.86) and taking into account that

$$K_{j,i} = G_j K_i$$

we obtain

$$K_{n+1} = H_{n+1} + \sum_{j=0}^{n} C_j^n L_{j+1} H_{n-j} + \sum_{j=1}^{n} C_j^n K_{j,1+n-j} - \frac{\partial W_{n+1}}{\partial t}.$$

Setting $m = n + 1$ and reorganizing the indices we obtain the equality (3.81). □

Remark 3.23 The formula of Kamel (3.81) allows us to compute the polynomials of the final Hamiltonian directly from the polynomials of the original Hamiltonian, but this may not represent a computational simplification because we have to compute the Poisson brackets successively to get $K_{j,i}$ through the formula (3.82). For example, for the first values of G_j, we have

$$G_1 = L_1, \quad G_2 = L_2 - L_1 L_1, \quad G_3 = L_3 - L_1(L_2 - L_1 L_1) - 2L_2 L_1.$$

We see that for G_3 we already need $L_1 L_1 L_1 g = \{\{\{g, W_1\}, W_1\}, W_1\}$.

The use of the Lie triangle to compute the polynomials of the final Hamiltonian based on the construction of the auxiliary polynomials may allow intermediate computations to be simplified more effectively.

Chapter 4
Spectral Decomposition of Hamiltonian Matrices

4.1 Introduction

In this chapter we study the normal form of a Hamiltonian matrix when the matrix is diagonalizable. For multiple eigenvalues we give the normal form only when the eigenvalues are purely imaginary numbers, the case relevant to the study of stability.

For the study of normal forms of Hamiltonian matrices here and in the next chapter we need some knowledge of symplectic spaces, which constitutes the first section of this chapter.

4.2 Symplectic Spaces

Let V be a vector space of dimension n over the field $\mathbf{F}$ ($\mathbf{F} = \mathbf{R}$ or $\mathbf{C}$). Let $\beta(\mathbf{x}, \mathbf{y})$ be a bilinear form on V with matrix $B = \left[\beta(\mathbf{u}_i, \mathbf{u}_j)\right]$ in a basis $\mathbf{u}_1, \ldots, \mathbf{u}_n$ of V. Note that B is symmetric or skew-symmetric according to β being a symmetric or skew-symmetric bilinear form.

If $\mathbf{v}_1, \ldots, \mathbf{v}_n$ is another basis of V and Q is the transition matrix, that is, $\mathbf{v}_j = \sum_i Q_{ij}\mathbf{u}_i$, then the matrix C of β on this basis is related to B by the equation $C = Q^T B Q$. Indeed,

$$
\begin{aligned}
C_{ij} = \beta(\mathbf{v}_i, \mathbf{v}_j) &= \beta\left(\sum_r Q_{ri}\mathbf{u}_r, \sum_s Q_{sj}\mathbf{u}_s\right) = \sum_{r,s} Q_{ri} B_{rs} Q_{sj} \\
&= \sum_s (Q^T B)_{is} Q_{sj} = (Q^T B Q)_{ij}.
\end{aligned}
$$

Therefore, the matrices $Q^T B Q$ with Q invertible represent the same bilinear form β.

H. E. Cabral, L. Brandão Dias, *Normal Forms and Stability of Hamiltonian Systems*, Applied Mathematical Sciences 218,
https://doi.org/10.1007/978-3-031-33046-9_4

Example 4.1 Any bilinear form in $\mathbf{R}^m$ is given in terms of the Euclidean inner product (1.2) by

$$\beta(\mathbf{x}, \mathbf{y}) = \langle \mathbf{x}, B\mathbf{y} \rangle,$$

for some $m \times m$ real matrix B.

Indeed, let $\beta_{ij} = \beta(\mathbf{e}_i, \mathbf{e}_j)$ be the value of β in the canonical basis vectors (1.3). Since $\mathbf{x} = x_1\mathbf{e}_1 + \cdots + x_m\mathbf{e}_m$ and $\mathbf{y} = y_1\mathbf{e}_1 + \cdots + y_m\mathbf{e}_m$, the bilinearity of β gives

$$\beta(\mathbf{x}, \mathbf{y}) = \sum_{i,j} x_i y_j \beta_{ij} = \sum_i x_i \Big(\sum_j \beta_{ij} y_j \Big) = \langle \mathbf{x}, B\mathbf{y} \rangle,$$

where B is the matrix $B = [\beta_{ij}]$. Notice that B is the matrix of β on the canonical basis.

Given a bilinear form β on V, we have a linear map from V to its dual space V^*, defined by

$$L_\beta : V \to V^*, \quad L_\beta(\mathbf{y}) = \beta(\cdot, \mathbf{y}) \tag{4.1}$$

Let $\mathbf{u}_1, \ldots, \mathbf{u}_n$ be a basis of V and $\mathbf{u}^1, \ldots, \mathbf{u}^n$ its dual basis, i.e., $\mathbf{u}^i(\mathbf{u}_j) = \delta_{ij}$, the Kronecker delta, that is, $\delta_{ii} = 1$ and $\delta_{ij} = 0$ for $i \neq j$.

Then the matrix $\mathcal{B} = (b_{ij})$ of L_β on these bases, which is defined by $L_\beta(\mathbf{u}_j) = \sum_{k=1}^n b_{kj}\mathbf{u}^k$, is the matrix $B = \big(\beta(\mathbf{u}_i, \mathbf{u}_j)\big)$ of the bilinear form β. Indeed,

$$\beta(\mathbf{u}_i, \mathbf{u}_j) = L_\beta(\mathbf{u}_j) \cdot \mathbf{u}_i = \sum_{k=1}^n b_{kj}\mathbf{u}^k \cdot \mathbf{u}_i = b_{ij}.$$

Proposition 4.1 *The linear map L_β defined in* (4.1) *is an isomorphism if, and only if, β is nondegenerate.*

Proof This is clear since a nondegenerate bilinear form β means that $\beta(\mathbf{x}, \mathbf{y}) = 0$ for all $\mathbf{x}$ implies $\mathbf{y} = 0$. □

If the bilinear form β on V is nondegenerate and skew-symmetric, then the dimension n of the space V is even, since $B^T = -B$ gives $\det B = (-1)^n \det B$.

Definition 4.2 Let ω be a nondegenerate skew-symmetric bilinear form on a vector space V. The pair (V, ω) is called a *symplectic vector space*.

If (V, ω) is a symplectic space, then the dimension of V is even.

Example 4.2 Let $l_1, l_2 \in V^*$ and write $\beta(\mathbf{u}, \mathbf{v}) = l_1(\mathbf{u})l_2(\mathbf{v}) - l_1(\mathbf{v})l_2(\mathbf{u})$. This is clearly a skew-symmetric bilinear form on V but it is degenerate if the dimension of V is greater than two. Indeed, $l_1^{-1}(0) \cap l_2^{-1}(0) \neq \{0\}$ if the dimension is greater than two, so for $\mathbf{u} \neq 0$ in this intersection, $\beta(\mathbf{u}, \mathbf{v}) = 0$ for all $\mathbf{v} \in V$. Therefore, β

is not a symplectic form on V. However, it is a symplectic form on V if dimension of V is two.

A most important symplectic form on the vector space $\mathbf{F}^{2n}$ is given in the following example.

Example 4.3 The vector space $\mathbf{F}^{2n}$ endowed with the bilinear form

$$\omega_J(\mathbf{x}, \mathbf{y}) = \mathbf{x}^T J\mathbf{y}, \tag{4.2}$$

where J is the standard symplectic matrix in a symplectic space given by Definition 1.7.

The bilinear form (4.2) is called the *standard symplectic form* or the *Poisson bracket* in $\mathbf{F}^{2n}$. We use also the notation $\{\mathbf{x}, \mathbf{y}\} = \omega_J(\mathbf{x}, \mathbf{y})$. If $\mathbf{F} = \mathbf{R}$ this is the symplectic product considered in Sect. 2.6.

Example 4.4 Let A be an $m \times m$ skew-symmetric matrix and write $\omega_A(\mathbf{x}, \mathbf{y}) = \langle A\mathbf{x}, \mathbf{y}\rangle$, for $\mathbf{x}, \mathbf{y} \in \mathbf{R}^m$, where $\langle\ ,\ \rangle$ is the Euclidean inner product (1.2) in $\mathbf{R}^m$. Then, ω_A is a skew-symmetric bilinear form on $\mathbf{R}^m$. If $m = 2n$ and A is an invertible matrix, then ω_A is a symplectic form on $\mathbf{R}^{2n}$ since if $\omega_A(\mathbf{x}, \mathbf{y}) = 0$ for all $\mathbf{y}$, then $A\mathbf{x} = 0$ so $\mathbf{x} = 0$ because A is invertible.

Definition 4.3 Let V be a vector space. The *annihilator* of a subspace U of V is the subspace of V^*

$$U^0 = \{f \in V^*;\ f(\mathbf{x}) = 0,\ \text{for all } \mathbf{x} \in U\}.$$

And the *annihilator* of a subspace U of V^* is the subspace of V

$$U_0 = \{\mathbf{x} \in V;\ f(\mathbf{x}) = 0,\ \text{for all } f \in U\}.$$

If $\mathbf{u}_1, \ldots, \mathbf{u}_k$ are elements of the vector space V we denote by $\left[\mathbf{u}_1, \ldots, \mathbf{u}_k\right]$ the subspace of V generated by them.

Consider a basis $\mathbf{v}_1, \ldots, \mathbf{v}_k, \ldots, \mathbf{v}_n$ of V and its dual basis $\mathbf{v}^1, \ldots, \mathbf{v}^k, \ldots, \mathbf{v}^n$ of V^*. We easily see that:

(1) If $U = \left[\mathbf{v}_1, \ldots, \mathbf{v}_k\right] \subset V$ then $U^0 = \left[\mathbf{v}^{k+1}, \ldots, \mathbf{v}^n\right]$, so the dimension of U^0 is $n - k$;
(2) If $U = \left[\mathbf{v}^1, \ldots, \mathbf{v}^k\right] \subset V^*$, then $U_0 = \left[\mathbf{v}_{k+1}, \ldots, \mathbf{v}_n\right]$, so the dimension of U_0 is $n - k$.

From this we see that $(U^0)_0 = U$ and $(U_0)^0 = U$.

Proposition 4.4 *For any subspace U of the vector space V, we have*

$$dim U + dim U^0 = dim V. \tag{4.3}$$

Proof Take a basis $\mathbf{v}_1, \ldots, \mathbf{v}_k$ of U and complete it with vectors $\mathbf{v}_{k+1}, \ldots, \mathbf{v}_n$ to form a basis of V and then use (1). □

If U is a subspace of the symplectic space (V, ω), its *symplectic orthogonal* is defined as the subspace of V

$$U^{\perp} = \{\mathbf{x} \in V;\ \omega(\mathbf{x}, \mathbf{y}) = 0,\ \text{for all } \mathbf{y} \in U\}.$$

Be careful with this notation because here, differently from the inner product case we may have $U \subset U^{\perp}$, for example, in the case when U is one dimensional.

Proposition 4.5 *Let $L_\omega : V \to V^*$ be the isomorphism defined in* (4.1). *Then, for any subspace $U \subset V$, we have:*
(a) $L_\omega(U^{\perp}) = U^0$; (b) $dim(U) + dim(U^{\perp}) = dim V$.

Proof Since $L_\omega(\mathbf{y})(\mathbf{x}) = \omega(\mathbf{x}, \mathbf{y})$, to prove (a) we just notice that

$$U^{\perp} = \{\mathbf{y} \in V;\ L_\omega(\mathbf{y})(\mathbf{x}) = 0,\ \text{for all } \mathbf{x} \in U\} = \{\mathbf{y} \in V;\ L_\omega(\mathbf{y}) \in U^0\} = L_\omega^{-1}(U^0).$$

Now, (b) follows from (4.3) since $\dim(U^{\perp}) = \dim(U^0)$ by (a). □

Let (V, ω) be a symplectic space. For subspaces U, W of V denote by $\{U, W\}$ the set

$$\{U, W\} = \{\omega(\mathbf{x}, \mathbf{y});\ \mathbf{x} \in U,\ \mathbf{y} \in W\}.$$

If $\{U, W\} = \{0\}$, we say that U and W are symplectically orthogonal.

Definition 4.6 A subspace U of V is said to be symplectic if ω restricted to U is nondegenerate.

Proposition 4.7 *Let U, W be subspaces of the symplectic space (V, ω).*

(a) *If $V = U \oplus W$ and $\{U, W\} = \{0\}$, then U and W are symplectic subspaces.*
(b) *If U is a symplectic subspace of V, then so is $U^{\perp}$ and $V = U \oplus U^{\perp}$.*

Proof To prove (a) let $\mathbf{x} \in U$ be such that $\omega(\mathbf{x}, \mathbf{y}) = 0$ for all $\mathbf{y} \in U$. Since $\{U, W\} = \{0\}$, we also have $\omega(\mathbf{x}, \mathbf{y}') = 0$ for all $\mathbf{y}' \in W$. As $V = U \oplus W$ it follows that $\omega(\mathbf{x}, \mathbf{z}) = 0$ for all $\mathbf{z} \in V$; hence, $\mathbf{x} = 0$. This shows that $\omega|_U$ is nondegenerate; hence, U is a symplectic subspace of V. Similarly, it is proved that W is a symplectic subspace of V.

We now prove (b). Let $\mathbf{x} \in U \cap U^{\perp}$. Then $\omega(\mathbf{x}, \mathbf{y}) = 0$ for all $\mathbf{y} \in U$ because $\mathbf{x} \in U^{\perp}$. As $\mathbf{x} \in U$ and $\omega(\mathbf{x}, \mathbf{y}) = 0$ for all $\mathbf{y} \in U$, we have $\mathbf{x} = 0$, because U is a symplectic subspace of V. Therefore, $U \cap U^{\perp} = \{0\}$; hence, $\dim(U + U^{\perp}) = \dim U + \dim U^{\perp} = \dim V$, so $U + U^{\perp} = V$. Since $U \cap U^{\perp} = \{0\}$, we have $U \oplus U^{\perp} = V$. It then follows from (a) that $U^{\perp}$ is a symplectic subspace. □

Definition 4.8 Let (V, ω) be a symplectic space. A *symplectic basis* of V is a basis $\mathbf{v}_1, \ldots, \mathbf{v}_{2n}$ in which the matrix of ω is J, so $\omega(\mathbf{v}_k, \mathbf{v}_l) = J_{kl}$.

Example 4.5 Construct a symplectic basis for the symplectic space $(\mathbf{R}^{2n}, \omega_A)$ of Example (4.4). Recall that $\omega_A(\mathbf{x}, \mathbf{y}) = \langle A\mathbf{x}, \mathbf{y}\rangle$. Denote by $(\mathbf{x}, \mathbf{y})$ the scalar product in $\mathbf{C}^n$ defined as the inner product (1.2) in $\mathbf{R}^n$. For real vectors we have $(\mathbf{x}, \mathbf{y}) = \langle \mathbf{x}, \mathbf{y} \rangle$. Let $\mathbf{x}_1$ be an eigenvector of A belonging to an eigenvalue λ_1. Then, $(A\mathbf{x}_1, \overline{\mathbf{x}}_1) = \lambda_1 \|\mathbf{x}_1\|^2 \neq 0$. Write the vector in terms of its real and imaginary parts, $\mathbf{x}_1 = \mathbf{r}_1 + i\mathbf{s}_1$. Then, $(A\mathbf{x}_1, \overline{\mathbf{x}}_1) = -2i\langle A\mathbf{r}_1, \mathbf{s}_1 \rangle$. Therefore, $\langle A\mathbf{r}_1, \mathbf{s}_1 \rangle$ is a nonzero real number. Also, notice that $\lambda_1 = i\alpha_1$ is a purely imaginary number. Let $\delta_1 = \text{sign}\langle A\mathbf{r}_1, \mathbf{s}_1 \rangle$ and $\kappa_1 = \frac{1}{\sqrt{|\langle A\mathbf{r}_1, \mathbf{s}_1 \rangle|}}$ and consider the real vectors

$$\mathbf{u}_1 = \kappa_1 \mathbf{r}_1, \quad \mathbf{v}_1 = \delta_1 \kappa_1 \mathbf{s}_1. \tag{4.4}$$

Then, $\omega_A(\mathbf{u}_1, \mathbf{v}_1) = 1$. Let $U_1 = [\mathbf{u}_1, \mathbf{v}_1]$ be the subspace generated by $\mathbf{u}_1, \mathbf{v}_1$. By the equality just proved, U_1 is a two-dimensional symplectic subspace of $\mathbf{R}^{2n}$ because $\omega_A|_{U_1}$ is nondegenerate. By Proposition 4.7(b) we have the direct sum $\mathbf{R}^{2n} = U_1 \oplus U_1^\perp$ with $U_1^\perp$ also a symplectic subspace of $\mathbf{R}^{2n}$. The equation $A\mathbf{x}_1 = \lambda_1 \mathbf{x}_1$ amounts to the equations $A\mathbf{u}_1 = -\alpha_1 \mathbf{v}_1$ and $A\mathbf{v}_1 = \alpha_1 \mathbf{u}_1$, hence $\langle \mathbf{u}_1, A\mathbf{z} \rangle = \alpha_1 \langle \mathbf{v}_1, \mathbf{z} \rangle$ and $\langle \mathbf{v}_1, A\mathbf{z} \rangle = -\alpha_1 \langle \mathbf{u}_1, \mathbf{z} \rangle$, so $A\mathbf{z} \in U_1^\perp$ for any $\mathbf{z} \in U_1^\perp$ and therefore this is an A-invariant subspace of $U_1^\perp$.

The same argument with an eigenvector $\lambda_2 = i\alpha_2$ of $A|_{U_1^\perp}$ shows the existence of a second pair of real vectors $\mathbf{u}_2, \mathbf{v}_2 \in U_1^\perp$ giving an A-invariant decomposition $U_1^\perp = U_2 \oplus U_2^\perp$ with $U_2^\perp$ also a symplectic subspace of $U_1^\perp$ and such that $\omega_A(\mathbf{u}_2, \mathbf{v}_2) = 1$. We proceed with $\mathbf{R}^{2n} = (U_1 \oplus U_1^\perp) \oplus (U_2 \oplus U_2^\perp)$ repeating the argument successively to find an A-invariant decomposition $\mathbf{R}^{2n} = (U_1 \oplus U_1^\perp) \oplus (U_2 \oplus U_2^\perp) \oplus \ldots \oplus (U_n \oplus U_n^\perp)$.

Then the vectors $\mathbf{u}_1, \mathbf{u}_2, \ldots, \mathbf{u}_k, \mathbf{v}_1, \mathbf{v}_2, \ldots, \mathbf{v}_k$ form a symplectic basis of $\mathbf{R}^{2n}$.

Proposition 4.9 *Every symplectic space has a symplectic basis.*

Proof By induction on dimension $2n$ of V. For $n = 1$ we can find vectors $\mathbf{u}_1, \mathbf{v}_1$ such that $\omega(\mathbf{u}_1, \mathbf{v}_1) = 1$, since ω is nondegenerate. This is a symplectic basis for V.

Now, assume $n > 1$. There exist vectors $\mathbf{u}_1, \mathbf{v}_1$ such that $\omega(\mathbf{u}_1, \mathbf{v}_1) = 1$. The subspace $U = [\mathbf{u}_1, \mathbf{v}_1]$ is symplectic and its generators $\mathbf{u}_1, \mathbf{v}_1$ form a symplectic basis for it. By Proposition 4.7(b) we have $V = U \oplus U^\perp$ with $U^\perp$ symplectic. But $\dim U^\perp = 2n - 2$; hence, by induction $U^\perp$ has a symplectic basis, $\mathbf{u}_2, \ldots, \mathbf{u}_n, \mathbf{v}_2, \ldots, \mathbf{v}_n$. Then, the vectors $\mathbf{u}_1, \mathbf{u}_2, \ldots, \mathbf{u}_n, \mathbf{v}_1, \mathbf{v}_2, \ldots, \mathbf{v}_n$ form a symplectic basis for V. □

Proposition 4.10 *Let the symplectic space U be a direct sum of two symplectic subspaces, $U = W_1 \oplus W_2$, with $\{W_1, W_2\} = \{0\}$. Let $\mathbf{u}_1, \ldots, \mathbf{u}_k, \mathbf{v}_1, \ldots, \mathbf{v}_k$ be a symplectic basis of W_1 and $\mathbf{z}_1, \ldots, \mathbf{z}_l, \mathbf{w}_1, \ldots, \mathbf{w}_l$ be a symplectic basis of W_2. Then,*

$$\mathbf{u}_1, \ldots, \mathbf{u}_k, \mathbf{z}_1, \ldots, \mathbf{z}_l, \mathbf{v}_1, \ldots, \mathbf{v}_k, \mathbf{w}_1, \ldots, \mathbf{w}_l$$

is a symplectic basis of U.

Proof By Definition 4.8 the vectors $\mathbf{x}_1, \ldots, \mathbf{x}_m, \mathbf{y}_1, \ldots, \mathbf{y}_m$ form a symplectic basis for the symplectic space (U, ω) if and only if $\omega(\mathbf{x}_i, \mathbf{x}_j) = 0$, $\omega(\mathbf{y}_i, \mathbf{y}_j) = 0$,

$\omega(\mathbf{x}_k, \mathbf{y}_l) = 0$ for $k \neq l$ and $\omega(\mathbf{x}_k, \mathbf{y}_k) = 1$, $k = 1, \ldots, m$. Using this information, the statement in the proposition is obvious since W_1 and W_2 are symplectically orthogonal. □

Definition 4.11 A linear mapping $L : V_1 \to V_2$ between two symplectic spaces (V_1, ω_1) and (V_2, ω_2) is called a *symplectic mapping* if

$$\omega_2(L\mathbf{u}, L\mathbf{v}) = \omega_1(\mathbf{u}, \mathbf{v}).$$

Since ω_1 is nondegenerate, the symplectic mapping L is injective. If V_1 and V_2 have the same dimension, then L is an isomorphism.

Example 4.6 Let $(\mathbf{R}^{2n}, \omega_A)$ be the symplectic space of Example 4.4. Let $L : \mathbf{R}^{2n} \to \mathbf{R}^{2n}$ be a linear operator and B its matrix in the canonical basis of $\mathbf{R}^{2n}$. Then, L is symplectic if, and only if, $B^T A B = A$, since $\omega_A(L\mathbf{x}, L\mathbf{y}) = \omega_A(\mathbf{x}, \mathbf{y})$ means that $\langle AB\mathbf{x}, B\mathbf{y}\rangle = \langle A\mathbf{x}, \mathbf{y}\rangle$, that is, $\langle B^T AB\mathbf{x}, \mathbf{y}\rangle = \langle A\mathbf{x}, \mathbf{y}\rangle$.

Notice that when A is the standard symplectic matrix J the above equation becomes $B^T J B = J$, so B is a symplectic matrix (see Definition 2.5).

Let us explore Example 4.6 further in the particular case of $n = 2$.

Example 4.7 Let the matrix A defining the symplectic space $(\mathbf{R}^4, \omega_A)$ be given by

$$A = \begin{bmatrix} 0 & -1 & 1 & -1 \\ 1 & 0 & -1 & 1 \\ -1 & 1 & 0 & -1 \\ 1 & 0 & 1 & 0 \end{bmatrix}.$$

Find an explicit symplectic basis for this space.

We first compute the eigenvalues of A, $\lambda = \pm i$ and $\lambda = \pm 2i$. Then we compute the eigenvectors, $\mathbf{u}_1 = \mathbf{r}_1 + i\mathbf{s}_1$ for $\lambda_1 = i$ and $\mathbf{u}_2 = \mathbf{r}_2 + i\mathbf{s}_2$ for $\lambda_2 = 2i$ finding

$$\mathbf{r}_1 = (1, 1, 0, -1),\ \ \mathbf{s}_1 = (0, 1, 1, 1) \quad \text{and} \quad \mathbf{r}_2 = (2, -1, 0, 1),\ \ \mathbf{s}_2 = (0, 1, -2, 1).$$

Computing $\delta_j = \text{sign}\langle A\mathbf{r}_j, \mathbf{s}_j\rangle$, $\kappa_j = 1\sqrt{|\langle A\mathbf{r}_j, \mathbf{s}_j\rangle|}$ and using (4.4) we find the vectors

$$\begin{aligned} \mathbf{u}_1 &= \frac{1}{\sqrt{3}}(1, 1, 0, -1), \quad & \mathbf{v}_1 &= \frac{1}{\sqrt{3}}(0, -1, -1, -1) \\ \mathbf{u}_2 &= \frac{1}{2\sqrt{3}}(2, -1, 0, 1), \quad & \mathbf{v}_2 &= \frac{1}{2\sqrt{3}}(0, -1, 2, -1), \end{aligned} \tag{4.5}$$

which give a symplectic basis $\mathbf{u}_1, \mathbf{u}_2, \mathbf{v}_1, \mathbf{v}_2$ of $(\mathbf{R}^4, \omega_A)$.

We use this symplectic basis of $(\mathbf{R}^4, \omega_A)$ to illustrate the property that the matrix of a symplectic operator in a symplectic basis is a symplectic matrix.

Example 4.8 Consider the linear mapping $L : \mathbf{R}^4 \to \mathbf{R}^4$ whose matrix in the canonical basis of $\mathbf{R}^4$ is

$$B = \begin{bmatrix} 1 & -2 & 0 & 0 \\ 1 & -1 & 0 & 0 \\ 0 & 0 & -1 & 2 \\ 0 & 0 & -1 & 1 \end{bmatrix}.$$

The matrix B satisfies the equation $B^T AB = A$, so by Example 4.6 L is a symplectic operator in $(\mathbf{R}^4, \omega_A)$. Notice that the matrix B is not symplectic. However, we will see that the matrix $[L]$ of L in the symplectic basis $\mathbf{u}_1, \mathbf{u}_2, \mathbf{v}_1, \mathbf{v}_2$ is a symplectic matrix.

Let $U = x_1\mathbf{u}_1 + x_2\mathbf{u}_2 + y_1\mathbf{v}_1 + y_2\mathbf{v}_2$. Then,

$$U = \frac{1}{\sqrt{3}}\Big(x_1 + x_2,\ x_1 - \frac{1}{2}x_2 - y_1 - \frac{1}{2}y_2,\ -y_1 + y_2,\ -x_1 + \frac{1}{2}x_2 - y_1 - \frac{1}{2}y_2\Big).$$

Setting $U = \frac{1}{\sqrt{3}}(a, b, c, d)$ and solving for x_i, y_i we get

$$x_1 = \frac{1}{3}(a + b - d), \quad x_2 = \frac{1}{3}(2a + b - d), \quad y_1 = -\frac{1}{3}(b + c + d),$$
$$y_2 = -\frac{1}{3}(b - 2c + d). \tag{4.6}$$

Computing L in the basis vectors (4.5) we find

$$L\mathbf{u}_1 = \frac{1}{\sqrt{3}}(-1, 0, -2, -1), \quad L\mathbf{u}_2 = \frac{1}{\sqrt{3}}\Big(2, \frac{3}{2}, 1, \frac{1}{2}\Big),$$

$$L\mathbf{v}_1 = \frac{1}{\sqrt{3}}(2, 1, -1, 0), \quad L\mathbf{v}_2 = \frac{1}{\sqrt{3}}\Big(1, \frac{1}{2}, -2, -\frac{3}{2}\Big).$$

Thus, taking a, b, c, d in U for each case $U = L\mathbf{u}_i$ and $U = L\mathbf{v}_i$ we conclude that

$$L\mathbf{u}_1 = -\mathbf{u}_2 + \mathbf{v}_1 - \mathbf{v}_2, \quad L\mathbf{u}_2 = \mathbf{u}_1 + \mathbf{u}_2 - \mathbf{v}_1, \quad L\mathbf{v}_1 = \mathbf{u}_1 + \mathbf{u}_2 - \mathbf{v}_2,$$
$$L\mathbf{v}_2 = \mathbf{u}_1 + \mathbf{v}_1 - \mathbf{v}_2.$$

Therefore, the matrix of the operator L on the basis $\mathbf{u}_1, \mathbf{u}_2, \mathbf{v}_1, \mathbf{v}_2$ is

$$[L] = \begin{bmatrix} 0 & 1 & 1 & 1 \\ -1 & 1 & 1 & 0 \\ 1 & -1 & 0 & 1 \\ -1 & 0 & -1 & -1 \end{bmatrix},$$

which is a symplectic matrix.

Let $\mathbf{v}_1, \ldots, \mathbf{v}_m$ be a basis of the vector space V over the field $\mathbf{F} = \mathbf{R}$ or $\mathbf{C}$. Then, we have the *coordinate isomorphism* $\mathcal{C} : V \to \mathbf{F}^m$ defined by,

$$\mathcal{C}(\mathbf{v}) = \mathbf{x}, \quad \text{that is} \quad \mathcal{C}(x_1\mathbf{v}_1 + \cdots + x_m\mathbf{v}_m) = (x_1, \ldots, x_m).$$

Proposition 4.12 *Let $L : V \to V$ be a linear operator with matrix A in the basis $\mathbf{v}_1, \ldots, \mathbf{v}_m$. If $\mathcal{C}(\mathbf{v}) = \mathbf{x}$, then $\mathcal{C}(L\mathbf{v}) = A\mathbf{x}$.*

Proof We have $L\mathbf{v}_j = \sum_i A_{ij}\mathbf{v}_i$, so if $\mathcal{C}(\mathbf{v}) = \mathbf{x}$, that is, $\mathbf{v} = \sum_j x_j\mathbf{v}_j$, we get

$$L\mathbf{v} = \sum_j x_j L\mathbf{v}_j = \sum_j x_j \Big(\sum_i A_{ij}\mathbf{v}_i\Big) = \sum_i \Big(\sum_j A_{ij}x_j\Big)\mathbf{v}_i = \sum_i (A\mathbf{x})_i\,\mathbf{v}_i;$$

hence, $A\mathbf{x} = \mathcal{C}(L\mathbf{v})$. □

Proposition 4.13 *If $\mathbf{v}_1, \ldots, \mathbf{v}_{2n}$ is a symplectic basis of the symplectic space (V, ω), then the coordinate isomorphism*

$$\mathcal{C} : V \to \mathbf{F}^{2n}, \quad \mathcal{C}\mathbf{v} = \mathbf{x} \tag{4.7}$$

is a symplectic mapping.

Proof As the basis $\mathbf{v}_1, \ldots, \mathbf{v}_{2n}$ is symplectic the matrix of the bilinear form ω on this basis is J. Therefore, if $\mathbf{u} = x_1\mathbf{v}_1 + \cdots + x_{2n}\mathbf{v}_{2n}$ and $\mathbf{v} = y_1\mathbf{v}_1 + \cdots + y_{2n}\mathbf{v}_{2n}$ then $\omega(\mathbf{u}, \mathbf{v}) = \sum_{k,l=1}^{2n} J_{kl}x_k y_l = \mathbf{x}^T J\mathbf{y}$. Since $\mathcal{C}(\mathbf{u}) = \mathbf{x}$ and $\mathcal{C}(\mathbf{v}) = \mathbf{y}$, then by (4.2) we have $\omega_J(\mathcal{C}\mathbf{u}, \mathcal{C}\mathbf{v}) = \mathbf{x}^T J\mathbf{y}$, hence we get

$$\omega_J(\mathcal{C}\mathbf{u}, \mathcal{C}\mathbf{v}) = \omega(\mathbf{u}, \mathbf{v}),$$

showing that $\mathcal{C}$ is symplectic. □

We also have the symplectic isomorphism defined by the standard symplectic matrix J,

$$J : (\mathbf{F}^{2n}, \omega_J) \to (\mathbf{F}^{2n}, \omega_J), \quad J(\mathbf{x}) = J\mathbf{x}. \tag{4.8}$$

Definition 4.14 Let (V, ω) be a symplectic space. A linear operator $L : V \to V$ is said to be *Hamiltonian* if it satisfies the equation

$$\omega(L\mathbf{u}, \mathbf{v}) + \omega(\mathbf{u}, L\mathbf{v}) = 0. \tag{4.9}$$

Example 4.9 Let $(\mathbf{R}^{2n}, \omega_A)$ be the symplectic space of Example 4.4. Let $L : \mathbf{R}^{2n} \to \mathbf{R}^{2n}$ be a linear operator and B its matrix in the canonical basis of $\mathbf{R}^{2n}$. Then, L is Hamiltonian if, and only if, $B^T A + AB = O$. Indeed, $\omega_A(L\mathbf{x}, \mathbf{y}) + \omega_A(\mathbf{x}, L\mathbf{y}) = 0$ means that $\langle AB\mathbf{x}, \mathbf{y}\rangle + \langle A\mathbf{x}, B\mathbf{y}\rangle = 0$, that is $\langle (AB + B^T A)\mathbf{x}, \mathbf{y}\rangle = 0$.

Notice that when A is the standard symplectic matrix J the matrix equation in this example becomes $B^T J + JB = 0$, so B is a Hamiltonian matrix (see item (3) of Proposition 2.26).

We consider the following example with $n = 2$ to illustrate a property of Hamiltonian operators similar to that of a symplectic operator.

Example 4.10 Consider the linear mapping $L : \mathbf{R}^4 \to \mathbf{R}^4$ whose matrix in the canonical basis of $\mathbf{R}^4$ is

$$B = \begin{bmatrix} 1 & 1 & 0 & 0 \\ 1 & -1 & 0 & 0 \\ 0 & 0 & 1 & 1 \\ 0 & 0 & 1 & -1 \end{bmatrix}.$$

The matrix B satisfies the equation $B^T A + BA = O$, so by the previous example L is a Hamiltonian operator in $(\mathbf{R}^4, \omega_A)$. Notice that B is not a Hamiltonian matrix. However, we will see that the matrix $[L]$ of L on the symplectic basis $\mathbf{u}_1, \mathbf{u}_2, \mathbf{v}_1, \mathbf{v}_2$ given in (4.5) is a Hamiltonian matrix.

Computing L in the basis vectors (4.5) we find

$$L\mathbf{u}_1 = \frac{1}{\sqrt{3}}(1, 0, 0, 1), \quad L\mathbf{u}_2 = \frac{1}{\sqrt{3}}\Big(1, \frac{3}{2}, 0, -\frac{1}{2}\Big),$$

$$L\mathbf{v}_1 = \frac{1}{\sqrt{3}}(0, 1, -1, 0), \quad L\mathbf{v}_2 = \frac{1}{\sqrt{3}}\Big(0, \frac{1}{2}, 1, \frac{3}{2}\Big).$$

Thus, referring to Example 4.8, taking a, b, c, d in U for each case $U = L\mathbf{u}_i$ and $U = L\mathbf{v}_i$ we conclude that

$$L\mathbf{u}_1 = \mathbf{u}_2 - \frac{1}{3}\mathbf{v}_1 - \frac{1}{3}\mathbf{v}_2, \; L\mathbf{u}_2 = \mathbf{u}_1 - \frac{1}{3}\mathbf{v}_1 - \frac{1}{3}\mathbf{v}_2, \; L\mathbf{v}_1 = \frac{1}{3}\mathbf{u}_1 - \frac{1}{3}\mathbf{u}_2 - \mathbf{v}_2,$$

$$L\mathbf{v}_2 = -\frac{1}{3}\mathbf{u}_1 + \frac{1}{3}\mathbf{u}_2 - \mathbf{v}_1.$$

Therefore, the matrix of the operator L in the basis $\mathbf{u}_1, \mathbf{u}_2, \mathbf{v}_1, \mathbf{v}_2$ is

$$[L] = \begin{bmatrix} 0 & 1 & \frac{1}{3} & -\frac{1}{3} \\ 1 & 0 & -\frac{1}{3} & \frac{1}{3} \\ -\frac{1}{3} & -\frac{1}{3} & 0 & -1 \\ -\frac{1}{3} & -\frac{1}{3} & -1 & 0 \end{bmatrix},$$

which is a Hamiltonian matrix.

The properties illustrated in Examples 4.8 and 4.10 hold in general and this is proved in the following proposition.

Proposition 4.15 *Let (V, ω) be a symplectic space. Then, the matrix of a linear operator $L : V \to V$ on a symplectic basis is Hamiltonian (symplectic) if, and only if, the operator is Hamiltonian (symplectic).*

Proof Let A be the matrix of $L : V \to V$ in the symplectic basis $\mathbf{v}_1, \ldots, \mathbf{v}_{2n}$. Let $\mathbf{x}, \mathbf{y}$ be arbitrary vectors of $\mathbf{F}^{2n}$ and let $\mathbf{x} = \mathcal{C}\mathbf{u}$ and $\mathbf{y} = \mathcal{C}\mathbf{v}$. By Proposition 4.12 $A\mathbf{x} = \mathcal{C}L\mathbf{u}$ and $A\mathbf{y} = \mathcal{C}L\mathbf{v}$. Using the fact that $\mathcal{C}$ is a symplectic mapping we have

$$
\begin{aligned}
\mathbf{x}^T(A^T J + JA)\mathbf{y} =& (A\mathbf{x})^T J\mathbf{y} + \mathbf{x}^T J(A\mathbf{y}) = \omega_J(A\mathbf{x}, \mathbf{y}) + \omega_J(\mathbf{x}, A\mathbf{y}) \\
=& \omega_J(\mathcal{C}L\mathbf{u}, \mathcal{C}\mathbf{v}) + \omega_J(\mathcal{C}\mathbf{u}, \mathcal{C}L\mathbf{v}) = \omega(L\mathbf{u}, \mathbf{v}) + \omega(\mathbf{u}, L\mathbf{v}).
\end{aligned}
$$

From this we see that $A^T J + JA = 0$, if, and only if, $\omega(L\mathbf{u}, \mathbf{v}) + \omega(\mathbf{u}, L\mathbf{v}) = 0$, for all $\mathbf{u}, \mathbf{v}$. This proves the proposition for the Hamiltonian case.

Analogously, on the one hand we have

$$
\mathbf{x}^T(A^T JA)\mathbf{y} = (A\mathbf{x})^T J(A\mathbf{y}) = \omega_J(A\mathbf{x}, A\mathbf{y}) = \omega_J(\mathcal{C}L\mathbf{u}, \mathcal{C}L\mathbf{v}) = \omega(L\mathbf{u}, L\mathbf{v}),
$$

and on the other hand we have $\mathbf{x}^T J\mathbf{y} = \omega_J(\mathbf{x}, \mathbf{y}) = \omega(\mathbf{u}, \mathbf{v})$.

Therefore, $A^T JA = J$, if, and only if, $\omega(L\mathbf{u}, L\mathbf{v}) = \omega(\mathbf{u}, \mathbf{v})$, for all $\mathbf{u}, \mathbf{v}$. This proves the proposition for the symplectic case. □

Remark 4.16 By this proposition we can define the operator L to be symplectic (Hamiltonian) if its matrix on some symplectic basis is symplectic (Hamiltonian).

Definition 4.17 Let (V, ω) be a symplectic space. A subspace U of V is said to be a *Lagrangian subspace* if its dimension is half that of V and ω restricted to U is zero.

So, if $\dim V = 2n$, then by definition:

$$
U \subset V \text{ is Lagrangian if and only if } \dim U = n \text{ and } \omega(\mathbf{x}, \mathbf{y}) = 0 \text{ for all } \mathbf{x}, \mathbf{y} \in U. \tag{4.10}
$$

Example 4.11 For $\mathbf{z}_1 = (\mathbf{x}_1, \mathbf{y}_1), \mathbf{z}_2 = (\mathbf{x}_2, \mathbf{y}_2) \in \mathbf{F}^{2n} = \mathbf{F}^n \times \mathbf{F}^n$ the symplectic form (4.2) is given by

$$
\omega_J(\mathbf{z}_1, \mathbf{z}_2) = \langle \mathbf{x}_1, \mathbf{y}_2 \rangle - \langle \mathbf{x}_2, \mathbf{y}_1 \rangle.
$$

We readily see that $U = \mathbf{F}^n \times \{0\}$ and $W = \{0\} \times \mathbf{F}^n$ are Lagrangian subspaces of $\mathbf{F}^{2n}$.

Notice that we have the decomposition $\mathbf{F}^{2n} = U \oplus W$ into Lagrangian subspaces.

Definition 4.18 A direct sum decomposition $V = U \oplus W$ of the symplectic space (V, ω) into Lagrangian subspaces U and W is called a *Lagrangian splitting* of V.

The Lagrangian splitting is not unique, as shown by the example $(\mathbf{R}^2, \omega_J)$ in which any two distinct lines through the origin gives a direct sum decomposition of the space $\mathbf{R}^2$ into Lagrangian subspaces.

Remark 4.19 If $L : V \to V$ is a symplectic operator and $U \subset V$ is a Lagrangian subspace, so is the subspace $L(U)$, since for $\mathbf{x}' = L(\mathbf{x}), \mathbf{y}' = L(\mathbf{y})$ in $L(U)$ we have $\omega(\mathbf{x}', \mathbf{y}') = \omega(L(\mathbf{x}), L(\mathbf{y})) = \omega(\mathbf{x}, \mathbf{y}) = 0$.

Proposition 4.20 *Given a Lagrangian subspace U of the symplectic space (V, ω), there is a Lagrangian splitting $V = U \oplus W$.*

Proof Let $W = \mathcal{C}^{-1}J\mathcal{C}(U)$, where $\mathcal{C} : V \to \mathbf{F}^{2n}$ and $J : \mathbf{F}^{2n} \to \mathbf{F}^{2n}$ are the isomorphisms (4.7) and (4.8). The composition $\mathcal{C}^{-1}J\mathcal{C}$ is a symplectic operator; hence, by the above Remark W is a Lagrangian subspace. Now, let $\mathbf{v} \in U \cap W$. Since $\mathbf{v} \in W$, we have $\mathcal{C}\mathbf{v} = J\mathcal{C}\mathbf{u}$ for some $\mathbf{u} \in U$ and since U is Lagrangian and $\mathbf{v} \in U$ also, we have

$$0 = \omega(\mathbf{v}, \mathbf{u}) = \omega_J(\mathcal{C}\mathbf{v}, \mathcal{C}\mathbf{u}) = \omega_J(J\mathcal{C}\mathbf{u}, \mathcal{C}\mathbf{u}) = (J\mathcal{C}\mathbf{u})^T J\mathcal{C}\mathbf{u} = \|J\mathcal{C}\mathbf{u}\|^2;$$

hence, $J\mathcal{C}\mathbf{u} = 0$, so $\mathbf{u} = 0$. Therefore, $U \cap W = \{0\}$; hence, $V = U \oplus W$ is a Lagrangian splitting. □

Example 4.12 Find all the Lagrangian subspaces of the symplectic space $(\mathbf{R}^4, \omega_A)$ of Example 4.7.

Let $\mathbf{u}, \mathbf{v}$ be generators of a 2-dimensional subspace U of $\mathbf{R}^4$. For any vectors $\mathbf{x} = \alpha\mathbf{u} + \beta\mathbf{v}$ and $\mathbf{y} = \gamma\mathbf{u} + \delta\mathbf{v}$ in U we compute $\omega_A(\mathbf{x}, \mathbf{y}) = (\alpha\delta - \beta\gamma)\langle A\mathbf{u}, \mathbf{v}\rangle$. Therefore, U is a Lagrangian subspace if, and only if, $\langle A\mathbf{u}, \mathbf{v}\rangle = 0$.

Example 4.13 In the symplectic space $(\mathbf{R}^4, \omega_A)$ of Example 4.7 consider the subspace U generated by the vectors $\mathbf{u}_1 = (1, 0, 1, 0)$ and $\mathbf{v}_1 = (0, 0, 2, 1)$. Show that it is a Lagrangian subspace and find a Lagrangian complement for it.

The vectors $\mathbf{u}_1$ and $\mathbf{v}_1$ are linearly independent, so the subspace U of $\mathbf{R}^4$ is 2-dimensional. Since $\langle A\mathbf{u}_1, \mathbf{v}_1\rangle = 0$, the subspace U is Lagrangian by the above example. We could now find the Lagrangian complement W as in the beginning of the proof of Proposition 4.20, but we can also proceed as follows. Let W be a subspace generated by two vectors $\mathbf{u}_2, \mathbf{v}_2$ which together with $\mathbf{u}_1, \mathbf{v}_1$ give a basis of $\mathbf{R}^4$ and are such that $\langle A\mathbf{u}_2, \mathbf{v}_2\rangle = 0$, for instance $\mathbf{u}_2 = (0, 1, 0, 1)$ and $\mathbf{v}_2 = (0, 0, 1, 0)$. Then, U_2 is a Lagrangian subspace and since $U \cap V = \{0\}$ we have $\mathbf{R}^4 = U \oplus V$, so V is a symplectic complement of U.

Example 4.14 Consider the real 4×4 Hamiltonian matrix $A = \begin{bmatrix} O & \Omega \\ -\Omega & O \end{bmatrix}$, where $\Omega = \begin{bmatrix} \omega_1 & 0 \\ 0 & \omega_2 \end{bmatrix}$ with $\omega_1 \neq \omega_2$. Find a two-dimensional A-invariant subspace of $\mathbf{R}^4$.

We look for linearly independent vectors $\mathbf{u} = (x, y, z, w)$, $\mathbf{v} = (\xi, \eta, \zeta, \chi) \in \mathbf{R}^4$ such that

$$A\mathbf{u} = \alpha\mathbf{u} + \beta\mathbf{v} \quad A\mathbf{v} = \gamma\mathbf{u} + \delta\mathbf{v}, \tag{4.11}$$

for some real numbers α, β, γ, δ. The Equations (4.11) define a homogeneous linear system in the unknown components of the vectors **u**, **v** and for this system to have nontrivial solutions it must be zero the determinant of its matrix of coefficients, that is, the matrix

$$M = \begin{bmatrix} \alpha I & -\Omega & \beta I & O \\ \Omega & \alpha I & O & \beta I \\ \gamma I & O & \delta I & -\Omega \\ O & \gamma I & \Omega & \delta I \end{bmatrix}.$$

Using the rule $\mathrm{Det}\begin{bmatrix} X & Y \\ Z & W \end{bmatrix} = \mathrm{Det}([WX - WYW^{-1}Z]$, we have

$$\begin{aligned}
\mathrm{Det}\, M &= \mathrm{Det}\left\{\begin{bmatrix} \delta I & -\Omega \\ \Omega & \delta I \end{bmatrix}\begin{bmatrix} \alpha I & -\Omega \\ \Omega & \alpha I \end{bmatrix} - \begin{bmatrix} \beta\gamma I & 0 \\ 0 & \beta\gamma I \end{bmatrix}\right\} \\
&= \mathrm{Det}\begin{bmatrix} (\alpha\delta - \beta\gamma)I - \Omega^2 & -(\alpha+\delta)\Omega \\ (\alpha+\delta)\Omega & (\alpha\delta - \beta\gamma)I - \Omega^2 \end{bmatrix} \\
&= \mathrm{Det}\left(\left((\alpha\delta - \beta\gamma)I - \Omega^2\right)^2 + (\alpha+\delta)^2\Omega^2\right) \\
&= \mathrm{Det}\left(\Omega^4 - 2\Delta\Omega^2 + \Delta^2 I + \tau^2\Omega^2\right),
\end{aligned}$$

where $\Delta = \alpha\delta - \beta\gamma$ and $\tau = \alpha + \delta$. The condition $\mathrm{Det}\, M = 0$ means that

$$\Delta^2 - 2\omega_1^2\Delta + (\omega_1^4 + \tau^2\omega_1^2) = 0 \quad \text{or} \quad \Delta^2 - 2\omega_2^2\Delta + (\omega_2^4 + \tau^2\omega_2^2) = 0.$$

Solving for Δ we get $\Delta = \omega_j^2 \pm \sqrt{-\tau^2\omega_j^2}$ for $j = 1$ or $j = 2$. Since Δ is real we must have $\tau = 0$ and then $\Delta = \omega_j^2$ for $j = 1$ or $j = 2$. For $\alpha = -\delta$, β and γ satisfying the condition $\Delta = \omega_1^2$ or $\Delta = \omega_2^2$ we have $\mathrm{Det}\, M = 0$ and the system (4.11) has non-trivial solutions $x, y, z, w, \xi, \eta, \zeta, \chi$. The corresponding vectors **u** and **v** are linearly independent because if, say, $\mathbf{v} = \sigma\mathbf{u}$, for some real σ, then $A\mathbf{u} = (\alpha + \beta\sigma)\mathbf{u}$ showing that A has a real eigenvalue, which is not the case since the eigenvalues of A are $\pm i\omega_1$ and $\pm i\omega_2$.

Therefore, the vectors **u**, **v** determine an A-invariant two-dimensional subspace U of $\mathbf{R}^4$.

We observe that $\{\mathbf{u}, \mathbf{v}\}$ cannot be zero, so U is not a Lagrangian subspace of $\mathbf{R}^4$. Let us prove that $\{\mathbf{u}, \mathbf{v}\} \neq 0$. From (4.11) we have $\{A\mathbf{u}, A\mathbf{v}\} = \Delta\{\mathbf{u}, \mathbf{v}\}$. But

$$\{A\mathbf{u}, A\mathbf{v}\} = \omega_1^2(x\zeta - z\xi) + \omega_2^2(y\chi - w\eta) \quad \text{and} \quad \{\mathbf{u}, \mathbf{v}\} = (x\zeta - z\xi) + (y\chi - w\eta),$$

so if $\{\mathbf{u}, \mathbf{v}\} = 0$, then since $\omega_1 \neq \omega_2$ we get

$$x\zeta - z\xi = 0 \quad \text{and} \quad y\chi - w\eta = 0. \tag{4.12}$$

Since $\Delta = \alpha\delta - \beta\gamma \neq 0$, α and γ cannot both be zero. Assume that $\gamma \neq 0$ and write the last four linear equations defined by the matrix M, namely:

$$\gamma x + \delta\xi - \omega_1\zeta = 0, \quad \gamma y + \delta\eta - \omega_2\chi = 0, \quad \gamma z + \omega_1\xi + \delta\zeta = 0,$$
$$\gamma w + \omega_2\eta + \delta\chi = 0.$$

Multiplying the first by ζ, the third by ξ and subtracting we get, using the first equation in (4.12)

$$\omega_1(\xi^2 + \zeta^2) = 0.$$

Multiplying the second by χ, the fourth by η and subtracting we get, using the second equation in (4.12)

$$\omega_2(\chi^2 + \eta^2) = 0.$$

Therefore, we get $\xi = \zeta = \chi = \eta = 0$ and since $\gamma \neq 0$ this implies $x = y = z = w = 0$, an absurd because the solution we are working with is nontrivial. The absurd came from the hypothesis $\{\mathbf{u}, \mathbf{v}\} = 0$.

Since $\{\mathbf{u}, \mathbf{v}\} \neq 0$, the subspace U is symplectic. By Proposition 4.5(b), we have a direct sum of symplectic subspaces $\mathbf{R}^4 = U \oplus U^\perp$.

We notice that $U^\perp$ is also invariant under A because if $\mathbf{z} \in U^\perp$, then $\{\mathbf{z}, \mathbf{u}\} = 0$ and $\{\mathbf{z}, \mathbf{v}\} = 0$ and since A is Hamiltonian we have $\{A\mathbf{z}, \mathbf{u}\} = -\{\mathbf{z}, A\mathbf{u}\} = 0$ and $\{A\mathbf{z}, \mathbf{v}\} = -\{\mathbf{z}, A\mathbf{v}\} = 0$.

We use this example to make a more general statement.

Example 4.15 Let A be any real 4×4 Hamiltonian matrix with purely imaginary eigenvalues $\pm i\omega_1, \pm i\omega_2$, with $\omega_1 \neq \omega_2$. Then, there is a direct sum decomposition $\mathbf{R}^4 = U \oplus V$, where U and V are symplectic subspaces of $\mathbf{R}^4$, which are invariant under A.

Let $H(\mathbf{z}) = \frac{1}{2}\mathbf{z}^T S\mathbf{z}$, with $S = -JA$ be the quadratic Hamiltonian defined by A. Since $\omega_1 \neq \omega_2$, by means of the linear symplectic transformation defined by the matrix (3.13) we obtain the normal form (3.14), which in our case has the matrix $\mathcal{A} = P^{-1}AP = \begin{bmatrix} O & \Omega \\ -\Omega & O \end{bmatrix}$, where $\Omega = \begin{bmatrix} \delta_1\omega_1 & 0 \\ 0 & \delta_2\omega_2 \end{bmatrix}$. By the previous example, we have an $\mathcal{A}$-invariant symplectic decomposition $\mathbf{R}^4 = \mathcal{U} \oplus \mathcal{V}$. Since P^{-1} is symplectic the subspaces $U = P^{-1}(\mathcal{U})$ and $V = P^{-1}(\mathcal{V})$ are symplectic and we get the A-invariant symplectic decomposition $A = U \oplus V$ because $A(U) = P^{-1}\mathcal{A}P\big(P^{-1}(\mathcal{U})\big) = P^{-1}\mathcal{A}(\mathcal{U}) \subset P^{-1}(\mathcal{U}) = U$. And the same is true of $A(V) \subset V$.

Proposition 4.21 *Given a Lagrangian splitting $V = U \oplus W$, let $\mathbf{u}_1, \ldots, \mathbf{u}_n$ be any basis of the Lagrangian subspace U. Then, there exists a unique basis $\mathbf{w}_1, \ldots, \mathbf{w}_n$ of W such that the vectors $\mathbf{u}_1, \ldots, \mathbf{u}_n, \mathbf{w}_1, \ldots, \mathbf{w}_n$ form a symplectic basis of V.*

If $V \subset \mathbf{C}^{2n}$ and ω is real on real vectors, then for a real basis of U the corresponding basis of W is also real.

Proof Consider the elements $\mathbf{w}^1, \ldots, \mathbf{w}^n$ of the dual space W^* of W given by $\mathbf{w}^i(\mathbf{x}) = \omega(\mathbf{u}_i, \mathbf{x})$, for $\mathbf{x} \in W$. They are linearly independent because if $\sum_i c_i \mathbf{w}^i = 0$, then $\omega(\sum_i c_i \mathbf{u}_i, \mathbf{x}) = 0$, for all $\mathbf{x} \in W$ and since this is also true for all $\mathbf{x} \in U$ because U is Lagrangian, it follows that it is true as well for all $\mathbf{x} \in V = U \oplus W$. But ω is nondegenerate, so $\sum_i c_i \mathbf{u}_i = 0$ and the linear independence of $\mathbf{u}_1, \ldots, \mathbf{u}_n$ implies $c_i = 0$ for all i. Hence, $\mathbf{w}^1, \ldots, \mathbf{w}^n$ are linearly independent. Therefore, they form a basis for W^*. Let $\mathbf{w}_1, \ldots, \mathbf{w}_n \in W$ be its dual basis. Then, $\omega(\mathbf{u}_i, \mathbf{w}_j) = \mathbf{w}^i(\mathbf{w}_j) = \delta_{ij}$, and this shows that $\mathbf{u}_1, \ldots, \mathbf{u}_n, \mathbf{w}_1, \ldots, \mathbf{w}_n$ is a symplectic basis of V. Uniqueness follows from the nondegeneracy of ω, because two such bases would lead to $\omega(\mathbf{u}_i, \tilde{\mathbf{w}}_j - \mathbf{w}_j) = 0$, for all i; hence, $\omega(\mathbf{x}, \tilde{\mathbf{w}}_j - \mathbf{w}_j) = 0$, for all $\mathbf{x} \in U$, and since this is also true for all $\mathbf{x} \in W$ because W is Lagrangian, it is also true for all $\mathbf{x} \in V$, so $\tilde{\mathbf{w}}_j = \mathbf{w}_j$.

To prove the last part assume the vectors $\mathbf{u}_1, \ldots, \mathbf{u}_n$ are real. Write $\mathbf{w}_j$ in terms of real vectors, $\mathbf{w}_j = \mathbf{w}_j^{(1)} + i\mathbf{w}_j^{(2)}$. Then, $\delta_{ij} = \omega(\mathbf{u}_i, \mathbf{w}_j) = \omega(\mathbf{u}_i, \mathbf{w}_j^{(1)}) + i\omega(\mathbf{u}_i, \mathbf{w}_j^{(2)})$, hence by the hypothesis on ω we have $\omega(\mathbf{u}_i, \mathbf{w}_j^{(2)}) = 0$, for all i. Therefore, $\omega(\mathbf{u}, \mathbf{w}_j^{(2)}) = 0$ for all $\mathbf{u} \in U$. Since W is Lagrangian, $\omega(\mathbf{z}, \mathbf{w}_j^{(2)}) = 0$ for all $\mathbf{z} \in W$. Consequently, $\omega(\mathbf{v}, \mathbf{w}_j^{(2)}) = 0$ for all $\mathbf{v} \in V$ and since ω is nondegenerate we get $\mathbf{w}_j^{(2)} = 0$. Therefore, $\mathbf{w}_j = \mathbf{w}_j^{(1)}$ is a real vector. □

Example 4.16 Consider the basis $\mathbf{u}_1 = (1, 0, 1, 0)$, $\mathbf{v}_1 = (0, 0, 2, 1)$ of U in the Lagrangian splitting $\mathbf{R}^4 = U \oplus V$ given in Example 4.13. Complete these vectors to a symplectic basis of $\mathbf{R}^4$.

Consider the functions $\mathbf{w}^1(\mathbf{x}) = \omega_A(\mathbf{u}_1, \mathbf{x})$ and $\mathbf{w}^2(\mathbf{x}) = \omega_A(\mathbf{u}_2, \mathbf{x})$ for $\mathbf{x} \in V$. Since $\omega_A(\mathbf{u}, \mathbf{x}) = \langle A\mathbf{u}, \mathbf{x}\rangle$, computing $A\mathbf{u}_1 = (1, 0, -1, 2)$ and $A\mathbf{v}_1 = (2, -2, -1, 2)$ we have for $\mathbf{x} = (x_1, x_2, x_3, x_4) \in \mathbf{R}^4$,

$$\mathbf{w}^1(\mathbf{x}) = x_1 - x_3 + 2x_4, \quad \mathbf{w}^2(\mathbf{x}) = x_1 - 2x_2 - x_3 + 2x_4.$$

The subspace V is generated by the vectors $\mathbf{u}_2 = (0, 1, 0, 1)$ and $\mathbf{v}_2 = (0, 0, 1, 0)$, see Example 4.13, so the vectors $\mathbf{x}$ in the subspace V are given by

$$\mathbf{x} = (0, a, b, a), \quad \text{for any} \quad a, b \in \mathbf{R}.$$

So, for $\mathbf{x} \in V$ we have $\mathbf{w}^1(\mathbf{x}) = -b + 2a$ and $\mathbf{w}^2(\mathbf{x}) = -b$. Let $\mathbf{w}_1, \mathbf{w}_2$ be the dual basis of $\mathbf{w}^1, \mathbf{w}^2$. Then, $\mathbf{w}^1(\mathbf{w}_1) = 1$ and $\mathbf{w}^1(\mathbf{w}_2) = 0$, which gives the values $a = \frac{1}{2}$ and $b = 0$, whereas $\mathbf{w}^2(\mathbf{w}_1) = 0$ and $\mathbf{w}^2(\mathbf{w}_2) = 1$, which gives the values $a = 0$ and $b = -1$. Therefore, the vectors of the basis of V are $\mathbf{w}_1 = (0, \frac{1}{2}, 0, \frac{1}{2})$ and $\mathbf{w}_2 = (0, 0, -1, 0)$ and we readily check that $\omega_A(\mathbf{u}_1, \mathbf{w}_1) = 1$ and $\omega_A(\mathbf{u}_2, \mathbf{w}_2) = 1$, as should be. Therefore, the vectors $\mathbf{u}_1, \mathbf{v}_1, \mathbf{w}_1, \mathbf{w}_2$ form a symplectic basis of $(\mathbf{R}^4, \omega_A)$.

Proposition 4.22 *Let $V = U \oplus W$ be a Lagrangian splitting and let $L : V \to V$ be a linear operator respecting this splitting, that is, $L(U) \subset U$ and $L(W) \subset W$. Let $\mathbf{u}_1, \dots, \mathbf{u}_n, \mathbf{w}_1, \dots, \mathbf{w}_n$ be a symplectic basis of V, with $\mathbf{u}_i \in U$ and $\mathbf{w}_j \in W$. Then, the matrix A of L on this basis is block-diagonal and is given by*

$$A = \begin{bmatrix} B & O \\ O & -B^T \end{bmatrix} \quad or \quad A = \begin{bmatrix} B & O \\ O & B^{-T} \end{bmatrix}, \tag{4.13}$$

according to whether L is a Hamiltonian or a symplectic operator.

Proof The matrix A has the block form $A = \begin{bmatrix} X & O \\ O & Y \end{bmatrix}$. Now use the fifth item in Propositions 2.6 and 2.26. □

4.3 The Spectral Decomposition of Hamiltonian Matrices

This section deals with the decomposition of a Hamiltonian matrix determined by the set of eigenvalues.

Definition 4.23 A polynomial $p(x) = a_0x^m + a_1x^{m-1} + \cdots + a_{m-1}x + a_m$ is said to be *even* if $p(x) = p(-x)$. It is said to be *reciprocal* if $p(x) = x^m p(\frac{1}{x})$.

An even polynomial $p(x)$ contains only even powers of x. If the polynomial $p(x)$ is reciprocal the powers of x equidistant from the extremes have the same coefficients, $a_k = a_{m-k}$.

Proposition 4.24 *The characteristic polynomial of a Hamiltonian matrix is even and that of a symplectic matrix is reciprocal.*

Proof We have already seen the Hamiltonian case in Proposition 2.23. Now, suppose that M is a symplectic matrix, $M^TJM = J$. Then, $M = J^{-1}M^{-T}J$; hence,

$$M - xI = J^{-1}(M^{-T} - xI)J = J^{-1}M^{-T}(I - xM)^TJ = -xJ^{-1}M^{-T}\left(M - \frac{1}{x}I\right)^T J.$$

Since a matrix and its transpose have the same determinant and $\det M = +1$ by the corollary to Proposition 2.31, then because we are dealing with matrices of even order $2n$, it follows from the above that

$$p(x) = \det(M - xI) = x^{2n}\det\big(M - \frac{1}{x}I\big) = x^{2n}p\left(\frac{1}{x}\right).$$

□

The statement about the symplectic matrix in this proposition is known as the Lyapunov–Poincaré theorem.

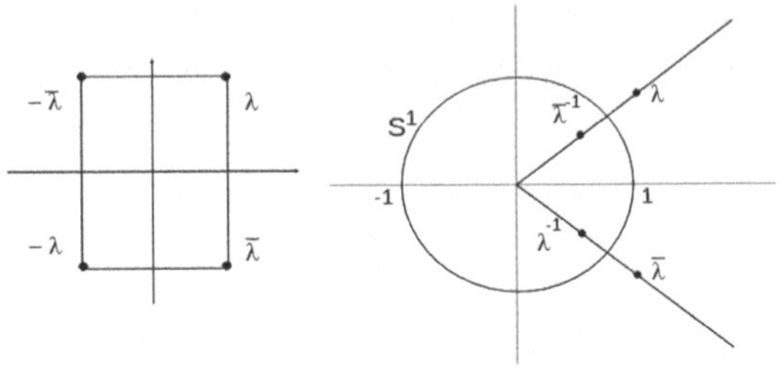

Fig. 4.1 Symmetry of the eigenvalues

It follows from this proposition that the eigenvalues of Hamiltonian and symplectic matrices come in pairs, namely $\lambda, -\lambda$ and λ, λ^{-1} respectively.

If the matrices are real we can even group them in quadruples

$$\lambda,\ -\lambda,\ \overline{\lambda},\ -\overline{\lambda} \text{ and } \lambda,\ \lambda^{-1},\ \overline{\lambda},\ \overline{\lambda}^{-1}$$

symmetric with respect to the real and imaginary axes in the Hamiltonian case and symmetric with respect to the unit circle in the symplectic case (see Fig. 4.1).

Proposition 4.25 *If A is a Hamiltonian matrix and M is symplectic, the matrix $B = M^{-1}AM$ is Hamiltonian.*

Proof Indeed, $A = JS$ with S symmetric and $M^{-1} = J^{-1}M^TJ$, so $B = J(M^TSM)$ is Hamiltonian, since M^TSM is symmetric. □

Given a complex square matrix A, the matrix in the collection of similar matrices $M^{-1}AM$, which has the simplest structure is its *Jordan canonical form*. In the class of Hamiltonian matrices we take the similar matrices only through symplectic M, so we expect the number of canonical forms to be greater than in the general case of real Jordan forms.

For example, for a 2×2 real Hamiltonian matrix $A = \begin{bmatrix} a & b \\ c & d \end{bmatrix}$ with eigenvalues i and $-i$, the complex Jordan canonical form $\begin{bmatrix} i & 0 \\ 0 & -i \end{bmatrix}$ can be written in real form in two ways, namely $A_1 = \begin{bmatrix} 0 & 1 \\ -1 & 0 \end{bmatrix}$ and $A_2 = \begin{bmatrix} 0 & -1 \\ 1 & 0 \end{bmatrix}$. This is essentially the same real Jordan normal form because if we consider the reflection $R = \begin{bmatrix} 0 & 1 \\ 1 & 0 \end{bmatrix}$ we see that $R^{-1}A_1R = A_2$.

However, the Hamiltonian matrices A_1 and A_2 are not similar *through a real symplectic matrix M*. Indeed, if $A_1 = M^{-1}A_2M$, then since $A_1 = J$ and $A_2 = -J$, we get $MJ = -JM$. Because M is symplectic this implies that $M^TM = -I$. This is not possible since the left-hand side is a positive definite matrix whereas the right-hand side is negative definite. Therefore, the Hamiltonian matrix A has the two nonsymplectically similar canonical forms A_1 and A_2.

To study the structure of the real canonical forms of a real Hamiltonian matrix A we will decompose the symplectic space $(\mathbf{C}^{2n}, \omega_J)$ into an A-invariant direct sum of symplectic subspaces that admit real symplectic bases.

Definition 4.26 A subspace U of $\mathbf{C}^n$ is a *complexification* if it has a real basis.

If $\mathbf{u}_1, \ldots, \mathbf{u}_k$ is a real basis of the subspace U of $\mathbf{C}^n$, then the subspace $U_{\mathbf{R}} = \{x_1\mathbf{u}_1 + \cdots + x_k\mathbf{u}_k;\ x_j \in \mathbf{R}\}$ of U is formed by real vectors and $U = \{\mathbf{x} + i\mathbf{y};\ \mathbf{x}, \mathbf{y} \in U_{\mathbf{R}}\}$, so U is the complexification of the real subspace $U_{\mathbf{R}}$.

Notice that $\mathbf{C}^n$ is a complexification as the canonical basis of $\mathbf{R}^n$ is a real basis of $\mathbf{C}^n$. Of course, $\mathbf{C}^n = \mathbf{R}^n + i\mathbf{R}^n$.

Proposition 4.27 *The subspace U is a complexification if, and only if, $\mathbf{u} \in U$ implies $\overline{\mathbf{u}} \in U$.*

Proof If U is a complexification it has a real basis $\mathbf{v}_1, \ldots, \mathbf{v}_k$; hence, for any $\mathbf{u} = z_1\mathbf{v}_1 + \cdots + z_k\mathbf{v}_k \in U$, we have $\overline{\mathbf{u}} = \overline{z}_1\mathbf{v}_1 + \cdots + \overline{z}_k\mathbf{v}_k \in U$. For the converse, let $\mathbf{v}_1, \ldots, \mathbf{v}_k$ be any basis of U. Then, the vectors $\mathbf{u}_j = \frac{1}{2}(\mathbf{v}_j + \overline{\mathbf{v}}_j)$ and $\mathbf{w}_j = \frac{1}{2i}(\mathbf{v}_j - \overline{\mathbf{v}}_j)$ are real vectors belonging to U and they generate this subspace, so from them we can extract a real basis; hence, U is a complexification. □

This proposition says that the subspace $U \subset \mathbf{C}^n$ is a complexification if, and only if, it is invariant under conjugation.

Denoting by $\eta(\lambda) = \ker(A - \lambda I)$ the eigenspace of the eigenvalue λ of the matrix A we have the following.

Lemma 4.28 *Let λ and μ be eigenvalues of a Hamiltonian matrix A with $\lambda + \mu \neq 0$. Then, $\{\eta(\lambda), \eta(\mu)\} = 0$, that is, eigenvectors corresponding to λ and μ are symplectically orthogonal (J-orthogonal).*

Proof Since A is Hamiltonian, $A^TJ + JA = 0$, so if $A\mathbf{u} = \lambda\mathbf{u}$ and $A\mathbf{v} = \mu\mathbf{v}$, we have

$$0 = \mathbf{u}^T(A^TJ + JA)\mathbf{v} = (A\mathbf{u})^TJ\mathbf{v} + \mathbf{u}^TJA\mathbf{v} = \lambda\mathbf{u}^TJ\mathbf{v} + \mu\mathbf{u}^TJ\mathbf{v} = (\lambda + \mu)\{\mathbf{u}, \mathbf{v}\}.$$

Since $\lambda + \mu \neq 0$ it follows that $\{\mathbf{u}, \mathbf{v}\} = 0$. □

Now let A be a real Hamiltonian matrix. For the moment we assume that all of its eigenvalues are distinct, so none of them is zero and also the matrix is diagonalizable. Later on we will consider the case of multiple eigenvalues but with the matrix A diagonalizable. In the next chapter we consider the general case of a nondiagonalizable matrix.

Let us group the eigenvalues of A in three classes, real, purely imaginary, and truly complex,

$$\pm\alpha_1, \ldots, \pm\alpha_k, \quad \pm i\beta_1, \ldots, \pm i\beta_r, \quad \pm\gamma_1 \pm i\delta_1, \ldots, \pm\gamma_s \pm i\delta_s$$

and consider the subspaces of $\mathbf{C}^{2n}$,

$$\begin{aligned}
U_j &= \eta(\alpha_j) \oplus \eta(-\alpha_j), \\
Y_j &= \eta(i\beta_j) \oplus \eta(-i\beta_j), \\
Z_j &= \big[\eta(\gamma_j + i\delta_j) \oplus \eta(\gamma_j - i\delta_j)\big] \oplus \big[\eta(-\gamma_j + i\delta_j) \oplus \eta(-\gamma_j - i\delta_j)\big],
\end{aligned} \tag{4.14}$$

and the direct sum of each kind of these subspaces,

$$U = U_1 \oplus \ldots \oplus U_k, \quad Y = Y_1 \oplus \ldots \oplus Y_r \quad \text{and} \quad Z = Z_1 \oplus \ldots \oplus Z_s.$$

Then we have the direct sum decomposition

$$\mathbf{C}^{2n} = U \oplus Y \oplus Z,$$

which is A-invariant because, in fact, each subspace U_j, Y_j, Z_j is invariant under A.

By Lemma 4.28, $\{U, Y \oplus Z\} = 0$, so by part (a) of Proposition 4.7 the subspaces U and $Y \oplus Z$ are symplectic. It then follows, by the same reason, that Y and Z are symplectic. And again by the same reason each of the subspaces U_j, Y_j, Z_j are symplectic. Using a symplectic basis of each one of the subspaces U, Y, Z, we can get a symplectic basis for $\mathbf{C}^{2n}$ using Proposition 4.10.

Similarly, if the symplectic space $V = V_1 \oplus \ldots \oplus V_k$ is a direct sum of two-dimensional subspaces mutually J-orthogonal and we choose symplectic bases $\mathbf{u}_1, \mathbf{v}_1; \mathbf{u}_2, \mathbf{v}_2, \ldots, \mathbf{u}_l, \mathbf{v}_l$ in each one of them, by Proposition 4.10 we get the symplectic basis $\mathbf{u}_1, \ldots, \mathbf{u}_l; \mathbf{v}_1, \ldots, \mathbf{v}_l$ of V. In this way we can construct a symplectic basis in U from symplectic bases in $U_1, \ldots, U_k$ and similarly for Y from symplectic bases in $Y_1, \ldots, Y_k$. And, also, from symplectic bases in the four-dimensional subspaces $Z_1, \ldots, Z_l$ we can construct a symplectic basis for Z using Proposition 4.10.

Notice that the A-invariant subspaces U_j, Y_j, Z_j of $\mathbf{C}^{2n}$ are complexifications because they are invariant under conjugation, so they have real bases. Therefore, if we can obtain a real symplectic basis in each invariant subspace U_j, Y_j, Z_j so as to get simple representations (normal forms) of the restrictions of A to these subspaces, we will find a real symplectic basis of $\mathbf{C}^{2n}$ which yields a real normal form of A.

4.4 Normal Form of Diagonalizable Real Hamiltonian Matrices

In this section we normalize the real Hamiltonian matrix assuming the eigenvalues to be distinct but not necessarily purely imaginary numbers. We then specialize to the purely imaginary case to show that we recover the linear normalization of

Chap. 3. We end the section with Theorem 4.39 and its corollary, which deal with the case of a diagonalizable real Hamiltonian matrix having multiple purely imaginary eigenvalues.

Lemma 4.29 *There is a real symplectic basis of* U_j *in which the matrix of* $A_{|U_j}$ *is of the form*

$$A = \begin{bmatrix} \alpha_j & 0 \\ 0 & -\alpha_j \end{bmatrix}, \quad \alpha_j \neq 0.$$

Proof Let $\mathbf{x}_j$ and $\mathbf{y}_j$ be real eigenvectors of $A_{|U_j}$ corresponding to the eigenvalues α_j and $-\alpha_j$. Then, they are linearly independent vectors of the two-dimensional symplectic subspace U_j; hence, $\gamma_j = \{\mathbf{x}_j, \mathbf{y}_j\} \neq 0$, otherwise $\{\ ,\ \}$ would be zero on U_j. Let $\mathbf{u}_j = \mathbf{x}_j$ and $\mathbf{v}_j = \frac{1}{\gamma_j}\mathbf{y}_j$. Then, $\mathbf{u}_j, \mathbf{v}_j$ is a real symplectic basis of U_j. Since $A\mathbf{u}_j = \alpha_j \mathbf{u}_j$, $A\mathbf{v}_j = -\alpha_j \mathbf{v}_j$, the matrix of A on this basis is that in the statement of the lemma. □

Lemma 4.30 *There is a real symplectic basis of* Y_j *in which the matrix of* $A_{|Y_j}$ *is one of the two matrices*

$$A = \begin{bmatrix} 0 & \beta_j \\ -\beta_j & 0 \end{bmatrix} \quad or \quad A = \begin{bmatrix} 0 & -\beta_j \\ \beta_j & 0 \end{bmatrix}.$$

Proof Let $\mathbf{x}_j = \mathbf{r}_j + i\mathbf{s}_j \in Y_j$ be an eigenvector of A corresponding to $i\beta_j$, so $A\mathbf{r}_j = -\beta_j \mathbf{s}_j$ and $A\mathbf{s}_j = \beta_j \mathbf{r}_j$. Since the vectors $\mathbf{r}_j + i\mathbf{s}_j$ and $\mathbf{r}_j - i\mathbf{s}_j$ are linearly independent, so are $\mathbf{r}_j$ and $\mathbf{s}_j$. These real vectors belong to Y_j because this subspace is invariant under complex conjugation. Therefore, $\mathbf{r}_j, \mathbf{s}_j$ form a real basis for Y_j but it may not be symplectic. However, $\{\mathbf{r}_j, \mathbf{s}_j\} \neq 0$ otherwise, $\{\ ,\ \}$ would be zero on Y_j.

If $\{\mathbf{r}_j, \mathbf{s}_j\} = \gamma_j^2 > 0$, $\gamma_j > 0$, take $\mathbf{u}_j = \gamma_j^{-1}\mathbf{r}_j$, $\mathbf{v}_j = \gamma_j^{-1}\mathbf{s}_j$. Then, $\{\mathbf{u}_j, \mathbf{v}_j\} = \gamma_j^{-2}\{\mathbf{r}_j, \mathbf{s}_j\} = 1$ and so $\mathbf{u}_j, \mathbf{v}_j$ is a real symplectic basis of Y_j. Since $A\mathbf{u}_j = -\beta_j \mathbf{v}_j$ and $A\mathbf{v}_j = \beta_j \mathbf{u}_j$, the matrix of $A_{|Y_j}$ on this basis is the first matrix in the statement of the lemma.

If $\{\mathbf{r}_j, \mathbf{s}_j\} = -\gamma_j^2 < 0$, $\gamma_j > 0$, take $\mathbf{u}_j = \gamma_j^{-1}\mathbf{s}_j$, $\mathbf{v}_j = \gamma_j^{-1}\mathbf{r}_j$. Then, $\{\mathbf{u}_j, \mathbf{v}_j\} = \gamma_j^{-2}\{\mathbf{s}_j, \mathbf{r}_j\} = 1$ and so $\mathbf{u}_j, \mathbf{v}_j$ is a real symplectic basis of Y_j. Since $A\mathbf{u}_j = \beta_j \mathbf{v}_j$ and $A\mathbf{v}_j = -\beta_j \mathbf{v}_j$, the matrix $A_{|Y_j}$ in this case is the second in the statement of the lemma. □

Lemma 4.31 *There is a real symplectic basis in* Z_j *in which the matrix of* $A_{|Z_j}$ *has the form*

$$A = \begin{bmatrix} B_j & O \\ O & -B_j^T \end{bmatrix},$$

where B_j *is a* 2×2 *real matrix with eigenvalues* $\gamma_j \pm i\delta_j$.

Proof First, we prove that the decomposition of Z_j in (4.14) is a Lagrangian splitting.

Indeed, let $\mathbf{v}_1 = a_1\mathbf{u} + b_1\bar{\mathbf{u}}$ and $\mathbf{v}_2 = a_2\mathbf{u} + b_2\bar{\mathbf{u}}$ be elements of the first summand, where $\mathbf{u}$ is an eigenvector of A corresponding to $\gamma_j + i\delta_j$ and $\bar{\mathbf{u}}$ the eigenvector corresponding to $\gamma_j - i\delta_j$. Then, we have $\omega(\mathbf{v}_1, \mathbf{v}_2) = \{\mathbf{v}_1, \mathbf{v}_2\} = (a_1b_2 - a_2b_1)\{\mathbf{u}, \bar{\mathbf{u}}\} = 0$, by Lemma 4.28, because the sum of the eigenvalues is $2\gamma_j \neq 0$. The same argument goes for the second summand. Therefore, the decomposition of Z_j in (4.14) is a Lagrangian splitting.

Now, each summand in that decomposition of Z_j is a complexification of a real subspace because it is invariant under complex conjugation so it has a real basis. Choose a real basis in the first summand and complete it with a real basis of the second to get a real symplectic basis of Z_j as in Proposition 4.21. Now use Proposition 4.22 to finish the argument. Notice that the eigenvalues of B_j are $\gamma_j + i\delta_j$ and $\gamma_j - i\delta_j$. □

Using these lemmas and Proposition 4.10 we find a symplectic basis of $\mathbf{R}^{2n}$ and the corresponding normal form of A.

Let us consider in more detail the case of purely imaginary eigenvalues.

First, notice that the number γ_j appearing in the proof of Lemma 4.30 is $\gamma_j = \sqrt{|\{\mathbf{r}_j, \mathbf{s}_j\}|}$. Let $\delta_j = \text{sign}\{\mathbf{r}_j, \mathbf{s}_j\}$ and set

$$\tilde{\mathbf{u}}_j = -\kappa_j \mathbf{s}_j \quad \text{and} \quad \tilde{\mathbf{v}}_j = \delta_j \kappa_j \mathbf{r}_j, \quad \text{with} \quad \kappa_j = \gamma_j^{-1}.$$

Then $\tilde{\mathbf{u}}_j, \tilde{\mathbf{v}}_j$ is a real symplectic basis of Y_j and we also have $A\tilde{\mathbf{u}}_j = -\delta_j\beta_j\tilde{\mathbf{v}}_j$ and $A\tilde{\mathbf{v}}_j = \delta_j\beta_j\tilde{\mathbf{u}}_j$.

Therefore, we get the following version of Lemma 4.30:

Lemma 4.32 *Let $\delta_j = sign\{\mathbf{r}_j, \mathbf{s}_j\}$, $\kappa_j^{-1} = \sqrt{|\{\mathbf{r}_j, \mathbf{s}_j\}|}$ and consider the vectors of Y_j*

$$\tilde{\mathbf{u}}_j = -\kappa_j \mathbf{s}_j \quad \textit{and} \quad \tilde{\mathbf{v}}_j = \delta_j \kappa_j \mathbf{r}_j.$$

Then, these vectors form a real symplectic basis of Y_j and the matrix of $A_{|Y_j}$ is given by

$$A = \begin{bmatrix} 0 & \delta_j\beta_j \\ -\delta_j\beta_j & 0 \end{bmatrix}.$$

Theorem 4.33 *Let A be a real Hamiltonian matrix with all its eigenvalues distinct and purely imaginary, $\pm i\omega_1, \ldots, \pm i\omega_n$, with $\omega_1, \ldots, \omega_n$ positive numbers. Let $\mathbf{r}_j + i\mathbf{s}_j$ be an eigenvector of A belonging to the eigenvalue $i\omega_j$ and let $\delta_j = sign\{\mathbf{r}_j, \mathbf{s}_j\}$, $\kappa_j^{-1} = \sqrt{|\{\mathbf{r}_j, \mathbf{s}_j\}|}$. Then, the following real matrix is symplectic*

$$P = Col\big[-\kappa_1\mathbf{s}_1, \ldots, -\kappa_n\mathbf{s}_n; \delta_1\kappa_1\mathbf{r}_1, \ldots, \delta_n\kappa_n\mathbf{r}_n\big] \tag{4.15}$$

and

$$P^{-1}AP = \begin{bmatrix} O & \Omega \\ -\Omega & O \end{bmatrix}, \tag{4.16}$$

where $\Omega = diag[\delta_1\omega_1, \ldots, \delta_n\omega_n]$.

Proof The spectral decomposition of A is $\mathbf{C}^{2n} = \eta(i\omega_1) \oplus \eta(-i\omega_1) \oplus \ldots \oplus \eta(i\omega_n)\oplus\eta(-i\omega_n)$. Let $\mathbf{u}_j, \mathbf{v}_j$ be the real symplectic basis of $Y_j = \eta(i\omega_j)\oplus\eta(-i\omega_j)$ constructed in Lemma 4.32. Then, the ordered sequence of vectors

$$\mathbf{u}_1, \mathbf{u}_2, \ldots, \mathbf{u}_n, \mathbf{v}_1, \mathbf{v}_2, \ldots, \mathbf{v}_n \tag{4.17}$$

is a real symplectic basis of $V = \mathbf{C}^{2n}$. From the lemma we have $A\mathbf{u}_j = -\delta_j\omega_j\mathbf{v}_j$ and $A\mathbf{v}_j = \delta_j\omega_j\mathbf{u}_j$, so the matrix (of the linear operator) A in the basis (4.17) is the matrix on the right-hand side of (4.16). But if P is the transition matrix from the canonical basis $\mathbf{e}_1, \ldots, \mathbf{e}_{2n}$ of $\mathbf{R}^{2n}$ to the symplectic basis (4.17) the matrix A on this basis is $P^{-1}AP$. Therefore, we have the equality (4.16).

Now, by definition of the transition matrix between two bases, we have

$$P\mathbf{e}_j = \mathbf{u}_j = -\kappa_j\mathbf{s}_j \quad \text{and} \quad P\mathbf{e}_{n+j} = \mathbf{v}_j = \delta_j\kappa_j\mathbf{r}_j,$$

so P is given by the expression (4.15), the matrix whose columns are $\mathbf{u}_1, \ldots, \mathbf{u}_n, \mathbf{v}_1, \ldots, \mathbf{v}_n$. As a transition matrix between symplectic bases P is a symplectic matrix. □

Remark 4.34 Notice that the matrix P in (4.15) is the same as that in (3.13).

Now we consider the case where the eigenvalues of A may not be distinct so, in particular, zero may appear among them.

For an eigenvalue λ of A, we have the A-invariant subspace $\eta_k(\lambda) = \ker(A - \lambda I)^k$. Clearly $\eta_k(\lambda) \subset \eta_{k+1}(\lambda)$, so

$$\eta^\dagger(\lambda) = \cup_{k\geq 1}\eta_k(\lambda) \tag{4.18}$$

is the largest, in the sense of inclusion, of these invariant subspaces.

Definition 4.35 The subspace $\eta^\dagger(\lambda)$ is called the generalized eigenspace of λ.

The dimension of $\eta^\dagger(\lambda)$ is the multiplicity of λ as a root of the characteristic polynomial.

Since A commutes with $(A - \lambda I)^k$, if $\mathbf{v} \in \eta_k(\lambda)$ then $(A - \lambda I)^k A\mathbf{v} = A(A - \lambda I)^k\mathbf{v} = 0$, hence $A\mathbf{v} \in \eta_k(\lambda)$. Therefore, each subspace $\eta_k(\lambda)$, so also $\eta^\dagger(\lambda)$, is A-invariant.

We need a lemma similar to Lemma 4.28.

Lemma 4.36 *Let λ and μ be eigenvalues of a Hamiltonian matrix A with $\lambda+\mu \neq 0$. Then, $\{\eta^\dagger(\lambda), \eta^\dagger(\mu)\} = 0$, that is, the generalized eigenspaces corresponding to λ and μ are J-orthogonal.*

Proof By induction. Notice that $\{\eta_1(\lambda), \eta_1(\mu)\} = 0$ by Lemma 4.28. We will prove that $\{\eta_k(\lambda), \eta_k(\mu)\} = 0$ implies $\{\eta_{k+1}(\lambda), \eta_{k+1}(\mu)\} = 0$ by first proving that $\{\eta_{k+1}(\lambda), \eta_k(\mu)\} = 0$.

First observe that $A^T J + JA = 0$ implies $\{\mathbf{u}, A\mathbf{v}\} = -\{A\mathbf{u}, \mathbf{v}\}$, which leads to the equality $\{\mathbf{u}, (A+\lambda I)^k \mathbf{v}\} = \{(-A+\lambda I)^k \mathbf{u}, \mathbf{v}\}$ for all $k \geq 1$, as is easily checked.

Now, for $\mathbf{u} \in \eta_{k+1}(\lambda)$ and $\mathbf{v} \in \eta_k(\mu)$ we will prove that $\{\mathbf{u}, \mathbf{v}\} = 0$. Write $(A-\mu I)^k = [A + \lambda I - (\lambda+\mu)I)]^k$ and use the binomial formula to get

$$(A-\mu I)^k = \sum_{j=0}^{k} C^k_{k-j}(-(\lambda+\mu))^{k-j}(A+\lambda I)^j.$$

Then, using the identity $\{\mathbf{u}, (A+\lambda I)^j \mathbf{v}\} = \{(-A+\lambda I)^j \mathbf{u}, \mathbf{v}\}$ we get

$$0 = \{\mathbf{u}, (A-\mu I)^k \mathbf{v}\} = \sum_{j=0}^{k} C^k_{k-j}(-(\lambda+\mu))^{k-j}\{(-A+\lambda I)^j \mathbf{u}, \mathbf{v}\}.$$

Since $\mathbf{u} \in \eta_{k+1}(\lambda)$ we have $(-A+\lambda I)^j \mathbf{u} = (-1)^j (A-\lambda I)^j \mathbf{u} \in \eta_k(\lambda)$ for every $j = 1, 2, \ldots, k$. Therefore, since $\mathbf{v} \in \eta_k(\mu)$, the assumption $\{\eta_k(\lambda), \eta_k(\mu)\} = 0$ implies that in the above sum all the terms are zero except that for $j = 0$. This means that we have $0 = (-(\lambda+\mu))^k \{\mathbf{u}, \mathbf{v}\}$ and since $\lambda + \mu \neq 0$ we get $\{\mathbf{u}, \mathbf{v}\} = 0$. This proves that $\{\eta_{k+1}(\lambda), \eta_k(\mu)\} = 0$. By the same argument we then show that this implies $\{\eta_{k+1}(\lambda), \eta_{k+1}(\mu)\} = 0$. □

We have a corresponding result for symplectic matrices, namely the following.

Lemma 4.37 *Let ρ and σ be eigenvalues of a symplectic matrix M with $\rho\sigma \neq 1$. Then, $\{\eta^\dagger(\rho), \eta^\dagger(\sigma)\} = 0$, that is, the generalized eigenspaces corresponding to ρ and σ are J-orthogonal.*

Proof First, we prove that $\{\eta_1(\rho), \eta_1(\sigma)\} = 0$. Indeed, since $M^T JM = J$ we have $\{M\mathbf{u}, M\mathbf{v}\} = \mathbf{u}^T M^T JM\mathbf{v} = \mathbf{u}^T J\mathbf{v} = \{\mathbf{u}, \mathbf{v}\}$. Therefore, if $\mathbf{u} \in \eta_1(\rho)$ and $\mathbf{v} \in \eta_1(\sigma)$ we get $\rho\sigma\{\mathbf{u}, \mathbf{v}\} = \{\mathbf{u}, \mathbf{v}\}$; hence, $\{\mathbf{u}, \mathbf{v}\} = 0$ since $\rho\sigma \neq 1$.

We observe that $M^T JM = J$ implies $\{\mathbf{u}, M\mathbf{v}\} = \{M^{-1}\mathbf{u}, \mathbf{v}\}$, which leads to the equality $\{\mathbf{u}, (M+\rho I)^k \mathbf{v}\} = \{(M^{-1}+\lambda I)^k \mathbf{u}, \mathbf{v}\}$ for all $k \geq 1$, as is easily checked.

Now, by induction, we assume that $\{\eta_k(\rho), \eta_k(\sigma)\} = 0$. With an argument similar to that for Lemma 4.36 we prove that for $\mathbf{u} \in \eta_{k+1}(\rho)$ and $\mathbf{v} \in \eta_k(\sigma)$ we have $\{\mathbf{u}, \mathbf{v}\} = 0$, so $\{\eta_{k+1}(\rho), \eta_k(\sigma)\} = 0$. Then, with the same argument, it is proved that $\{\eta_{k+1}(\rho), \eta_{k+1}(\sigma)\} = 0$. □

Back to Hamiltonian matrices. We will group the eigenvalues of A into four classes: the zero eigenvalue and the nonzero real, purely imaginary, and truly complex eigenvalues

$$0;\ \pm\alpha_1, \ldots, \pm\alpha_k; \quad \pm i\beta_1; \ldots \pm i\beta_r, \quad \pm\gamma_1 \pm i\delta_1, \ldots, \pm\gamma_s \pm i\delta_s \tag{4.19}$$

and consider the A-invariant subspaces of $\mathbf{C}^{2n}$,

$$X = \eta^\dagger(0), \quad U = U_1 \oplus \ldots \oplus U_k, \quad Y = Y_1 \oplus \ldots \oplus Y_r \quad \text{and} \quad Z = Z_1 \oplus \ldots \oplus Z_s,$$

where U_j, Y_j, Z_j are the subspaces given in (4.14) with eigenspaces replaced by generalized eigenspaces. Then we have the A-invariant direct sum decomposition

$$\mathbf{C}^{2n} = X \oplus U \oplus Y \oplus Z. \tag{4.20}$$

By Lemma 4.36, $\{X, U \oplus Y \oplus Z\} = 0$, so by Proposition 4.7(a) the subspaces X and $U \oplus Y \oplus Z$ are symplectic. It then follows, for the same reason, that U and $Y \oplus Z$ are symplectic and also that Y and Z are symplectic. And again for the same reason each of the subspaces U_j, Y_j, Z_j are symplectic.

From symplectic bases of $U_1, \ldots, U_k$ we get a symplectic basis of $U = U_1 \oplus \ldots \oplus U_k$ using Proposition 4.10 and similarly for $Y = Y_1 \oplus \ldots \oplus Y_r$ and $Z = Z_1 \oplus \ldots \oplus Z_s$. Considering a symplectic basis for each one of the mutually J-orthogonal symplectic subspaces X, U, Y, Z, we get a symplectic basis for $\mathbf{C}^{2n}$ using Proposition 4.10.

The construction of these symplectic bases will be the goal of the next chapter.

To conclude this section we consider the normal form of a real Hamiltonian matrix in the special case, where A has multiple purely imaginary eigenvalues and is diagonalizable.

First, we prove the following lemma.

Lemma 4.38 *Let $i\beta$ be a purely imaginary eigenvalue with multiplicity k of the real Hamiltonian matrix A. Let the restriction of A to the subspace $U = \eta^\dagger(i\beta) \oplus \eta^\dagger(-i\beta)$ be diagonalizable. Then, there is a basis $\mathbf{w}_1, \ldots, \mathbf{w}_k, \overline{\mathbf{w}}_1 \ldots, \overline{\mathbf{w}}_k$ of U formed by eigenvectors of A belonging to the eigenvalues $i\beta$ and $-i\beta$ such that $\{\mathbf{w}_1, \overline{\mathbf{w}}_1\} \neq 0$.*

Proof Consider a basis of U formed by eigenvectors $\mathbf{w}_1, \ldots, \mathbf{w}_k$ of A belonging to $i\beta$, and the corresponding eigenvectors $\overline{\mathbf{w}}_1 \ldots, \overline{\mathbf{w}}_k$ of A belonging to $-i\beta$.

If $\{\mathbf{w}_1, \overline{\mathbf{w}}_1\} \neq 0$ we are done. If not, it may happen that $\{\mathbf{w}_j, \overline{\mathbf{w}}_j\} \neq 0$ for some j, in which case we take $\mathbf{w}_j$ as $\mathbf{w}_1$. If $\{\mathbf{w}_j, \overline{\mathbf{w}}_j\} = 0$ for all j, there is some $j \geq 2$ such that $\{\mathbf{w}_1, \overline{\mathbf{w}}_j\} \neq 0$; otherwise, $\mathbf{w}_1$ would be symplectically orthogonal to all the basis vectors, because $\eta^\dagger(i\beta)$ is a Lagrangian subspace of U by Lemma 4.37. Since the symplectic form $\{\ ,\ \}$ is nondegenerate on U we would get $\mathbf{w}_1 = 0$, a contradiction.

Now, $\mathbf{w}_1 + \mathbf{w}_j$ is an eigenvector of A belonging to $i\beta$ and

$$\{\mathbf{w}_1 + \mathbf{w}_j, \overline{\mathbf{w}}_1 + \overline{\mathbf{w}}_j\} = \{\mathbf{w}_1, \overline{\mathbf{w}}_j\} + \{\mathbf{w}_j, \overline{\mathbf{w}}_1\}.$$

If the right-hand side of this equality is nonzero, we take $\mathbf{w}_1 + \mathbf{w}_j$ as the vector $\mathbf{w}_1$. If it is zero we consider the eigenvector $\mathbf{w}_1 + i\mathbf{w}_j$ of A belonging to $i\beta$ and compute

$$\{\mathbf{w}_1 + i\mathbf{w}_j, \overline{\mathbf{w}}_1 - i\overline{\mathbf{w}}_j\} = -i\big(\{\mathbf{w}_1, \overline{\mathbf{w}}_j\} - \{\mathbf{w}_j, \overline{\mathbf{w}}_1\}\big).$$

This time, the right-hand side of this equality is nonzero; otherwise, together with $\{\mathbf{w}_1, \overline{\mathbf{w}}_j\} + \{\mathbf{w}_j, \overline{\mathbf{w}}_1\} = 0$ we would get $\{\mathbf{w}_1, \overline{\mathbf{w}}_j\} = 0$, a contradiction. In this case we take $\mathbf{w}_1 + i\mathbf{w}_j$ as the vector $\mathbf{w}_1$.

The above discussion shows that we can assume that $\{\mathbf{w}_1, \overline{\mathbf{w}}_1\} \neq 0$. □

Theorem 4.39 *Let $i\beta$ be a purely imaginary eigenvalue with multiplicity k of the real Hamiltonian matrix A. Let the restriction A_β of A to the subspace $U = \eta^\dagger(i\beta) \oplus \eta^\dagger(-i\beta)$ be diagonalizable. Then, there is a real symplectic basis of U such that the matrix of A_β on this basis is $\begin{bmatrix} O & B_\delta \\ -B_\delta & O \end{bmatrix}$, where $B_\delta = diag\big[\delta_1\beta, \dots, \delta_k\beta\big]$, with $\delta_j = \pm 1$ determined in the process of finding the symplectic basis.*

Proof Let $\mathbf{w}_1, \dots, \mathbf{w}_k, \overline{\mathbf{w}}_1 \dots, \overline{\mathbf{w}}_k$ be a basis of U as in the above lemma. Let $\mathbf{w}_1 = \mathbf{r}_1 + i\mathbf{s}_1$. Then $\{\mathbf{r}_1, \mathbf{s}_1\} \neq 0$ because $\{\mathbf{w}_1, \overline{\mathbf{w}}_1\} \neq 0$. Let $\delta_1 = \text{sign}\{\mathbf{r}_1, \mathbf{s}_1\}$, $\kappa_1 = 1/\sqrt{|\{\mathbf{r}_1, \mathbf{s}_1\}|}$ and consider the vectors $\mathbf{u}_1 = \kappa_1\mathbf{r}_1, \mathbf{v}_1 = \delta_1\kappa_1\mathbf{s}_1$. Then, the vectors $\mathbf{u}_1$, $\mathbf{v}_1$ form a real symplectic basis for the subspace U_1 of U generated by the vectors $\mathbf{w}_1, \overline{\mathbf{w}}_1$. By Proposition 4.7(b) the subspace $U_1^\perp$ is symplectic and since A is Hamiltonian, $U = U_1 \oplus U_1^\perp$ is an A-invariant decomposition. Restricted to $U_1^\perp$, A has $i\beta$ as an eigenvalue of multiplicity $k - 1$. Repeat the argument with a basis $\mathbf{w}_2, \dots, \mathbf{w}_k, \overline{\mathbf{w}}_2 \dots, \overline{\mathbf{w}}_k$ of $U_1^\perp$ such that $\{\mathbf{w}_2, \overline{\mathbf{w}}_2\} \neq 0$. This gives a real symplectic basis $\mathbf{u}_2$, $\mathbf{v}_2$ for the subspace of U generated by the vectors $\mathbf{w}_2, \overline{\mathbf{w}}_2$. Then, $\mathbf{u}_1, \mathbf{u}_2, \mathbf{v}_1, \mathbf{v}_2$ is a symplectic basis of the subspace $U_1 \oplus U_2$ and we have an A-invariant decomposition of $U = \big(U_1 \oplus U_2\big) \oplus \big(U_1 \oplus U_2\big)^\perp$. Repeating the argument successively we find a real symplectic basis $\mathbf{u}_1, \mathbf{u}_2, \dots, \mathbf{u}_k, \mathbf{v}_1, \mathbf{v}_2, \dots, \mathbf{v}_k$ of U, where

$$\mathbf{u}_j = \kappa_j\mathbf{r}_j, \quad \mathbf{v}_j = \delta_j\kappa_j\mathbf{s}_j, \qquad (j = 1, \dots, k) \tag{4.21}$$

with $\delta_j = \text{sign}\{\mathbf{r}_j, \mathbf{s}_j\}$ and $\kappa_j = 1/\sqrt{|\{\mathbf{r}_j, \mathbf{s}_j\}|}$, for $j = 1, \dots, k$.

Since $\mathbf{w}_j = \mathbf{r}_j + i\mathbf{s}_j$ and $A\mathbf{w}_j = i\beta\mathbf{w}_j$, we have $A\mathbf{u}_j = -\delta_j\beta\mathbf{v}_j$, $A\mathbf{v}_j = \delta_j\beta\mathbf{u}_j$. Therefore, the matrix of the restriction A_β of A to U on this basis is that stated in this Theorem. □

Corollary 4.39.1 *Let $x_1, \dots, x_k; y_1, \dots, y_k$ be the coordinates of $\mathbf{z} \in U$ on the basis (4.21). Let $S = -JA$ and let $H = \frac{1}{2}\mathbf{x}^T S\mathbf{x}$ be the Hamiltonian function defined by S. Then, the expression of the induced Hamiltonian on U is*

$$H(\mathbf{z}) = \frac{1}{2}\beta \sum_{s=1}^{k} \delta_s (x_s^2 + y_s^2). \tag{4.22}$$

Proof Let $\mathbf{z} = \sum_{l=1}^{k} \big(x_l \mathbf{u}_l + y_l \mathbf{v}_l\big)$. Then, $S\mathbf{z} = \beta \sum_{s=1}^{k} \delta_s \big(x_s J \mathbf{v}_s - y_s J \mathbf{u}_s\big)$, so we get

$$\langle \mathbf{z}, S\mathbf{z} \rangle = \beta \sum_{l,s=1}^{k} \delta_s \big[x_l x_s \langle \mathbf{u}_l, J\mathbf{v}_s \rangle - x_l y_s \langle \mathbf{u}_l, J\mathbf{u}_s \rangle + y_l x_s \langle \mathbf{v}_l, J\mathbf{v}_s \rangle - y_l y_s \langle \mathbf{v}_l, J\mathbf{u}_s \rangle \big].$$

Since $\langle \mathbf{x}, J\mathbf{y} \rangle = \{\mathbf{x}, \mathbf{y}\}$ and the basis (4.21) is symplectic the above equation reduces to $\langle \mathbf{z}, S\mathbf{z} \rangle = \beta \sum_{s=1}^{k} \delta_s (x_s^2 + y_s^2)$, which proves (4.22). □

Chapter 5
The General Linear Normalization

5.1 Introduction

In the previous chapter we discussed the normalization of a real Hamiltonian matrix when all the eigenvalues are distinct. We also studied the case when the matrix has multiple eigenvalues but they are purely imaginary numbers and the matrix is diagonalizable. Now we discuss the problem in the general case. The problem of describing the normal forms of Hamiltonian and symplectic matrices was first solved by Williamson in a series of papers in the 1930s (see [59–61]).

We group the eigenvalues as in (4.19) and consider the A-invariant direct sum decomposition (4.20) with $X = \eta^\dagger(0)$ and the subspaces U_j, Y_j, Z_j defined in (4.14) with eigenspaces replaced by generalized eigenspaces.

Normalizing A restricted to X requires a lot of work and this is done in the next section. To normalize A on Y we use part of what will be developed for the nilpotent case of $A\big|_X$ and this is done in Sect. 5.3.

Now for U and Z we profit from the existence of real Jordan bases for A restricted to U_j and Z_j. For this part we follow the article by Laub and Meyer [35]

Let $\alpha = \alpha_j$ be one of the real eigenvalues. Take a real basis of $\eta^\dagger(\alpha)$, which yields the real Jordan canonical form of $A|_{\eta^\dagger(\alpha)}$, namely, with δ_i equals 1 or 0

$$A_j = \begin{bmatrix} \alpha & & & & & \\ \delta_2 & \alpha & & & & \\ & \delta_3 & \alpha & & & \\ & & & \ddots & & \\ & & & \delta_{k-1} & \alpha & \\ & & & & \delta_k & \alpha \end{bmatrix}, \quad \text{e.g.,} \quad A_j = \left[\begin{array}{ccc|cc|c} \alpha & & & & & \\ 1 & \alpha & & & & \\ & 1 & \alpha & & & \\ \hline & & 0 & \alpha & & \\ & & & 1 & \alpha & \\ \hline & & & & 0 & \alpha \end{array}\right].$$

H. E. Cabral, L. Brandão Dias, *Normal Forms and Stability of Hamiltonian Systems*, Applied Mathematical Sciences 218,
https://doi.org/10.1007/978-3-031-33046-9_5

As $\eta^\dagger(\alpha) \oplus \eta^\dagger(-\alpha)$ is an A-invariant Lagrangian splitting, then by Proposition 4.21 we can complete this basis of $\eta^\dagger(\alpha)$ with a basis of $\eta^\dagger(-\alpha)$ so as to get a symplectic basis of $\eta^\dagger(\alpha) \oplus \eta^\dagger(-\alpha)$. As the matrix A is Hamiltonian, its restriction to $\eta^\dagger(\alpha) \oplus \eta^\dagger(-\alpha)$ when expressed in this symplectic basis has the form (see Proposition 4.22)

$$A^*_{\alpha_j} = \begin{bmatrix} A_j & O \\ O & -A_j^T. \end{bmatrix}, \tag{5.1}$$

which is the normal form of $A|_{\eta^\dagger(\alpha_j)\oplus\eta^\dagger(-\alpha_j)}$. Doing this for each eigenvalue α_j and using Proposition 4.10 we get the normal form of $A|_U$.

Now let us consider the A-invariant decomposition $Z_j = Z_j^{(1)} \oplus Z_j^{(2)}$, where

$$Z_j^{(1)} = \eta^\dagger(\gamma+i\delta)\oplus\eta^\dagger(\gamma-i\delta) \quad \text{and} \quad Z_j^{(2)} = \eta^\dagger(-\gamma+i\delta)\oplus\eta^\dagger(-\gamma-i\delta). \tag{5.2}$$

The subspaces (5.2) are invariant under conjugation so they are complexifications; hence, they have real bases.

Take a real basis of $Z_j^{(1)}$ that yields the real Jordan canonical form of A restricted to $Z_j^{(1)}$, namely

$$A_j = \begin{bmatrix} B_{j1} & & & \\ & B_{j2} & & \\ & & \ddots & \\ & & & B_{jk_j} \end{bmatrix}, \quad \text{with} \quad B_{js} = \begin{bmatrix} D_j & & & \\ I_2 & D_j & & \\ & & \ddots & \\ & & I_2 & D_j \end{bmatrix}$$

,where $D_j = \begin{bmatrix} \gamma_j & \delta_j \\ -\delta_j & \gamma_j \end{bmatrix}$, $I_2 = \begin{bmatrix} 1 & 0 \\ 0 & 1 \end{bmatrix}$ and the number of blocks D_j in B_{js} decreases as s increases.

By Lemma 4.36, $Z_j = Z_j^{(1)} \oplus Z_j^{(2)}$ is a Lagrangian splitting, so by Proposition 4.21 we can complete this real basis of $Z_j^{(1)}$ with a basis of $Z_j^{(2)}$ so as to get a real symplectic basis of Z_j. As the matrix A is Hamiltonian its restriction to Z_j when expressed in this symplectic basis has the form

$$A_j^* = \begin{bmatrix} A_j & O \\ O & -A_j^T. \end{bmatrix}, \tag{5.3}$$

which is the normal form of $A\big|_{Z_j}$. Doing this for each j and using Proposition 4.10 we get the normal form of $A|_Z$.

Now we consider the normalization for the more complicated cases of the spaces X and Y. Here, although we continue using ideas from Laub and Meyer [35] and from the book by Meyer and Offin [41], we present the subject from a different approach.

5.2 The Real Nilpotent Case

First, the nilpotent case $X = \eta^{\dagger}(0)$. All the eigenvalues of $A|_X$ are equal to zero.

Let $A : V \to V$ be a linear operator on the finite dimensional real vector space V.

Definition 5.1 A vector $\mathbf{x} \in V$ has nilpotent index $s+1$ denoted $\mathrm{nil}(\mathbf{x}) = s+1$ if $A^s\mathbf{x} \neq 0$ and $A^{s+1}\mathbf{x} = 0$. The operator A is nilpotent of index $s+1$ if $A^s \neq 0$ and $A^{s+1} = 0$.

If A is nilpotent of index $s+1$, then $\mathrm{nil}(\mathbf{x}) \leq s+1$ for every $\mathbf{x} \in V$ with equality holding for some $\mathbf{x}$.

Let $A : V \to V$ be a nilpotent operator. Then the space V is a direct sum of the form (see Appendix III of [30])

$$V = Z(\mathbf{v}_1, A) \oplus \ldots Z(\mathbf{v}_r, A) \oplus L, \tag{5.4}$$

where L is contained in the kernel of A and $Z(\mathbf{v}_j, A)$ is the cyclic subspace of A,

$$Z(\mathbf{v}_j, A) = \left[\mathbf{v}_j, A\mathbf{v}_j, \ldots, A^{s_j}\mathbf{v}_j\right], \qquad s_j + 1 = \mathrm{nil}(\mathbf{v}_j).$$

Now let V be a symplectic space and assume that A is Hamiltonian. In the above direct sum decomposition the summands need not be symplectic subspaces of V. For instance, $s_j + 1 = \mathrm{nil}(\mathbf{v}_j)$ is an odd number, then $Z(\mathbf{v}_j, A)$ has an odd dimension, so it cannot be a symplectic subspace. Our goal is to construct a direct sum decomposition of V by A-invariant symplectic subspaces of V.

For a symplectic cyclic subspace we have the following result.

Proposition 5.2 *If $Z(\mathbf{v}, A)$ is a symplectic subspace, then $\{A^m\mathbf{v}, \mathbf{v}\} \neq 0$, where $m + 1 = \mathrm{nil}(\mathbf{v})$.*

Proof Since $Z(\mathbf{v}, A) = [\mathbf{v}, A\mathbf{v}, \ldots, A^m\mathbf{v}]$ is symplectic m is odd. Let $m = 2r - 1$ and consider the basis $\mathcal{B} = \left\{\mathbf{v}, A\mathbf{v}, \ldots, A^r\mathbf{v}; A^{r+1}\mathbf{v}, \ldots, A^{2r-2}\mathbf{v}, A^{2r-1}\mathbf{v}\right\}$ of $Z(\mathbf{v}, A)$. The symplectic form ω restricted to this subspace has the following matrix, on this basis

$$[\omega]_{\mathcal{B}} = \begin{bmatrix} X & Y \\ -Y^T & O \end{bmatrix},$$

where

$$Y = \begin{bmatrix} \{\mathbf{v}, A^{r+1}\mathbf{v}\} & \{\mathbf{v}, A^{r+2}\mathbf{v}\} & \ldots & \{\mathbf{v}, A^{m-1}\mathbf{v}\} & \{\mathbf{v}, A^m\mathbf{v}\} \\ \{A\mathbf{v}, A^{r+1}\mathbf{v}\} & \{A\mathbf{v}, A^{r+2}\mathbf{v}\} & \ldots & \{A^m\mathbf{v}, \mathbf{v}\} & 0 \\ \{A^2\mathbf{v}, A^{r+1}\mathbf{v}\} & \{A^2\mathbf{v}, A^{r+2}\mathbf{v}\} & \ldots & 0 & 0 \\ \vdots & & \ldots & & \vdots \\ \{A^m\mathbf{v}, \mathbf{v}\} & 0 & \ldots & 0 & 0 \end{bmatrix}.$$

Since $Z(\mathbf{v}, A)$ is symplectic the restriction of ω is nondegenerate; hence, the matrix $[\omega]_{\mathcal{B}}$ has a nonzero determinant. But this determinant is equal to

$$\pm \left(\operatorname{Det} Y\right)^2 = \pm\{\mathbf{v}, A^m\mathbf{v}\}^{2(r+1)}.$$

Therefore, $\{A^m\mathbf{v}, \mathbf{v}\} \neq 0$. □

5.2.1 Two Main Theorems in the Nilpotent Context

The two theorems of this section are the fundamental results that will be used to construct the normal form of a real Hamiltonian matrix in the nilpotent case.

Theorem 5.3 *Let U_1 be an A-invariant symplectic subspace of V. Let $\mathbf{u} \in U_1$ be an element of nilpotent index $m+1$ such that $\{A^m\mathbf{u}, \mathbf{u}\} \neq 0$.*

$Z(\mathbf{u}, A) = \left[\mathbf{u}, A\mathbf{u}, \ldots, A^m\mathbf{u}\right]$ is a symplectic subspace of V and an explicit symplectic basis can be constructed for it.

Example 5.1 Simplify the quadratic Hamiltonian

$$H = \frac{1}{2}(y_1^2 + y_2^2) - \frac{1}{2}x_1^2 + \frac{1}{2}x_2^2 + x_2 y_1$$

by means of a linear symplectic transformation.

If S is the Hessian matrix of H, we compute the Hamiltonian matrix $A = JS$ and its powers

$$A = \begin{bmatrix} 0 & 1 & 1 & 0 \\ 0 & 0 & 0 & 1 \\ 1 & 0 & 0 & 0 \\ 0 & -1 & -1 & 0 \end{bmatrix}, \quad A^2 = \begin{bmatrix} 1 & 0 & 0 & 1 \\ 0 & -1 & -1 & 0 \\ 0 & 1 & 1 & 0 \\ -1 & 0 & 0 & -1 \end{bmatrix}, \quad A^3 = \begin{bmatrix} 0 & 0 & 0 & 0 \\ -1 & 0 & 0 & -1 \\ 1 & 0 & 0 & 1 \\ 0 & 0 & 0 & 0 \end{bmatrix}, \quad A^4 = 0,$$

so A is nilpotent. For the vector $\mathbf{u} = (1, 0, 0, 0)$ we have $A\mathbf{u} = (0, 0, 1, 0)$, $A^2\mathbf{u} = (1, 0, 0, -1)$ $A^3\mathbf{u} = (0, -1, 1, 0)$. The vector $\mathbf{u}$ has a nilpotent index equal to $s + 1 = 4$ and we compute $\{A^3\mathbf{u}, \mathbf{u}\} = -1$. By Theorem 5.3, the cyclic subspace $Z(\mathbf{u}, A) = [\mathbf{u}, A\mathbf{u}, A^2\mathbf{u}, A^3\mathbf{u}]$ is symplectic.

In the proof of the theorem we will see how to get a symplectic basis for $Z(\mathbf{u}, A)$. But in this case we easily find a symplectic basis. Indeed, we already computed $\{\mathbf{u}, A^3\mathbf{u}\} = 1$. Now we compute $\{\mathbf{u}, A\mathbf{u}\} = 1$ and $\{A\mathbf{u}, A^2\mathbf{u}\} = -1$. We readily see that the vectors

$$\mathbf{u}_1 = \mathbf{u}, \quad \mathbf{u}_2 = A\mathbf{u} - A^3\mathbf{u}, \quad \mathbf{v}_1 = A^3\mathbf{u}, \quad \mathbf{v}_2 = -A^2\mathbf{u}$$

form a symplectic basis $\mathbf{R}^4 = Z(\mathbf{u}, A)$.

The transition matrix P from the canonical basis of $\mathbf{R}^4$, which is symplectic to the above symplectic basis is symplectic. It is given by

$$P = \begin{bmatrix} 1 & 0 & 0 & -1 \\ 0 & 1 & -1 & 0 \\ 0 & 0 & 1 & 0 \\ 0 & 0 & 0 & 1 \end{bmatrix}$$

and defines the symplectic linear transformation

$$x_1 = \xi_1 - \eta_2, \quad x_2 = \xi_2 - \eta_1, \quad y_1 = \eta_1, \quad y_2 = \eta_2$$

that transforms the given Hamiltonian into the new Hamiltonian

$$\mathcal{H} = -\frac{1}{2}\xi_1^2 + \frac{1}{2}\xi_2^2 + \xi_1\eta_2. \tag{5.5}$$

Another application of this theorem is the following.

Example 5.2 Simplify the quadratic Hamiltonian

$$H = -\frac{1}{2}(y_1^2 + y_2^2) - \frac{1}{2}(x_1^2 + x_2^2) + x_1 y_2 + x_2 y_1$$

by means of a linear symplectic transformation.

In this case we find

$$A = \begin{bmatrix} 0 & 1 & -1 & 0 \\ 1 & 0 & 0 & -1 \\ 1 & 0 & 0 & -1 \\ 0 & 1 & -1 & 0 \end{bmatrix} \quad \text{and} \quad A^2 = 0.$$

For $\mathbf{u}_1 = (1, 0, 0, 0)$ and $\mathbf{u}_2 = (0, 1, 0, 0)$ we have $A\mathbf{u}_1 = (0, 1, 1, 0)$, $A\mathbf{u}_2 = (1, 0, 0, 1)$ and we readily check that $\{A\mathbf{u}_1, \mathbf{u}_1\} = -1$, $\{A\mathbf{u}_2, \mathbf{u}_2\} = -1$. Since the nilpotent index of $\mathbf{u}_1$ and $\mathbf{u}_2$ is two, by Theorem 5.3 the subspaces $Z(\mathbf{u}_1, A) = [\mathbf{u}_1, A\mathbf{u}_1]$ and $Z(\mathbf{u}_2, A) = [\mathbf{u}_2, A\mathbf{u}_2]$ are symplectic subspaces of $\mathbf{R}^4$. Clearly, $\mathbf{u}_1, A\mathbf{u}_1$ is a symplectic basis of $Z(\mathbf{u}_1, A)$ and $\mathbf{u}_2, A\mathbf{u}_2$ is a symplectic basis of $Z(\mathbf{u}_2, A)$. Since $\{\mathbf{u}_1, \mathbf{u}_2\} = 0$ and $\{A\mathbf{u}_1, A\mathbf{u}_2\} = 0$ we get the symplectic basis $\mathbf{u}_1, \mathbf{u}_2, A\mathbf{u}_1, A\mathbf{u}_2$ of $\mathbf{R}^4 = Z(\mathbf{u}_1, A) \oplus Z(\mathbf{u}_2, A)$.

The transition matrix P from the canonical basis of $\mathbf{R}^4$ to this basis is the symplectic matrix

$$P = \begin{bmatrix} 1 & 0 & 0 & 1 \\ 0 & 1 & 1 & 0 \\ 0 & 0 & 1 & 0 \\ 0 & 0 & 0 & 1 \end{bmatrix},$$

which defines the symplectic linear transformation

$$x_1 = \xi_1 + \eta_2, \quad x_2 = \xi_2 + \eta_1, \quad y_1 = \eta_1, \quad y_2 = \eta_2$$

that transforms the given Hamiltonian into the new Hamiltonian

$$\mathcal{H} = -\frac{1}{2}(\xi_1^2 + \xi_2^2). \tag{5.6}$$

In the next example we consider a three-degree of freedom Hamiltonian.

Example 5.3 Simplify the quadratic Hamiltonian

$$H = \frac{1}{2}(x_2^2 + y_2^2) + x_1x_3 + x_1y_1 + x_2y_2 + x_3y_3 + y_1y_3.$$

by means of a linear symplectic transformation.

The Hamiltonian matrix $A = J\operatorname{Hess} H$ is the 6×6 matrix

$$A = \begin{bmatrix} I & I' \\ -I' & -I \end{bmatrix}, \quad \text{where} \quad I' = \begin{bmatrix} 0 & 0 & 1 \\ 0 & 1 & 0 \\ 1 & 0 & 0 \end{bmatrix},$$

and I is the 3×3 identity matrix. Since $I'I' = I$, we find $A^2 = 0$ and we also find that the characteristic polynomial of A is $p(\lambda) = \lambda^6$. Therefore, $\eta^\dagger(0) = \operatorname{Ker}(A - 0I)^6 = \mathbf{R}^6$.

For $\mathbf{x} = (x_1, x_2, x_3, x_4, x_5, x_6) \in \mathbf{R}^6$, we have

$$A\mathbf{x} = (x_1 + x_6, x_2 + x_5, x_3 + x_4, -(x_3 + x_4), -x_2 - x_5, -x_1 - x_6)$$

and $\{A\mathbf{x}, \mathbf{x}\} = (x_2 + x_5)^2 + 2(x_1 + x_6)(x_3 + x_4)$.

Notice that every vector has index at most two. For the vector $\mathbf{u}_1 = (0, 1, 0, 0, 0, 0)$ we have $\operatorname{nil}(\mathbf{u}_1) = 2$ and $U_1 = \{A\mathbf{u}_1, \mathbf{u}_1\} = 1$. By Theorem 5.3 the subspace $Z(\mathbf{u}_1, A)$ is symplectic. Therefore, $\mathbf{R}^6 = U_1 \oplus U_1^\perp$ with $U_1^\perp$ an A-invariant symplectic subspace. The vector $\mathbf{u}_2 = (1, 0, \frac{1}{2}, 0, 0, 0) \in U_1^\perp$ since $\{\mathbf{u}_1, \mathbf{u}_2\} = 0$ and $\{A\mathbf{u}_1, \mathbf{u}_2\} = 0$. Now $\{A\mathbf{u}_2, \mathbf{u}_2\} = 1$ and since $\mathbf{u}_2$ has nilpotent index 2 the subspace $U_2 = Z(\mathbf{u}_2, A)$ is a symplectic and $U_1^\perp = U_2 \oplus U_2^\perp$, with $U_2^\perp$ a symplectic subspace.

One more example.

Example 5.4 We have $A^2 = 0$ for the following Hamiltonian matrix

$$A = \begin{bmatrix} 0 & 0 & 0 & 0 & 0 & 0 \\ 1 & 0 & 0 & 0 & 0 & 0 \\ 0 & 0 & 0 & 0 & 0 & 0 \\ 1 & 0 & 0 & 0 & -1 & 0 \\ 0 & 0 & 0 & 0 & 0 & 0 \\ 0 & 0 & 1 & 0 & 0 & 0 \end{bmatrix}. \tag{5.7}$$

For $\mathbf{x} = (x_1, x_2, x_3, x_4, x_5, x_6)$ we have

$$A\mathbf{x} = (0, x_1, 0, x_1 - x_5, 0, x_3) \quad \text{and} \quad \{A\mathbf{x}, \mathbf{x}\} = 2x_1x_5 - x_1^2 - x_3^2. \tag{5.8}$$

Taking $\mathbf{u} = (0, 0, 1, 0, 0, 0)$ we have $\text{nil}(\mathbf{u}) = 2$ and $\{A\mathbf{u}, \mathbf{u}\} = -1$; hence, by Theorem 5.3,

$$U = Z(\mathbf{u}, A) = [\mathbf{u}, A\mathbf{u}]$$

is a symplectic subspace of $\mathbf{R}^6$ and we have the A-invariant direct sum decomposition $\mathbf{R}^6 = U \oplus U^\perp$ with $U^\perp$ a symplectic subspace of $\mathbf{R}^6$. A symplectic basis for U is given by the vectors $\mathbf{u}_1 = \mathbf{u}$, $\mathbf{v}_1 = -A\mathbf{u}$.

For $\mathbf{x} = (y_1, y_2, y_3, y_4, y_5, y_6)$ we have

$$\{\mathbf{x}, \mathbf{y}\} = x_1y_4 + x_2y_5 + x_3y_6 - x_4y_1 - x_5y_2 - x_6y_3 \text{ and } \{A\mathbf{x}, \mathbf{y}\} = 2x_1y_5 - x_1y_1 - x_3y_3.$$

The vector $\mathbf{y} \in U^\perp$ if, and only if, $\{\mathbf{u}, \mathbf{y}\} = 0$ and $\{A\mathbf{u}, \mathbf{y}\} = 0$ which means that $y_3 = y_6 = 0$; hence,

$$U^\perp = \{(y_1, y_2, 0, y_4, y_5, 0)\}.$$

For $\mathbf{y}_* = (1, 0, 0, 0, 0, 0) \in U^\perp$ we have $\text{nil}(\mathbf{y}) = 2$ and $\{A\mathbf{y}_*, \mathbf{y}_*\} = -1$. By Theorem 5.3, $Z(\mathbf{y}_*, A) = [\mathbf{y}_*, A\mathbf{y}_*]$ is a symplectic subspace of $U^\perp$ and a symplectic basis for it is formed by the vectors $\mathbf{u}_2 = \mathbf{y}_*$, $\mathbf{v}_2 = -A\mathbf{y}_*$. We have the A-invariant symplectic decomposition $U^\perp = Z(\mathbf{y}_*, A) \oplus Z(\mathbf{y}_*, A)^\perp$.

Now, $\mathbf{y} \in U^\perp \cap \oplus Z(\mathbf{y}_*, A)^\perp$ if, and only if, $\{\mathbf{y}_*, \mathbf{y}\} = 0$ and $\{A\mathbf{y}_*, \mathbf{y}\} = 0$, that is, $y_1 = y_5$ and $y_4 = 0$, so $Z(\mathbf{y}_*, A)^\perp = \{(y_1, y_2, 0, 0, y_1, 0)\}$. For $\mathbf{u}_3 = (1, 0, 0, 0, 1, 0)$ and $\mathbf{v}_3 = (0, -1, 0, 0, 0, 0)$ we have $\{\mathbf{u}_3, \mathbf{v}_3\} = 1$, so these two vectors form a symplectic basis of $Z(\mathbf{y}_*, A)^\perp$.

Therefore, we have the symplectic A-invariant direct sum decomposition

$$\mathbf{R}^6 = Z(\mathbf{u}, A) \oplus Z(\mathbf{y}_*, A) \oplus Z(\mathbf{y}_*, A)^\perp,$$

and since these three subspaces are pairwise symplectically orthogonal, then by Proposition 4.10 the vectors $\mathbf{u}_1, \mathbf{u}_2, \mathbf{u}_3, \mathbf{v}_1, \mathbf{v}_2, \mathbf{v}_3$ form a symplectic basis for it.

Now, we have $A\mathbf{u}_1 = -\mathbf{v}_1$, $A\mathbf{u}_2 = -\mathbf{v}_2$, $A\mathbf{u}_3 = -\mathbf{v}_3$, $A\mathbf{v}_1 = 0$, $A\mathbf{v}_2 = 0$, $A\mathbf{v}_3 = 0$, so the matrix A on this basis is

$$A_* = \begin{bmatrix} 0 & 0 & 0 & 0 & 0 & 0 \\ 0 & 0 & 0 & 0 & 0 & 0 \\ 0 & 0 & 0 & 0 & 0 & 0 \\ -1 & 0 & 0 & 0 & 0 & 0 \\ 0 & -1 & 0 & 0 & 0 & 0 \\ 0 & 0 & -1 & 0 & 0 & 0 \end{bmatrix}.$$

To give the expressions of the Hamiltonians defined by the matrices A and A_* we compute the symmetric matrices $S = -JA$, $S_* = -JA_*$ and we find

$$H = -\frac{1}{2}(x_1^2 + x_3^2) + x_1 y_1 \quad \text{and} \quad \mathcal{H} = \frac{1}{2}(\xi_1^2 + \xi_2^2 + \xi_3^2).$$

Now we deal with the case where $\{A^s\mathbf{x}, \mathbf{x}\} = 0$ for all $\mathbf{x}$ with $s + 1 = \text{nil}(\mathbf{x})$. This implies that the number r in the direct sum decomposition (5.4) is at least 2.

Indeed, if $V = Z(\mathbf{v}, A) \oplus L$ is symplectic, then since $\{Z(\mathbf{v}, A), L\} = 0$, both $Z(\mathbf{v}, A)$ and L are symplectic subspaces, by Proposition 4.7(a). If $m + 1 = \text{nil}(\mathbf{v})$, then $\{A^m\mathbf{v}, \mathbf{v}\} \neq 0$, by Proposition 5.2, contrary to the assumption above.

Let us first prove the following lemma.

Lemma 5.4 *Let U be a $2q$-dimensional A-invariant symplectic subspace of V such that $\{A^s\mathbf{x}, \mathbf{x}\} = 0$ for all $\mathbf{x} \in U$, where $s + 1 = \text{nil}(\mathbf{x})$. Then, there are vectors $\mathbf{u}, \mathbf{v} \in U$ with the same nilpotent index $m + 1 \leq q$ such that $\{A^m\mathbf{u}, \mathbf{v}\} \neq 0$ and $Z(\mathbf{u}, A) \cap Z(\mathbf{v}, A) = \{0\}$.*

Proof Consider as in (5.4), the direct sum decomposition

$$U = Z(\mathbf{u}_1, A) \oplus \ldots \oplus Z(\mathbf{u}_r, A) \oplus L, \quad \text{with} \quad r \geq 2 \tag{5.9}$$

where $L \subset \ker A$ and $Z(\mathbf{u}_j, A) = \left[\mathbf{u}_j, A\mathbf{u}_j, \ldots, A^{k_j - 1}\mathbf{u}_j\right]$ with $k_j = \text{nil}(\mathbf{u}_j)$. We take $k_1 \geq \ldots \geq k_r$. Then, since A is a Hamiltonian matrix we have for $s \geq 1$

$$\{A^{k_1-1}\mathbf{u}_1, A^s\mathbf{u}_j\} = \pm\{\mathbf{u}_1, A^{k_1-1+s}\mathbf{u}_j\} = 0, \quad \text{for} \quad j = 1, \ldots, r.$$

This will imply that there exists an index $j \geq 2$ such that $\{A^{k_1-1}\mathbf{u}_1, \mathbf{u}_j\} \neq 0$.

Indeed, since $\{A^{k_1-1}\mathbf{u}_1, \mathbf{u}_1\} = 0$ if we also had $\{A^{k_1-1}\mathbf{u}_1, \mathbf{u}_j\} = 0$ for $j \geq 2$ then we would have $\{A^{k_1-1}\mathbf{u}_1, \mathbf{x}\} = 0$ for any $\mathbf{x} \in Z(\mathbf{u}_j, A)$ for $j \geq 1$ and therefore $\{A^{k_1-1}\mathbf{u}_1, \mathbf{x}\} = 0$ for all $\mathbf{x}$ in the symplectic space U, which cannot happen because $A^{k_1-1}\mathbf{u}_1 \neq 0$.

Let $l \geq 2$ be the first index such that $\{A^{k_1-1}\mathbf{u}_1, \mathbf{u}_l\} \neq 0$ and let $\mathbf{u} = A^{k_1-k_l}\mathbf{u}_1$. Then, $A^{k_l}\mathbf{u} = A^{k_1}\mathbf{u}_1 = 0$ and $A^{k_l-1}\mathbf{u} = A^{k_1-1}\mathbf{u}_1 \neq 0$, because $k_1 = \text{nil}(\mathbf{u}_1)$. Therefore, $k_l = \text{nil}(\mathbf{u})$. Take $\mathbf{v} = \mathbf{u}_l$. Then, $\mathbf{u}$ and $\mathbf{v}$ have nilpotent index $m + 1 = k_l$ and $\{A^m\mathbf{u}, \mathbf{v}\} = \{A^{k_1-1}\mathbf{u}_1, \mathbf{u}_l\} \neq 0$.

Since $\mathbf{u} = A^{k_1-k_l}\mathbf{u}_1 \in Z(\mathbf{u}_1, A)$, we have $Z(\mathbf{u}, A) \subset Z(\mathbf{u}_1, A)$ and the direct sum decomposition (5.9) implies that $Z(\mathbf{u}, A) \cap Z(\mathbf{v}, A) = \{0\}$. □

Notice that in the proof the condition $r \geq 2$ is essential. The following example shows that the lemma is false if $r = 1$.

Example 5.5 Check that for the Hamiltonian matrix $A = \begin{bmatrix} 0 & 1 & 0 & 0 \\ 0 & 0 & 0 & 1 \\ 0 & 0 & 0 & 0 \\ 0 & 0 & -1 & 0 \end{bmatrix}$ we have

$$\{A^s\mathbf{x}, \mathbf{x}\} = 0, \quad \text{for any} \quad \mathbf{x} \text{ with } \operatorname{nil}(\mathbf{x}) = s + 1,$$

and $\{A\mathbf{x}, \mathbf{y}\} = 0$, for any $\mathbf{x}, \mathbf{y} \in U = \mathbf{R}^4$, with $\operatorname{nil}(\mathbf{x}) = \operatorname{nil}(\mathbf{y}) = 2$.

Computing the powers of A

$$A^2 = \begin{bmatrix} 0 & 0 & 0 & 1 \\ 0 & 0 & -1 & 0 \\ 0 & 0 & 0 & 0 \\ 0 & 0 & 0 & 0 \end{bmatrix}, \quad A^3 = \begin{bmatrix} 0 & 0 & -1 & 0 \\ 0 & 0 & 0 & 0 \\ 0 & 0 & 0 & 0 \\ 0 & 0 & 0 & 0 \end{bmatrix} \quad \text{and} \quad A^4 = 0$$

we see that for $\mathbf{x} = (x_1, x_2, x_3, x_4) \in \mathbf{R}^4$ we have

$$A\mathbf{x} = (x_2, x_4, 0, -x_3), \quad A^2\mathbf{x} = (x_4, -x_3, 0, 0) \quad \text{and} \quad A^3\mathbf{x} = (-x_3, 0, 0, 0);$$

hence, $\operatorname{nil}(\mathbf{x}) = 3$ if $x_3 = 0$ and $x_4 \neq 0$, $\operatorname{nil}(\mathbf{x}) = 2$ if $x_3 = x_4 = 0$ and $x_2 \neq 0$, and that $\operatorname{nil}(\mathbf{x}) = 1$ if $x_2 = x_3 = x_4 = 0$ and $x_1 \neq 0$. Since $J\mathbf{x} = (x_3, x_4, -x_1, -x_2)$, we see that

$$\{A\mathbf{x}, \mathbf{x}\} = 2x_2x_3 + x_4^2, \quad \{A^2\mathbf{x}, \mathbf{x}\} = 0, \quad \{A^3\mathbf{x}, \mathbf{x}\} = -x_3^2,$$

so $\{A^s\mathbf{x}, \mathbf{x}\} = 0$, for $s = 1, 2, 3, 4$.

Since $\{A\mathbf{x}, \mathbf{y}\} = x_2y_3 + x_4y_4 + x_3y_2$, we have

$$\{A\mathbf{x}, \mathbf{y}\} = 0 \quad \text{for any vectors} \quad \mathbf{x}, \mathbf{y} \quad \text{of index 2.}$$

Notice that for $\mathbf{u} = (0, 0, 1, 0)$ we have $A\mathbf{u} = (0, 0, 0, -1)$, $A^2\mathbf{u} = (0, -1, 0, 0)$ and $A^3\mathbf{u} = (-1, 0, 0, 0)$; hence, $\mathbf{R}^4 = Z(\mathbf{u}, A)$ and the direct sum decomposition (5.9) in this case has a single term, that is, $r = 1$.

Now we can state the second theorem in which the existence of vectors $\mathbf{u}, \mathbf{v}$ with the required conditions is guaranteed by the lemma above

Theorem 5.5 *Let U_2 be a $2q$-dimensional A-invariant symplectic subspace of V such that $\{A^s\mathbf{x}, \mathbf{x}\} = 0$ for all $\mathbf{x} \in U_2$ with $s + 1 = \operatorname{nil}(\mathbf{x})$. Let $\mathbf{u}, \mathbf{v} \in U_2$ be vectors with a nilpotent index $m + 1 \leq q$ such that $\{A^m\mathbf{u}, \mathbf{v}\} \neq 0$. Then, the subspace $Z(\mathbf{u}, A) \oplus Z(\mathbf{v}, A)$ is symplectic and an explicit symplectic basis can be constructed for it.*

The process of finding the symplectic base is described in the proofs of Theorem 5.3 and Theorem 5.5. The process is summarized in the Remark after the proof of each one of these theorems.

Let us see an application of Theorem 5.5.

Example 5.6 Consider the Hamiltonian matrix A and its powers given below

$$A = \begin{bmatrix} 0 & 1 & 0 & 0 & 0 & 0 \\ 0 & 0 & 0 & 0 & 0 & 1 \\ 0 & 1 & 0 & 0 & 1 & 0 \\ 0 & 0 & 0 & 0 & 0 & 0 \\ 0 & 0 & 0 & -1 & 0 & -1 \\ 0 & 0 & 0 & 0 & 0 & 0 \end{bmatrix}, \quad A^2 = \begin{bmatrix} 0 & 0 & 0 & 0 & 0 & 1 \\ 0 & 0 & 0 & 0 & 0 & 0 \\ 0 & 0 & 0 & -1 & 0 & 0 \\ 0 & 0 & 0 & 0 & 0 & 0 \\ 0 & 0 & 0 & 0 & 0 & 0 \\ 0 & 0 & 0 & 0 & 0 & 0 \end{bmatrix}, \quad A^3 = 0. \tag{5.10}$$

For all nonzero $\mathbf{x}$, $\text{nil}(\mathbf{x}) \le 3$. For $\mathbf{x} = (x_1, x_2, x_3, x_4, x_5, x_6)$ we have,

$$A\mathbf{x} = (x_2, x_6, x_2 + x_5, 0, -x_4 - x_6, 0) \quad \text{and} \quad A^2\mathbf{x} = (x_6, 0, -x_4, 0, 0, 0).$$

Of course, $\{A^0\mathbf{x}, \mathbf{x}\} = 0$, $\{A^2\mathbf{x}, \mathbf{x}\} = 0$, whereas

$$\{A\mathbf{x}, \mathbf{x}\} = 2(x_2x_4 + x_2x_6 + x_4x_6).$$

If $\text{nil}(\mathbf{x}) = 2$, then $x_4 = x_6 = 0$; hence, $\{A\mathbf{x}, \mathbf{x}\} = 0$. Therefore, $\{A^s\mathbf{x}, \mathbf{x}\} = 0$ for all $\mathbf{x}$ with $\text{nil}(\mathbf{x}) = s + 1$.

By Lemma 5.4, there are vectors $\mathbf{u}, \mathbf{v} \in \mathbf{R}^6$ with $\text{nil}(\mathbf{u}) = \text{nil}(\mathbf{v}) = m + 1 \le 3$ such that $\{A^m\mathbf{u}, \mathbf{v}\} \ne 0$ and by Theorem 5.5, the subspace $U = Z(\mathbf{u}, A) \oplus Z(\mathbf{u}, A)$ of $\mathbf{R}^6$ is symplectic.

For $\mathbf{y} = (y_1, y_2, y_3, y_4, y_5, y_6)$ we compute

$$\begin{aligned} \{\mathbf{x}, \mathbf{y}\} &= x_1y_4 + x_2y_5 + x_3y_6 - x_4y_1 - x_5y_2 - x_6y_3, \\ \{A\mathbf{x}, \mathbf{y}\} &= x_2y_4 + x_4y_2 + x_2y_6 + x_6y_2 + x_5y_6 + x_6y_5. \end{aligned} \tag{5.11}$$

If $\text{nil}(\mathbf{x}) = 1$, then $A\mathbf{x} = 0$ so $x_2 = x_4 = x_5 = x_6 = 0$. If $\text{nil}(\mathbf{x}) = 2$, then $A^2\mathbf{x} = 0$ so $x_4 = x_6 = 0$. Therefore, from (5.11) we see that if $\text{nil}(\mathbf{u}) = \text{nil}(\mathbf{v}) = m + 1$ with $m = 0$ or $m = 1$, then $\{A^m\mathbf{u}, \mathbf{v}\} = 0$.

Therefore, $\{A^2\mathbf{u}, \mathbf{v}\} \ne 0$. In this case $Z(\mathbf{u}, A) = [\mathbf{u}, A\mathbf{u}, A^2\mathbf{u}]$ and U has dimension 6, so $U = \mathbf{R}^6$. For $\mathbf{u}$ and $\mathbf{v}$ we can take $\mathbf{u} = (0, 0, 0, 0, 0, 1)$ and $\mathbf{v} = (0, 0, 0, 1, 0, 0)$, so we have $\{A^2\mathbf{u}, \mathbf{v}\} = 1$.

In the proof of Theorem 5.5 it is shown that vectors $\mathbf{u}, \mathbf{v}$ of the nilpotent index $m + 1$ such that $\{A^m\mathbf{u}, \mathbf{v}\} \ne 0$ determine two vectors $\mathbf{f}, \mathbf{g} \in Z(\mathbf{u}, A) \oplus Z(\mathbf{v}, A)$ satisfying the conditions (5.21) and (5.22). In the present case these conditions are the following:

$$\begin{aligned} &\{\mathbf{f}, \mathbf{f}\} = 0, \quad \{\mathbf{g}, \mathbf{g}\} = 0, \quad \{\mathbf{f}, \mathbf{g}\} = 0, \quad \{A\mathbf{f}, \mathbf{f}\} = 0, \\ &\{A\mathbf{f}, \mathbf{g}\} = 0, \quad \{A\mathbf{g}, \mathbf{g}\} = 0, \quad \{A^2\mathbf{f}, \mathbf{f}\} = 0, \quad \{A^2\mathbf{g}, \mathbf{g}\} = 0, \end{aligned}$$

and $\{A^2\mathbf{f}, \mathbf{g}\} = 1$. From these equalities we readily see that the following vectors

$$\mathbf{u}_1 = \mathbf{f}, \ \mathbf{u}_2 = A\mathbf{f}, \ \mathbf{u}_3 = A^2\mathbf{f}, \ \mathbf{v}_1 = A^2\mathbf{g}, \ \mathbf{v}_2 = -A\mathbf{g}, \ \mathbf{v}_3 = \mathbf{g}$$

form a symplectic basis $\mathcal{B}$ of U. Since

$$A\mathbf{u}_1 = \mathbf{u}_2,\ A\mathbf{u}_2 = \mathbf{u}_3,\ A\mathbf{u}_3 = 0,\ A\mathbf{v}_1 = 0,\ A\mathbf{v}_2 = -\mathbf{v}_1,\ A\mathbf{v}_3 = -\mathbf{v}_2,$$

the matrix of the linear map $\mathbf{x} \mapsto A\mathbf{x}$ on this basis is

$$[A]_{\mathcal{B}} = \begin{bmatrix} 0 & 0 & 0 & 0 & 0 & 0 \\ 1 & 0 & 0 & 0 & 0 & 0 \\ 0 & 1 & 0 & 0 & 0 & 0 \\ 0 & 0 & 0 & 0 & -1 & 0 \\ 0 & 0 & 0 & 0 & 0 & -1 \\ 0 & 0 & 0 & 0 & 0 & 0 \end{bmatrix}.$$

The Hamiltonian defined by the original matrix is

$$H = x_2 y_1 + x_2 y_3 + y_2 y_3$$

and the transformed Hamiltonian, which is defined by the matrix $[A]_{\mathcal{B}}$ is

$$\mathcal{H} = \xi_1 \eta_2 + \xi_2 \eta_3.$$

To prove Theorems 5.3 and 5.5 we need some preliminaries, which are developed in the lemmas of the next section.

5.2.2 *The Basic Lemmas*

Let U be an A-invariant subspace of V and let $A : U \to U$ be nilpotent of index $s + 1$.

Let $\mathcal{A}$ be the commutative algebra generated by the linear operator $A : U \to U$. The elements of $\mathcal{A}$ are the polynomials in A with real coefficients and degree less than or equal to s, $\Phi = \alpha_0 I + \alpha_1 A + \cdots + \alpha_s A^s$.

Definition 5.6 The conjugate Φ^* of the element $\Phi = \alpha_0 I + \alpha_1 A + \cdots + \alpha_s A^s$ in $\mathcal{A}$ is the element

$$\Phi^* = \alpha_0 I - \alpha_1 A + \cdots + (-1)^s \alpha_s A^s. \tag{5.12}$$

We say that Φ is *even* if $\Phi^* = \Phi$ and that Φ is *odd* if $\Phi^* = -\Phi$.

Notice that Φ even means that it contains only even powers of A and Φ odd means that it contains only odd powers of A.

Lemma 5.7

(1) *The element* $\Phi = \alpha_0 I + \alpha_1 A + \cdots + \alpha_s A^s$ *of* $\mathcal{A}$ *is nonsingular, that is, it has an inverse, if, and only if,* $\alpha_0 \neq 0$.
(2) $(\Phi\Psi)^* = \Phi^*\Psi^*$.
(3) *If* Φ *is nonsingular it has a square root; precisely, there is a* $\Psi \in \mathcal{A}$ *such that* $\Psi^2 = \text{sign}(\alpha_0)\Phi$.
(4) *Let* Φ *be nonsingular. If it is even, then so is its square root* Ψ.

Proof If $\Phi = \alpha_0 I + \alpha_1 A + \cdots + \alpha_s A^s$ and $\Psi = \beta_0 I + \beta_1 A + \cdots + \beta_s A^s$, then

$$\Phi\Psi = \alpha_0\beta_0 I + (\alpha_0\beta_1 + \alpha_1\beta_0)A + (\alpha_0\beta_2 + \alpha_1\beta_1 + \alpha_2\beta_0)A^2 + (\alpha_0\beta_3 + \alpha_1\beta_2 + \alpha_2\beta_1 + \alpha_3\beta_0)A^3 + \cdots , \tag{5.13}$$

(1) We see from (5.13) that the coefficients of Ψ in the equation $\Phi\Psi = I$ can be determined if, and only if, $\alpha_0 \neq 0$.
(2) Also from (5.13) we see that changing the signs of the coefficients of odd powers of A both in Φ and Ψ simply changes the signs of the corresponding terms in the product $\Phi\Psi$, so $(\Phi\Psi)^* = \Phi^*\Psi^*$.
(3) Comparing the coefficients in the equation $\Psi^2 = \text{sign}(\alpha_0)\Phi$, we see that the coefficients of Ψ are given by, with $\epsilon = \text{sign}(\alpha_0)$,

$$\beta_0^2 = \epsilon\alpha_0,\quad \beta_1 = \frac{1}{2\beta_0}\epsilon\alpha_1,\quad \beta_2 = \frac{1}{2\beta_0}(\epsilon\alpha_2 - \beta_1^2),\quad \beta_3 = \frac{1}{2\beta_0}(\epsilon\alpha_3 - 2\beta_1\beta_2)$$
$$\beta_4 = \frac{1}{2\beta_0}(\epsilon\alpha_4 - 2\beta_1\beta_3 + \beta_2^2),\quad \beta_5 = \frac{1}{2\beta_0}(\epsilon\alpha_5 - 2\beta_1\beta_4 + 2\beta_2\beta_3), \ldots . \tag{5.14}$$

If $\alpha_0 \neq 0$ all the coefficients β_j can be determined guaranteeing the existence of Ψ.
(4) We see from the expressions (5.14) that the coefficients of Ψ with odd subscripts are zero if the same occurs for the coefficients of Φ so Ψ is even if Φ is even.

□

For $\mathbf{x}, \mathbf{y} \in U$, consider the element of $\mathcal{A}$, defined by the polynomial in A,

$$\Omega(\mathbf{x}, \mathbf{y}) = \{A^s\mathbf{x}, \mathbf{y}\}I + \{A^{s-1}\mathbf{x}, \mathbf{y}\}A + \cdots + \{\mathbf{x}, \mathbf{y}\}A^s. \tag{5.15}$$

The mapping $\Omega : U \times U \to \mathcal{A}$ is clearly **R**-bilinear on the vector space U, but we want to consider it as a mapping on U when vectors of U are multiplied by elements of the ring $\mathcal{A}$, that is, when U is viewed as a module over $\mathcal{A}$. In this context, it is linear in the first variable but not in the second, as the item (2) of the lemma below shows.

Lemma 5.8 *The following properties hold:*

(1) $\Omega(\mathbf{x}, \mathbf{y}) = (-1)^{s+1}\Omega(\mathbf{y}, \mathbf{x})^*$;
(2) $\Omega(\mathbf{x}, \mathbf{y})$ *is* $\mathcal{A}$*-linear in the first variable and* $\mathcal{A}^*$*-linear in the second, that is,* $\Omega(\mathbf{x}, \mathbf{y})$ *is additive in* $\mathbf{x}$ *and* $\mathbf{y}$*, and moreover*

$$\Omega(\Phi\mathbf{x}, \mathbf{y}) = \Phi\Omega(\mathbf{x}, \mathbf{y}) \quad \text{and} \quad \Omega(\mathbf{x}, \Psi\mathbf{y}) = \Psi^*\Omega(\mathbf{x}, \mathbf{y}). \tag{5.16}$$

Proof (1) Since A is Hamiltonian, $\{A\mathbf{x}, \mathbf{y}\} = -\{\mathbf{x}, A\mathbf{y}\}$. Then, we have

$$\begin{aligned}\Omega(\mathbf{x}, \mathbf{y}) &= \sum_{l=0}^{s}\{A^{s-l}\mathbf{x}, \mathbf{y}\}A^l = \sum_{l=0}^{s}(-1)^{s-l}\{\mathbf{x}, A^{s-l}\mathbf{y}\}A^l \\ &= (-1)^{s+1}\sum_{l=0}^{s}(-1)^l\{A^{s-l}\mathbf{y}, \mathbf{x}\}A^l = (-1)^{s+1}\Omega(\mathbf{y}, \mathbf{x})^*.\end{aligned}$$

To prove (2), first use the equality $A^{s+1} = 0$ in the transformation of the right-hand side of the equation $\Omega(A\mathbf{x}, \mathbf{y}) = \sum_{l=0}^{s}\{A^{s-l}(A\mathbf{x}), \mathbf{y}\}A^l$ to get

$$\Omega(A\mathbf{x}, \mathbf{y}) = A\sum_{l=1}^{s}\{A^{s-(l-1)}\mathbf{x}, \mathbf{y}\}A^{l-1} = A\sum_{l'=0}^{s}\{A^{s-l'}\mathbf{x}, \mathbf{y}\}A^{l'} = A\Omega(\mathbf{x}, \mathbf{y}).$$

Then, for any positive integer p, we have $\Omega(A^p\mathbf{x}, \mathbf{y}) = A^p\Omega(\mathbf{x}, \mathbf{y})$. The $\mathcal{A}$-linearity in the first variable then follows from the corresponding $\mathbf{R}$-linearity. Now, the $\mathcal{A}^*$-linearity in the second variable follows from the $\mathcal{A}$-linearity in the first variable, item (1) just proved and item 2 of Lemma 5.7. □

5.2.3 *Proof of Theorem 5.3*

Let $U = Z(\mathbf{u}, A)$. Then, U is an A-invariant subspace of U_1 and $A : U \to U$ is nilpotent of index $m + 1$. Let $\alpha_0 = \{A^m\mathbf{u}, \mathbf{u}\} \neq 0$. First, we prove that there exists a vector $\mathbf{e} \in U$ such that

$$\{A^m\mathbf{e}, \mathbf{e}\} = \operatorname{sign}(\alpha_0) \quad \text{and} \quad \{A^j\mathbf{e}, \mathbf{e}\} = 0, \quad \text{for } j = 1, 2, \ldots, m-1. \tag{5.17}$$

Since $\alpha_0 \neq 0$, then by (1) and (3) of Lemma 5.7 $\Phi = \Omega(\mathbf{u}, \mathbf{u})$ is nonsingular and has a square root Ψ, that is, $\Psi^2 = \operatorname{sign}(\alpha_0)\Phi$. As $\{A^m\mathbf{u}, \mathbf{u}\} \neq 0$, m is odd since A is Hamiltonian. By Lemma 5.8, $\Omega(\mathbf{u}, \mathbf{u}) = (-1)^{m+1}\Omega(\mathbf{u}, \mathbf{u})^*$; hence, Φ is even.

Therefore, by item (4) of Lemma 5.7 $\Psi^* = \Psi$. Now, let $\mathbf{e} = \Psi^{-1}\mathbf{u}$. By (2) of Lemma 5.8 and (3) of Lemma 5.7 we have

$$\Phi = \Omega(\mathbf{u}, \mathbf{u}) = \Omega(\Psi\mathbf{e}, \Psi\mathbf{e}) = \Psi\Psi^*\Omega(\mathbf{e}, \mathbf{e}) = \Psi^2\Omega(\mathbf{e}, \mathbf{e}) = \text{sign}(\alpha_0)\Phi\Omega(\mathbf{e}, \mathbf{e});$$

hence, since Φ is nonsingular we get $\Omega(\mathbf{e}, \mathbf{e}) = \text{sign}(\alpha_0)I$. From this equality and the definition (5.15) of Ω we get

$$\big[\{A^m\mathbf{e}, \mathbf{e}\} - \text{sign}(\alpha_0)\big]I + \{A^{m-1}\mathbf{e}, \mathbf{e}\}A + \{A^{m-2}\mathbf{e}, \mathbf{e}\}A^2 + \cdots + \{A\mathbf{e}, \mathbf{e}\}A^{m-1} = 0,$$

which gives the equalities (5.17).

Set $m = 2l - 1$ and consider the ordered sequence of $2l$ vectors

$$\mathbf{u}_1 = \mathbf{e},\ \mathbf{u}_2 = A\mathbf{e},\ \ldots,\ \mathbf{u}_l = A^{l-1}\mathbf{e};\ \mathbf{v}_1 = A^{2l-1}\mathbf{e},\ \ldots,\ \mathbf{v}_{l-1} = A^{l+1}\mathbf{e},\ \mathbf{v}_l = A^l\mathbf{e}. \tag{5.18}$$

Using (5.17) and the fact that A is Hamiltonian, hence $\{\mathbf{x}, A\mathbf{y}\} = -\{A\mathbf{x}, \mathbf{y}\}$, we compute for $1 \le i < j \le l$,

$$\begin{aligned}\{\mathbf{u}_i, \mathbf{u}_j\} &= \{A^{i-1}\mathbf{e}, A^{j-1}\mathbf{e}\}\\ &= (-1)^{i-1}\{\mathbf{e}, A^{i+j-2}\mathbf{e}\} = 0, \quad \text{because} \quad 1 \le i + j - 2 \le m - 1,\end{aligned}$$

and similarly,

$$\begin{aligned}\{\mathbf{v}_i, \mathbf{v}_j\} &= \{A^{2l-i}\mathbf{e}, A^{2l-j}\mathbf{e}\}\\ &= (-1)^{2l-i}\{\mathbf{e}, A^{4l-i-j}\mathbf{e}\} = 0 \quad \text{because} \quad 4l - i - j > m + 1.\end{aligned}$$

Now, for $1 \le i, j \le l$, we have

$$\{\mathbf{u}_i, \mathbf{v}_j\} = \{A^{i-1}\mathbf{e}, A^{2l-j}\mathbf{e}\} = (-1)^{i-1}\{\mathbf{e}, A^{2l+i-j-1}\mathbf{e}\} = (-1)^{i-1}\{\mathbf{e}, A^{m+i-j}\mathbf{e}\},$$

and we see that $\{\mathbf{u}_i, \mathbf{v}_j\} = 0$ for $i > j$, because $A^{m+1} = 0$ and for $i < j$ by (5.17), whereas for $i = j$ we get

$$\{\mathbf{u}_j, \mathbf{v}_j\} = (-1)^{j-1}\{\mathbf{e}, A^m\mathbf{e}\} = (-1)^j\{A^m\mathbf{e}, \mathbf{e}\} = (-1)^j\text{sign}(\alpha_0).$$

From the above we see that if we multiply the vectors $\mathbf{v}_j$ by $(-1)^j\text{sign}(\alpha_0)$ we get from (5.18) a symplectic basis for the subspace U, which is therefore a symplectic subspace of U_1. □

Remark 5.9 From the proof of the theorem we see that the process to construct the symplectic basis of $Z(\mathbf{u}, A)$ is the following.

Find a square root Ψ of $\Phi = \Omega(\mathbf{u}, \mathbf{u})$ and take $\mathbf{e} = \Psi^{-1}\mathbf{u}$. Then, $m = 2l - 1$ and the symplectic basis is given by

$$\mathbf{u}_1 = \mathbf{e},\ \mathbf{u}_2 = A\mathbf{e},\ \ldots,\ \mathbf{u}_l = A^{l-1}\mathbf{e};\ \mathbf{v}_1 = -\epsilon A^{2l-1}\mathbf{e},$$
$$\mathbf{v}_2 = \epsilon A^{2l-2}\mathbf{e},\ \ldots,\ \mathbf{v}_l = (-1)^l \epsilon A^l \mathbf{e},$$

with $\epsilon = \text{sign}(\alpha_0)$, where $\alpha_0 = \{A^m\mathbf{u}, \mathbf{u}\}$.

5.2.4 *Proof of Theorem 5.5*

Let $W = Z(\mathbf{u}, A) \oplus Z(\mathbf{v}, A)$. Then, the restriction of A to W is a nilpotent linear operator $A : W \to W$ of index $m + 1$. We consider the algebra $\mathcal{A} = \{\alpha_0 I + \alpha_1 A + \cdots + \alpha_m A^m;\ \alpha_j \in \mathbf{R}\}$. Since $\{A^m\mathbf{u}, \mathbf{v}\} \neq 0$ the element $\Omega(\mathbf{u}, \mathbf{v})$ of $\mathcal{A}$ is nonsingular and we can assume, without loss of generality[1] that $\Omega(\mathbf{u}, \mathbf{v}) = I$. By item (1) of Lemma 5.8, we then have $\Omega(\mathbf{v}, \mathbf{u}) = I$ or $\Omega(\mathbf{v}, \mathbf{u}) = -I$ according to m being odd or even. We now look for a vector $\mathbf{f} \in W$ such that $\Omega(\mathbf{f}, \mathbf{f}) = 0$, taking it in the form

$$\mathbf{f} = \mathbf{u} + \Phi\mathbf{v}, \qquad \Phi \in \mathcal{A}. \tag{5.19}$$

In view of Lemma 5.8 this leads to the equality

$$\Omega(\mathbf{u}, \mathbf{u}) + \Phi\Omega(\mathbf{v}, \mathbf{u}) + \Phi^*\Omega(\mathbf{u}, \mathbf{v}) + \Phi\Phi^*\Omega(\mathbf{v}, \mathbf{v}) = 0.$$

We will look for Φ such that $\Phi^* = (-1)^{m+1}\Phi$, that is, ϕ is even if m is odd and Φ is odd if m is even. Then, the above equality leads to

$$\Phi = \frac{1}{2}\big[\Omega(\mathbf{u}, \mathbf{u}) - \Phi^2\Omega(\mathbf{v}, \mathbf{v})\big] \quad \text{or} \quad \Phi = -\frac{1}{2}\big[\Omega(\mathbf{u}, \mathbf{u}) + \Phi^2\Omega(\mathbf{v}, \mathbf{v})\big], \tag{5.20}$$

according to m being even or odd.

Now, for any $\mathbf{z} \in W$, we have $\Omega(\mathbf{z}, \mathbf{z}) = (-1)^{m+1}\Omega(\mathbf{z}, \mathbf{z})^*$, so $\Omega(\mathbf{z}, \mathbf{z})$ is even if m is odd and it is odd if m is even. Since $(\Phi^2)^* = (\Phi^*)^2$ we have Φ^2 even whether Φ is even or odd. Therefore, the right-hand side of each equation in (5.20) is even or odd according to m being odd or even. Consequently, we can solve the Eqs. (5.20) for Φ recursively by comparing the coefficients of equal powers of A on both sides.[2]

Once Φ is determined we have $\mathbf{f}$ and then we define

$$\mathbf{g} = \mathbf{v} - \frac{1}{2}\Omega(\mathbf{v}, \mathbf{v})\mathbf{f}.$$

---

[1] We have $\Psi\Omega(\mathbf{u}, \mathbf{v}) = I$ for some $\Psi \in \mathcal{A}$, so $\Omega(\mathbf{u}, \Psi^*\mathbf{v}) = I$ and we can just replace $\mathbf{v}$ by $\Psi^*\mathbf{v}$.

[2] If m is even then $\Phi = \alpha_1 A + \cdots$ and $\Omega(\mathbf{u}, \mathbf{u}) = \{A^{m-1}\mathbf{u}, \mathbf{u}\}A + \cdots$ whereas if m is odd, then $\Phi = \alpha_0 I + \alpha_2 A^2 + \cdots$ and $\Omega(\mathbf{u}, \mathbf{u}) = \{A^{m-2}\mathbf{u}, \mathbf{u}\}A^2 + \cdots$, $\Omega(\mathbf{v}, \mathbf{v}) = \{A^{m-2}\mathbf{v}, \mathbf{v}\}A^2 + \cdots$, since $\{A^m\mathbf{v}, \mathbf{v}\} = 0$. Comparison of coefficients in (5.20) allows us to find Φ in both cases.

We observe that $\mathbf{f}$ can be taken so that[3] $\Omega(\mathbf{f}, \mathbf{v}) = I$, hence $\Omega(\mathbf{f}, \mathbf{g}) = \Omega(\mathbf{f}, \mathbf{v}) = I$.

Now, by (2) of Lemma 5.8, we have using the fact that $\Omega(\mathbf{f}, \mathbf{f}) = 0$,

$$\Omega(\mathbf{g}, \mathbf{g}) = \Omega(\mathbf{v}, \mathbf{v}) - \frac{1}{2}\Omega(\mathbf{v}, \mathbf{v})\Omega(\mathbf{f}, \mathbf{v}) - \frac{1}{2}\Omega(\mathbf{v}, \mathbf{v})^*\Omega(\mathbf{v}, \mathbf{f}).$$

But $\Omega(\mathbf{v}, \mathbf{v})^*\Omega(\mathbf{v}, \mathbf{f}) = (-1)^{m+1}\Omega(\mathbf{v}, \mathbf{v})(-1)^{m+1}\Omega(\mathbf{f}, \mathbf{v})^* = \Omega(\mathbf{v}, \mathbf{v})$, because $\Omega(\mathbf{f}, \mathbf{v}) = I$. Therefore, $\Omega(\mathbf{g}, \mathbf{g}) = 0$. Now, see (5.15), the equalities $\Omega(\mathbf{f}, \mathbf{f}) = 0$ and $\Omega(\mathbf{g}, \mathbf{g}) = 0$ give the relations

$$\{A^i\mathbf{f}, \mathbf{f}\} = 0, \quad \{A^i\mathbf{g}, \mathbf{g}\} = 0, \qquad \text{for} \quad i = 0, 1, \ldots, m, \tag{5.21}$$

whereas $\Omega(\mathbf{f}, \mathbf{g}) = I$ gives

$$\{A^m\mathbf{f}, \mathbf{g}\} = 1 \qquad \text{and} \qquad \{A^i\mathbf{f}, \mathbf{g}\} = 0, \quad \text{for} \quad i = 0, 1, \ldots, m-1. \tag{5.22}$$

Using these relations one easily verifies that the ordered sequence of vectors of W

$$\begin{aligned} &\mathbf{u}_1 = \mathbf{f}, \ \mathbf{u}_2 = A\mathbf{f}, \ \ldots, \ \mathbf{u}_m = A^{m-1}\mathbf{f}, \ \mathbf{u}_{m+1} = A^m\mathbf{f}; \\ &\mathbf{v}_1 = (-1)^m A^m\mathbf{g}, \ \mathbf{v}_2 = (-1)^{m-1}A^{m-1}\mathbf{g}, \ \ldots, \ \mathbf{v}_m = (-1)^{m-(m-1)}A\mathbf{g}, \ \mathbf{v}_{m+1} = \mathbf{g} \end{aligned} \tag{5.23}$$

form a symplectic basis for W, which is therefore a symplectic subspace of U_1. □

Remark 5.10 The process of constructing the symplectic basis in the case of Theorem 5.5 is a little more complicated in view of the two steps described in footnotes [(1)] and [(3)]. The step of solving the Eqs. (5.20) to get Φ corresponds to the one for finding the square root Ψ in Remark 5.9. Now, once found Φ we have $\mathbf{f} = \mathbf{u} + \Phi\mathbf{v}$ and then we define $\mathbf{g} = \mathbf{v} - \frac{1}{2}\Omega(\mathbf{v}, \mathbf{v})\mathbf{f}$. With the vectors $\mathbf{f}$ and $\mathbf{g}$ thus constructed we get the symplectic basis (5.23).

5.2.5 The Normal Form in the Nilpotent Context

We now use the theorems of Sect. 5.2.1 to find the normal form of a Hamiltonian nilpotent matrix.

Consider the basis constructed in Theorem 5.3 for $A : U \to U = Z(\mathbf{u}, A)$ (see Remark 5.9). On this basis we have

$$\begin{aligned} &A\mathbf{u}_1 = \mathbf{u}_2, \ A\mathbf{u}_2 = \mathbf{u}_3, \ \ldots, \ A\mathbf{u}_{l-1} = \mathbf{u}_l, \ A\mathbf{u}_l = \delta_l \mathbf{v}_l; \\ &A\mathbf{v}_1 = 0, \ A\mathbf{v}_2 = -\mathbf{v}_1, \ldots, A\mathbf{v}_{l-1} = -\mathbf{v}_{l-2}, A\mathbf{v}_l = -\mathbf{v}_{l-1}, \end{aligned}$$

---

[3] Since $\{A^m\mathbf{v}, \mathbf{v}\} = 0$, by (5.19) we have $\{A^m\mathbf{v}, \mathbf{f}\} = \{A^m\mathbf{v}, \mathbf{u}\} \neq 0$ so $\Omega(\mathbf{v}, \mathbf{f})$l hence, $\Omega(\mathbf{f}, \mathbf{v})$ is also nonsingular. Take $\Theta \in \mathcal{A}$ such that $\Theta\Omega(\mathbf{f}, \mathbf{v}) = I$ and then replace $\mathbf{f}$ by $\Theta\mathbf{f}$.

where $\delta_l = (-1)^l \epsilon = \pm 1$. Therefore, the matrix of $A : U \to U$ on this basis is the following

$$A^* = \begin{bmatrix} N & O \\ D & -N^T \end{bmatrix}, \tag{5.24}$$

where the $l \times l$ matrices N and D are given by

$$N = \begin{bmatrix} 0 & & & \\ 1 & 0 & & \\ & 1 & & \\ & & \ddots & \\ & & 1 & 0 \end{bmatrix} \quad \text{and} \quad D = D_l = \begin{bmatrix} 0 & & & \\ 0 & 0 & & \\ 0 & 0 & & \\ & & \ddots & \\ & & 0 & \delta_l \end{bmatrix}. \tag{5.25}$$

For the basis (5.23) constructed in Theorem 5.5 for $A : W \to W = Z(\mathbf{u}, A) \oplus Z(\mathbf{v}, A)$ we have

$$A\mathbf{u}_1 = \mathbf{u}_2, \; Au_2 = \mathbf{u}_3, \; \ldots, \; A\mathbf{u}_m = \mathbf{u}_{m+1}, \; A\mathbf{u}_{m+1} = 0;$$
$$A\mathbf{v}_1 = 0, \; A\mathbf{v}_2 = -\mathbf{v}_1, \ldots, A\mathbf{v}_m = -\mathbf{v}_{m-1}, \; A\mathbf{v}_{m+1} = -\mathbf{v}_m.$$

Therefore, the matrix of $A : W \to W$ on this basis has exactly the same structure (5.24) but now N and D are matrices of order $m + 1$ and $D = 0$, the zero matrix.

By repeated applications of Theorems 5.3 and 5.5 we can get a direct sum decomposition of V into A-invariant symplectic subspaces and construct a symplectic basis for V, which gives the normal form of the nilpotent Hamiltonian operator $A : V \to V$.

Indeed, if there is a vector $\mathbf{v}_1 \in V$ such that $\{A^{s_1}\mathbf{v}_1, \mathbf{v}_1\} \neq 0$, $s_1 = \mathrm{nil}(\mathbf{v}_1) - 1$, we apply Theorem 5.3 to get the symplectic subspace $V_1 = Z(\mathbf{v}_1, A)$ and construct the symplectic basis for it, obtaining the matrix of $A|_{V_1}$ as described above.

Now we write $V = V_1 \oplus W_1$, where $W_1 = V_1^{\perp}$. By Proposition 4.7(b) W_1 is an A-invariant symplectic subspace of V. If there is a vector $\mathbf{v}_2 \in W_1$ such that $\{A^{s_2}\mathbf{v}_2, \mathbf{v}_2\} \neq 0$, where $s_2 = \mathrm{nil}(\mathbf{v}_2) - 1$, we again apply Theorem 5.3 to get the symplectic subspace $V_2 = Z(\mathbf{v}_2, A)$, construct the symplectic basis for it, and find the matrix of $A|_{V_2}$.

Next, we get $V = V_1 \oplus V_2 \oplus W_2$, where $W_2 = (V_1 \oplus V_2)^{\perp}$ is an A-invariant symplectic subspace of V. We proceed with this process until we get

$$V = V_1 \oplus V_2 \oplus \ldots \oplus V_k \oplus W_k,$$

where $W_k = (V_1 \oplus V_2 \oplus \ldots \oplus V_k)^{\perp}$ is such that $\{A^s\mathbf{v}, \mathbf{v}\} = 0$, for all $\mathbf{v} \in W_k$.

At this point we apply Theorem 5.5 successively to get A-invariant symplectic subspaces $V_{k+1}, \ldots, V_t$ of V with the corresponding symplectic bases and obtain the matrices $A\big|_{V_j}$. We remark that in the final direct sum decomposition

$$V = V_1 \oplus V_2 \oplus \ldots \oplus V_k \oplus V_{k+1}, \ldots, V_{k+s} \oplus L,$$

the subspace L, which is contained in the kernel of A, is also symplectic.

The normal form of A is therefore a block diagonal matrix formed by blocks of the form (5.24), where in the first k blocks $D = D_l$ and in the next s blocks, $D = 0$. The block corresponding to L is of course the zero matrix.

It is worth mentioning that the normal form of a nilpotent real Hamiltonian matrix is given in terms of the nilpotent matrices of maximal rank having the number 1 in the subdiagonal, that is, matrices N as in (5.25).

5.3 Case of Purely Imaginary Eigenvalues

Finally, we consider the normalization of A restricted to the space Y,

$$Y = \eta^{\dagger}(i\beta_1) \oplus \eta^{\dagger}(-i\beta_1) \oplus \ldots \oplus \eta^{\dagger}(i\beta_n) \oplus \eta^{\dagger}(-i\beta_n). \tag{5.26}$$

Let $\pm i\beta$ be one of the pairs of eigenvalues $\pm i\beta_j$ of the real Hamiltonian matrix A. Then, as we have seen at the end of Sect. 4.3 the subspace

$$E(\beta) = \eta^{\dagger}(i\beta) \oplus \eta^{\dagger}(-i\beta) \tag{5.27}$$

is symplectic. We consider now the following linear operators on the generalized eigenspaces of $i\beta$ and $-i\beta$,

$$B = A - i\beta I : \eta^{\dagger}(i\beta) \to \eta^{\dagger}(i\beta) \quad \text{and} \quad \overline{B} = A + i\beta I : \eta^{\dagger}(-i\beta) \to \eta^{\dagger}(-i\beta), \tag{5.28}$$

which define the linear operator on $E(\beta)$,

$$\tilde{B} : E(\beta) \to E(\beta), \quad \tilde{B}(\mathbf{x} + \mathbf{y}) = B(\mathbf{x}) + \overline{B}(\mathbf{y}), \tag{5.29}$$

where $\mathbf{x} \in \eta^{\dagger}(i\beta)$ and $\mathbf{y} \in \eta^{\dagger}(-i\beta)$.

We observe that B and $\overline{B}$ are nilpotent operators of the same index equal to the multiplicity of the eigenvalue $i\beta$. Notice also that, in view of Lemma 4.36, $\eta^{\dagger}(i\beta)$ is a Lagrangian subspace of the space $E(\beta)$ in (5.27) because for $\mathbf{u}, \mathbf{v} \in \eta^{\dagger}(i\beta)$ we have $\{\mathbf{u}, \mathbf{v}\} = 0$ since $i\beta + i\beta \neq 0$. Also, $\eta^{\dagger}(-i\beta)$ is a Lagrangian subspace of $E(\beta)$. Therefore, the direct sum (5.27) is a Lagrangian splitting of $E(\beta)$.

We will first find a symplectic basis of $E(\beta)$ formed by complex vectors and subsequently from this basis we construct a real symplectic basis.

Remark 5.11 Since $E(\beta) = \eta^\dagger(i\beta) \oplus \eta^\dagger(-i\beta)$ is a Lagrangian splitting and the subspace $\eta^\dagger(i\beta)$ is $\tilde{B}$-invariant, one is tempted to take a Jordan basis for $\tilde{B}$: $\eta^\dagger(i\beta) \to \eta^\dagger(i\beta)$ and complete it with a basis of $\eta^\dagger(-i\beta)$ to form a symplectic basis of $E(\beta)$. This would be fine if $\tilde{B} : E(\beta) \to E(\beta)$ were a Hamiltonian operator, which is not the case. So we have to proceed differently and the process is not straightforward.

In the process below we will find symplectic subspaces of $E(\beta)$ restricted to which the operator $\tilde{B}$ is Hamiltonian. The following identity

$$\{B\mathbf{x}, \mathbf{y}\} = -\{\mathbf{x}, \overline{B}\mathbf{y}\}, \quad \text{for all} \quad \mathbf{x} \in \eta^\dagger(i\beta),\ \mathbf{y} \in \eta^\dagger(-i\beta) \tag{5.30}$$

will be used frequently. To prove it, notice that since A is Hamiltonian we have $\{A\mathbf{x}, \mathbf{y}\} = -\{\mathbf{x}, A\mathbf{y}\}$, for all $\mathbf{x}, \mathbf{y}$ so for $\mathbf{x} \in \eta^\dagger(i\beta)$ and $\mathbf{y} \in \eta^\dagger(-i\beta)$ we have

$$\{B\mathbf{x}, \mathbf{y}\} + \{\mathbf{x}, \overline{B}\mathbf{y}\} = \{A\mathbf{x} - i\beta\mathbf{x}, \mathbf{y}\} + \{\mathbf{x}, A\mathbf{y} + i\beta\mathbf{y}\} = \{A\mathbf{x}, \mathbf{y}\} + \{\mathbf{x}, A\mathbf{y}\} = 0,$$

which proves (5.30).

5.3.1 The First Main Theorem in the Imaginary Case

In this Section we prove two theorems that will be the tools for constructing the normal forms in the case of the space Y.

The first theorem is the following.

Theorem 5.12 *Let* $\mathbf{u} \in \eta^\dagger(i\beta)$ *be a vector of nilpotent index* $s + 1$ *such that* $\{B^s\mathbf{u}, \overline{\mathbf{u}}\} \neq 0$. *Then,* $Z(\mathbf{u}, B) \oplus Z(\overline{\mathbf{u}}, \overline{B})$ *is a symplectic subspace of* $E(\beta)$ *and an explicit basis can be constructed for it.*

Let us illustrate the use of this theorem by means of the following example.

Example 5.7 Simplify the following quadratic Hamiltonian

$$H = -2x_1^2 - x_2^2 + \frac{3}{2}y_1^2 + \frac{5}{2}y_2^2 - 3x_1x_2 + x_1y_1 - x_1y_2 + x_2y_1 - x_2y_2 - 4y_1y_2,$$

by means of a linear symplectic transformation.

The Hamiltonian matrix $A = J\mathrm{Hess}H$ defined by H is

$$A = \begin{bmatrix} 1 & 1 & 3 & -4 \\ -1 & -1 & -4 & 5 \\ 4 & 3 & -1 & 1 \\ 3 & 2 & -1 & 1 \end{bmatrix}.$$

This matrix has a pair of double eigenvalues $\lambda = \pm i$. Now, $A\mathbf{x} = i\mathbf{x}$ if, and only if, $\mathbf{x} = x_3(i, -i, 1, 1)$, so the eigenspace of the eigenvalue i of A is one-dimensional; hence, A is nondiagonalizable.

Let $E(\beta) = \eta^\dagger(i\beta) \oplus \eta^\dagger(-i\beta)$, with $\beta = 1$, and consider B and $\overline{B}$ as in (5.28). The subspace $\eta^\dagger(i\beta)$ is generated by two vectors, $\mathbf{w}_1, \mathbf{w}_2$ where $A\mathbf{w}_2 = i\beta\mathbf{w}_2$ and $A\mathbf{w}_1 = i\beta\mathbf{w}_1 + \mathbf{w}_2$. For $\mathbf{w}_2 = (i, -i, 1, 1)$ we get $\mathbf{w}_1 = (5+3i, -7-3i, 1+2i, 1)$ and we compute $\{B\mathbf{w}_1, \overline{\mathbf{w}}_1\} = 4$. Taking $\mathbf{u} = \mathbf{w}_1$ we have $\mathbf{u} \in \eta^\dagger(i\beta)$ and $\{B\mathbf{u}, \overline{\mathbf{u}}\} = 4 \neq 0$. By Theorem 5.12 $Z(\mathbf{u}, B) \oplus Z(\overline{\mathbf{u}}, \overline{B})$ is a symplectic subspace of $E(\beta)$.

In the proof of Theorem 5.12 it is shown that from the vector $\mathbf{u}$ one obtains a vector $\mathbf{e} \in \eta^\dagger(i\beta)$, which in our present case satisfies the equalities

$$\{\mathbf{e}, \overline{\mathbf{e}}\} = 0 \quad \text{and} \quad \{B\mathbf{e}, \overline{\mathbf{e}}\} = 1. \tag{5.31}$$

It follows that the vectors

$$\mathbf{u}_1 = \mathbf{e}, \qquad \mathbf{u}_2 = B\mathbf{e}, \qquad \mathbf{v}_1 = -\overline{B}\overline{\mathbf{e}}, \qquad \mathbf{v}_2 = \overline{\mathbf{e}} \tag{5.32}$$

form a symplectic basis of $Z(\mathbf{u}, B) \oplus Z(\overline{\mathbf{u}}, \overline{B})$. Indeed, $\{\mathbf{u}_1, \mathbf{u}_2\} = 0$, $\{\mathbf{v}_1, \mathbf{v}_2\} = 0$ because $\eta^\dagger(i\beta)$ and $\eta^\dagger(-i\beta)$ are Lagrangian subspaces of $E(\beta)$, $\{\mathbf{u}_2, \mathbf{v}_2\} = \{B\mathbf{e}, \overline{\mathbf{e}}\} = 1$ and by (5.30) we have $\{\mathbf{u}_1, \mathbf{v}_1\} = -\{\mathbf{e}, \overline{B}\overline{\mathbf{e}}\} = \{B\mathbf{e}, \overline{\mathbf{e}}\} = 1$.

Now we have $B\mathbf{u}_1 = \mathbf{u}_2$, $B\mathbf{u}_2 = 0$, $\overline{B}\mathbf{v}_1 = 0$, $\overline{B}\mathbf{v}_2 = \mathbf{v}_1$ and since $B = A - i\beta I$ on $\eta^\dagger(i\beta)$ and $\overline{B} = A + i\beta I$ on $\eta^\dagger(-i\beta)$ we get

$$A\mathbf{u}_1 = i\beta\mathbf{u}_1 + \mathbf{u}_2, \quad A\mathbf{u}_2 = i\beta\mathbf{u}_2, \quad A\mathbf{v}_1 = -i\beta\mathbf{v}_1, \quad A\mathbf{v}_2 = -\mathbf{v}_1 - i\beta\mathbf{v}_2, \tag{5.33}$$

so the matrix A in this basis, that is, the matrix of the linear operator $\mathbf{x} \mapsto A\mathbf{x}$ in the symplectic basis $\mathbf{u}_1, \mathbf{u}_2, \mathbf{v}_1, \mathbf{v}_2$ is the following complex Hamiltonian matrix

$$[A] = \begin{bmatrix} i\beta & 0 & 0 & 0 \\ 1 & i\beta & 0 & 0 \\ 0 & 0 & -i\beta & -1 \\ 0 & 0 & 0 & -i\beta \end{bmatrix}.$$

To get a real Hamiltonian matrix we have to find a real symplectic basis. The general procedure to pass from the complex basis to a real symplectic basis is described by (5.63) and (5.64) in Sect. 5.5. In the present case, this process gives the vectors

$$\mathbf{u}_1^* = \frac{1}{\sqrt{2}}(\mathbf{u}_1 + \mathbf{v}_2), \quad \mathbf{u}_2^* = \frac{1}{\sqrt{2}i}(\mathbf{u}_2 + \mathbf{v}_1), \quad \mathbf{v}_1^* = \frac{1}{\sqrt{2}}(-\mathbf{u}_2 + \mathbf{v}_1),$$
$$\mathbf{v}_2^* = \frac{1}{\sqrt{2}i}(\mathbf{u}_1 - \mathbf{v}_2). \tag{5.34}$$

From the fact that (5.36) is a symplectic basis it is readily shown that the vectors $\mathbf{u}_1^*, \mathbf{u}_2^*, \mathbf{v}_1^*, \mathbf{v}_2^*$ form a symplectic basis. Moreover, (5.36) shows that $\overline{\mathbf{u}}_1 = \mathbf{v}_2, \overline{\mathbf{u}}_2 = -\mathbf{v}_1, \overline{\mathbf{v}}_1 = -\mathbf{u}_2, \overline{\mathbf{v}}_2 = \mathbf{u}_1$, so we have

$$\mathbf{u}_1^* = \sqrt{2}\mathrm{Re}\,\mathbf{u}_1, \quad \mathbf{u}_2^* = \sqrt{2}\mathrm{Im}\,\mathbf{u}_2, \quad \mathbf{v}_1^* = \sqrt{2}\mathrm{Re}\,\mathbf{v}_1, \quad \mathbf{v}_2^* = -\sqrt{2}\mathrm{Im}\,\mathbf{v}_2;$$

hence, $\mathbf{u}_1^*, \mathbf{u}_2^*, \mathbf{v}_1^*, \mathbf{v}_2^*$ is a real symplectic basis of $E(\beta)$.

Using (5.38) and (5.34) and we have

$$A\mathbf{u}_1^* = \frac{1}{\sqrt{2}}\big(A\mathbf{u}_1 + A\mathbf{v}_2\big) = \frac{1}{\sqrt{2}}\big(i\beta(\mathbf{u}_1 - \mathbf{v}_2)\big) + (\mathbf{u}_2 - \mathbf{v}_1) = -\mathbf{v}_1^* - \beta\mathbf{v}_2^*.$$

Similarly, we get $A\mathbf{u}_2^* = -\beta\mathbf{v}_1^*$, $A\mathbf{v}_1^* = \beta\mathbf{u}_2^*$, and $A\mathbf{v}_2^* = \beta\mathbf{u}_1^* + \mathbf{u}_2^*$.

Therefore, in the real symplectic basis (5.34) the matrix A becomes

$$A^* = \begin{bmatrix} 0 & 0 & 0 & \beta \\ 0 & 0 & \beta & 1 \\ -1 & -\beta & 0 & 0 \\ -\beta & 0 & 0 & 0 \end{bmatrix}.$$

Computing the symmetric matrix $S^* = -JA^*$ we find that the Hamiltonian $\mathcal{H} = \frac{1}{2}\zeta^T S^* \zeta$, in the coordinates ξ_i, η_i of the symplectic basis $\mathbf{u}_1^*, \mathbf{u}_2^*, \mathbf{v}_1^*, \mathbf{v}_2^*$ is given by

$$\mathcal{H} = \frac{1}{2}\xi_1^2 + \frac{1}{2}\eta_2^2 + \beta(\xi_1\xi_2 + \eta_1\eta_2).$$

The linear symplectic transformation that leads the original Hamiltonian H to the Hamiltonian $\mathcal{H}$ is given by the transition matrix from the canonical basis of $\mathbf{R}^4$ to the symplectic $\mathbf{u}_1^*, \mathbf{u}_2^*, \mathbf{v}_1^*, \mathbf{v}_2^*$ and to get it we must know explicitly the vector $\mathbf{e}$ whose construction is described in the beginning of the proof of Theorem 5.12.

The following example deals with a three-degree of freedom Hamiltonian.

Example 5.8 Simplify the following quadratic Hamiltonian

$$H = -\frac{1}{2}x_3^2 - \frac{1}{2}y_3^2 + x_1x_2 - x_1y_3 - x_2y_1 + y_1y_2,$$

by means of a symplectic linear transformation.

The Hamiltonian matrix $A = J\mathrm{Hess}H$ defined by H is

$$A = \begin{bmatrix} 0 & -1 & 0 & 0 & 1 & 0 \\ 0 & 0 & 0 & 1 & 0 & 0 \\ -1 & 0 & 0 & 0 & 0 & -1 \\ 0 & -1 & 0 & 0 & 0 & 1 \\ -1 & 0 & 0 & 1 & 0 & 0 \\ 0 & 0 & 1 & 0 & 0 & 0 \end{bmatrix}.$$

The characteristic polynomial of A is $p(\lambda) = (\lambda^2 + 1)^3$, so the matrix has a pair of triple eigenvalues $\lambda = \pm i$.

Let $E(\beta) = \eta^\dagger(i\beta) \oplus \eta^\dagger(-i\beta)$, with $\beta = 1$, and consider the operators B and $\overline{B}$ in (5.28). The matrix of B is given by

$$B = \begin{bmatrix} -i & -1 & 0 & 0 & 1 & 0 \\ 0 & -i & 0 & 1 & 0 & 0 \\ -1 & 0 & -i & 0 & 0 & -1 \\ 0 & -1 & 0 & -i & 0 & 1 \\ -1 & 0 & 0 & 1 & -i & 0 \\ 0 & 0 & 1 & 0 & 0 & -i \end{bmatrix}.$$

Let us describe the subspace $\eta^\dagger(i\beta) = \mathrm{Ker}B^3$, where $B = A - i\beta I$. We compute

$$B^2 = \begin{bmatrix} -2 & 2i & 0 & 0 & -2i & 0 \\ 0 & -2 & 0 & -2i & 0 & 1 \\ 2i & 1 & -2 & 0 & -1 & 2i \\ 0 & 2i & 1 & -2 & 0 & -2i \\ 2i & 0 & 0 & -2i & -2 & 1 \\ -1 & 0 & -2i & 0 & 0 & -2 \end{bmatrix}, \quad B^3 = \begin{bmatrix} 4i & 4 & 0 & 0 & -4 & 0 \\ 0 & 4i & 1 & -4 & 0 & -3i \\ 5 & -3i & 4i & 0 & 3i & 4 \\ -1 & 4 & -3i & 4i & 0 & -5 \\ 4 & 0 & 1 & -4 & 4i & -3i \\ 3i & 1 & -4 & 0 & -1 & 4i \end{bmatrix}.$$

Looking for the vectors $\mathbf{x} = (x_1, x_2, x_3, x_4, x_5, x_6)$ such that $B^3\mathbf{x} = 0$ we find

$$\mathbf{x} = (-8x_2 - 8ix_4 + 4x_6,\ x_2,\ -4ix_2 + 4x_4 + 3ix_6,\ x_4,\ (1-8i)x_2 + 8x_4 + 4ix_6,\ x_6).$$

For such vectors we compute

$$B^2\mathbf{x} = (0, -2x_2 - 2ix_4 + x_6, 0, -2ix_2 + 2x_4 + ix_6, -2x_2 - 2ix_4 + x_6, 0)$$

and for real numbers x_2, x_4, x_6 we find that[4]

$$\{B^2\mathbf{x}, \overline{\mathbf{x}}\} = -8i\left[(2x_2 - x_6)^2 + 4x_4^2\right].$$

Choosing $x_2 = \frac{1}{4}$, $x_4 = x_6 = 0$, we get the vector $\mathbf{u} = (-2, \frac{1}{4}, -i, 0, \frac{1}{4} - 2i, 0) \in \eta^\dagger(i\beta)$ with $\mathrm{nil}(\mathbf{u}) = 3$ and $\{B^2\mathbf{u}, \overline{\mathbf{u}}\} = -2i \neq 0$.

By Theorem 5.12 the direct sum $Z(\mathbf{u}, B) \oplus Z(\overline{\mathbf{u}}, \overline{B})$ is a symplectic subspace of $E(\beta) = \eta^\dagger(i\beta) \oplus \eta^\dagger(-i\beta)$. Since $Z(\mathbf{u}, B) = [\mathbf{u}, B\mathbf{u}, B^2\mathbf{u}]$ is three-dimensional we

[4] It is no coincidence that $\{B^2\mathbf{x}, \overline{\mathbf{x}}\}$ is a purely imaginary number; this will always happen for $\{B^s\mathbf{x}, \overline{\mathbf{x}}\}$ if s is an even number. In the present case we have, using (5.30)

$$\{B^2\mathbf{x}, \overline{\mathbf{x}}\} = -\{B\mathbf{x}, \overline{B\mathbf{x}}\} = \{\overline{B\mathbf{x}}, B\mathbf{x}\} = \overline{\{B\mathbf{x}, \overline{B\mathbf{x}}\}} = -\overline{\{B^2\mathbf{x}, \overline{\mathbf{x}}\}}$$

so $\{B^2\mathbf{x}, \overline{\mathbf{x}}\}$ is a purely imaginary number.

have

$$\eta^\dagger(i\beta) \oplus \eta^\dagger(-i\beta) = Z(\mathbf{u}, B) \oplus Z(\overline{\mathbf{u}}, \overline{B}).$$

Let us find a symplectic basis for this space. In this case, the vector $\mathbf{e} \in \eta^\dagger(i\beta)$ constructed in the proof of Theorem 5.12 satisfies the equalities

$$\{B^2\mathbf{e}, \overline{\mathbf{e}}\} = i, \quad \{B\mathbf{e}, \overline{\mathbf{e}}\} = 0 \quad \text{and} \quad \{\mathbf{e}, \overline{\mathbf{e}}\} = 0. \tag{5.35}$$

It follows that the vectors

$$\mathbf{u}_1 = \mathbf{e}, \quad \mathbf{u}_2 = B\mathbf{e}, \quad \mathbf{u}_3 = B^2\mathbf{e}, \quad \mathbf{v}_1 = -i\overline{B}^2\overline{\mathbf{e}}, \quad \mathbf{v}_2 = i\overline{B}\overline{\mathbf{e}}, \quad \mathbf{v}_3 = -i\overline{\mathbf{e}} \tag{5.36}$$

form a symplectic basis of $Z(\mathbf{u}, B) \oplus Z(\overline{\mathbf{u}}, \overline{B})$.

Now we have

$$B\mathbf{u}_1 = \mathbf{u}_2, \quad B\mathbf{u}_2 = \mathbf{u}_3, \quad B\mathbf{u}_3 = 0, \quad \overline{B}\mathbf{v}_1 = 0, \quad \overline{B}\mathbf{v}_2 = -\mathbf{v}_1, \quad \overline{B}\mathbf{v}_3 = -\mathbf{v}_2, \tag{5.37}$$

and since $B = A - i\beta I$ on $\eta^\dagger(i\beta)$ and $\overline{B} = A + i\beta I$ on $\eta^\dagger(-i\beta)$ we get

$$\begin{aligned} A\mathbf{u}_1 &= i\beta\mathbf{u}_1 + \mathbf{u}_2, \quad A\mathbf{u}_2 = i\beta\mathbf{u}_2 + \mathbf{u}_3, \quad A\mathbf{u}_3 = i\beta\mathbf{u}_3, \\ A\mathbf{v}_1 &= -i\beta\mathbf{v}_1, \quad A\mathbf{v}_2 = -\mathbf{v}_1 - i\beta\mathbf{v}_2, \quad A\mathbf{v}_3 = -\mathbf{v}_2 - i\beta\mathbf{v}_3 \end{aligned} \tag{5.38}$$

so the matrix A on this basis, that is, the matrix of the linear operator $\mathbf{x} \mapsto A\mathbf{x}$ on the symplectic basis $\mathbf{u}_1, \mathbf{u}_2, \mathbf{u}_3, \mathbf{v}_1, \mathbf{v}_2, \mathbf{v}_3$ is the following complex Hamiltonian matrix

$$[A] = \begin{bmatrix} i\beta & 0 & 0 & 0 & 0 & 0 \\ 1 & i\beta & 0 & 0 & 0 & 0 \\ 0 & 1 & i\beta & 0 & 0 & 0 \\ 0 & 0 & 0 & -i\beta & -1 & 0 \\ 0 & 0 & 0 & 0 & -i\beta & -1 \\ 0 & 0 & 0 & 0 & 0 & -i\beta \end{bmatrix}.$$

To get a real Hamiltonian matrix we have to find a real symplectic basis. In this case, $\mathrm{nil}(\mathbf{u}) = s + 1$ means $s = 2$ and the general procedure to pass from the complex basis to a real symplectic basis described by (5.61) in Sect. 5.5 gives the vectors

$$\begin{aligned} \mathbf{u}_1^* &= \frac{1}{\sqrt{2}}(\mathbf{u}_1 + i\mathbf{v}_3), \quad \mathbf{u}_2^* = \frac{1}{\sqrt{2}}(\mathbf{u}_2 - i\mathbf{v}_2), \quad \mathbf{u}_3^* = \frac{1}{\sqrt{2}}(\mathbf{u}_3 + i\mathbf{v}_1) \\ \mathbf{v}_1^* &= \frac{1}{\sqrt{2}}(i\mathbf{u}_3 + \mathbf{v}_1), \quad \mathbf{v}_2^* = \frac{1}{\sqrt{2}}(-i\mathbf{u}_2 + \mathbf{v}_2), \quad \mathbf{v}_3^* = \frac{1}{\sqrt{2}}(i\mathbf{u}_1 + \mathbf{v}_3). \end{aligned} \tag{5.39}$$

Since the vectors (5.36) form a symplectic basis the vectors $\mathbf{u}_1^*, \mathbf{u}_2^*, \mathbf{u}_3^*, \mathbf{v}_1^*, \mathbf{v}_2^*, \mathbf{v}_3^*$ also form a symplectic basis. Moreover, (5.36) shows that

$$\overline{\mathbf{u}}_1 = \mathbf{v}_2, \quad \overline{\mathbf{u}}_2 = -\mathbf{v}_1, \quad \overline{\mathbf{u}}_3 = i\mathbf{v}_1, \quad \overline{\mathbf{v}}_1 = -\mathbf{u}_2, \quad \overline{\mathbf{v}}_2 = \mathbf{u}_1, \quad \overline{\mathbf{v}}_3 = i\mathbf{u}_1,$$

so we have

$$\mathbf{u}_j^* = \sqrt{2}\mathrm{Re}\,\mathbf{u}_j, \quad \mathbf{v}_j^* = \sqrt{2}\mathrm{Re}\,\mathbf{v}_, \quad \text{for } j = 1, 2, 3;$$

hence, $\mathbf{u}_1^*, \mathbf{u}_2^*, \mathbf{u}_3^*, \mathbf{v}_1^*, \mathbf{v}_2^*, \mathbf{v}_3^*$ is a real symplectic basis of $E(\beta)$.

Using (5.34) and (5.38), we have

$$\begin{aligned} A\mathbf{u}_1^* &= \frac{1}{\sqrt{2}}\left(A\mathbf{u}_1 + iA\mathbf{v}_3\right) = \frac{1}{\sqrt{2}}\left(i\beta\mathbf{u}_1 + \mathbf{u}_2 - i\mathbf{v}_2 + \beta\mathbf{v}_3\right) \\ &= \frac{1}{\sqrt{2}}(\mathbf{u}_2 - i\mathbf{v}_2) + \beta\frac{1}{\sqrt{2}}(i\mathbf{u}_1 + \mathbf{v}_3) = \mathbf{u}_2^* + \beta\mathbf{v}_3^*. \end{aligned}$$

Similarly, we get

$$A\mathbf{u}_2^* = \mathbf{u}_3^* - \beta\mathbf{v}_2^*, \quad A\mathbf{u}_3^* = \beta\mathbf{v}_1^*, \quad A\mathbf{v}_1^* = -\beta\mathbf{u}_3^*, \quad A\mathbf{v}_2^* = -\mathbf{v}_1^* + \beta\mathbf{u}_2^*,$$
$$A\mathbf{v}_3^* = -\mathbf{v}_2^* - \beta\mathbf{u}_1^*.$$

Therefore, in the real symplectic basis (5.39) the matrix A becomes

$$A^* = \begin{bmatrix} 0 & 0 & 0 & 0 & 0 & -\beta \\ 1 & 0 & 0 & 0 & \beta & 0 \\ 0 & 1 & 0 & -\beta & 0 & 0 \\ 0 & 0 & \beta & 0 & -1 & 0 \\ 0 & -\beta & 0 & 0 & 0 & -1 \\ \beta & 0 & 0 & 0 & 0 & 0 \end{bmatrix}.$$

Computing the symmetric matrix $S^* = -JA^*$ we find that the Hamiltonian $\mathcal{H} = \frac{1}{2}\zeta^T S^* \zeta$, in the coordinates ξ_i, η_i of the symplectic basis $\mathbf{u}_1^*, \mathbf{u}_2^*, \mathbf{u}_3^*, \mathbf{v}_1^*, \mathbf{v}_2^*, \mathbf{v}_3^*$ is given by

$$\mathcal{H} = \frac{1}{2}\beta\xi_2^2 + \frac{1}{2}\beta\eta_2^2 - \beta\xi_1\xi_3 + \xi_1\eta_2 + \xi_2\eta_3 - \beta\eta_1\eta_3.$$

The linear symplectic transformation that leads the original Hamiltonian H to the Hamiltonian $\mathcal{H}$ is given by the transition matrix from the canonical basis of $\mathbf{R}^6$ to the symplectic basis $\mathbf{u}_1^*, \mathbf{u}_2^*, \mathbf{u}_3^*, \mathbf{v}_1^*, \mathbf{v}_2^*, \mathbf{v}_3^*$ and to get it we must know explicitly the vector $\mathbf{e}$ whose construction is described in the beginning of the proof of Theorem 5.12.

Remark 5.13 It seems that the Hamiltonian $\mathcal{H}$ in this example is not simpler than the original Hamiltonian H. Indeed, to make the computations simple we started with a Hamiltonian of a simple structure. However, the final Hamiltonian $\mathcal{H}$ would be the same for any Hamiltonian matrix having the characteristic polynomial of A. For instance, for any symplectic matrix P, the matrix $\mathcal{A} = P^{-1}AP$ is a Hamiltonian matrix with the characteristic polynomial of A and $\mathcal{A}$ could be a matrix of a complicated structure. See Remark 5.19 after the proof of Theorem 5.12.

In the next example, we consider a four-degree of freedom Hamiltonian.

Example 5.9 Simplify the following quadratic four-degree of freedom Hamiltonian

$$H = \frac{1}{2}(y_1^2+y_2^2+y_3^2+y_4^2)+\sqrt{3}x_1y_2+2x_1y_4-\sqrt{3}x_2y_1+\sqrt{3}x_3y_4-2x_4y_1-\sqrt{3}x_4y_3.$$

The Hamiltonian matrix defined by H is the 8×8 matrix

$$\mathcal{A} = \begin{bmatrix} A & I \\ O & A \end{bmatrix}, \quad \text{where} \quad A = \begin{bmatrix} 0 & -\sqrt{3} & 0 & -2 \\ \sqrt{3} & 0 & 0 & 0 \\ 0 & 0 & 0 & -\sqrt{3} \\ 0 & 0 & \sqrt{3} & 0 \end{bmatrix}.$$

The characteristic polynomial of A is $p_A(x) = (x^2+1)(x^2+9)$ and that of $\mathcal{A}$ is $p_{\mathcal{A}}(x) = (x^2+1)^2(x^2+9)^2$. The eigenvalues of $\mathcal{A}$ are $\pm i$ and $\pm 3i$, each with multiplicity two, so we consider $i\beta_1$ and $i\beta_2$ with $\beta_1 = 1$ and $\beta_2 = 3$. We have the $\mathcal{A}$-invariant direct sum decomposition

$$\mathbf{C}^8 = E(\beta_1) \oplus E(\beta_2),$$

where $E(\beta_k) = \eta^\dagger(i\beta_k) \oplus \eta^\dagger(-i\beta_k)$ is a four-dimensional symplectic subspace of $\mathbf{C}^8$. Consider the linear operator

$$\tilde{\mathcal{B}_1} : E(\beta_1) \to E(\beta_1),$$

with $\mathcal{B}_1 = \mathcal{A} - i\beta_1 I : \eta^\dagger(i\beta_1) \to \eta^\dagger(i\beta_1)$ and $\overline{\mathcal{B}}_1 = \mathcal{A} + i\beta_1 I : \eta^\dagger(-i\beta_1) \to \eta^\dagger(-i\beta_1)$ as we have defined in (5.27) and (5.28). Let us see the structure of $\eta^\dagger(i\beta_1) = \operatorname{Ker}\mathcal{B}_1^2$. We have

$$\mathcal{B}_1 = \begin{bmatrix} -i & -\sqrt{3} & 0 & -2 \\ \sqrt{3} & -i & 0 & 0 \\ 0 & 0 & -i & -\sqrt{3} \\ 2 & 0 & \sqrt{3} & -i \end{bmatrix} \quad \text{and} \quad \mathcal{B}_1^2 = \begin{bmatrix} -8 & 2i\sqrt{3} & -2\sqrt{3} & 4i \\ -2i\sqrt{3} & -4 & 0 & -2\sqrt{3} \\ -2\sqrt{3} & 0 & -4 & 2i\sqrt{3} \\ -4i & -2\sqrt{3} & -2i\sqrt{3} & -8 \end{bmatrix}.$$

For $\mathbf{z} = (x_1, x_2, x_3, x_4; y_1, y_2, y_3, y_4) \in \eta^\dagger(i\beta_1)$, that is $\mathcal{B}_1^2\mathbf{z} = 0$, we find

$$\mathbf{z} = \left(x_1, -i\sqrt{3}x_1, -\sqrt{3}x_1, -ix_1; y_1, -i\sqrt{3}y_1, -\sqrt{3}y_1, iy_1\right)$$

and we compute $\{\mathcal{B}_1\mathbf{z}, \overline{\mathbf{z}}\} = 8y_1^2$. Taking for $\mathbf{z}_*$ the vector $\mathbf{z}$ with $x_1 = 0$ and $y_1 = 1$ we have $\mathbf{z}_* \in \eta^\dagger(i\beta_1)$, $\operatorname{nil}(\mathbf{z}_*) = 2$ and $\{\mathcal{B}_1\mathbf{z}_*, \overline{\mathbf{z}}_*\} = 8$, so by Theorem 5.12 we have the symplectic subspace $U_* = Z(\mathbf{z}_*, \mathcal{B}_1) \oplus Z(\overline{\mathbf{z}}_*, \overline{\mathcal{B}})$ of $E(\beta_1)$. Since $Z(\mathbf{z}_*, \mathcal{B}_1) = [\mathbf{z}_*, \mathcal{B}_1\mathbf{z}_*]$ is two-dimensional, the subspace U_* is four-dimensional; hence, $U_* = E(\beta_1)$.

To construct a symplectic basis for this space we observe that the vector $\mathbf{e}_1 \in \eta^\dagger(i\beta_1)$ constructed in the proof of Theorem 5.12 satisfies the equalities

$$\{B\mathbf{e}_1, \overline{\mathbf{e}}_1\} = 1, \qquad \{\mathbf{e}_1, \overline{\mathbf{e}}_1\} = 0. \tag{5.40}$$

It follows that the vectors

$$\mathbf{u}_1 = \mathbf{e}_1, \quad \mathbf{u}_2 = B\mathbf{e}_1, \quad \mathbf{v}_1 = -\overline{B}\overline{\mathbf{e}}_1, \quad \mathbf{v}_2 = \overline{\mathbf{e}}_1$$

form a symplectic basis of $Z(\mathbf{u}, \mathcal{B}_1) \oplus Z(\overline{\mathbf{u}}, \overline{\mathcal{B}_1})$. We have

$$\mathcal{B}\mathbf{u}_1 = \mathbf{u}_2, \;\; \mathcal{B}\mathbf{u}_2 = 0 \;\; \overline{\mathcal{B}}\mathbf{v}_1 = 0, \;\; \overline{\mathcal{B}}\mathbf{v}_2 = -\mathbf{v}_1, \tag{5.41}$$

and therefore for $\mathcal{A}$ restricted to $E(\beta_1)$ we have

$$\mathcal{A}\mathbf{u}_1 = i\beta_1\mathbf{u}_1 + \mathbf{u}_2, \quad \mathcal{A}\mathbf{u}_2 = i\beta_1\mathbf{u}_2, \quad \mathcal{A}\mathbf{v}_1 = -i\beta_1\mathbf{v}_1, \quad \mathcal{A}\mathbf{v}_2 = -\mathbf{v}_1 - i\beta_1\mathbf{v}_2 \tag{5.42}$$

To find a real symplectic basis we observe that $s = 1$ is odd and $\{\mathcal{B}_1\mathbf{z}_*, \overline{\mathbf{z}}_*\} = 8$ is a positive real number, so the vectors $\mathbf{u}_j^*, \mathbf{v}_j^*$ in (5.63) and (5.64) of the symplectic basis are given by

$$\mathbf{u}_1^* = \frac{1}{\sqrt{2}}(\mathbf{u}_1 + \mathbf{v}_2), \quad \mathbf{u}_2^* = \frac{1}{\sqrt{2}i}(\mathbf{u}_2 + \mathbf{v}_1), \quad \mathbf{v}_1^* = \frac{1}{\sqrt{2}}(-\mathbf{u}_2 + \mathbf{v}_1),$$
$$\mathbf{v}_2^* = \frac{1}{\sqrt{2}i}(\mathbf{u}_1 - \mathbf{v}_2).$$

Since $\overline{\mathbf{u}}_1 = \mathbf{v}_2, \overline{\mathbf{u}}_2 = -\mathbf{v}_1, \overline{\mathbf{v}}_1 = -\mathbf{u}_2, \overline{\mathbf{v}}_2 = \mathbf{u}_1$, we have

$$\mathbf{u}_1^* = \sqrt{2}\mathrm{Re}\,\mathbf{u}_1, \quad \mathbf{u}_2^* = \sqrt{2}\mathrm{Re}\,\mathbf{u}_1, \quad \mathbf{v}_1^* = \sqrt{2}\mathrm{Re}\,\mathbf{v}_1, \quad \mathbf{v}_2^* = \sqrt{2}\mathrm{Re}\,\mathbf{v}_2,$$

so $\mathbf{u}_1^*, \mathbf{u}_2^*, \mathbf{v}_1^*, \mathbf{v}_2^*$ is a real symplectic basis. Now, using (5.44) we have

$$\mathcal{A}\mathbf{u}_1^* = \frac{1}{\sqrt{2}}(\mathcal{A}\mathbf{u}_1 + \mathcal{A}\mathbf{v}_2) = \beta_1\frac{1}{\sqrt{2}i}(-\mathbf{u}_1 + \mathbf{v}_2) + \frac{1}{\sqrt{2}}(\mathbf{u}_2 - \mathbf{v}_1) = -\beta_1\mathbf{v}_2^* - \mathbf{v}_1^*,$$

and similarly we find

$$\mathcal{A}\mathbf{u}_2^* = -\beta_1\mathbf{v}_1^*, \quad \mathcal{A}\mathbf{v}_1^* = \beta_1\mathbf{u}_2^*, \quad \mathcal{A}\mathbf{v}_2^* = \beta_1\mathbf{u}_1^* + \mathbf{u}_2^*.$$

Therefore, the matrix of the restriction of $\mathcal{A}$ to $E(\beta_1)$ on this basis is

$$\mathcal{A}_*^{(1)} = \begin{bmatrix} 0 & 0 & 0 & \beta_1 \\ 0 & 0 & \beta_1 & 1 \\ -1 & -\beta_1 & 0 & 0 \\ -\beta_1 & 0 & 0 & 0 \end{bmatrix}.$$

Now, we repeat the above analysis for the operator $\mathcal{B}_2 = \mathcal{A} - i\beta_2 I : \eta^\dagger(i\beta_2) \to \eta^\dagger(i\beta_2)$. Let us see the structure of $\eta^\dagger(i\beta_1) = \mathrm{Ker}\mathcal{B}_1^2$. We have

$$\mathcal{B}_2 = \begin{bmatrix} -3i & -\sqrt{3} & 0 & -2 \\ \sqrt{3} & -3i & 0 & 0 \\ 0 & 0 & -3i & -\sqrt{3} \\ 2 & 0 & \sqrt{3} & -3i \end{bmatrix} \quad \text{and} \quad \mathcal{B}_2^2 = \begin{bmatrix} -16 & 6i\sqrt{3} & -2\sqrt{3} & 12i \\ -6i\sqrt{3} & -12 & 0 & -2\sqrt{3} \\ -2\sqrt{3} & 0 & -12 & 6i\sqrt{3} \\ -12i & -2\sqrt{3} & -6i\sqrt{3} & -16 \end{bmatrix}.$$

For $\mathbf{z} = (x_1, x_2, x_3, x_4; y_1, y_2, y_3, y_4) \in \eta^\dagger(i\beta_2)$, that is $\mathcal{B}_2^2\mathbf{z} = 0$, we find

$$\mathbf{z} = \frac{1}{3}\left(3x_1, -i\sqrt{3}x_1, \sqrt{3}x_1, -3ix_1; 3y_1, -i\sqrt{3}y_1, \sqrt{3}y_1, -3iy_1\right)$$

and we compute $\{\mathcal{B}_2\mathbf{z}, \overline{\mathbf{z}}\} = 8y_1^2$. Taking for $\mathbf{z}_{**}$ the vector $\mathbf{z}$ with $x_1 = 0$ and $y_1 = \sqrt{3}$ we have $\mathbf{z}_{**} \in \eta^\dagger(i\beta_2)$, $\mathrm{nil}(\mathbf{z}_{**}) = 2$ and $\{\mathcal{B}_2\mathbf{z}_{**}, \overline{\mathbf{z}}_{**}\} = 8$, so by Theorem 5.12 we have the symplectic subspace $U_{**} = Z(\mathbf{z}_{**}, \mathcal{B}_2) \oplus Z(\overline{\mathbf{z}}_{**}, \overline{\mathcal{B}_2})$ of $E(\beta_2)$. Since $Z(\mathbf{z}_{**}, \mathcal{B}_2) = [\mathbf{z}_{**}, \mathcal{B}_2\mathbf{z}_{**}]$ is two dimensional, the subspace U_{**} is four dimensional; hence, $U_{**} = E(\beta_2)$.

Using a vector $\mathbf{e}_2$ satisfying equations similar to (5.40) we find the symplectic basis

$$\mathbf{u}_3 = \mathbf{e}_2, \quad \mathbf{u}_4 = \mathcal{B}_2\mathbf{e}_2, \quad \mathbf{v}_3 = -\overline{\mathcal{B}_2}\overline{\mathbf{e}}_2, \quad \mathbf{v}_4 = \overline{\mathbf{e}}_2$$

of $U_{**} = Z(\mathbf{z}_{**}, \mathcal{B}_2) \oplus Z(\overline{\mathbf{z}_{**}}, \overline{\mathcal{B}_2})$. We have

$$\mathcal{B}_2\mathbf{u}_3 = \mathbf{u}_4, \quad \mathcal{B}_2\mathbf{u}_4 = 0 \quad \overline{\mathcal{B}_2}\mathbf{v}_3 = 0, \quad \overline{\mathcal{B}_2}\mathbf{v}_4 = -\mathbf{v}_3, \tag{5.43}$$

and therefore for $\mathcal{A}$ restricted to $E(\beta_2)$ we have

$$\mathcal{A}\mathbf{u}_3 = i\beta_2\mathbf{u}_3 + \mathbf{u}_4, \quad \mathcal{A}\mathbf{u}_4 = i\beta_2\mathbf{u}_4, \quad \mathcal{A}\mathbf{v}_3 = -i\beta_2\mathbf{v}_3, \quad \mathcal{A}\mathbf{v}_4 = -\mathbf{v}_3 - i\beta_2\mathbf{v}_4 \tag{5.44}$$

A real symplectic basis $\mathbf{u}_3^*, \mathbf{u}_4^*, \mathbf{v}_3^*, \mathbf{v}_4^*$ is constructed exactly as for the case β_1 and we find that the matrix of the restriction of $\mathcal{A}$ to $E(\beta_2)$ on this basis is

$$\mathcal{A}_*^{(2)} = \begin{bmatrix} 0 & 0 & 0 & \beta_2 \\ 0 & 0 & \beta_2 & 1 \\ -1 & -\beta_2 & 0 & 0 \\ -\beta_2 & 0 & 0 & 0 \end{bmatrix}.$$

Since $i\beta_1 \neq i\beta_2$ the subspaces U_* and U_{**} are symplectically orthogonal and we can use Proposition 4.10 to construct the real symplectic basis of $\mathbf{R}^8 = U_* \oplus U_{**}$. On this basis the matrix becomes

$$\mathcal{A}_* = \begin{bmatrix} 0 & 0 & 0 & 0 & 0 & \beta_1 & 0 & 0 \\ 0 & 0 & 0 & 0 & \beta_1 & 1 & 0 & 0 \\ 0 & 0 & 0 & 0 & 0 & 0 & 0 & \beta_2 \\ 0 & 0 & 0 & 0 & 0 & 0 & \beta_2 & 1 \\ -1 & -\beta_1 & 0 & 0 & 0 & 0 & 0 & 0 \\ -\beta_1 & 0 & 0 & 0 & 0 & 0 & 0 & 0 \\ 0 & 0 & -1 & -\beta_2 & 0 & 0 & 0 & 0 \\ 0 & 0 & -\beta_2 & 0 & 0 & 0 & 0 & 0 \end{bmatrix}.$$

Computing the symmetric matrix $\mathcal{S}^* = -J\mathcal{A}^*$, which defines the new Hamiltonian $\mathcal{H}$, we find the expression

$$\mathcal{H} = \frac{1}{2}\xi_1^2 + \frac{1}{2}\xi_3^2 + \frac{1}{2}\eta_2^2 + \frac{1}{2}\eta_4^2 + \xi_1\xi_2 + \eta_1\eta_2 + 3(\xi_3\xi_4 + \eta_3\eta_4).$$

Remark 5.14 Here again $\mathcal{H}$ is not much simpler than the original Hamiltonian H, but see the comments in Remark 5.13 and also Remark 5.19.

To prove Theorem 5.12, just as in the real nilpotent case, we need to develop some preliminary machinery, which is done in the next subsection.

5.3.2 Basic Lemmas in the Imaginary Case

Let U be a finite dimensional complex vector space and let $\mathcal{B}$ be the commutative algebra generated by the nilpotent operator $B : U \to U$ of nilpotent index $s+1$. The elements of $\mathcal{B}$ are the polynomials in B with complex coefficients and degree less than or equal to s, $\Phi = \alpha_0 I + \alpha_1 B + \cdots + \alpha_s B^s$.

Definition 5.15 To $\Phi = \alpha_0 I + \alpha_1 B + \cdots + \alpha_s B^s \in \mathcal{B}$ we assign the element

$$\Phi^* = \overline{\alpha}_0 I - \overline{\alpha}_1 B + \overline{\alpha}_2 B^2 + \cdots + (-1)^s \overline{\alpha}_s B^s. \tag{5.45}$$

Lemma 5.16

(1) $\Phi = \alpha_0 I + \alpha_1 B + \cdots + \alpha_s B^s \in \mathcal{B}$ *is nonsingular if, and only if,* $\alpha_0 \neq 0$.
(2) $(\Phi\Psi)^* = \Phi^*\Psi^*$.
(3) *Let* $\Phi \in \mathcal{B}$ *be nonsingular and satisfy* $\Phi = (-1)^{s+1}\Phi^*$. *Then, there exists a nonsingular element* $\Psi \in \mathcal{B}$ *such that* $\Psi = (-1)^{s+1}\Psi^*$ *and* $\Psi\Psi^* = \epsilon\Phi$, *where* $\epsilon = \operatorname{sign}(\alpha_0)$ *if* s *is odd and* $\epsilon = -i\operatorname{sign}(\alpha_0/i)$ *if* s *is even.*

Proof The proof of (1) is exactly equal to that in Lemma 5.7. For (2) the argument is similar to the case of that lemma, with the only change due to the conjugation of coefficients.

Let us prove (3). We want to find Ψ such that $\Psi = (-1)^{s+1}\Psi^*$ and $\Psi\Psi^* = \epsilon\Phi$. Since $\Phi = (-1)^{s+1}\Phi^*$, we get $\Psi^2 = (-1)^{s+1}\Psi\Psi^* = (-1)^{s+1}\epsilon\Phi$; hence,

$$\Psi^2 = \epsilon\Phi^*. \tag{5.46}$$

Now we consider the identities (5.14) for the coefficients of Ψ^2 with the α's and β's as complex numbers and the coefficients of Φ^*, not those of Φ. From the first equation in (5.14) we get $\beta_0^2 = \epsilon\overline{\alpha}_0$ and we will take $\beta_0 = \sqrt{(-1)^{s-1}|\alpha_0|}$. Now, from $\Phi = (-1)^{s+1}\Phi^*$ we see that $\alpha_0 = (-1)^{s+1}\overline{\alpha}_0$, so α_0 is real if s is odd and α_0 is purely imaginary if s is even. Therefore, we have $\epsilon = \operatorname{sign}(\alpha_0)$ if s is odd and and $\epsilon = -i\operatorname{sign}(\alpha_0/i)$ if s is even. Once found β_0 all the other coefficients of Ψ are uniquely determined from the identities (5.14). □

For $\mathbf{x}, \mathbf{y} \in U$ consider the element $\Omega(\mathbf{x}, \mathbf{y})$ of $\mathcal{B}$ defined by the polynomial in B,

$$\Omega(\mathbf{x}, \mathbf{y}) = \{B^s\mathbf{x}, \overline{\mathbf{y}}\}I + \{B^{s-1}\mathbf{x}, \overline{\mathbf{y}}\}B + \cdots + \{\mathbf{x}, \overline{\mathbf{y}}\}B^s. \tag{5.47}$$

The mapping $\Omega : U \times U \to \mathcal{B}$ is clearly $\mathbf{R}$-bilinear, but we want to consider it as a mapping on U with this set viewed as a module over the ring $\mathcal{B}$. As before, we have

Lemma 5.17 *The following properties hold:*

(1) $\Omega(\mathbf{x}, \mathbf{y}) = (-1)^{s+1}\Omega(\mathbf{y}, \mathbf{x})^*$*;*
(2) $\Omega(\mathbf{x}, \mathbf{y})$ *is $\mathcal{B}$-linear in the first variable and $\mathcal{B}^*$-linear in the second, that is,* $\Omega(\mathbf{x}, \mathbf{y})$ *is additive in* $\mathbf{x}$ *and* $\mathbf{y}$*, and moreover*

$$\Omega(\Phi\mathbf{x}, \mathbf{y}) = \Phi\Omega(\mathbf{x}, \mathbf{y}) \quad \textit{and} \quad \Omega(\mathbf{x}, \Psi\mathbf{y}) = \Psi^*\Omega(\mathbf{x}, \mathbf{y}).$$

Proof (1) Using (5.30) we have

$$\begin{aligned}
\Omega(\mathbf{x}, \mathbf{y}) &= \sum_{l=0}^{s}\{B^{s-l}\mathbf{x}, \overline{\mathbf{y}}\}B^l = \sum_{l=0}^{s}(-1)^{s-l}\{\mathbf{x}, \overline{B}^{s-l}\overline{\mathbf{y}}\}B^l \\
&= (-1)^{s+1}\sum_{l=0}^{s}(-1)^l\{\overline{B}^{s-l}\overline{\mathbf{y}}, \mathbf{x}\}B^l \\
&= (-1)^{s+1}\sum_{l=0}^{s}(-1)^l\overline{\{B^{s-l}\mathbf{y}, \overline{\mathbf{x}}\}}B^l = (-1)^{s+1}\Omega(\mathbf{y}, \mathbf{x})^*.
\end{aligned}$$

The proof of (2) is exactly as before. For the $\mathcal{B}^*$-linearity we make use of item (1) just proved and property (2) of Lemma 5.16. □

5.3.3 Proof of Theorem 5.12

First we prove that there exists a vector $\mathbf{e} \in U = Z(\mathbf{u}, B)$ such that

$$\{B^s\mathbf{e}, \overline{\mathbf{e}}\} = \overline{\epsilon} \quad \text{and} \quad \{B^j\mathbf{e}, \overline{\mathbf{e}}\} = 0, \quad \text{for } j = 0, 1, 2, \ldots, s-1, \tag{5.48}$$

where $\epsilon = \operatorname{sign}(\alpha_0)$ if s is odd and $\epsilon = i \operatorname{sign}(\alpha_0/i)$, if s is even, with $\alpha_0 = \{B^s\mathbf{u}, \overline{\mathbf{u}}\}$.

Since $\alpha_0 = \{B^s\mathbf{u}, \overline{\mathbf{u}}\} \neq 0$, then by (1) of Lemma 5.16 the element $\Phi = \Omega(\mathbf{u}, \mathbf{u})$ of $\mathcal{B}$ is nonsingular. Also by item (1) of Lemma 5.17, $\Omega(\mathbf{u}, \mathbf{u}) = (-1)^{s+1}\Omega(\mathbf{u}, \mathbf{u})^*$, that is, $\Phi = (-1)^{s+1}\Phi^*$. Therefore, by (3) of Lemma 5.16 there exists $\Psi \in \mathcal{B}$ such that $\Psi = (-1)^{s+1}\Psi^*$ and $\Psi\Psi^* = \epsilon\Phi$, where $\epsilon = \operatorname{sign}(\alpha_0)$ if s is odd and $\epsilon = i\operatorname{sign}(\alpha_0/i)$, if s is even.

Let $\mathbf{e} = \Psi^{-1}\mathbf{u}$. Then, $\mathbf{e} \in U$ and by (2) of Lemma 5.17 we have

$$\Phi = \Omega(\mathbf{u}, \mathbf{u}) = \Omega(\Psi\mathbf{e}, \Psi\mathbf{e}) = \Psi\Psi^*\Omega(\mathbf{e}, \mathbf{e}) = \epsilon\Phi\Omega(\mathbf{e}, \mathbf{e}).$$

Since Φ is nonsingular and $\overline{\epsilon}\epsilon = 1$, we get $\Omega(\mathbf{e}, \mathbf{e}) = \overline{\epsilon}I$. From this equality we get

$$\big[\{B^s\mathbf{e}, \overline{\mathbf{e}}\} - \overline{\epsilon}\big]I + \{B^{s-1}\mathbf{e}, \overline{\mathbf{e}}\}B + \{B^{s-2}\mathbf{e}, \overline{\mathbf{e}}\}B^2 + \cdots + \{B\mathbf{e}, \overline{\mathbf{e}}\}B^{s-1} + \{\mathbf{e}, \overline{\mathbf{e}}\}B^s = 0,$$

which gives the equalities (5.48).

The vectors $\mathbf{e}, B\mathbf{e}, \ldots, B^s\mathbf{e}, \overline{\mathbf{e}}, \overline{B}\overline{\mathbf{e}}, \ldots, \overline{B}^s\overline{\mathbf{e}}$ form a basis for $U \oplus \overline{U}$.[5] Now, consider the sequence of vectors.

$$\mathbf{u}_1 = \mathbf{e},\ \mathbf{u}_2 = B\mathbf{e},\ \ldots,\ \mathbf{u}_{s+1} = B^s\mathbf{e};\ \tilde{\mathbf{v}}_1 = \overline{B}^s\overline{\mathbf{e}},\ \tilde{\mathbf{v}}_2 = \overline{B}^{s-1}\overline{\mathbf{e}},\ \ldots,\ \tilde{\mathbf{v}}_{s+1} = \overline{\mathbf{e}}. \tag{5.49}$$

Because $U = Z(\mathbf{u}, B)$ and $\overline{U} = Z(\overline{\mathbf{u}}, \overline{B})$ are contained in the Lagrangian subspaces $\eta(i\beta)$ and $\eta(-i\beta)$ respectively, we have $\{\mathbf{u}_i, \mathbf{u}_j\} = 0$ and $\{\tilde{\mathbf{v}}_i, \tilde{\mathbf{v}}_j\} = 0$ for all i, j, whereas using (5.30) we compute for $1 \le i, j \le s+1$,

$$\{\mathbf{u}_i, \tilde{\mathbf{v}}_j\} = \{B^{i-1}\mathbf{e}, \overline{B}^{s-j+1}\overline{\mathbf{e}}\} = (-1)^{s-j+1}\{B^{s+i-j}\mathbf{e}, \overline{\mathbf{e}}\}.$$

Therefore, by (5.48) we see that $\{\mathbf{u}_i, \tilde{\mathbf{v}}_j\} = 0$ for $i \neq j$, whereas $\{\mathbf{u}_j, \tilde{\mathbf{v}}_j\} = (-1)^{s-j+1}\overline{\epsilon}$.

So by taking $\mathbf{v}_j = (-1)^{s-j+1}\epsilon\tilde{\mathbf{v}}_j$ we get from (5.49) a symplectic basis for the subspace $Z(\mathbf{u}, B) \oplus Z(\overline{\mathbf{u}}, \overline{B})$ given by the vectors $\mathbf{u}_1, \ldots, \mathbf{u}_{s+1}, \mathbf{v}_1, \ldots, \mathbf{v}_{s+1}$. This subspace is therefore symplectic. □

[5] Indeed, applying B^s to $a_1\mathbf{e} + a_2B\mathbf{e} + \cdots + a_sB^s\mathbf{e} = 0$ we get $a_1B^s\mathbf{e} = 0$, hence, $a_1 = 0$. Repeating this operation successively applying B^{s-j}, we obtain $a_1 = \cdots = a_{s+1} = 0$, showing that these vectors are linearly independent. It follows that the $s+1$ vectors $\overline{\mathbf{e}}, \overline{B}\overline{\mathbf{e}}, \ldots, \overline{B}^s\overline{\mathbf{e}} \in \eta^\dagger(-i\beta)$ are also linearly independent and since $\eta^\dagger(i\beta) \cap \eta^\dagger(-i\beta) = \{0\}$, the $2(s+1)$ vectors form a basis of $U \oplus \overline{U}$.

Remark 5.18 To construct the symplectic basis in this case we set $\Phi = \Omega(\mathbf{u}, \mathbf{u})$ and find the square root Ψ using (5.46). This requires us to solve (5.14) with the α's and β's as complex numbers and using the coefficients of Φ^* instead of those of Φ. Next we find the vector $\mathbf{e} = \Psi^{-1}\mathbf{u}$. Then, the symplectic basis of $Z(\mathbf{u}, B) \oplus Z(\overline{\mathbf{u}}, \overline{B})$ is given by

$$\mathbf{u}_1, \mathbf{u}_2, \ldots, \mathbf{u}_{s+1}; \mathbf{v}_1, \mathbf{v}_2, \ldots, \mathbf{v}_{s+1},$$

with

$$\mathbf{u}_j = B^{j-1}\mathbf{e}, \quad \mathbf{v}_j = (-1)^{s-j+1}\epsilon \overline{B}^{s-j+1}\overline{\mathbf{e}}, \quad j = 1, \ldots, s+1. \tag{5.50}$$

We recall that $\epsilon = \operatorname{sign}(\alpha_0)$ if s is odd and $\epsilon = i \operatorname{sign}(\alpha_0/i)$, if s is even with $\alpha_0 = \{B^s\mathbf{u}, \overline{\mathbf{u}}\}$.

Remark 5.19 Let $\mathcal{A} = P^{-1}AP$, where A is a Hamiltonian matrix of order $2n$ and P a symplectic matrix. The matrices $\mathcal{A}$ and A have the same characteristic polynomial. Let $i\beta$ be an eigenvalue of A with multiplicity $k \geq 2$. Then, $\mathcal{B} = P^{-1}BP$ where $\mathcal{B} = \mathcal{A} - i\beta I$ and $B = A - i\beta I$. For any vector $\mathbf{x} \in \mathbf{C}^{2n}$ we have $\operatorname{nil}_{\mathcal{B}}(\mathbf{x}) = \operatorname{nil}_B(\mathbf{x})$ and because P is symplectic we also have $\{\mathcal{B}^s\mathbf{x}, \overline{\mathbf{x}}\} = \{B^s\mathbf{x}, \overline{\mathbf{x}}\}$. We have $\eta^{\dagger}_{\mathcal{B}}(i\beta) = \eta^{\dagger}_B(i\beta)$; let us denote this subspace by W. Let $m + 1$ be the common nilpotent index of $\mathcal{B}$ and B and denote by $\Phi_{\mathcal{B}}$ and Φ_B the elements of the corresponding commutative algebras. Then, we have

$$\Phi_{\mathcal{B}} = P^{-1}\Phi_B P, \quad \Phi^*_{\mathcal{B}} = P^{-1}\Phi^*_B P$$

and if $\Psi_B\Psi^*_B = \epsilon\Phi_B$, then $\Psi_{\mathcal{B}}\Psi^*_{\mathcal{B}} = \epsilon\Phi_{\mathcal{B}}$.

Now reporting to the proof of Theorem 5.12, for $\mathbf{e} = \Psi_B^{-1}\mathbf{u}$ let $\mathbf{u}^* = P^{-1}\mathbf{u}$ and $\mathbf{e}^* = P^{-1}\mathbf{e}$. Then, $\mathbf{e}^* = \Psi_{\mathcal{B}}^{-1}\mathbf{u}^*$ and $\{\mathcal{B}^j\mathbf{e}^*, \overline{\mathbf{e}}^*\} = \{B^j\mathbf{e}, \overline{\mathbf{e}}\}$, for any j. Thus, the same construction (5.49) to find the symplectic basis $\mathbf{u}_1, \ldots, \mathbf{u}_s, \mathbf{v}_1, \ldots, \mathbf{v}_s$ can be done with $\mathcal{B}$ and $\mathbf{e}^*$ to yield a symplectic basis $\mathbf{u}^*_1, \ldots, \mathbf{u}^*_s, \mathbf{v}^*_1, \ldots, \mathbf{v}^*_s$ for $W \oplus \overline{W}$. Since

$$\mathbf{u}^*_j = \mathcal{B}^{j-1}\mathbf{e}^* = P^{-1}B^{j-1}P(P^{-1}\mathbf{e}) = P^{-1}B^{j-1}\mathbf{e} = P^{-1}\mathbf{u}_j,$$

and similarly $\mathbf{v}^*_j = P^{-1}\mathbf{u}_j$, we see that the symplectic matrix P^{-1} carries the symplectic $\mathbf{u}_1, \ldots, \mathbf{u}_s, \mathbf{v}_1, \ldots, \mathbf{v}_s$ to the symplectic basis $\mathbf{u}^*_1, \ldots, \mathbf{u}^*_s, \mathbf{v}^*_1, \ldots, \mathbf{v}^*_s$.

Now, if $[b_{ij}]$ is the matrix of B on the basis $\mathbf{u}_j, \mathbf{v}_j$ and $[b^*_{ij}]$ is the matrix of $\mathcal{B}$ in the basis $\mathbf{u}^*_j, \mathbf{v}^*_j$, then $\mathcal{B}\mathbf{u}^*_j = \sum_i(b^*_{ij}\mathbf{u}^*_i + b^*_{n+i,j}\mathbf{v}^*_i)$. Since $\mathcal{B} = P^{-1}BP$ and $\mathbf{u}^*_j = P^{-1}\mathbf{u}_j$, $\mathbf{v}^*_j = P^{-1}\mathbf{v}_j$, we get for $1 \leq i, j \leq n$,

$$P^{-1}B\mathbf{u}_j = P^{-1}\sum_i(b^*_{ij}(\mathbf{u}_i + b^*_{n+i,j}\mathbf{v}_i)).$$

But $B\mathbf{u}_j = \sum_i (b_{ij}\mathbf{u}_i + b_{n+i,j}\mathbf{v}_i)$; hence, $b^*_{ij} = b_{ij}$ and $b^*_{n+i,j} = b_{n+i,j}$. We have equalities for all the other entries, so the two matrices are the same. It follows that the matrices A and $\mathcal{A}$ in the corresponding bases are the same, so the final Hamiltonian has the same expression in the coordinates of the corresponding bases.

5.3.4 Second Main Theorem in the Imaginary Case

Assume that there is a vector $\mathbf{u}_1 \in \eta^\dagger(i\beta)$ such that $\text{nil}(\mathbf{u}_1) = s_1 + 1$ and $\{B^{s_1}\mathbf{u}_1, \overline{\mathbf{u}}_1\} \neq 0$. Then, by Theorem 5.3.1 $W_1 = Z(\mathbf{u}_1, B) \oplus Z(\overline{\mathbf{u}}_1, \overline{B})$ is a symplectic subspace of $E(\beta)$ and we have the $\tilde{B}$-invariant symplectic decomposition $E(\beta) = W_1 \oplus W_1^\perp$. A basis for W_1 is

$$\mathbf{u}_1, B\mathbf{u}_1, \ldots, B^{s_1}\mathbf{u}_1, \overline{\mathbf{u}}_1, \overline{B}\overline{\mathbf{u}}_1, \ldots, \overline{B}^{s_1}\overline{\mathbf{u}}_1.$$

If there is a vector $\mathbf{u}_2 \in \eta^\dagger(i\beta) \cap W_1^\perp$ such that $\text{nil}(\mathbf{u}_2) = s_2 + 1$ and $\{B^{s_2}\mathbf{u}_2, \overline{\mathbf{u}}_2\} \neq 0$ then $W_2 = Z(\mathbf{u}_2, B) \oplus Z(\overline{\mathbf{u}}_2, \overline{B})$ is a symplectic subspace of the symplectic space $W_1^\perp$ and we have the $\tilde{B}$-invariant symplectic decomposition $W_1^\perp = W_2 \oplus W_2^\perp$. A basis for W_2 is

$$\mathbf{u}_2, B\mathbf{u}_2, \ldots, B^{s_2}\mathbf{u}_2, \overline{\mathbf{u}}_2, \overline{B}\overline{\mathbf{u}}_2, \ldots, \overline{B}^{s_2}\overline{\mathbf{u}}_2.$$

This gives the $\tilde{B}$-invariant symplectic decomposition

$$E(\beta) = (W_1 \oplus W_2) \oplus (W_1 \oplus W_2)^\perp.$$

If there is a vector $\mathbf{u}_3 \in \eta^\dagger(i\beta) \cap (W_1 \oplus W_2)^\perp$ such that $\text{nil}(\mathbf{u}_3) = s_3 + 1$ and $\{B^{s_3}\mathbf{u}_3, \overline{\mathbf{u}}_3\} \neq 0$ we can find a $\tilde{B}$-invariant symplectic decomposition of $(W_1 \oplus W_2)^\perp$ and get

$$E(\beta) = (W_1 \oplus W_2 \oplus W_3) \oplus (W_1 \oplus W_2 \oplus W_3)^\perp.$$

We proceed in this way until we arrive at a subspace $W = W_1 \oplus W_2 \oplus \cdots \oplus W_k$ giving a $\tilde{B}$-invariant symplectic decomposition

$$E(\beta) = W \oplus W^\perp \tag{5.51}$$

such that $\{B^s\mathbf{u}, \overline{\mathbf{u}}\} = 0$, for all $\mathbf{u} \in \eta^\dagger(i\beta) \cap W^\perp$. We notice that a basis for W is given by the vectors

$$\mathbf{u}_1, B\mathbf{u}_1, \ldots, B^{s_1}\mathbf{u}_1, \overline{\mathbf{u}}_1, \overline{B}\overline{\mathbf{u}}_1, \ldots, \overline{B}^{s_1}\overline{\mathbf{u}}_1, \ldots, \mathbf{u}_k, B\mathbf{u}_k, \ldots, B^{s_k}\mathbf{u}_k, \overline{\mathbf{u}}_k, \overline{B}\overline{\mathbf{u}}_k, \ldots, \overline{B}^{s_k}\overline{\mathbf{u}}_k. \tag{5.52}$$

Lemma 5.20 *Let W be the subspace in the decomposition* (5.51). *Let* $\mathbf{u} \in \eta^\dagger(i\beta) \cap W^\perp$ *with* $\mathrm{nil}(\mathbf{u}) = s+1$. *Then, there is a vector* $\mathbf{v} \in \eta^\dagger(i\beta) \cap W^\perp$ *with* $\mathrm{nil}(\mathbf{v}) \geq s+1$ *such that* $\{B^s\mathbf{u}, \overline{\mathbf{v}}\} \neq 0$.

Proof Since $\tilde{B}(W^\perp) \subset W^\perp$, $\tilde{B}\mathbf{u} \in W^\perp$; hence, $\tilde{B}^s\mathbf{u} \in W^\perp$. But $\mathbf{u} \in \eta^\dagger(i\beta)$ so, see definition (5.29) of $\tilde{B}$, $\tilde{B}\mathbf{u} = B\mathbf{u}$; hence, $B^s\mathbf{u} \in W^\perp$. Since $W^\perp$ is symplectic and $B^s\mathbf{u} \neq 0$, there is a vector $\mathbf{x}+\mathbf{y} \in W^\perp$ with $\mathbf{x} \in \eta^\dagger(i\beta)$ and $\mathbf{y} \in \eta^\dagger(-i\beta)$ such that $\{B^s\mathbf{u}, \mathbf{x}+\mathbf{y}\} \neq 0$. Since $\eta^\dagger(i\beta)$ is Lagrangian, $\{B^s\mathbf{u}, \mathbf{x}\} = 0$, so $\{B^s\mathbf{u}, \mathbf{y}\} \neq 0$. Let $\mathbf{v} = \overline{\mathbf{y}}$. Then, $\mathbf{v} \in \eta^\dagger(i\beta)$ and $\{B^s\mathbf{u}, \overline{\mathbf{v}}\} \neq 0$.

By (5.30) and $\{B^s\mathbf{u}, \overline{\mathbf{v}}\} \neq 0$ we have $\{\mathbf{u}, \overline{B}^s\overline{\mathbf{v}}\} \neq 0$; hence, $\overline{B}^s\overline{\mathbf{v}} \neq 0$, so $\overline{\mathbf{v}}$ has a nilpotent index at least $s+1$ and therefore $\mathrm{nil}(\mathbf{v}) \geq s+1$ also.

It remains to prove that $\mathbf{v} \in W^\perp$. We prove this using the basis (5.52) of W. The equalities $\{\mathbf{v}, B^l\mathbf{u}_i\} = 0$ hold because $\eta^\dagger(i\beta)$ is Lagrangian. On the other hand, since $\mathbf{x}+\mathbf{y} \in W^\perp$, we get

$$\{\mathbf{v}, \overline{B}^l\overline{\mathbf{u}}_i\} = \overline{\{\mathbf{y}, B^l\mathbf{u}_i\}} = \overline{\{\mathbf{x}+\mathbf{y}, B^l\mathbf{u}_i\}} = 0.$$

This shows that $\mathbf{v}$ is symplectically orthogonal to the basis vectors of W; hence, $\mathbf{v} \in W^\perp$. □

The second theorem is the following:

Theorem 5.21 *Let W be the subspace in the decomposition* (5.51). *Let* $\mathbf{u} \in \eta^\dagger(i\beta) \cap W^\perp$ *be an element of maximal*[6] *nilpotent index* $m+1$. *Then,* $\mathbf{u}$ *determines two vectors* $\mathbf{f}, \mathbf{g} \in \eta^\dagger(i\beta) \cap W^\perp$ *with the same nilpotent index* $m+1$ *such that*

$$\mathcal{Z}_{\mathbf{fg}} = Z(\mathbf{f}, B) \oplus Z(\mathbf{g}, B) \oplus Z(\overline{\mathbf{f}}, \overline{B}) \oplus Z(\overline{\mathbf{g}}, \overline{B}) \tag{5.53}$$

is a symplectic subspace of $W^\perp$ *and an explicit symplectic basis can be constructed for it.*

Proof By Lemma 5.20 there exists a vector $\mathbf{v} \in \eta^\dagger(i\beta) \cap W^\perp$ such that $\{B^m\mathbf{u}, \overline{\mathbf{v}}\} \neq 0$ which also has nilpotent index $m+1$.

Then, the element $\Omega(\mathbf{u}, \mathbf{v}) = \{B^m\mathbf{u}, \overline{\mathbf{v}}\}I + \cdots + \{\mathbf{u}, \overline{\mathbf{v}}\}B^m$ of the algebra $\mathcal{B} = \{\alpha_0 I + \alpha_1 B + \cdots + \alpha_m B^m;\ \alpha_j \in \mathbf{C}\}$ is nonsingular and we can assume, without loss of generality[7] that $\Omega(\mathbf{u}, \mathbf{v}) = I$. By item (1) of Lemma 5.17, we then have $\Omega(\mathbf{v}, \mathbf{u}) = (-1)^{m+1}I$.

We now look for a vector $\mathbf{f} \in \eta^\dagger(i\beta) \cap W^\perp$ such that $\Omega(\mathbf{f}, \mathbf{f}) = 0$, taking it in the form $\mathbf{f} = \mathbf{u} + \Phi\mathbf{v}$, with $\Phi \in \mathcal{B}$ such that $\Phi^* = (-1)^{m+1}\Phi$. Proceeding exactly as we did for the Eq. (5.19) we are led to the equalities (5.20) from which we can determine Φ. Once determined Φ we consider the element $\mathbf{g} = \mathbf{v} - \frac{1}{2}\Omega(\mathbf{v}, \mathbf{v})\mathbf{f}$ of

[6] By this we mean that $B^m\mathbf{u} \neq 0$ and $B^{m+1}\mathbf{x} = 0$, for all $\mathbf{x} \in \eta^\dagger(i\beta) \cap W^\perp$.

[7] See footnote [(1)].

$\eta^\dagger(i\beta)\cap W^\perp$. Moreover,[8] we can take $\mathbf{f}$ so that $\Omega(\mathbf{f},\mathbf{v}) = I$; hence, also $\Omega(\mathbf{f},\mathbf{g}) = I$. We then get $\Omega(\mathbf{g},\mathbf{g}) = 0$, whether m is even or odd. Now, the equalities $\Omega(\mathbf{f},\mathbf{f}) = 0$ and $\Omega(\mathbf{g},\mathbf{g}) = 0$ give the relations

$$\{B^i\mathbf{f},\bar{\mathbf{f}}\} = 0, \quad \{B^i\mathbf{g},\bar{\mathbf{g}}\} = 0, \qquad \text{for} \quad i = 0, 1, \ldots, m, \tag{5.54}$$

whereas the equality $\Omega(\mathbf{f},\mathbf{g}) = I$ gives

$$\{B^m\mathbf{f},\bar{\mathbf{g}}\} = 1 \qquad \text{and} \qquad \{B^i\mathbf{f},\bar{\mathbf{g}}\} = 0, \quad \text{for} \quad i = 0, 1, \ldots, m-1. \tag{5.55}$$

Consider the $4(m+1)$ vectors (the vectors in the first row are vectors in $\eta^\dagger(i\beta)\cap W^\perp$, those in the second row are vectors in $\eta^\dagger(-i\beta)\cap W^\perp$)

$$\begin{array}{c} \mathbf{f},\ B\mathbf{f}\ \ldots,\ B^{m-1}\mathbf{f},\ B^m\mathbf{f};\ \ \mathbf{g},\ B\mathbf{g},\ \ldots,\ B^{m-1}\mathbf{g},\ B^m\mathbf{g}; \\ \overline{B}^m\bar{\mathbf{g}},\ \overline{B}^{m-1}\bar{\mathbf{g}},\ \ldots,\ \overline{B}\bar{\mathbf{g}},\ \bar{\mathbf{g}};\ \ \overline{B}^m\bar{\mathbf{f}},\ \overline{B}^{m-1}\bar{\mathbf{f}},\ \ldots,\ \overline{B}\bar{\mathbf{f}},\ \bar{\mathbf{f}}. \end{array} \tag{5.56}$$

The vectors of the first row give a basis for the subspace $Z(\mathbf{f},B)\oplus Z(\mathbf{g},B)$[9] and those of the second row give a basis for $Z(\bar{\mathbf{f}},\overline{B})\oplus Z(\bar{\mathbf{g}},\overline{B})$. Since $\eta^\dagger(i\beta)\cap\eta^\dagger(-i\beta) = \{0\}$ these $4(m+1)$ vectors give a basis for $\mathcal{Z}_{\mathbf{fg}}$.

From this basis we can get a symplectic basis for $\mathcal{Z}_{\mathbf{fg}}$ so this is a symplectic subspace. Indeed, denote by $\mathbf{u}_1,\ldots,\mathbf{u}_{m+1};\mathbf{u}_{m+2},\ldots,\mathbf{u}_{2(m+1)}$ the vectors of the first row with the last $m+1$ vectors multiplied respectively by $(-1)^1,\ldots,(-1)^m$, $(-1)^{m+1}$ and likewise denote by $\mathbf{v}_1,\ldots,\mathbf{v}_{m+1};\mathbf{v}_{m+2},\ldots,\mathbf{v}_{2(m+1)}$ the vectors of the second row with the first $m+1$ vectors multiplied respectively by $(-1)^m,(-1)^{m-1},\ldots,(-1)^1,(-1)^0$. Then using the relations (5.54) and (5.55) we easily check that

$$\mathbf{u}_1,\ \ldots,\ \mathbf{u}_{2(m+1)};\ \ \mathbf{v}_1,\ \ldots,\ \mathbf{v}_{2(m+1)} \tag{5.57}$$

form a symplectic basis for $\mathcal{Z}_{\mathbf{fg}}$. Notice that the symplectic product $\{\mathbf{z}_i,\mathbf{z}_j\}$ of any two vectors of the first row or of the second row in (5.56) is zero because $\eta^\dagger(i\beta)$ and $\eta^\dagger(-i\beta)$ are Lagrangian subspaces of $\eta^\dagger(i\beta)\oplus\eta^\dagger(-i\beta)$.

Finally, we have $\mathcal{Z}_{\mathbf{fg}}\subset W^\perp$ because all the vectors (5.56) are in $W^\perp$. □

Remark 5.22 To construct the symplectic basis in this case, first take a vector $\mathbf{u}\in\eta^\dagger(i\beta)\cap W^\perp$ of a maximal nilpotent index $m+1$, then find a vector $\mathbf{v}\in\eta^\dagger(i\beta)\cap W^\perp$ such that $\{B^m\mathbf{u},\bar{\mathbf{v}}\}\neq 0$ and consider $\Omega(\mathbf{u},\mathbf{v}) = \{B^m\mathbf{u},\bar{\mathbf{v}}\}I + \cdots + \{\mathbf{u},\bar{\mathbf{v}}\}B^m$. Next,

[8] Since the subspace W in the decomposition (5.51) satisfies the condition $\{B^m\mathbf{x},\bar{\mathbf{x}}\} = 0$, for all $\mathbf{x}\in\eta^\dagger(i\beta)\cap W^\perp$ and $\mathbf{v}\in\eta^\dagger(i\beta)\cap W^\perp$, we have $\{B^m\mathbf{v},\bar{\mathbf{v}}\} = 0$; hence, $\Omega(\mathbf{f},\mathbf{v}) = \{B^m\mathbf{u},\bar{\mathbf{v}}\}I + \cdots$ is nonsingular, so $\Psi\Omega(\mathbf{f},\mathbf{v}) = I$, for some $\Psi\in\mathcal{B}$. Replace $\mathbf{f}$ with $\Psi\mathbf{f}$.

[9] If $a_0\mathbf{f} + a_1B\mathbf{f} + \cdots a_mB^m\mathbf{f} + b_0\mathbf{g} + b_1B\mathbf{g} + \cdots + b_mB^m\mathbf{g} = 0$ taking the symplectic product by $\bar{\mathbf{g}}$ and using (5.54) and (5.55) we get $a_m = 0$. Applying B and taking the symplectic product again by $\bar{\mathbf{g}}$ we get $a_{m-1} = 0$ and so on.

find a solution Φ of Eqs. (5.20) satisfying the condition $\Phi^* = (-1)^{m+1}\Phi$ and take $\mathbf{f} = \mathbf{u} + \Phi\mathbf{v}$ and $\mathbf{g} = \mathbf{v} - \frac{1}{2}\Omega(\mathbf{v}, \mathbf{v})\mathbf{f}$. Then we get the symplectic basis (5.57) from (5.56) in the way described in the final stage of the proof of Theorem 5.21.

Example 5.10 Let H be an n-degree of freedom real Hamiltonian whose matrix A has a purely imaginary eigenvalue $i\beta$ of multiplicity $k \geq 2$ and consider the $2k$-dimensional A-invariant symplectic subspace $E(\beta) = \eta^\dagger(i\beta) \oplus \eta^\dagger(-i\beta)$ of $\mathbf{C}^{2n}$.

Suppose that for A restricted to $E(\beta)$ we have, for any s, $\{A^s\mathbf{x}, \overline{\mathbf{x}}\} = 0$ for all vectors $\mathbf{x} \in \eta^\dagger(i\beta)$ with $\text{nil}(\mathbf{x}) = s + 1$.

Let $\mathbf{u} \in \eta^\dagger(i\beta)$ be a vector with a maximal nilpotent index $m+1$. By Lemma 5.20 there is vector $\mathbf{v} \in \eta^\dagger(i\beta)$ of the same nilpotent index $m+1$ such that $\{B^m\mathbf{u}, \overline{\mathbf{v}}\} \neq 0$. Theorem 5.21 then says that the direct sum (5.53) is a symplectic subspace of $E(\beta)$.

In the proof of Theorem 5.21 it is shown that the vectors $\mathbf{u}$, $\mathbf{v}$ determine two other vectors $\mathbf{f}, \mathbf{g} \in \eta^\dagger(i\beta)$ of nilpotent index $m + 1$ satisfying the equalities (5.54) and (5.55). These equalities imply that, up to an adjustment of signs, the sequence of $4(m + 1)$ vectors (5.56) furnishes a symplectic basis for $E(\beta)$.

Example 5.16 in Sect. 5.7 shows that a symplectic subspace of the form (5.53) cannot arise in the case of a 4×4 Hamiltonian matrix. Therefore, $m \geq 1$ so dim $E(\beta) \geq 8$ and the Hamiltonian must have at least four degrees of freedom.

5.4 The Complex Normal Form

5.4.1 *Normal Form of A in the Subspace* $Z(\mathbf{u}, B) \oplus Z(\overline{\mathbf{u}}, \overline{B})$

Using the basis (5.50) constructed in the proof of Theorem 5.12 we have

$$B\mathbf{u}_1 = \mathbf{u}_2, \quad B\mathbf{u}_2 = \mathbf{u}_3, \quad \ldots, \quad B\mathbf{u}_s = \mathbf{u}_{s+1}, \quad B\mathbf{u}_{s+1} = 0,$$

$$\overline{B}\mathbf{v}_1 = 0, \quad \overline{B}\mathbf{v}_2 = -\mathbf{v}_1, \quad \ldots, \quad \overline{B}\mathbf{v}_s = -\mathbf{v}_{s-1}, \quad \overline{B}\mathbf{v}_{s+1} = -\mathbf{v}_s.$$

Remark 5.23 The matrix of $\tilde{B} : Z(\mathbf{u}, B) \oplus Z(\overline{\mathbf{u}}, \overline{B}) \to Z(\mathbf{u}, B) \oplus Z(\overline{\mathbf{u}}, \overline{B})$ in the symplectic basis $\mathbf{u}_1, \ldots, \mathbf{u}_{s+1}, \mathbf{v}_1, \ldots, \mathbf{v}_{s+1}$ is Hamiltonian, so this operator is Hamiltonian by Proposition 4.15.

Since $A = B + i\beta I$ in $\eta^\dagger(i\beta)$ and $A = \overline{B} - i\beta I$ in $\eta^\dagger(-i\beta)$, we get

$$A\mathbf{u}_1 = i\beta\mathbf{u}_1 + \mathbf{u}_2, \; A\mathbf{u}_2 = i\beta\mathbf{u}_2 + \mathbf{u}_3, \; \ldots, \; A\mathbf{u}_s = i\beta\mathbf{u}_s + \mathbf{u}_{s+1}, \; A\mathbf{u}_{s+1} = i\beta u_{s+1},$$

$$A\mathbf{v}_1 = -i\beta\mathbf{v}_1, \; A\mathbf{v}_2 = -i\beta\mathbf{v}_2 - \mathbf{v}_1, \; \ldots, \; A\mathbf{v}_s = -i\beta\mathbf{v}_s - \mathbf{v}_{s-1}, \; A\mathbf{v}_{s+1} = -i\beta\mathbf{v}_{s+1} - \mathbf{v}_s.$$

Therefore, the matrix of the restriction of A to the subspace $Z(\mathbf{u}, B) \oplus Z(\overline{\mathbf{u}}, \overline{B})$ of $E(\beta) = \eta^\dagger(i\beta) \oplus \eta^\dagger(-i\beta)$ in the symplectic basis $\mathbf{u}_1, \ldots, \mathbf{u}_{s+1}, \mathbf{v}_1, \ldots, \mathbf{v}_{s+1}$ is

$$A_\beta = \begin{bmatrix} N_\beta & O \\ & \\ O & -N_\beta^T \end{bmatrix}, \qquad N_\beta = \begin{bmatrix} i\beta & & & & & \\ 1 & i\beta & & & & \\ & 1 & i\beta & & & \\ & & & \ddots & & \\ & & & & & \\ & & & & 1 & i\beta & \\ & & & & & 1 & i\beta \end{bmatrix}. \tag{5.58}$$

5.4.2 Normal Form of A in the Subspace $\mathcal{Z}_{\mathbf{fg}}$

For the basis vectors (5.57) of the subspace $\mathcal{Z}_{\mathbf{fg}}$ constructed in Theorem 5.21 we have

$$\begin{aligned} B\mathbf{u}_1 &= \mathbf{u}_2,\ B\mathbf{u}_2 = \mathbf{u}_3,\ \ldots,\ B\mathbf{u}_m = \mathbf{u}_{m+1},\ B\mathbf{u}_{m+1} = 0; \\ B\mathbf{u}_{m+2} &= -\mathbf{u}_{m+3},\ B\mathbf{u}_{m+3} = -\mathbf{u}_{m+4}, \ldots, B\mathbf{u}_{2m+1} = -\mathbf{u}_{2m+2},\ B\mathbf{u}_{2m+2} = 0, \\ \overline{B}\mathbf{v}_1 &= 0,\ \overline{B}\mathbf{v}_2 = -\mathbf{v}_1,\ \ldots, \overline{B}\mathbf{v}_m = -\mathbf{v}_{m-1},\ \overline{B}\mathbf{v}_{m+1} = -\mathbf{v}_m; \\ \overline{B}\mathbf{v}_{m+2} &= 0,\ \overline{B}\mathbf{v}_{m+3} = \mathbf{v}_{m+2},\ \ldots, \overline{B}\mathbf{v}_{2m+1} = \mathbf{v}_{2m},\ \overline{B}\mathbf{v}_{2m+2} = \mathbf{v}_{2m+1}, \end{aligned}$$

and since $A = B + i\beta I$ in $\eta^\dagger(i\beta)$ and $A = \overline{B} - i\beta I$ in $\eta^\dagger(-i\beta)$, we easily find the matrix of A in the basis (5.57), namely

$$A_\beta = \begin{bmatrix} N_\beta & O & & \\ O & N_\beta^- & & \\ & & -N_\beta^T & O \\ & & O & -(N_\beta^-)^T \end{bmatrix}, \tag{5.59}$$

where N_β is the matrix in (5.58) and N_β^- is a matrix with the same structure but with -1 in place of 1.

By repeated application of Theorem 5.21 we get a symplectic decomposition of $W^\perp$, which, together with the decomposition of W found previously, gives a symplectic decomposition of the space $\eta^\dagger(i\beta) \oplus \eta^\dagger(-i\beta)$.

Doing this for each eigenvalue $i\beta_j$ we find a symplectic direct sum decomposition of the space Y in (5.26) together with symplectic bases constructed as above for each eigenvalue and this furnishes the complex normal form of A restricted to Y.

5.5 Construction of Real Symplectic Bases

The subspace W in (5.51) is clearly invariant under conjugation. But the same is also true of $W^\perp$ because if $\mathbf{z} \in W^\perp$, using the basis vectors (5.52) of W we have

$$\{\overline{\mathbf{z}}, B^l \mathbf{u}_i\} = \overline{\{\mathbf{z}, \overline{B}^l \overline{\mathbf{u}}_i\}} = 0 \quad \text{and} \quad \{\overline{\mathbf{z}}, \overline{B}^l \overline{\mathbf{u}}_i\} = \overline{\{\mathbf{z}, B^l \mathbf{u}_i\}} = 0,$$

so $\overline{\mathbf{z}} \in W^\perp$. Therefore, the symplectic spaces W and $W^\perp$ are complexifications of real spaces and so they have real bases. We have constructed complex symplectic bases for these spaces (see Remarks 5.18 and 5.22). In this section we see how to get real symplectic bases from the complex bases.

We remind the reader that for the nilpotent case of the real subspace $X = \eta^\dagger(0)$ and for the subspace U of nonzero real eigenvalues, the constructed symplectic bases are real bases. The same is also true for the subspace Z corresponding to truly complex eigenvalues. We now consider the case of the subspace Y of purely imaginary eigenvalues.

Real Basis from the Complex Basis of $Z(\mathbf{u}, B) \oplus Z(\overline{\mathbf{u}}, \overline{B})$
Recall from Remark 5.18 that the symplectic basis of $Z(\mathbf{u}, B) \oplus Z(\overline{\mathbf{u}}, \overline{B})$ is given by

$$\mathbf{u}_j = B^{j-1}\mathbf{e}, \quad \mathbf{v}_j = \epsilon(-1)^{s-j+1}\overline{B}^{s-j+1}\overline{\mathbf{e}}, \qquad j = 1, \ldots, s+1$$

with $\epsilon = \pm i$ if s is even and $\epsilon = \pm 1$ if s is odd. Notice that

$$\overline{\mathbf{u}}_j = \overline{\epsilon}(-1)^{j-1}\mathbf{v}_{s-j+2}, \quad \overline{\mathbf{v}}_j = \overline{\epsilon}(-1)^{s-j+1}\mathbf{u}_{s-j+2} \tag{5.60}$$

Now, consider the case of s even, so $\overline{\epsilon} = -\epsilon$. Then, $\overline{\mathbf{u}}_j = \epsilon(-1)^j \mathbf{v}_{s-j+2}$, $\overline{\mathbf{v}}_j = \epsilon(-1)^j \mathbf{u}_{s-j+2}$. Therefore, defining

$$\mathbf{u}_j^* = \frac{1}{\sqrt{2}}\big(\mathbf{u}_j + \epsilon(-1)^j \mathbf{v}_{s-j+2}\big) \text{ and } \mathbf{v}_j^* = \frac{1}{\sqrt{2}}\big(\epsilon(-1)^j \mathbf{u}_{s-j+2} + \mathbf{v}_j\big). \tag{5.61}$$

These are real vectors because $\mathbf{u}_j^* = \sqrt{2}\mathrm{Re}\,\mathbf{u}_j$, $\mathbf{v}_j^* = \sqrt{2}\mathrm{Re}\,\mathbf{v}_j$.

From (5.61) we see that the transition matrix from the basis $\mathbf{u}_1, \ldots, \mathbf{u}_{s+1}$; $\mathbf{v}_1, \ldots, \mathbf{v}_{s+1}$ to the vectors $\mathbf{u}_1^*, \ldots, \mathbf{u}_{s+1}^*$; $\mathbf{v}_1^*, \ldots, \mathbf{v}_{s+1}^*$ is the matrix of order $2(s+1)$

$$P = \frac{1}{\sqrt{2}}\begin{bmatrix} I & \epsilon C \\ \epsilon C & I \end{bmatrix}, \quad C = \begin{bmatrix} 0 & 0 & & 0 & -1 \\ 0 & 0 & & 1 & 0 \\ & & \ddots & & \\ 0 & 1 & & 0 & 0 \\ -1 & 0 & & 0 & 0 \end{bmatrix}. \tag{5.62}$$

Since the matrix C is symmetric and $(\epsilon C)^T(\epsilon C) = -I$, the matrix P is symplectic by Proposition 2.6(e). Since $\mathbf{u}_1, \ldots, \mathbf{u}_{s+1}; \mathbf{v}_1, \ldots, \mathbf{v}_{s+1}$ is a symplectic basis and the transition matrix P is symplectic the real vectors $\mathbf{u}_1^*, \ldots, \mathbf{u}_{s+1}^*; \mathbf{v}_1^*, \ldots, \mathbf{v}_{s+1}^*$ defined in (5.61) also form a symplectic basis of $Z(\mathbf{u}, B) \oplus Z(\overline{\mathbf{u}}, \overline{B})$.

Next, we assume that s is odd. In this case we define the vectors $\mathbf{u}_j^*$ and $\mathbf{v}_j^*$ considering the cases when j is odd and when j is even.

For j odd we define

$$\mathbf{u}_j^* = \frac{1}{\sqrt{2}}\big(\mathbf{u}_j + \epsilon(-1)^{j-1}\mathbf{v}_{s-j+2}\big), \quad \mathbf{v}_j^* = \frac{1}{\sqrt{2}}\big(-\epsilon(-1)^{j-1}\mathbf{u}_{s-j+2} + \mathbf{v}_j\big), \tag{5.63}$$

and for j even we define

$$\mathbf{u}_j^* = \frac{1}{\sqrt{2}i}\big(\mathbf{u}_j + \epsilon\mathbf{v}_{s-j+2}\big), \quad \mathbf{v}_j^* = \frac{1}{\sqrt{2}i}\big(\epsilon\mathbf{u}_{s-j+2} - \mathbf{v}_j\big). \tag{5.64}$$

For any $j = 1, \ldots, s+1$ these are real vectors. Indeed, using (5.60) and the fact that now $\epsilon = \pm 1$ is real, we have

$$\mathbf{u}_j^* = \sqrt{2}\mathrm{Re}\mathbf{u}_j, \quad \mathbf{v}_j^* = \sqrt{2}\mathrm{Re}\mathbf{v}_j, \quad \text{if } j \text{ is odd.}$$

$$\mathbf{u}_j^* = \sqrt{2}\mathrm{Im}\mathbf{u}_j, \quad \mathbf{v}_j^* = -\sqrt{2}\mathrm{Im}\mathbf{v}_j, \quad \text{if } j \text{ is even.}$$

From (5.63) and (5.64) we see that the transition matrix P is now given by

$$P = \frac{1}{\sqrt{2}}\begin{bmatrix} A_1 & A_2 \\ A_3 & A_4 \end{bmatrix}, \tag{5.65}$$

where

$$A_1 = \begin{bmatrix} 1 & & & & \\ & -i & & & \\ & & \cdots & & \\ & & & 1 & \\ & & & & -i \end{bmatrix}, \quad A_2 = -\epsilon\begin{bmatrix} & & & & i \\ & & & 1 & \\ & & \cdots & & \\ & i & & & \\ 1 & & & & \end{bmatrix},$$

$$A_3 = \epsilon\begin{bmatrix} & & & & -i \\ & & & 1 & \\ & & \cdots & & \\ & -i & & & \\ 1 & & & & \end{bmatrix}, \quad A_4 = \begin{bmatrix} 1 & & & & \\ & i & & & \\ & & \cdots & & \\ & & & 1 & \\ & & & & i \end{bmatrix}.$$

We compute

$$A_1^T A_3 = A_2^T A_4 = \epsilon \begin{bmatrix} & & & & -i \\ & & & -i & \\ & & \cdots & & \\ & -i & & & \\ -i & & & & \end{bmatrix},$$

so $A_1^T A_3$ and $A_2^T A_4$ are symmetric and since $\epsilon^2 = 1$, we find $A_1^T A_4 - A_3^T A_2 = 2I$. By Proposition 2.6(e) P is a symplectic matrix; hence, the vectors $\mathbf{u}_1^*, \ldots, \mathbf{u}_{s+1}^*; \mathbf{v}_1^*, \ldots, \mathbf{v}_{s+1}^*$ defined in (5.63) and (5.64) form a real symplectic basis of $Z(\mathbf{u}, B) \oplus Z(\overline{\mathbf{u}}, \overline{B})$.

Real Basis from the Complex Basis of $\mathcal{Z}_{\mathbf{fg}}$

In the case of Theorem 5.21, the symplectic basis is given by

$$\begin{aligned} \mathbf{u}_j &= B^{j-1}\mathbf{f}, \ \ \mathbf{v}_j = (-1)^{m-j+1}\overline{B}^{m-j+1}\overline{\mathbf{g}}, \quad j = 1, \ldots, m+1 \\ \mathbf{u}_j &= (-1)^{j-m-1}B^{j-m-2}\mathbf{g}, \ \ \mathbf{v}_j = \overline{B}^{2m-j+2}\overline{\mathbf{f}}, \quad j = m+2, \ldots, 2(m+1) \end{aligned} \tag{5.66}$$

and we then have

$$\begin{aligned} \overline{\mathbf{u}}_j &= \mathbf{v}_{2m-j+3}, \quad \overline{\mathbf{v}}_j = -\mathbf{u}_{2m-j+3}, \quad \text{if} \quad j \le m+1, \\ \overline{\mathbf{u}}_j &= -\mathbf{v}_{2m-j+3}, \quad \overline{\mathbf{v}}_j = \mathbf{u}_{2m-j+3}, \quad \text{if} \quad j \ge m+2. \end{aligned} \tag{5.67}$$

Therefore, setting

$$\begin{aligned} \mathbf{u}_j^* &= \frac{1}{\sqrt{2}}\big(\mathbf{u}_j + \mathbf{v}_{2m-j+3}\big), \quad \mathbf{v}_j^* = \frac{1}{\sqrt{2}}\big(-\mathbf{u}_{2m-j+3} + \mathbf{v}_j\big), \quad \text{for} \quad j \le m+1, \\ \mathbf{u}_j^* &= \frac{1}{\sqrt{2}i}\big(\mathbf{u}_j + \mathbf{v}_{2m-j+3}\big), \quad \mathbf{v}_j^* = \frac{1}{\sqrt{2}i}\big(\mathbf{u}_{2m-j+3} - \mathbf{v}_j\big), \quad \text{for} \quad j \ge m+2 \end{aligned} \tag{5.68}$$

we see that $\mathbf{u}_j^*$ and $\mathbf{v}_j^*$ are real vectors as we readily check that

$$\mathbf{u}_j^* = \sqrt{2}\mathrm{Re}\,\mathbf{u}_j, \quad \mathbf{v}_j^* = \sqrt{2}\mathrm{Re}\,\mathbf{v}_j, \qquad \text{for} \quad j \le m+1,$$

$$\mathbf{u}_j^* = \sqrt{2}\mathrm{Im}\,\mathbf{u}_j, \quad \mathbf{v}_j^* = -\sqrt{2}\mathrm{Im}\,\mathbf{v}_j, \qquad \text{for} \quad j \ge m+2.$$

The transition matrix from the basis $\mathbf{u}_1, \ldots, \mathbf{u}_{2(m+1)}; \mathbf{v}_1, \ldots, \mathbf{v}_{2(m+1)}$ to the vectors $\mathbf{u}_1^*, \ldots, \mathbf{u}_{2(m+1)}^*; \mathbf{v}_1^*, \ldots, \mathbf{v}_{2(m+1)}^*$ is given by

$$P = \frac{1}{\sqrt{2}}\begin{bmatrix} A_1 & A_2 \\ A_3 & A_4 \end{bmatrix}, \tag{5.69}$$

where now

$$A_1 = \begin{bmatrix} 1 & & & & & \\ & \ddots & & & & \\ & & 1 & & & \\ & & & -i & & \\ & & & & \ddots & \\ & & & & & -i \end{bmatrix}, \quad A_2 = \begin{bmatrix} & & & & & -i \\ & & & & \iddots & \\ & & & -i & & \\ & & -1 & & & \\ & \iddots & & & & \\ -1 & & & & & \end{bmatrix},$$

$$A_3 = \begin{bmatrix} & & & & & -i \\ & & & & \iddots & \\ & & & -i & & \\ & & 1 & & & \\ & \iddots & & & & \\ 1 & & & & & \end{bmatrix}, \quad A_4 = \begin{bmatrix} 1 & & & & & \\ & \ddots & & & & \\ & & 1 & & & \\ & & & i & & \\ & & & & \ddots & \\ & & & & & i \end{bmatrix},$$

We find that $A_1^T A_3 = A_2^T A_4$ are symmetric and $A_1^T A_4 - A_3^T A_2 = 2I$, so by Proposition 2.6(e) P is a symplectic matrix; therefore, the vectors $\mathbf{u}_1^*, \ldots, \mathbf{u}_{2(m+1)}^*; \mathbf{v}_1^*, \ldots, \mathbf{v}_{2(m+1)}^*$ defined in (5.63) and (5.64) form a real symplectic basis.

5.6 Real Normal Forms of Hamiltonian Matrices

We have seen in Sect. 5.4 that the complex normal form of a nondiagonalizable real Hamiltonian matrix is a block diagonal matrix where the diagonal blocks are matrices of the form (5.58) and (5.59).

From the complex bases giving these complex normal forms we have constructed in Sect. 5.5 real symplectic bases for the spaces $Z(\mathbf{u}, B) \oplus Z(\overline{\mathbf{u}}, \overline{B})$ and $\mathcal{Z}_{\mathbf{fg}}$ and found the transition matrix P from the complex to the real bases.

The matrix A_β^* of the linear operator defined by the real Hamiltonian matrix A expressed on the real basis is real. It is obtained from the complex normal form A_β by conjugation with the transition matrix P, namely

$$A_\beta^* = P^{-1} A_\beta P.$$

For the case of the matrix (5.58) with s even we have, denoting $\sigma = i\epsilon = \pm 1$,

$$
A^*_\beta = \begin{bmatrix}
0 & & & & & & & & & & & -\sigma\beta \\
1 & 0 & & & & & & & & & \sigma\beta & \\
 & & \cdots & & & & & & & \cdots & & \\
 & & & 1 & 0 & & & \sigma\beta & & & & \\
 & & & & 1 & 0 & -\sigma\beta & & & & & \\
 & & & & & \sigma\beta & 0 & -1 & & & & \\
 & & & & -\sigma\beta & & & 0 & -1 & & & \\
 & & \cdots & & & & & & & \cdots & & \\
0 & -\sigma\beta & & & & & & & & & 0 & -1 \\
\sigma\beta & & & & & & & & & & & 0
\end{bmatrix}, \tag{5.70}
$$

and for s odd,

$$
A^*_\beta = \epsilon \left[\begin{array}{cccccc|ccccc}
 & & & & & & & & & & \beta \\
 & & & & & & & & & \beta & 1 \\
 & & O & & & & & & & -1 & \\
 & & & & & & & & \cdots & & \\
 & & & & & & & \beta & -1 & & \\
 & & & & & & \beta & 1 & & & \\
\hline
 & & & & -1 & -\beta & & & & & \\
 & & & 1 & -\beta & & & & & & \\
 & & & -\beta & & & & & & & \\
 & & \cdots & & & & & & & O & \\
 & 1 & & & & & & & & & \\
-1 & -\beta & & & & & & & & & \\
-\beta & & & & & & & & & &
\end{array}\right]. \tag{5.71}
$$

For the case of the matrix (5.59), we get

$$
A^*_\beta = \begin{bmatrix}
N & O & O & C_\beta \\
O & -N & C_\beta & O \\
O & -C_\beta & -N^T & O \\
-C_\beta & O & O & N^T
\end{bmatrix}, \tag{5.72}
$$

where N is the subdiagonal nilpotent matrix of maximal rank of order $m+1$ and C_β is the counter diagonal matrix of order $m+1$ defined by β, that is, all the elements of the counter diagonal are equal to β. The matrix O is the zero square matrix of order $m+1$.

Therefore, the real normal form of a real Hamiltonian matrix is a block diagonal matrix formed by blocks that are matrices of the form (5.70), (5.71) and (5.72).

We note that the real normal forms so obtained coincide with those given in Laub-Meyer [35] except for the matrix (5.72), which differs from theirs by conjugation through one of the matrices

$$P_1 = \begin{bmatrix} I & O & O & O \\ O & O & O & I \\ O & O & I & O \\ O & -I & O & O \end{bmatrix}, \quad P_2 = \begin{bmatrix} I & O & O & O \\ O & O & O & -I \\ O & O & -I & O \\ O & -I & O & O \end{bmatrix},$$

the first being symplectic, the second symplectic with multiplier -1.

5.7 Further Examples

The following properties of Hamiltonian matrices will be used in the examples

Proposition 5.24 *Let $T : V \to V$ be a Hamiltonian linear operator on the symplectic subspace V of $\mathbf{C}^{2n}$.*

(1) *If k is even, then $\{T^k\mathbf{v}, \mathbf{v}\} = 0$ for all $\mathbf{v} \in V$.*
(2) *If $\{T\mathbf{v}, \mathbf{v}\} = 0$ for all $\mathbf{v} \in V$, then $T = 0$.*

Proof

(1) Let $k = 2s$, then $\{T^k\mathbf{v}, \mathbf{v}\} = (-1)^s\{T^s\mathbf{v}, T^s\mathbf{v}\} = 0$.
(2) Let M be the matrix of T on a symplectic basis of V. Then, M is a Hamiltonian matrix. Therefore, writing in block form $M = \begin{bmatrix} A & B \\ C & D \end{bmatrix}$ we have that B and C are symmetric and $A^T + D = O$. Let $\mathbf{z}$ be the coordinate vector of $\mathbf{v}$. Then, the hypothesis in item (2) translates into $\{M\mathbf{z}, \mathbf{z}\} = 0$ for all $\mathbf{z} \in \mathbf{C}^{2l}$, where $2l$ is the dimension of V. So if $\mathbf{z} = (\mathbf{x}, \mathbf{y})^T$ we have

$$\{M\mathbf{z}, \mathbf{z}\} = \langle A\mathbf{x} + B\mathbf{y}, \mathbf{y}\rangle - \langle C\mathbf{x} + D\mathbf{y}, \mathbf{x}\rangle = 0, \quad \text{for all} \quad \mathbf{x}, \mathbf{y}.$$

Setting $\mathbf{x} = 0$, we get $\langle B\mathbf{y}, \mathbf{y}\rangle = 0$, for all $\mathbf{y}$. Taking $\mathbf{y} = \mathbf{e}_i + \mathbf{e}_j$, a sum of two canonical vectors of $\mathbf{R}^l$, then using the fact that B is symmetric, we get $\langle B\mathbf{e}_i, \mathbf{e}_j\rangle = 0$; hence, $B_{ij} = 0$, so $B = O$. Taking $\mathbf{y} = 0$ we conclude that $C = O$. Since $D^T = -A$, we have $\{M\mathbf{z}, \mathbf{z}\} = 2\langle A\mathbf{x}, \mathbf{y}\rangle$. Therefore, $\langle A\mathbf{x}, \mathbf{y}\rangle = 0$ for all $\mathbf{x}, \mathbf{y}$, so taking $\mathbf{y} = A\mathbf{x}$ we get $\|A\mathbf{x}\|^2 = 0$ for all $\mathbf{x}$; hence, $A = O$. This proves that $M = O$, so $T = 0$ □

In the examples that follow we construct not only the real normal form of the Hamiltonian matrix and therefore the real normal form of the quadratic Hamiltonian but we also construct the real symplectic transformation that leads to the normal form. This point is very important because, for instance, when we study the stability of an equilibrium point of a Hamiltonian system we start by normalizing the quadratic part of the Hamiltonian expanded around the equilibrium. The canonical linear transformation used to find the normal form of the quadratic part affects the Hamiltonian as a whole, so it has to be known.

Example 5.11 Let $T : W \to W$ be a nilpotent Hamiltonian operator in the two dimensional symplectic space W. Then, the possible normal forms of its matrix are

$$\begin{bmatrix} 0 & 0 \\ 0 & 0 \end{bmatrix} \quad \text{and} \quad \begin{bmatrix} 0 & \delta \\ 0 & 0 \end{bmatrix}, \qquad \text{where} \quad \delta = \pm 1. \tag{5.73}$$

By Proposition 5.24 (2), either $T = 0$ or $\{T\mathbf{v}, \mathbf{v}\} \neq 0$ for some vector $\mathbf{v} \in W$. Let $\delta = \text{sign}\{T\mathbf{v}, \mathbf{v}\}$, $\kappa = 1/\sqrt{|\{T\mathbf{v}, \mathbf{v}\}|}$ and take the vectors $\mathbf{u}_1 = \kappa T\mathbf{v}$, $\mathbf{v}_1 = \delta\kappa\mathbf{v}$. Then, $\{\mathbf{u}_1, \mathbf{v}_1\} = 1$, so $\mathbf{u}_1, \mathbf{v}_1$ is a symplectic basis of W and $T\mathbf{u}_1 = 0$, $T\mathbf{v}_1 = \delta\mathbf{u}_1$ showing that the normal form is the second matrix in (5.73).

The next example deals with a matrix with a zero eigenvalue.

Example 5.12 Let A be a 4×4 real Hamiltonian matrix with eigenvalues $\pm i\beta$ and 0. Find its normal forms.

The subspaces $\eta(i\beta)$ and $\eta(-i\beta)$ have dimension 1 and since 0 is a double eigenvalue the subspace $\eta^\dagger(0)$ is two dimensional, so we have the A-invariant direct sum decomposition $\mathbf{C}^4 = Y \oplus \eta^\dagger(0)$, where $Y = \eta(i\beta) \oplus \eta(-i\beta)$. By Lemma 4.36 the subspaces Y and $\eta^\dagger(0)$ are J-orthogonal, so they are both symplectic subspaces of $\mathbf{C}^4$.

Consider $A_1 = A_{|Y}$. Let $\mathbf{v}$ be an eigenvector of A for $i\beta$. Since A is real $\overline{\mathbf{v}}$ is an eigenvector of A for $-i\beta$ and $\mathbf{v}$, $\overline{\mathbf{v}}$ are linearly independent. Let $\mathbf{v} = \mathbf{r} + i\mathbf{s}$. Since Y is invariant under conjugation the vectors $\mathbf{r}$ and $\mathbf{s}$ belong to Y and are linearly independent; hence, they form a real basis for Y. Let $\delta_1 = \text{sign}\{\mathbf{r}, \mathbf{s}\}$ and $\kappa_1 = 1/\sqrt{|\{\mathbf{r}, \mathbf{s}\}|}$. Now consider the vectors $\mathbf{u}_1 = -\kappa_1\mathbf{s}$, $\mathbf{v}_1 = \delta_1\kappa_1\mathbf{r}$ of Y. We have $\{\mathbf{u}_1, \mathbf{v}_1\} = 1$, so $\mathbf{u}_1, \mathbf{v}_1$ form a real symplectic basis of Y. Since $A\mathbf{v} = i\beta\mathbf{v}$ we have $A\mathbf{r} = -\beta\mathbf{s}$ and $A\mathbf{s} = \beta\mathbf{r}$, hence $A\mathbf{u}_1 = -\delta_1\beta\mathbf{v}_1$ and $A\mathbf{v}_1 = \delta_1\beta\mathbf{u}_1$.

By the previous example there are symplectic bases $\mathbf{u}_2$, $\mathbf{v}_2$ of $\eta^\dagger(0)$ giving the normal forms (5.73) of $A_0 = A_{|\eta^\dagger(0)}$, where we take $\delta = \delta_2$.

The vectors $\mathbf{u}_1, \mathbf{u}_2, \mathbf{v}_1, \mathbf{v}_2$ form a real symplectic basis of $\mathbf{C}^4$ and the matrix A on this basis, the real normal form A^* of A, is one of the following matrices

$$\begin{bmatrix} 0 & 0 & \delta_1\beta & 0 \\ 0 & 0 & 0 & 0 \\ -\delta_1\beta & 0 & 0 & 0 \\ 0 & 0 & 0 & 0 \end{bmatrix} \quad \text{or} \quad \begin{bmatrix} 0 & 0 & \delta_1\beta & 0 \\ 0 & 0 & 0 & \delta_2 \\ -\delta_1\beta & 0 & 0 & 0 \\ 0 & 0 & 0 & 0 \end{bmatrix}, \qquad (\delta_1 = \pm 1, \delta_2 = \pm 1).$$

The corresponding Hamiltonian functions, defined by the matrix $S = -JA^*$, are given by

$$H_1 = \frac{1}{2}\delta_1\beta(x_1^2 + y_1^2) \quad \text{or} \quad H_2 = \frac{1}{2}\delta_1\beta(x_1^2 + y_1^2) + \frac{1}{2}\delta_2 y_2^2, \tag{5.74}$$

where $\delta_1 = \pm 1, \ \ \delta_2 = \pm 1$.

When the matrix A in the above example is given explicitly the value of β and the signs δ_1 and δ_2 are well determined. This is the case in the following example.

Example 5.13 Find a normal form of the following Hamiltonian:

$$H = \frac{1}{2}(p_1^2 + p_2^2) - \frac{3}{8}q_1^2 + \frac{1}{2}q_2^2 + q_2 p_1$$

and the symplectic linear transformation that leads to it.

The matrix of the linear system defined by H is

$$A = \begin{bmatrix} 0 & 1 & 1 & 0 \\ 0 & 0 & 0 & 1 \\ \frac{3}{4} & 0 & 0 & 0 \\ 0 & -1 & -1 & 0 \end{bmatrix}, \quad \text{so} \quad A^2 = \begin{bmatrix} \frac{3}{4} & 0 & 0 & 1 \\ 0 & -1 & -1 & 0 \\ 0 & \frac{3}{4} & \frac{3}{4} & 0 \\ -\frac{3}{4} & 0 & 0 & -1 \end{bmatrix}.$$

The eigenvalues of A are $0, 0, i\beta, -i\beta$ with $\beta = \frac{1}{2}$, so $\mathbf{C}^4 = Y \oplus \eta^\dagger(0)$ with $Y = \eta(i\beta) \oplus \eta(-i\beta)$.

First we consider $A_1 = A|_Y$. An eigenvector $\mathbf{u} = \mathbf{r} + i\mathbf{s}$ of A belonging to $i\beta$ is found with $\mathbf{r} = (0, 1, -\frac{3}{4}, 0)$ and $\mathbf{s} = (-\beta, 0, 0, \beta)$. We compute $\{\mathbf{r}, \mathbf{s}\} = \frac{1}{4}\beta = \frac{1}{8}$, and therefore $\delta_1 = \text{sign}\{\mathbf{r}, \mathbf{s}\} = +1$ and $\kappa_1 = 2\sqrt{2}$.

A real symplectic basis of $Y = \eta^\dagger(i\beta) \oplus \eta^\dagger(-i\beta)$ is $\mathbf{u}_1 = -\kappa_1\mathbf{s}$, $\mathbf{v}_1 = \delta_1\kappa_1\mathbf{r}$.

Now we consider $A_0 = A|_{\eta^\dagger(0)}$ and recall that $\eta^\dagger(0)$ is the kernel of A^2. The real vector $\mathbf{v} = (1, 0, 0, -\frac{3}{4})$ belongs to $\eta^\dagger(0)$ and $A\mathbf{v} = (0, -\frac{3}{4}, \frac{3}{4}, 0)$. We compute $\{A\mathbf{v}, \mathbf{v}\} = -\frac{3}{16}$, so see Example 5.11, $\delta_2 = -1$ and a symplectic basis for $\eta^\dagger(0)$ is $\mathbf{u}_2 = \kappa_2 T\mathbf{v}$, $\mathbf{v}_2 = \delta\kappa_2\mathbf{v}$, where $\kappa_2 = 4/\sqrt{3}$.

Therefore, the matrix of the linear operator $A : Y \oplus \eta^\dagger(0) \to Y \oplus \eta^\dagger(0)$ on the symplectic basis of $\mathbf{C}^4$ given by $\mathbf{u}_1, \mathbf{u}_2, \mathbf{v}_1, \mathbf{v}_2$ is the second matrix A^* of the previous example with $\beta = \frac{1}{2}$ and $\delta_1 = +1$, $\delta_2 = -1$.

Let P be the transition matrix from the symplectic basis of $\mathbf{C}^4$ formed by the canonical vectors $\mathbf{e}_1, \mathbf{e}_2, \mathbf{e}_3, \mathbf{e}_4$ to the symplectic basis $\mathbf{u}_1, \mathbf{u}_2, \mathbf{v}_1, \mathbf{v}_2$. Then, P is symplectic and the canonical linear transformation $(q_1, q_2, p_1, p_2)^T = P(x_1, x_2, y_1, y_2)^T$ takes the Hamiltonian H of the example to the Hamiltonian

$$\mathcal{H}(\boldsymbol{\zeta}) = \frac{1}{4}(x_1^2 + y_1^2) - \frac{1}{2}y_2^2.$$

The Hamiltonian considered in the next example is that already worked out in Example 5.2, but here we find the symplectic basis using the technique prescribed in the proof of Theorem 5.3.

Example 5.14 Find a linear symplectic transformation that leads the following Hamiltonian

$$H = \frac{1}{2}(p_1^2 + p_2^2) - \frac{1}{2}q_1^2 + \frac{1}{2}q_2^2 + q_2 p_1$$

to a normal form.

In Example 5.2 we found the Hamiltonian matrix A of the linear system defined by H and its powers, A^2, A^3 and saw that $A^4 = 0$. For the vector $\mathbf{u} = (1, 0, 0, 0)$ we saw that

$$A\mathbf{u} = (0, 0, 1, 0), \ A^2\mathbf{u} = (1, 0, 0, -1) \ A^3\mathbf{u} = (0, -1, 1, 0).$$

Since $\{A^3\mathbf{u}, \mathbf{u}\} = -1$ the cyclic subspace $Z(\mathbf{u}, A) = [\mathbf{u}, A\mathbf{u}, A^2\mathbf{u}, A^3\mathbf{u}]$ is symplectic by Theorem 5.3. In that example we used the vectors $\mathbf{u}$, $A\mathbf{u}$, $A^2\mathbf{u}$ and $A^3\mathbf{u}$ to construct a symplectic basis and found the transformed Hamiltonian (5.6).

Let us now compute the symplectic basis for this space as described in Remark 5.9. We have to find the vector $\mathbf{e} = \Psi^{-1}\mathbf{u}$, where Ψ is a square root of $\Phi = \Omega(\mathbf{u}, \mathbf{u})$ that is, see Lemma 5.7, $\Psi^2 = \epsilon\Phi$, where $\epsilon = \text{sign}(\{A^3\mathbf{u}, \mathbf{u}\})$.

We compute $\{\mathbf{u}, \mathbf{u}\} = 0$, $\{A\mathbf{u}, \mathbf{u}\} = -1$ and $\{A^2\mathbf{u}, \mathbf{u}\} = 0$, hence $\Phi = -I - A^2$. As $\epsilon = -1$, Ψ is determined from the equation $\Psi^2 = I + A^2$. We find $\Psi = I + \frac{1}{2}A^2$, which has the inverse $\Psi^{-1} = I - \frac{1}{2}A^2$. So in terms of the canonical basis vectors $\mathbf{e}_1, \mathbf{e}_2, \mathbf{e}_3, \mathbf{e}_4$ of $\mathbf{R}^4$, we get

$$\mathbf{e} = \Psi^{-1}\mathbf{u} = \frac{1}{2}(\mathbf{e}_1 + \mathbf{e}_4), \ \ A\mathbf{e} = \frac{1}{2}(\mathbf{e}_2 + \mathbf{e}_3), \ \ A^2\mathbf{e} = \mathbf{e}_1 - \mathbf{e}_4, \ \ A^3\mathbf{e} = -\mathbf{e}_2 + \mathbf{e}_3.$$

The symplectic basis for $Z(\mathbf{u}, A)$ (see Remark 5.9) is $\mathbf{u}_1 = \mathbf{e}$, $\mathbf{u}_2 = A\mathbf{e}$, $\mathbf{v}_1 = A^3\mathbf{e}$, $\mathbf{v}_2 = -A^2\mathbf{e}$, so

$$A\mathbf{u}_1 = \mathbf{u}_2, \quad A\mathbf{u}_2 = -\mathbf{v}_2, \quad A\mathbf{v}_1 = 0, \quad A\mathbf{v}_2 = -\mathbf{v}_1.$$

Therefore, the matrix of A on this basis, its real normal form, is

$$A_N = \begin{bmatrix} 0 & 0 & 0 & 0 \\ 1 & 0 & 0 & 0 \\ 0 & 0 & 0 & -1 \\ 0 & -1 & 0 & 0 \end{bmatrix}.$$

The symplectic linear transformation that takes H to the new Hamiltonian $\mathcal{H}$ is defined by the transition matrix from the symplectic basis $\mathbf{e}_1, \mathbf{e}_2, \mathbf{e}_3, \mathbf{e}_4$ to the symplectic basis $\mathbf{u}_1, \mathbf{u}_2, \mathbf{v}_1, \mathbf{v}_2$ and is given by

$$q_1 = \frac{1}{2}X_1 - Y_2, \ q_2 = \frac{1}{2}X_2 - Y_1, \ p_1 = \frac{1}{2}X_2 + Y_1, \ p_2 = \frac{1}{2}X_1 + Y_2.$$

Substituting these coordinates into H or computing the symmetric matrix $S = -JA_N$ that defines the transformed Hamiltonian we find the new Hamiltonian,

$$\mathcal{H} = \frac{1}{2}X_2^2 + X_1Y_2, \tag{5.75}$$

which is the normal form of H.

The Hamiltonian in this example is the same one that appears on page 182 of the book [40], where it is given the expression $H_0 = \frac{1}{2}\eta_1^2 - \xi_1\xi_2$ for its normal form, obtained through the symplectic transformation

$$q_1 = -\xi_1 - \frac{1}{2}\xi_2,\ q_2 = -\frac{1}{2}\eta_1 + \eta_2,\ p_1 = -\frac{1}{2}\eta_1 - \eta_2,\ p_2 = \xi_1 - \frac{1}{2}\xi_2.$$

We note that by means of the symplectic transformation $X_1 = \xi_2,\ X_2 = \eta_1,\ Y_1 = \eta_2,\ Y_2 = -\xi_1$ we pass from our normalized Hamiltonian to the Hamiltonian $H_0 = \frac{1}{2}\eta_1^2 - \xi_1\xi_2$.

Remark 5.25 Through the symplectic transformation

$$\xi_1 = X_1, \quad \xi_2 = X_2, \quad \eta_1 = \frac{1}{2}X_2 + Y_1, \quad \eta_2 = \frac{1}{2}X_1 + Y_2,$$

we can pass from the Hamiltonian (5.6) found in Example 5.2 to the above Hamiltonian (5.75). Notice that the latter has a simpler form than the former and this shows the importance of constructing the symplectic basis using the scheme given in the proof of Theorem 5.3.

The next example deals with the diagonalizable case of a pair of double purely imaginary eigenvalues.

Example 5.15 Let A be a real 4×4 Hamiltonian matrix with a pair of double purely imaginary eigenvalues $\pm i\beta$ and assume it is diagonalizable. Find the real normal form of the Hamiltonian function defined by A.

In the proof of Theorem 4.39 we constructed a symplectic basis $\mathbf{u}_1, \mathbf{u}_2, \mathbf{v}_1, \mathbf{v}_2$ for the space $\mathbf{C}^2 = \eta^\dagger(i\beta) \oplus \eta^\dagger(-i\beta)$ in which the matrix A assumes the normal form

$$A^* = \begin{bmatrix} 0 & 0 & \delta_1\beta & 0 \\ 0 & 0 & 0 & \delta_2\beta \\ -\delta_1\beta & 0 & 0 & 0 \\ 0 & -\delta_2\beta & 0 & 0 \end{bmatrix}.$$

Therefore, the normal form of the Hamiltonian function defined by A is

$$H = \frac{1}{2}\delta_1\beta(x_1^2 + y_1^2) + \frac{1}{2}\delta_2\beta(x_2^2 + y_2^2), \qquad (\delta_1 = \pm 1,\ \delta_2 = \pm 1). \tag{5.76}$$

The example that follows treats the nondiagonalizable case of a pair of double purely imaginary eigenvalues.

Example 5.16 Let A be a real 4×4 Hamiltonian matrix with a pair of double purely imaginary eigenvalues $\pm i\beta$ and assume that it is nondiagonalizable. Find the complex normal form of A and the symplectic linear transformation that leads to it. Construct the real symplectic basis from the complex basis and find the corresponding real normal form.

Consider the A-invariant direct sum decomposition $Y = \eta^\dagger(i\beta) \oplus \eta^\dagger(-i\beta)$ in which the two subspaces are Lagrangian. Consider now the mappings

$$B = A - i\beta I : \eta^\dagger(i\beta) \to \eta^\dagger(i\beta) \quad \text{and} \quad \overline{B} = A + i\beta I : \eta^\dagger(-i\beta) \to \eta^\dagger(-i\beta). \tag{5.77}$$

Let $\mathbf{w}_2$ be an eigenvector of A corresponding to $i\beta$ and $\mathbf{w}_1$ an associate vector, that is, $A\mathbf{w}_2 = i\beta\mathbf{w}_2$ and $A\mathbf{w}_1 = i\beta\mathbf{w}_1 + \mathbf{w}_2$. Then $B\mathbf{w}_1 = \mathbf{w}_2$, hence $\mathbf{w}_1, \mathbf{w}_2$ form a basis of $\eta^\dagger(i\beta)$ and notice that $B\mathbf{w}_2 = 0$. Since A is real, $\overline{\mathbf{w}}_2$ and $\overline{\mathbf{w}}_1$ are the eigenvector of A and the associate vector corresponding to the eigenvalue $-i\beta$. Then, also $\overline{B}\overline{\mathbf{w}}_1 = \overline{\mathbf{w}}_2$ and $\overline{B}\overline{\mathbf{w}}_2 = 0$.

For the basis $\mathbf{w}_1, \mathbf{w}_2, \overline{\mathbf{w}}_1, \overline{\mathbf{w}}_2$ of Y we have $\{B\mathbf{w}_1, \mathbf{w}_1\} = 0$ and $\{B\mathbf{w}_1, \mathbf{w}_2\} = 0$ because $\eta^\dagger(i\beta)$ is a Lagrangian subspace of the symplectic space Y. Now, using (5.30) $\{B\mathbf{w}_1, \overline{\mathbf{w}}_2\} = \{B\mathbf{w}_1, \overline{B}\overline{\mathbf{w}}_1\} = -\{B^2\mathbf{w}_1, \overline{\mathbf{w}}_1\} = 0$. It follows that $\{B\mathbf{w}_1, \overline{\mathbf{w}}_1\} \neq 0$, otherwise $B\mathbf{w}_1$ would be symplectically orthogonal to the basis $\mathbf{w}_1, \mathbf{w}_2, \overline{\mathbf{w}}_1, \overline{\mathbf{w}}_2$ of Y, hence $B\mathbf{w}_1 = 0$, that is, $\mathbf{w}_2 = 0$, which is not the case.

Therefore, $\mathbf{u} = \mathbf{w}_1 \in \eta^\dagger(i\beta)$ has nilpotent index $s + 1 = 2$ and satisfies the condition $\{B\mathbf{u}, \overline{\mathbf{u}}\} \neq 0$. By Theorem 5.12 the space $Z(\mathbf{u}, B) \oplus Z(\overline{\mathbf{u}}, \overline{B})$ is symplectic.

We find a symplectic basis for it (see Remark 5.18). Since $B^2 = 0$ the algebra $\mathcal{B}$ is formed by the polynomials $aI + bB$ with complex coefficients a and b.

Let $\Phi = \Omega(\mathbf{u}, \mathbf{u}) = aI + bB$. Then, see (5.45), $\Phi^* = \overline{a}I - \overline{b}B$. As $s = 1$, $\Phi = \Phi^*$ by Lemma 5.17(1); hence, a is a real number and b is a purely imaginary number.

By Lemma 5.16(3) there is a $\Psi = cI + dB \in \mathcal{B}$ such that $\Psi^* = \Psi$ and $\Psi\Psi^* = \epsilon\Phi$ with $\epsilon = \operatorname{sign} a$. The equality $\Psi^2 = \epsilon\Phi$ gives $c^2 = \epsilon a$ and $2cd = \epsilon b$ so we can get Ψ with the choice $c = \sqrt{|a|}$ and $d = \epsilon b/2c$.

Now let $\mathbf{e} = \Psi^{-1}\mathbf{u}$. Then, the complex symplectic basis (see (5.50)) is given by:

$$\mathbf{u}_1 = \mathbf{e}, \quad \mathbf{u}_2 = B\mathbf{e}; \quad \mathbf{v}_1 = -\epsilon\,\overline{B\mathbf{e}}, \quad \mathbf{v}_2 = \epsilon\,\overline{\mathbf{e}}. \tag{5.78}$$

With these vectors we see from (5.77) that

$$A\mathbf{u}_1 = i\beta\mathbf{u}_1 + \mathbf{u}_2, \quad A\mathbf{u}_2 = i\beta\mathbf{u}_2, \quad A\mathbf{v}_1 = -i\beta\mathbf{v}_1, \quad A\mathbf{v}_2 = -\mathbf{v}_1 - i\beta\mathbf{v}_2, \tag{5.79}$$

so the matrix A in the basis (5.78), that is, the complex normal form of A, is

$$A^*_{\mathbf{C}} = \begin{bmatrix} i\beta & 0 & 0 & 0 \\ 1 & i\beta & 0 & 0 \\ 0 & 0 & -i\beta & -1 \\ 0 & 0 & 0 & -i\beta \end{bmatrix}.$$

As $s = 1$ is odd, we use (5.63) and (5.64) to construct the real symplectic basis, finding

$$\mathbf{u}_1^* = \frac{1}{\sqrt{2}}(\mathbf{u}_1 + \epsilon\mathbf{v}_2),\ \mathbf{u}_2^* = \frac{1}{\sqrt{2}i}(\mathbf{u}_2 + \epsilon\mathbf{v}_1),\ \mathbf{v}_1^* = \frac{1}{\sqrt{2}}(-\epsilon\mathbf{u}_2 + \mathbf{v}_1),$$
$$\mathbf{v}_2^* = \frac{1}{\sqrt{2}i}(\epsilon\mathbf{u}_1 - \mathbf{v}_2).$$

We now can use the transition matrix between the above symplectic bases to find the real normal form, but maybe it is more straightforward to compute

$$A\mathbf{u}_1^* = -\epsilon\mathbf{v}_1^* - \epsilon\beta\mathbf{v}_2^*, \quad A\mathbf{u}_2^* = -\epsilon\beta\mathbf{v}_1^*, \quad A\mathbf{v}_1^* = \epsilon\beta\mathbf{u}_2^*, \quad A\mathbf{v}_2^* = \epsilon\beta\mathbf{u}_1^* + \epsilon\mathbf{u}_2^*.$$

Therefore, the real normal form of A is given by

$$A_{\mathbf{R}}^* = \epsilon \begin{bmatrix} 0 & 0 & 0 & \beta \\ 0 & 0 & \beta & 1 \\ -1 & -\beta & 0 & 0 \\ -\beta & 0 & 0 & 0 \end{bmatrix}.$$

Computing $S^* = -JA_{\mathbf{R}}^*$ we get the normalized Hamiltonian, $\mathcal{H}(\mathbf{z}) = \frac{1}{2}\mathbf{z}^T S^* \mathbf{z}$, namely

$$\mathcal{H}(x_1, x_2, y_1, y_2) = \frac{1}{2}\epsilon(x_1^2 + y_2^2) + \epsilon\beta(x_1x_2 + y_1y_2).$$

Notice that the following vectors constructed from the symplectic basis (5.78)

$$\tilde{\mathbf{u}}_1 = \frac{1}{\sqrt{2}}(\mathbf{u}_1 + \epsilon\mathbf{v}_2),\ \tilde{\mathbf{u}}_2 = \frac{1}{\sqrt{2}i}(-\mathbf{u}_1 + \epsilon\mathbf{v}_2),\ \tilde{\mathbf{v}}_1 = \frac{1}{\sqrt{2}}(-\epsilon\mathbf{u}_2 + \mathbf{v}_1),$$
$$\tilde{\mathbf{v}}_2 = \frac{1}{\sqrt{2}i}(\epsilon\mathbf{u}_2 + \mathbf{v}_1)$$

also form a real symplectic basis (the transition matrix is symplectic). Using (5.79) we compute

$$A\tilde{\mathbf{u}}_1 = \beta\tilde{\mathbf{u}}_2 - \epsilon\tilde{\mathbf{v}}_1, \quad A\tilde{\mathbf{u}}_2 = -\tilde{\mathbf{u}}_1 - \epsilon\tilde{\mathbf{v}}_2, \quad A\tilde{\mathbf{v}}_1 = \beta\tilde{\mathbf{v}}_2, \quad A\tilde{\mathbf{v}}_2 = -\beta\tilde{\mathbf{v}}_1.$$

Therefore, on the symplectic basis $\tilde{\mathbf{u}}_1, \tilde{\mathbf{u}}_2, \tilde{\mathbf{v}}_1, \tilde{\mathbf{v}}_2$ the matrix A has the normal form

$$A^* = \begin{bmatrix} 0 & -\beta & 0 & 0 \\ \beta & 0 & 0 & 0 \\ -\epsilon & 0 & 0 & -\beta \\ 0 & -\epsilon & \beta & 0 \end{bmatrix}.$$

The corresponding Hamiltonian function is

$$H = \frac{1}{2}\epsilon(x_1^2 + x_2^2) + \beta(x_1 y_2 - x_2 y_1). \tag{5.80}$$

Notice that (5.80) is the Sokol'ski normal form given in (3.11) with $\beta = -\omega$ and $\epsilon = \delta$.

In the next example we describe all the normal forms of a nilpotent real Hamiltonian matrix of order 4.

Example 5.17 Let A be a 4×4 real nonzero nilpotent Hamiltonian matrix. Find its possible normal forms.

Either there is a vector $\mathbf{u}$ such that $\{A^3\mathbf{u}, \mathbf{u}\} \neq 0$ or $\{A^3\mathbf{x}, \mathbf{x}\} = 0$ for all $\mathbf{x}$. In the former case $\mathbf{R}^4 = Z(\mathbf{u}, A)$ is symplectic by Theorem 5.3 and a symplectic basis can be constructed as in Remark 5.9, the matrix being that in (5.24) with N of order 2.

In the latter case, either $\{A\mathbf{x}, \mathbf{x}\} = 0$ for all $\mathbf{x}$ of nilpotent index 2 or there is a vector $\mathbf{u}$ of index 2 such that $\{A\mathbf{u}, \mathbf{u}\} \neq 0$. In the first case by Lemma 5.4, there are vectors $\mathbf{u}, \mathbf{v}$ of the same nilpotent index $m + 1 \leq 2$ such that $\{A^m\mathbf{u}, \mathbf{v}\} \neq 0$. Then, by Proposition 5.5, the subspace $W = Z(\mathbf{u}, A) \oplus Z(\mathbf{v}, A)$ is symplectic and a symplectic basis can be constructed for it as in Remark 5.10. Since $m = 1$, $W = \mathbf{R}^4$ and the matrix is that in (5.24) with N of order 2 and $\delta_l = 0$. This gives two more possible normal forms

In the second case, by Theorem 5.3 $W = Z(\mathbf{u}, A)$ is a two-dimensional symplectic subspace and $\mathbf{R}^4 = W \oplus W^\perp$ with $W^\perp$ a two-dimensional symplectic subspace too. By Example 5.11 the canonical forms of the matrices of the restrictions $A : W \to W$ and $A : W^\perp \to W^\perp$ are those mentioned there with the possibility of the zero matrix only for the latter. If $\mathbf{e} \in W$ and $\mathbf{f} \in W^\perp$ are the vectors generating $W = Z(\mathbf{u}, A)$ and $W^\perp = Z(\mathbf{v}, A)$, then by Proposition 4.10 the possible symplectic bases for $W \oplus W^\perp$ are

$$\mathbf{e}, \mathbf{f}, A\mathbf{e}, A\mathbf{f}; \qquad \mathbf{e}, A\mathbf{f}, A\mathbf{e}, \mathbf{f}; \qquad A\mathbf{e}, \mathbf{f}, \mathbf{e}, A\mathbf{f}; \qquad A\mathbf{e}, A\mathbf{f}, \mathbf{e}, \mathbf{f},$$

so the corresponding canonical forms for A are

$$\begin{bmatrix} 0&0&0&0 \\ 0&0&0&0 \\ 1&0&0&0 \\ 0&1&0&0 \end{bmatrix}; \quad \begin{bmatrix} 0&0&0&0 \\ 0&0&0&1 \\ 1&0&0&0 \\ 0&0&0&0 \end{bmatrix}; \quad \begin{bmatrix} 0&0&1&0 \\ 0&0&0&0 \\ 0&0&0&0 \\ 0&1&0&0 \end{bmatrix}; \quad \begin{bmatrix} 0&0&1&0 \\ 0&0&0&1 \\ 0&0&0&0 \\ 0&0&0&0 \end{bmatrix}$$

together with the two matrices

$$\begin{bmatrix} 0&0&0&0 \\ 0&0&0&0 \\ 1&0&0&0 \\ 0&0&0&0 \end{bmatrix}; \quad \begin{bmatrix} 0&0&1&0 \\ 0&0&0&0 \\ 0&0&0&0 \\ 0&0&0&0 \end{bmatrix}$$

corresponding to the possibility of $A_{W^\perp} = 0$. We see that there are nine possible normal forms for the matrix A of order 4, but is it not possible that some pair of these six 4×4 Hamiltonian matrices are symplectically equivalent, that is, conjugated through a real symplectic matrix?

Let us check one case, that of the Hamiltonian matrices $M_1 = \begin{bmatrix} O & O \\ I & O \end{bmatrix}$ and $M_2 = \begin{bmatrix} O & I \\ O & O \end{bmatrix}$.

Let $P = \begin{bmatrix} A & B \\ C & D \end{bmatrix}$ be a real symplectic matrix. Since $P^{-1} = J^{-1}P^T J = \begin{bmatrix} D^T & -B^T \\ -C^T & A^T \end{bmatrix}$ we compute $P^{-1}M_1 P = \begin{bmatrix} -B^T A & -B^T B \\ A^T A & A^T D \end{bmatrix}$ and the equality $P^{-1}M_1 P = M_2$ gives $B^T B = -I$, so for all vectors $\mathbf{x}$ we have $\|\mathbf{x}\|^2 = \langle \mathbf{x}, \mathbf{x} \rangle = -\langle B^T B\mathbf{x}, \mathbf{x} \rangle = -\langle B\mathbf{x}, B\mathbf{x} \rangle = -\|B\mathbf{x}\|^2$, obviously impossible, so M_1 and M_2 are not real symplectically equivalent.

We leave as an exercise to prove that any two of the above last six Hamiltonian matrices are not real symplectically equivalent except for the pair consisting of the second and third matrices. Therefore, there are altogether eight possible normal forms of A.

The next example deals with 6×6 Hamiltonian matrix.

Example 5.18 Let A be a 6×6 real Hamiltonian matrix with double purely imaginary eigenvalues $\pm i\beta$ and 0. Find the complex normal forms of A and real symplectic normal forms too.

Since 0 is also a double eigenvalue, each of the subspaces $\eta^\dagger(0)$, $\eta^\dagger(i\beta)$ and $\eta^\dagger(-i\beta)$ is two-dimensional, so we have the A-invariant direct sum decomposition $\mathbf{C}^6 = Y \oplus \eta^\dagger(0)$, where $Y = \eta^\dagger(i\beta) \oplus \eta^\dagger(-i\beta)$. By Lemma 4.36 the subspaces $\eta^\dagger(0)$ and Y are J-orthogonal, so by Proposition 4.7(a) they are both symplectic subspaces of $\mathbf{C}^6$.

The normal form of $A_0 = A_{|\eta^\dagger(0)}$ is either $\begin{bmatrix} 0 & 0 \\ 0 & 0 \end{bmatrix}$ or $\begin{bmatrix} 0 & \delta \\ 0 & 0 \end{bmatrix}$ with $\delta = \pm 1$ (see Example 5.11).

Now we consider $A_1 = A_{|Y}$. Notice that $Y = \eta^\dagger(i\beta) \oplus \eta^\dagger(-i\beta)$ is an A-invariant Lagrangian splitting of Y. Let $B = A - i\beta I$ and $\overline{B} = A + i\beta I$.

If A_1 is diagonalizable, then $A = i\beta I$ on $\eta^\dagger(i\beta)$ and $A = -i\beta I$ on $\eta^\dagger(-i\beta)$. The matrix of A_1 on any basis of Y is the diagonal matrix, $A_1 = \text{diag}\big[i\beta, i\beta, -i\beta, -i\beta\big]$.

Now suppose that A_1 is not diagonalizable. Let us prove that there exists a vector $\mathbf{u} \in \eta^\dagger(i\beta)$ such that $\{B\mathbf{u}, \overline{\mathbf{u}}\} \neq 0$.

Indeed, there exists $\mathbf{u} \in \eta^\dagger(i\beta)$ such that $B\mathbf{u}$ and $\mathbf{u}$ are linearly independent; otherwise, A_1 would be diagonalizable. In particular, $B\mathbf{u} \neq 0$. Now $\{B\mathbf{u}, \mathbf{u}\} = 0$ because $\eta^\dagger(i\beta)$ is Lagrangian and $\{B\mathbf{u}, \overline{B\mathbf{u}}\} = -\{B^2\mathbf{u}, \mathbf{u}\} = 0$. If $\{B\mathbf{u}, \overline{\mathbf{u}}\} = 0$ also, then $B\mathbf{u}$ would be J-orthogonal to the basis $\mathbf{u}$, $B\mathbf{u}$, $\overline{\mathbf{u}}$, $\overline{B\mathbf{u}}$ of Y; hence, $B\mathbf{u} = 0$, a contradiction.

We have thus proved that $\alpha_0 = \{B\mathbf{u}, \overline{\mathbf{u}}\} \neq 0$.

Now we take $\mathbf{e} = \Psi^{-1}\mathbf{u}$, where (see the proof of Theorem 5.12) $\Psi\Psi^* = \epsilon\Phi$, with $\Phi(\mathbf{u}) = \Omega(\mathbf{u}, \mathbf{u})$ and $\epsilon = \text{sign}(\alpha_0)$, because the nilpotent index $m + 1 = 2$. Then, the symplectic basis of Y is $\mathbf{u}_1 = \mathbf{e}$, $\mathbf{u}_2 = B\mathbf{e}$; $\mathbf{v}_1 = -\epsilon\overline{B\mathbf{e}}$, $\mathbf{v}_2 = \epsilon\overline{\mathbf{e}}$ giving the standard normal form of A_1,

$$\begin{bmatrix} i\beta & 0 & 0 & 0 \\ 1 & i\beta & 0 & 0 \\ 0 & 0 & -i\beta & -1 \\ 0 & 0 & 0 & -i\beta \end{bmatrix}.$$

Using Proposition 4.10 we combine the symplectic bases of Y and $\eta^\dagger(0)$ to get symplectic bases of $Y \oplus \eta^\dagger(0)$ and so find the complex normal forms of A,

$$A^*_\beta = \begin{bmatrix} i\beta & 0 & 0 & 0 & 0 & 0 \\ 1 & i\beta & 0 & 0 & 0 & 0 \\ 0 & 0 & 0 & 0 & 0 & d \\ 0 & 0 & 0 & -i\beta & -1 & 0 \\ 0 & 0 & 0 & 0 & -i\beta & 0 \\ 0 & 0 & 0 & 0 & 0 & 0 \end{bmatrix}, \quad A^*_\beta = \begin{bmatrix} i\beta & 0 & 0 & 0 & 0 & 0 \\ 1 & i\beta & 0 & 0 & 0 & 0 \\ 0 & 0 & 0 & 0 & 0 & d \\ 0 & 0 & 0 & -i\beta & -1 & 0 \\ 0 & 0 & 0 & 0 & -i\beta & 0 \\ 0 & 0 & 1 & 0 & 0 & 0 \end{bmatrix},$$

where $d = 0, +1$ or -1.

Constructing now convenient real symplectic bases from the complex symplectic bases we can find real normal forms of A.

Another example with a 6×6 Hamiltonian matrix.

Example 5.19 Let A be a 6×6 real nonzero nilpotent Hamiltonian matrix. Find the number of possible normal forms.

If there is a vector $\mathbf{u}$ such that $\{A^5\mathbf{u}, \mathbf{u}\} \neq 0$, then $\mathbf{R}^6 = Z(\mathbf{u}, A)$ is symplectic by Theorem 5.3 and a symplectic basis can be constructed as in Remark 5.9(a), the matrix being that in (5.24) with N of order 3. We have in this case two possible matrices.

Suppose that $\{A^5\mathbf{x}, \mathbf{x}\} = 0$ for all $\mathbf{x}$. By item (1) of Proposition 5.24 $\{A^4\mathbf{x}, \mathbf{x}\} = 0$ for all $\mathbf{x}$. Now either there is a vector $\mathbf{u}$ of nilpotent index 4 such that $\{A^3\mathbf{u}, \mathbf{u}\} \neq 0$ or $\{A^3\mathbf{x}, \mathbf{x}\} = 0$ for all $\mathbf{x}$ of the nilpotent index 4. In the first case the four-dimensional subspace $W = Z(\mathbf{u}, A)$ is symplectic by Theorem 5.3 and $A|_W$ has two possible normal forms, depending on the sign of δ_l in (5.25). Now $\mathbf{R}^6 = W \oplus W^\perp$, with $\dim W^\perp = 2$. The matrix of $A|_{W^\perp}$ can be the zero matrix or the nonzero matrix of Example 5.11. Since $\delta = \pm 1$, we have 3 possible normal forms for $A|_{W^\perp}$. Together with the two possibilities for the matrix of $A|_W$, we get 6 possible normal forms for A in this first case.

In the second case, either there is a vector $\mathbf{u}$ of nilpotent index 2 such that $\{A\mathbf{u}, \mathbf{u}\} \neq 0$ or $\{A\mathbf{x}, \mathbf{x}\} = 0$ for all $\mathbf{x}$ of nilpotent index 2. If $\{A\mathbf{u}, \mathbf{u}\} \neq 0$ the two-dimensional subspace $W = Z(\mathbf{u}, A)$ is symplectic by Theorem 5.3 and contributes with two normal forms. Now $\mathbf{R}^6 = W \oplus W^\perp$ with dim $W^\perp = 4$. The number of possible normal forms for $A\big|_{W^\perp}$ is eight according to the Example 5.17, nine if we include the zero matrix. Therefore, in this second case we have 18 possible normal forms of A.

If $\{A\mathbf{x}, \mathbf{x}\} = 0$ for all $\mathbf{x}$ with nilpotent index 2 we are in the case where $\{A^s\mathbf{x}, \mathbf{x}\} = 0$ for all $\mathbf{x}$ with a nilpotent index $s + 1$. By Lemma 5.4 there are vectors $\mathbf{u}, \mathbf{v}$ with the same nilpotent index $m + 1 \leq 3$ such that $\{A^m\mathbf{u}, \mathbf{v}\} \neq 0$. Then, $W = Z(\mathbf{u}, A) \oplus Z(\mathbf{v}, A)$ is a symplectic subspace and it contributes with one possibility, because $\delta_l = 0$ in (5.25). If $m = 2$, then $W = \mathbf{R}^6$ and this gives one possibility. If $m = 1$, then $\mathbf{R}^6 = W \oplus W^\perp$ with dimension of $W^\perp$ equal to two so there are 3 possible normal forms for $A\big|_{W^\perp}$; hence, we get 3 possibilities. So in this case there are four more possibilities.

Therefore, there are altogether $2+6+18+4 = 30$ possible canonical forms for a nonzero nilpotent Hamiltonian matrix of order 6.

Finally, an example with a Hamiltonian matrix of higher order.

Example 5.20 Find a real normal form of the Hamiltonian matrix

$$A = \begin{bmatrix} K & I \\ U & K \end{bmatrix},$$

where I is the identity matrix of order n and U, K the following real $n \times n$ matrices

$$U = \begin{bmatrix} a & c & & & \\ c & b & & & \\ & & 0 & & \\ & & & \ddots & \\ & & & & 0 \end{bmatrix} \quad \text{and} \quad K = \begin{bmatrix} 0 & 1 & & & \\ -1 & 0 & & & \\ & & 0 & & \\ & & & \ddots & \\ & & & & 0 \end{bmatrix}.$$

An eigenvalue λ of A and corresponding eigenvectors $\mathbf{v} = (\mathbf{x}, \mathbf{y})$ are given by

$$\mathbf{y} = (\lambda I - K)\mathbf{x}, \quad \text{and} \quad (\lambda I - K)\mathbf{y} = U\mathbf{x},$$

so $\mathbf{x}$ is a vector in the kernel of the matrix

$$M_\lambda = (\lambda I - K)^2 - U = \begin{bmatrix} \lambda^2 - (a+1) & -(2\lambda + c) & & & \\ 2\lambda - c & \lambda^2 - (b+1) & & & \\ & & \lambda^2 & & \\ & & & \ddots & \\ & & & & \lambda^2 \end{bmatrix}. \tag{5.81}$$

and λ is a root of the equation

$$\lambda^{2(n-2)}\left[\lambda^4-(a+b-2)\lambda^2+(a+1)(b+1)-c^2\right]=0.$$

Therefore, we have zero as an eigenvalue of multiplicity $2(n-2)$ and the other four eigenvalues as the roots of the equation

$$\lambda^4-(a+b-2)\lambda^2+(a+1)(b+1)-c^2=0. \tag{5.82}$$

The necessary and sufficient condition for the roots of (5.82) to be distinct purely imaginary numbers is that

$$a+b-2<0,\ \ (a+1)(b+1)-c^2>0,\ \ \Delta=(a+b-2)^2-4[(a+1)(b+1)-c^2]>0. \tag{5.83}$$

When a, b, c satisfy these inequalities, let $\lambda=\pm i\beta_j$, $j=1,2$ with $\beta_j>0$ be these four eigenvalues. Eigenvectors of $i\beta_j$ can be taken to be $\mathbf{v}_j=\mathbf{r}_j+i\mathbf{s}_j$, $j=1,2$, where

$$\begin{aligned}
\mathbf{r}_1&=\left(c,\,-(\beta_1^2+a+1),\,0,\ldots,0;\ -\beta_1^2+a+1,\,c,\,0,\ldots,0\right),\\
\mathbf{s}_1&=\left(2\beta,\,0,\,0,\ldots,0;\ \beta_1c,\,-\beta_1(\beta_1^2+a-1),\,0,\ldots,0\right)\\
\mathbf{r}_2&=\left(\beta_2^2+b+1,\,-c,\,0,\ldots,0;\ c,\,-\beta_2^2+b+1,\,0,\ldots,0\right),\\
\mathbf{s}_2&=\left(0,\,2\beta,\,0,\,0,\ldots,0;\ \beta_2(\beta_2^2+b-1),\,-\beta_2c,\,0,\ldots,0\right).
\end{aligned} \tag{5.84}$$

Using the $2n\times 2n$ standard symplectic matrix J, we compute

$$\begin{aligned}
(\mathbf{r}_1, J\mathbf{s}_1)&=\beta_1\left[c^2+\beta_1^4+2(a+1)\beta_1^2+a^2-2a-3\right]\\
(\mathbf{r}_2, J\mathbf{s}_2)&=\beta_2\left[c^2+\beta_2^4+2(b+1)\beta_2^2+b^2-2b-3\right].
\end{aligned} \tag{5.85}$$

Since we have $A\mathbf{r}_j=-\beta_j\mathbf{s}_j$, $A\mathbf{s}_j=\beta_j\mathbf{r}_j$, we get the A-invariant bi-dimensional symplectic subspaces $W_j=\eta(i\beta_j)\oplus\eta(-i\beta_j)$, with symplectic bases $(j=1,2)$ (see the proof of Lemma 4.30)

$$\mathbf{u}_j=\gamma^{-1}\mathbf{r}_j,\ \mathbf{v}_j=\gamma^{-1}\mathbf{s}_j,\quad \text{if}\quad (\mathbf{r}_j, J\mathbf{s}_j)=\gamma_j^2>0$$

or

$$\mathbf{u}_j=\gamma^{-1}\mathbf{s}_j,\ \mathbf{v}_j=\gamma^{-1}\mathbf{r}_j,\quad \text{if}\quad -(\mathbf{r}_j, J\mathbf{s}_j)=\gamma_j^2>0.$$

The subspaces W_1 and W_2 are symplectically orthogonal so the subspace $W=W_1\oplus W_2$ is symplectic. It is A-invariant and has the symplectic basis $\mathbf{u}_1, \mathbf{u}_2, \mathbf{v}_1, \mathbf{v}_2$. Also, its symplectic orthogonal complement $W^{\perp}$ is an A-invariant symplectic space of dimension $2(n-2)$. Moreover, $A\big|_{W^{\perp}}$ has only zero eigenvalues, so it is nilpotent.

The subspace $W^{\perp}$ consists of the vectors $\mathbf{z}$ symplectically orthogonal to the vectors $\mathbf{r}_1, \mathbf{s}_1, \mathbf{r}_2, \mathbf{s}_2$, that is, the vectors of the form

$$\mathbf{z} = (0, 0, x_3, \ldots, x_n;\ 0, 0, , y_3, \ldots, y_n).$$

Now we compute

$$A^2 = \begin{bmatrix} U + K^2 & 2K \\ UK + KU & U + K^2 \end{bmatrix},$$

and observe that

$$UK + KU = \begin{bmatrix} 0 & a+b & & & \\ -(a+b) & 0 & & & \\ & & 0 & & \\ & & & \ddots & \\ & & & & 0 \end{bmatrix}, \quad U + K^2 = \begin{bmatrix} a-1 & c & & & \\ c & b-1 & & & \\ & & 0 & & \\ & & & \ddots & \\ & & & & 0 \end{bmatrix}.$$

Therefore, $A^2\mathbf{z} = 0$ for all $\mathbf{z} \in W^{\perp}$. But for $\mathbf{z} \in W^{\perp}$ we have

$$A\mathbf{z} = (0, 0, y_3, \ldots, y_n;\ 0, 0, 0, \ldots, 0), \quad \text{hence} \quad \{A\mathbf{z}, \mathbf{z}\} = y_3^2 + \cdots + y_n^2,$$

so $\{A\mathbf{z}, \mathbf{z}\} \neq 0$ if $y_i \neq 0$ for some $i \geq 3$. Since $\mathbf{z}$ has a nilpotent index 2 and $\{A\mathbf{z}, \mathbf{z}\} \neq 0$ the subspace $Z(\mathbf{z}, A) = [\mathbf{z}, A\mathbf{z}]$ is symplectic.

Taking $\mathbf{z}_i = (\mathbf{e}_i, \mathbf{e}_i)$, $i = 3, 4, \ldots, n$, we have $\{A\mathbf{z}_i, \mathbf{z}_i\} = 1$. Hence, $A\mathbf{z}_i, \mathbf{z}_i$ is a symplectic basis for $Z(\mathbf{z}_i, A) = [\mathbf{z}_i, A\mathbf{z}_i]$. As the subspaces $W, Z(\mathbf{z}_3, A), \ldots, Z(\mathbf{z}_n, A)$ are pairwise symplectically orthogonal, we obtain the A-invariant symplectic decomposition of $\mathbf{R}^{2n}$,

$$\mathbf{R}^{2n} = W \oplus Z(\mathbf{z}_3, A) \oplus Z(\mathbf{v}_4, A) \oplus \ldots \oplus Z(\mathbf{z}_n, A).$$

We can get a symplectic basis for the space $\mathbf{R}^{2n}$ combining the bases of the subspaces in the sequence

$$\mathbf{u}_1, \mathbf{u}_2, A\mathbf{z}_1, A\mathbf{z}_2, \ldots, A\mathbf{z}_n; \mathbf{v}_1, \mathbf{v}_2, \mathbf{z}_1, \mathbf{z}_2, \ldots, \mathbf{z}_n.$$

The matrix A in this basis, its normal form, is given by (the blank entries are all zero)

$$
A = \left[\begin{array}{ccccc|ccccc}
0 & & & & & \beta_1 & & & & \\
 & 0 & & & & & \beta_2 & & & \\
 & & 0 & & & & & 1 & & \\
 & & & \ddots & & & & & \ddots & \\
 & & & & 0 & & & & & 1 \\
\hline
-\beta_1 & & & & & 0 & & & & \\
 & -\beta_2 & & & & & 0 & & & \\
 & & 0 & & & & & 0 & & \\
 & & & \ddots & & & & & \ddots & \\
 & & & & 0 & & & & & 0
\end{array}\right],
$$

in the case $(\mathbf{r}_j, J\mathbf{s}_j) > 0$, $j = 1, 2$.

The quadratic Hamiltonian defined by $S = -JA$, $H(\mathbf{z}) = \frac{1}{2}\mathbf{z}^T S\mathbf{z}$, $\mathbf{z} = (\mathbf{x}, \mathbf{y})$ is given by

$$
H(\mathbf{x}, \mathbf{y}) = \frac{1}{2}\beta_1(x_1^2 + y_1^2) + \frac{1}{2}\beta_2(x_2^2 + y_2^2) + y_3^2 + \cdots + y_n^2.
$$

Chapter 6
Stability of Equilibria

6.1 Introduction

In Sect. 2.7 we defined the stability of an equilibrium of an autonomous ordinary differential equation $\dot{\mathbf{x}} = \mathbf{f}(\mathbf{x})$ and stated some stability results. Now we consider again this question but also include the case of a non-autonomous differential equation.

Let Ω be a region (open and connected set) in $\mathbf{R}^n$ and consider the differential equation

$$\dot{\mathbf{x}} = \mathbf{f}(\mathbf{x}, t), \tag{6.1}$$

with $\mathbf{f} : \Omega \times [0, \infty) \to \mathbf{R}^n$ continuous and locally Lipschitzian with respect to $\mathbf{x}$, so the existence and uniqueness of solutions of (6.1) are guaranteed.

We say that $\mathbf{x}^* \in \Omega$ is an equilibrium of this equation if $f(\mathbf{x}^*, t) = 0$ for all t. This of course includes the autonomous case. We define stable and unstable equilibria exactly as we did for the Eq. (2.112).

Given a differentiable function $V : \Omega \to \mathbf{R}$, define $V_t(\mathbf{x}) = \langle \nabla V(\mathbf{x}), \mathbf{f}(\mathbf{x}, t)\rangle$, where $\langle\ ,\ \rangle$ is the Euclidean inner product in $\mathbf{R}^n$. Notice that for a solution $\mathbf{x}(t)$ of (6.1) we have $V_t(\mathbf{x}(t)) = \frac{d}{dt} V(\mathbf{x}(t))$, by the chain rule.

6.2 Chetaev's Theorem

Theorem 6.1 (Chetaev) *Let* $\mathbf{x}^*$ *be an equilibrium of the Eq.* (6.1)*, where* $\mathbf{f}(\mathbf{x}, t)$ *is* T*-periodic in* t*. Suppose that there exist a function* $V : \Omega \to \mathbf{R}$ *and an open subset* $A \subset \Omega$ *such that the following properties hold:*

H. E. Cabral, L. Brandão Dias, *Normal Forms and Stability of Hamiltonian Systems*, Applied Mathematical Sciences 218,
https://doi.org/10.1007/978-3-031-33046-9_6

(1) $\mathbf{x}^* \in \partial A$, *the boundary of* A; (2) $V(\mathbf{x}) > 0$, *for* $\mathbf{x} \in A$;
(3) $V(\mathbf{x}) = 0$, *for* $\mathbf{x} \in \partial A \cap \Omega$; (4) $V_t(\mathbf{x}) > 0$, *for* $(\mathbf{x}, t) \in A \times [0, \infty)$.

Then, the equilibrium $\mathbf{x}^*$ *is unstable.*

Proof Fix $\epsilon > 0$ such that $\overline{B_\epsilon(\mathbf{x}^*)} \subset \Omega$. By condition (1), for any $\delta > 0$ there exists a point $\mathbf{x} \in B_\delta(\mathbf{x}^*) \cap A$. Let $\mathbf{x}(t)$ be the solution of the Eq. (6.1) with $\mathbf{x}(0) = \mathbf{x}$. We will show that either $\mathbf{x}(t)$ is not defined for all $t > 0$ or $\|\mathbf{x}(t)\| \geq \epsilon$ for some $t > 0$. This will prove the instability of the equilibrium $\mathbf{x}^*$.

Suppose that the solution $\mathbf{x}(t)$ is defined for all $t > 0$. If $\mathbf{x}(t)$ does not remain inside A for all $t > 0$ it has to leave this set at some first time $t_0 > 0$. As long as $(\mathbf{x}(t), t) \in A \times [0, t_0]$, we have $\frac{d}{dt}V(\mathbf{x}(t)) = V_t(\mathbf{x}(t)) > 0$, by condition (4); hence, $V(\mathbf{x}(t)) \geq V(\mathbf{x}(0)) = V_t(\mathbf{x}) > 0$, for $0 \leq t \leq t_0$. Therefore, by (3) the solution $\mathbf{x}(t)$ cannot leave A through the boundary $\partial A \cap \Omega$, so to leave A it first has to cross the boundary of the ball and consequently we must have $\|\mathbf{x}(t_0)\| > \epsilon$.

Suppose now that $\mathbf{x}(t)$ remains inside A, for all $t > 0$. Integrating $\frac{d}{dt}V(\mathbf{x}(t)) = V_t(\mathbf{x}(t))$ between 0 and t, we get

$$V(\mathbf{x}(t)) = V(\mathbf{x}) + \int_0^t V_s(\mathbf{x}(s))\,ds \tag{6.2}$$

so by (4) we have $V(\mathbf{x}(t)) > V(\mathbf{x})$, for $t > 0$.

Let $A_\mathbf{x} = \{\mathbf{y} \in A;\ \ V(\mathbf{y}) \geq V(\mathbf{x})\}$ and consider the bounded set $K = \overline{B_\epsilon(\mathbf{x}^*)} \cap A_\mathbf{x}$. This set is closed because it obviously contains the limits of its convergent sequences. As K is also bounded it is a compact subset of $\mathbf{R}^n$ by the Heine–Borel theorem. Since $K \subset A$, the function $V_t(\mathbf{x}) = \langle \nabla V(\mathbf{x}), \mathbf{f}(\mathbf{x}, t)\rangle$ is positive on $K \times [0, \infty)$, hence has a positive minimum α on the compact set $K \times [0, T]$. Since the function $\mathbf{f}(\mathbf{x}, t)$ is T-periodic in t, $V_t(\mathbf{x})$ is bounded from below by α all over $K \times [0, \infty)$. Because $V(\mathbf{x}(t)) > V(\mathbf{x})$, for all $t > 0$, we have $\mathbf{x}(t) \in A_\mathbf{x}$, for all $t > 0$. Now assume that $\mathbf{x}(t)$ remains in the ball $\overline{B_\epsilon(\mathbf{x}^*)}$ for $t > 0$. Then, $\mathbf{x}(t)$ lies in K for all $t > 0$.

Therefore, $V_t(\mathbf{x}(t)) \geq \alpha$, for all $t > 0$ and (6.2) implies that $V(\mathbf{x}(t)) \to \infty$ as $t \to \infty$, a contradiction, since V is bounded on K. This contradiction comes from the assumption that $\mathbf{x}(t)$ was defined for all $t > 0$ and remained inside the ball $\overline{B_\epsilon(\mathbf{x}^*)}$ for all $t > 0$. Therefore, either the solution $\mathbf{x}(t)$ is not defined for all $t > 0$ or $\|\mathbf{x}(t)\| \geq \epsilon$, for some $t > 0$. □

Example 6.1 (Instability of Stationary Rotation About the Mean Axis) In Example 2.21 we proved that the stationary rotations of a free rigid body about the minor and major axes of the ellipsoid of inertia are Lyapunov stable. Let us now prove that stationary rotations about the intermediary axis are unstable.

Recall from (2.38) the equations of motion for the free rigid body about its fixed center of mass, namely

$$I_1\dot{p} = (I_2 - I_3)qr, \quad I_2\dot{q} = (I_3 - I_1)pr, \quad I_3\dot{r} = (I_1 - I_2)pq,$$

where $\mathbf{\Omega} = p\mathbf{e}_1 + q\mathbf{e}_2 + r\mathbf{e}_3$. After the translation $x = p$, $y = q - b$, $z = r$, this system assumes the form

$$I_1\dot{x} = (I_2 - I_3)(y + b)z, \quad I_2\dot{y} = (I_2 - I_1)xz, \quad I_3\dot{z} = (I_1 - I_2)x(y + b),$$

and it is easy to see that for this system the function $v(x, y, z) = -xz$ satisfies the hypotheses of Chetaev's theorem, with respect to the open subset $A = \{(x, y, z);\ xz < 0\}$. Therefore, the equilibrium is unstable.

6.3 Moser's Invariant Curve Theorem

An important result for the study of stability of an equilibrium of an autonomous two-degree of freedom Hamiltonian system is Moser's invariant curve theorem, [45]. Consider the mapping in the plane defined in terms of polar coordinates r, θ by the rotation of angle $\alpha(r)$,

$$\theta_1 = \theta + \alpha(r), \quad r_1 = r. \tag{6.3}$$

We assume that $\alpha(r)$ is a smooth function of r with $\alpha'(r) \neq 0$ in the disk $0 \leq r \leq b$. This mapping preserves every circle $r = r^* \leq b$ rotating its points by the angle $\alpha(r^*)$, which justifies the name *twist mapping* for it. The question arises whether a smooth perturbation M of the twist mapping, 2π-periodic in θ (see Fig. 6.1).

$$\theta_1 = \theta + \alpha(r) + f(\theta, r), \quad r_1 = r + g(\theta, r) \tag{6.4}$$

preserves some closed curves near the circles $r = const$. The mere assumption that f and g are small does not ensure the existence of closed invariant curves as the choice $g(\theta, r) \equiv \epsilon$ shows since the iterations of a point by M increase the value of r. However, if the mapping M is area preserving in the disk $0 \leq r \leq b$ and f and g are sufficiently differentiable, then there are invariant closed curves near the

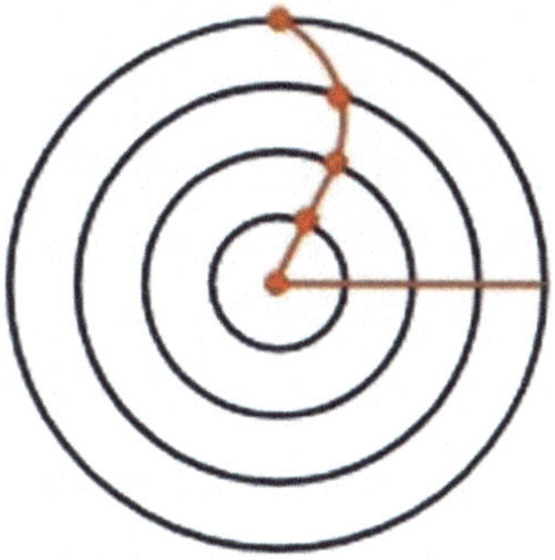

Fig. 6.1 Twist mapping

majority of the circles $r = r^*$. This is the content of Moser's twist theorem, which we state precisely below. First, we notice that if the mapping M is area preserving in the disk $0 \leq r \leq b$ and has the origin as a fixed point, then it has the *intersection property*: every closed curve Γ near a circle $r = r^*$ inside the disk intersects its image curve, that is $\Gamma \cap M\Gamma \neq \emptyset$.

In his original proof (see [45]), Moser requires the functions f and g to be at least of class C^{333}. His proof in the analytic context appears in sections 32 and 33 of the book [55]. Much effort has been spent on diminishing the order of differentiability of f and g. See the Introduction to chapters V and VI of [29] for more information on that subject.

We will consider the theorem in the analytic context and follow the exposition in [55]. So $f(\theta, r)$ and $g(\theta, r)$ will be real analytic functions in the annulus $0 \leq a_0 \leq r \leq b_0$ and 2π-periodic in θ. Let $x = \theta$ and with $\gamma = |\alpha(b_0) - \alpha(a_0)|$ take y defined by $y = \dfrac{\alpha(r)}{\gamma}$. Then, the mapping (6.5) assumes the form

$$x_1 = x + \gamma y + f(x, y), \quad y_1 = y + g(x, y) \tag{6.5}$$

with f and g usually different from the functions in (6.4) but still real analytic functions of x, y and 2π-periodic in x. The variable y ranges over an interval $a \leq y \leq b$ of length $b - a = 1$ and by narrowing the annulus $a_0 \leq r \leq b_0$ we can also assume that $\gamma \leq 1$.

We extend f, g to complex analytic functions defined in an open subset $\mathcal{D} \times \mathcal{D}'$ of $\mathbf{C}^2$,

$$\mathcal{D}: \quad |\Im x| < r_0, \quad y \in \mathcal{D}', \tag{6.6}$$

with $\mathcal{D}'$ an open neighborhood of the interval $a \leq x \leq b$ in the complex plane. We take $0 < r_0 \leq 1$ and assume that every closed curve $y = \phi(x) = \phi(x + 2\pi)$ intersects its image under the mapping (6.5).

For the proof of the following lemma see Section 33 of [55].

Lemma 6.2 *For $\mu > 1$ and $c_0 > 0$ sufficiently small there is in any interval of length $\gamma \leq 1$ a number ω satisfying the inequalities*

$$\left|\frac{\omega}{2\pi} q - p\right| \geq \gamma \frac{c_0}{q^{\mu}}, \qquad (p, q = 1, 2, \ldots). \tag{6.7}$$

Theorem 6.3 (Moser) *Take ω satisfying* (6.7). *Then, under the above hypotheses, for each $\epsilon > 0$ there is a positive number δ, depending on ϵ and $\mathcal{D}$ but not on γ such that if*

$$|f| + |g| < \gamma\delta$$

there is mapping of the form

$$\Phi: \quad x = \xi + u(\xi), \quad y = v(\xi), \tag{6.8}$$

with u and v 2π-periodic real analytic functions in the complex domain $|\Im\xi| < r_0/2$. The mapping (6.8), restricted to the real axis, parametrizes a curve Γ in the plane, $(x, y) = \Phi(\xi)$, such that $\Phi^{-1}M\Phi(\xi) = \xi + \omega$. So M on the invariant curve Γ corresponds to the rotation by the angle ω, $R(e^{i\xi}) = e^{i(\xi+\omega)}$. Moreover, the functions u, v satisfy the inequality

$$|u| + |v - \gamma^{-1}\omega| < \epsilon.$$

Remark 6.4 The condition $|v - \gamma^{-1}\omega| < \epsilon$ says that r is near a certain value r^* so the invariant curve is close to a circle with a rotation number close to ω, since $|\alpha(r) - \omega| < \gamma\epsilon$.

The proof of this theorem is lengthy and quite delicate. It can be seen in Sections 32 and 33 of [55].

Following [41] we present two variations of this theorem, which are more suitable for our purposes. We will state them in the next two theorems. In preparation for their proofs let us find estimates on the original functions $f(\theta, r)$ and $g(\theta, r)$ in the mapping (6.4) to guarantee the existence of invariant curves of that mapping. Again with $x = \theta$ and $y = \dfrac{\alpha(r)}{\gamma}$ we have $r = \alpha^{-1}(\gamma y)$, so instead of the expression (6.5) of the mapping M we have

$$x_1 = x + \gamma y + \tilde{f}(x, y), \quad y_1 = y + \tilde{g}(x, y),$$

with $\tilde{f}(x, y) = f(\theta, r)$ and $\tilde{g}(x, y)$ to be found shortly. From $r_1 = r + g(\theta, r)$ we get

$$\gamma y_1 = \alpha\big(r + g(\theta, r)\big) = \alpha(r) + \alpha'(r)g(\theta, r) + O\big(g(\theta, r)^2\big),$$

so $\tilde{g}(x, y) = \dfrac{1}{\gamma}g(\theta, r)\big[\alpha'(r) + O(g(\theta, r))\big]$.[1]

With ω satisfying (6.7) and the mapping M satisfying the intersection property, Theorem 6.3 says that for any $\epsilon > 0$ there is a positive number δ such that if $|\tilde{f}| + |\tilde{g}| < \gamma\delta$, there is a closed curve Γ: $x = \xi + u(\xi)$, $y = v(\xi)$ such that $\Phi^{-1}M\Phi(\xi) = \xi + \omega$ and this invariant curve is near a circle $r = r^*$ with the angle of rotation of the twist mapping ϵ-close to ω, $|\alpha(r^*) - \omega| < \gamma\epsilon$.

Let A be the maximum value of $|\alpha'(r)|$ in the annulus $0 \le a_0 \le r \le b_0$. Then $A > 0$ and by narrowing this annulus we can assume $|O\big(g(\theta, r)\big)| \le A$, so $\tilde{g}(x, y)| \le \dfrac{2A}{\gamma}|g(\theta, r)|$.

So, if we take f and g satisfying the inequalities

[1] The notation $O\big(g(\theta, r)^n\big)$ means a function $p(\theta, r)g(\theta, r)^n$ with $|p(\theta, r)| \le B$ for some constant B.

$$|f| \leq \frac{1}{2}\gamma\delta, \quad |g| \leq \frac{\gamma^2}{4A}\delta,$$

we have $|\tilde{f}| + |\tilde{g}| < \gamma\delta$, which guarantees the existence of the invariant curve Γ.

After this discussion we can prove the following version of the invariant curve theorem.

Theorem 6.5 *Let the real functions $\alpha(r)$ and $f(\theta, r)$ and $g(\theta, r)$ be analytic in the annulus $0 \leq a_0 \leq r \leq b_0$ with $\alpha'(r) \neq 0$, and $f(\theta, r) = r^{l+1} f_1(\theta, r)$, $g(\theta, r) = r^{l+1} g_1(\theta, r)$ of order $O(r^{l+1})$ with respect to r. Suppose that the mapping*

$$M: \quad \theta_1 = \theta + \alpha(r) + r^{l+1} f_1(\theta, r), \quad r_1 = r + r^{l+1} g_1(\theta, r)$$

has the intersection property and take ω satisfying condition (6.7)*. Then, given $\epsilon > 0$ there exists $r_0(\epsilon)$, $0 < r_0 \leq b_0$ such that for $r \leq r_0$ the mapping M has an invariant curve Γ defined by $\theta = \xi + u(\xi)$, $\alpha(r) = \gamma v(\xi)$ which is near a circle $r = r^*$ with the angle of rotation ϵ-close to ω, $|\alpha(r^*) - \omega| < \gamma\epsilon$.*

Proof With $x = \theta$ and $y = \frac{\alpha(r)}{\gamma}$ as before, we have

$$x_1 = x + \gamma y + \tilde{f}(x, y), \quad y_1 = y + \tilde{g}(x, y),$$

where from the above discussion

$$\tilde{f}(x, y) = r^{l+1} f_1(\theta, r), \quad \tilde{g}(x, y) = \frac{1}{\gamma} r^{l+1} g_1(\theta, r)\big(\alpha'(r) + O(g_1(\theta, r))\big). \tag{6.9}$$

By Theorem 6.3, for any $\epsilon > 0$ there is a number δ such that if $|\tilde{f}| + |\tilde{g}| < \gamma\delta$, an invariant closed curve Γ: $x = \xi + u(\xi)$, $y = v(\xi)$ exists, $\Phi^{-1} M \Phi(\xi) = \xi + \omega$, and Γ is near a circle $r = r^*$ with the angle of rotation ϵ-close to ω, that is $|\alpha(r^*) - \omega| < \gamma\epsilon$.

With A the maximum value of $|\alpha'(r)|$ in the annulus $0 \leq a_0 \leq r \leq b_0$ and $|O\big(g_1(\theta, r)\big)| \leq A$ we have from the second equation in (6.9) $|\tilde{g}(x, y)| \leq \frac{2A}{\gamma} r^{l+1} |g_1(\theta, r)|$. Let D be an upper bound for $|f_1|$ and $|g_1|$ and take $r_0 > 0$ such that

$$r_0^{l+1} D\Big(1 + \frac{2A}{\gamma}\Big) < \frac{1}{2}\gamma\delta.$$

Then, if $r \leq r_0$ from (6.9) we see that $|\tilde{f}| + |\tilde{g}| < \gamma\delta$, which implies the existence of the invariant curve Γ. □

Theorem 6.6 *Consider the mapping M given in action-angle variables (I, ϕ) by*

$$\phi_1 = \phi + \alpha(I) + \epsilon^{l+1} f(\phi, I), \quad I_1 = I + \epsilon^{l+1} g(\phi, I),$$

where α, f, g are real analytic in the annulus $0 \leq I_0 \leq I \leq I_1$ and $\alpha'(I) \neq 0$. The functions f and g are 2π-periodic in ϕ and the mapping M has the intersection property.

Then, given ω satisfying the condition (6.7), for any $\eta > 0$ there exists $\epsilon_0 = \epsilon_0(\eta)$ such that for $\epsilon \leq \epsilon_0$ the mapping has an invariant curve Γ defined by $\theta = \xi + u(\xi)$, $\alpha(I) = \gamma v(\xi)$ which is near a circle $I = I^*$ with the angle of rotation η-close to ω, $|\alpha(r^*) - \omega| < \gamma\eta$.

Proof Look at I, ϕ as polar coordinates in the plane. The proof is like that of the previous theorem. With $x = \theta$ and $y = \frac{\alpha(I)}{\gamma}$, where $\gamma = |\alpha(I_1) - \alpha(I_0)|$ we have

$$x_1 = x + \gamma y + \tilde{f}(x, y), \quad y_1 = y + \tilde{g}(x, y),$$

where

$$\tilde{f}(x, y) = \epsilon^{l+1} f(\phi, I), \quad \tilde{g}(x, y) = \frac{1}{\gamma}\epsilon^{l+1} g(\phi, I)\big(\alpha'(I) + O(g)\big). \tag{6.10}$$

By Theorem 6.3, for any $\eta > 0$ there is a number δ such that if $|\tilde{f}| + |\tilde{g}| < \gamma\delta$, an invariant closed curve Γ: $x = \xi + u(\xi)$, $y = v(\xi)$ exists, $\Phi^{-1} M \Phi(\xi) = \xi + \omega$, and Γ is near a circle $I = I^*$ with the angle of rotation η-close to ω, that is $|\alpha(I^*) - \omega| < \gamma\eta$.

With A the maximum value of $|\alpha'(I)|$ in the annulus $0 \leq a_0 \leq I \leq b_0$ and $|O\big(g(\phi, I)\big)| \leq A$ we have from the second equation in (6.9) $|\tilde{g}(x, y)| \leq \frac{2A}{\gamma}\epsilon^{l+1}|g(\phi, I)|$. Let D be an upper bound for $|f|$ and $|g|$ and take $\epsilon_0 > 0$ such that

$$\epsilon_0^{l+1} D\Big(1 + \frac{2A}{\gamma}\Big) < \frac{1}{2}\gamma\delta.$$

Then, if $\epsilon \leq \epsilon_0$ we have $|\tilde{f}| + |\tilde{g}| < \gamma\delta$, which implies the existence of the invariant curve Γ. □

6.4 Two Stability Lemmas

Let us go back to Sect. 3.6 where we considered the normalization of a two-degree of freedom Hamiltonian system with resonant frequencies $p\omega_1 - q\omega_2 = 0$, p, q positive integers. There, we showed that the normalized resonant term H_m is

$$H_m = H_m(I_1, I_2, p\phi_1 + q\phi_2)$$

a polynomial of degree m in $\sqrt{I_1}$, $\sqrt{I_2}$ with the angles ϕ_1, ϕ_2 appearing only in the combination $p\phi_1 + q\phi_2$; the degree is m because $2l_1 + sp + 2l_2 + sq = k_1 + l_1 + k_2 + l_2 = m$.

Example 6.2 (Cherry's Hamiltonian and Other Resonant Cases) Specific examples of Hamiltonians of this type are the following

$$H{=}2I_1 - I_2 + I_1^{1/2} I_2 \cos(\phi_1 + 2\phi_2), \quad \text{(Cherry's Hamiltonian, [19])}$$

$$H{=}2\omega I_1 - \omega I_2 + \delta I_1^{1/2} \cos(\phi_1 + 2\phi_2) + \big((I_1 + I_2)^2\big),$$

$$H{=}3\omega I_1 - \omega I_2 + \delta I_1^{1/2} I_2^{3/2} \cos(\phi_1 + 3\phi_2) + \big((I_1 + I_2)^{5/2}\big),$$

presented in Section 8 of [11], where their stability was studied.

Normalizing the Hamiltonian H up to H_m the remainder is an analytic function given by a power series in $\sqrt{I_1}$, $\sqrt{I_2}$ where the coefficients are finite sums of sines and co-sines of the angles ϕ_1, ϕ_2.

In the level set $H = 0$ we can express $I_2 = I_2(I_1, \phi_1, \phi_2)$ in a neighborhood of the origin and the analysis of the system $\dot{\mathbf{z}} = J\nabla H(\mathbf{z})$ on $H = 0$ gives a hint of why we consider below the stability lemmas for time-dependent periodic Hamiltonian systems with one-degree of freedom (see [12]).

Lemma 6.7 *Let $K(r, \phi, t) = \Psi(\phi)r^n + O\left(r^{n+\frac{1}{2}}\right)$ be an analytic function of $\sqrt{r}, \phi, t$, τ-periodic in ϕ and T-periodic in t; $n = \frac{m}{2}$, with $m \geq 3$ an integer. Then, if $\Psi(\phi) \neq 0$, for all ϕ, the origin $r = 0$ is a stable equilibrium for the Hamiltonian system defined by K, in the sense that given $\epsilon > 0$, there exists a $\delta > 0$ such that if $r(0) < \delta$, then $r(t)$ is defined for all $t > 0$ and $r(t) < \epsilon$. If Ψ has a simple zero, that is, a value ϕ^* such that $\Psi(\phi^*) = 0$ and $\Psi'(\phi^*) \neq 0$, the equilibrium is unstable.*

Proof We prove the first part on the assumption that $\Psi(\phi) > 0$ for all ϕ; the proof for $\Psi(\phi) < 0$ is made by a slight modification. Define the analytic function $I = I(k)$ for $k > 0$ by

$$I = \frac{1}{2\pi}\int_0^\tau r(k, \phi)d\phi, \qquad \text{where} \qquad r(k, \phi) = \frac{k^{1/n}}{\Psi(\phi)^{1/n}}.$$

Since $I = \dfrac{k^{1/n}}{2\pi}\displaystyle\int_0^\tau \frac{1}{\Psi(\phi)^{1/n}}d\phi$, we get

$$k(I) = \beta^n I^n \ \text{ where } \beta = \left(\frac{1}{2\pi}\int_0^\tau \frac{1}{\Psi(\phi)^{1/n}}d\phi\right)^{-1}.$$

Now consider the analytic function

$$S(I, \phi) = \int_0^\phi r(k, \theta)d\theta.$$

Notice that

$$S(I, \phi) = \beta I G(\phi) \quad \text{where} \quad G(\phi) = \int_0^\phi \frac{1}{\Psi(\theta)^{1/n}}d\theta.$$

The function S generates a canonical transformation $r = r(I, W)$, $\phi = \phi(I, W)$ by the relations

$$W = \frac{\partial S}{\partial I} = \beta G(\phi), \quad r = \frac{\partial S}{\partial \phi} = \beta I G'(\phi),$$

and the original Hamiltonian $K(r, \phi, t) = \Psi(\phi) r^n + O\big(r^{n+\frac{1}{2}}\big)$ is taken to the new Hamiltonian

$$K(I, W, t) = \beta^n I^n + O\big(I^{n+\frac{1}{2}}\big). \tag{6.11}$$

Indeed, Ψ is periodic in ϕ and different from zero for all ϕ; hence, $|\Psi|$ is bounded away from zero, so $G'(\phi) = \frac{1}{\Psi(\phi)^{1/n}}$ is bounded. Therefore, $r = \beta I G' \leq bI$; hence, $O\big(r^{n+\frac{1}{2}}\big) = O\big(I^{n+\frac{1}{2}}\big)$.

The function $K(I, W, t)$ is analytic in $\sqrt{I}$, W, t. Notice that W is a true angular variable because since $\Psi(\phi)$ is τ-periodic we have

$$\begin{aligned} W(\phi + \tau) &= \beta \int_0^{\phi+\tau} \frac{1}{\Psi(\theta)^{1/n}} d\theta = \beta \int_0^{\tau} \frac{1}{\Psi(\theta)^{1/n}} d\theta + \beta \int_{\tau}^{\phi+\tau} \frac{1}{\Psi(\theta)^{1/n}} d\theta \\ &= 2\pi + W(\phi). \end{aligned}$$

Therefore, $K(I, W, t)$ is 2π-periodic in W and, of course, T-periodic in t.

Now, we move from the variable I, near the origin, to a variable J in the interval $1 \leq J \leq 2$. We do that by using a small positive parameter σ and defining the transformation

$$L: \quad I = \gamma \sigma J, \quad W = \psi, \tag{6.12}$$

where γ is the number related to β through the equation $\beta^n \gamma^{n-1} = 1/n$. This relation simplifies the expression of the Hamiltonian $\mathcal{K}$ given below. This transformation is symplectic with multiplier $\gamma\sigma$, so the previous Hamiltonian is transformed to the following:

$$\mathcal{K}(J, \psi, t) = \frac{1}{n} \sigma^{n-1} J^n + O\big(\sigma^{n-\frac{1}{2}}\big),$$

where $\mathcal{K}(J, \psi, t)$ is analytic in (J, ψ, t), 2π-periodic in W and T-periodic in t. The corresponding system of Hamiltonian differential equations, with $\epsilon = \sqrt{\sigma}$, is

$$\frac{dJ}{dt} = O\big(\epsilon^{2n-1}\big), \quad \frac{d\psi}{dt} = -\epsilon^{2n-2} J^{n-1} + O\big(\epsilon^{2n-1}\big),$$

where the right-hand sides are analytic in J, ψ, t, ϵ, 2π-periodic in ψ and T-periodic in t, with $1 \leq J \leq 2$. The flow of this system gives the period map $(J_1, \psi_1) = \Phi(J, \psi, T, \epsilon)$

$$J_1 = J + \epsilon^{2n-1} F_1(J, \psi, \epsilon),$$
$$\psi_1 = \psi - \epsilon^{2n-2} T J^{n-1} + \epsilon^{2n-1} F_2(J, \psi, \epsilon) \tag{6.13}$$

defined in the region $1 \leq J \leq 2$, $\psi \in \mathbf{R}$, $0 < \epsilon < \epsilon_0$ with the functions F_1, F_2 being 2π-periodic in ψ.

The flow of the Hamiltonian K in (6.11) is area preserving on the disk $0 \leq I \leq 2\gamma\sigma$ and has the origin as a fixed point; hence, its period mapping has the intersection property. The transformation L conjugates this period mapping to the period mapping (6.13), so this one also has the intersection property.

Therefore, by the invariant curve theorem in the version of Theorem 6.6, for small σ there exist invariant curves $J = J(\psi) = J(\psi + 2\pi)$ close to circles centered at the origin inside the ring $1 < J < 2$. Since $I = \gamma\sigma J$, the corresponding curves $I = I(\psi)$ can be taken inside small neighborhoods of the origin (by taking σ sufficiently small). In the three-dimensional space (I, ψ, t), identifying the sections $t = 0$ and $t = T$, we get a torus formed by solution curves that begin on the closed curve $I = I(\psi)$. By continuity of solutions, if $I(0)$ is small, then $I(t)$ remains small for t in the interval $0 \leq t \leq T$. So these tori will have a very thin section if $I(0)$ is small. By uniqueness of solutions any solution $(I(t), \psi(t))$ that starts at a point inside the region bounded by the curve $I = I(\psi)$ cannot cross the torus, and therefore $I(t)$ remains small for all t in its domain of definition. As the solution stays inside a compact set it is defined for all t. This proves the stability of the origin $r = 0$.

To prove the instability statement we use Chetaev's theorem. Assume that $\Psi(\phi^*) = 0$ and $\Psi'(\phi^*) > 0$. The proof can easily be adapted to the case $\Psi'(\phi^*) < 0$. Choose $\delta > 0$ so that

$$\Psi(\phi) \neq 0 \quad \text{and} \quad \Psi'(\phi) > 0 \quad \text{for} \quad 0 < |\phi - \phi^*| \leq \delta.$$

Let $\Phi = \dfrac{\pi}{2\delta}(\phi - \phi^*)$ and consider the function

$$V(r, \phi) = r^n \cos \Phi, \tag{6.14}$$

defined in a region $\Omega = \{(r, \phi)\} \subset B$, where $B \times [0, \infty)$ is the domain of the function $K(r, \phi, t)$. Also consider the open subset of Ω defined by

$$A = \{(r, \phi) : \ r > 0, \quad \phi^* - \delta < \phi < \phi^* + \delta\} \cap \Omega.$$

By the definition of Φ the function V is positive in the set A and vanishes on $\partial A \cap \Omega$; moreover, the equilibrium $r = 0$ belongs to ∂A. Therefore, the conditions (1), (2), and (3) of Theorem 6.1 are verified. To conclude that the equilibrium $r = 0$ is unstable it suffices to verify the condition (4) of that theorem. Let

$$\mathbf{f}(r, \phi, t) = (K_\phi(r, \phi, t), -K_r(r, \phi, t))$$

and $V_t(r, \phi) = \langle \nabla V(r, \phi), \mathbf{f}(r, \phi, t)\rangle$. Since

$$K_\phi = r^n \Psi'(\phi) + O\big(r^{n+\frac{1}{2}}\big) \quad \text{and} \quad K_r = nr^{n-1}\Psi(\phi) + O\big(r^{n-\frac{1}{2}}\big)$$

we compute

$$V_t(r, \phi) = nr^{2n-1}\Big[\Psi'(\phi)\cos\Phi + \frac{\pi}{2\delta}\Psi(\phi)\sin\Phi\Big] + O\big(r^{2n-\frac{1}{2}}\big). \tag{6.15}$$

Since $\Psi'(\phi)$ and $\cos\Phi$ are positive for $|\phi - \phi^*| < \delta$, we have $\Psi'(\phi)\cos\Phi > 0$ for $|\phi - \phi^*| < \delta$. Now $\sin\Phi$ and $\Psi(\phi)$ are positive on the interval $0 < \phi - \phi^* < \delta$ and negative on the interval $-\delta < \phi - \phi^* < 0$, so we have $\Psi(\phi)\sin\Phi > 0$ for $0 < |\phi - \phi^*| < \delta$. Therefore, $\Psi'(\phi)\cos\Phi + \frac{\pi}{2\delta}\Psi(\phi)\sin\Phi > 0$ for $|\phi - \phi^*| < \delta$; hence, $V_t(r, \phi) > 0$ for $|\phi - \phi^*| < \delta$. We conclude, by the periodicity of $\mathbf{f}$ in t that V_t is positive in $A \times [0, \infty)$ if r is sufficiently small. So by Chetaev's theorem, the equilibrium is unstable. □

Example 6.3 (Stability in the Absence of Third-Order Resonance) Let us study the stability of the equilibrium $x = 0, y = 0$ of the one-degree of freedom analytic Hamiltonian, 2π-periodic in time t, defined by

$$H = \frac{1}{2}(x^2 + y^2) + x^4 + \lambda y^4 + O\Big(\sqrt{x^2 + y^2}^{\,9/2}\Big), \tag{6.16}$$

where λ is a positive constant.

Making a scaling, $x \mapsto \epsilon x,\ \ y \mapsto \epsilon y$, which is ϵ^2-symplectic mapping, we can assume that the Hamiltonian has the form

$$H = \frac{1}{2}(x^2 + y^2) + \frac{\epsilon^2}{2}(x^4 + \lambda y^4) + O\Big(\sqrt{x^2 + y^2}^{\,9/2}\Big).$$

Through the symplectic transformation given by the rotation

$$x = X\cos t + Y\sin t, \qquad y = -X\sin t + Y\cos t,$$

we eliminate the quadratic term of H and get the new 2π-periodic Hamiltonian

$$\mathcal{H}(X, Y, t) = \frac{\epsilon^2}{2}\Big[(X\cos t + Y\sin t)^4 + \lambda(-X\sin t + Y\cos t)^4\Big] + O\Big(\sqrt{X^2 + Y^2}^{\,9/2}\Big).$$

Notice that this Hamiltonian has the form

$$\mathcal{H} = \mathcal{H}_0 + \frac{\epsilon}{1!}\mathcal{H}_1 + \frac{\epsilon^2}{2!}\mathcal{H}_2 + O\Big(\sqrt{X^2 + Y^2}^{\,9/2}\Big). \tag{6.17}$$

where $\mathcal{H}_0 = 0,\ \mathcal{H}_1 = 0$, and

$$\mathcal{H}_2 = (X\cos t + Y\sin t)^4 + \lambda(-X\sin t + Y\cos t)^4. \tag{6.18}$$

We now use Deprit–Hori method to find a symplectic mapping $(X, Y) \mapsto (u, v)$ generated by a 2π-periodic function

$$\mathcal{W}(X, v, t) = \mathcal{W}_1 + \frac{\epsilon}{1!}\mathcal{W}_2 + \frac{\epsilon^2}{2!}\mathcal{W}_3 + \dots$$

that transforms (6.17) to the following Hamiltonian, autonomous up to order ϵ^2,

$$\mathcal{K}(u, v, t) = \mathcal{K}_0 + \frac{\epsilon}{1!}\mathcal{K}_1 + \frac{\epsilon^2}{2!}\mathcal{K}_2 + O\left(\sqrt{u^2 + v^2}^{9/2}\right), \tag{6.19}$$

Since $\mathcal{H}_0 = 0$ and $\mathcal{H}_1 = 0$, the Deprit–Hori method gives (see Example 3.80, $\mathcal{K}_0 = 0,\ \mathcal{K}_1 = 0$ and

$$\mathcal{K}_2 = \mathcal{H}_2 - \frac{\partial \mathcal{W}_2}{\partial t}.$$

Therefore,

$$\mathcal{W}_2(t) = \mathcal{W}_2(0) + \int_0^t (\mathcal{H}_2 - \mathcal{K}_2)\, dt.$$

From the requirements that $\mathcal{W}_2$ is 2π-periodic and $\mathcal{K}_2$ is constant we get

$$\mathcal{K}_2 = \frac{1}{2\pi}\int_0^{2\pi} \mathcal{H}_2 dv. \tag{6.20}$$

Since $\int_0^{2\pi} \cos^4 t\, dt = \int_0^{2\pi} \sin^4 t\, dt = \frac{3}{4}\pi,\ \int_0^{2\pi} \cos^2 t \sin^2 t\, dt = \frac{1}{4}\pi$ and

$$X = u + \dots, Y = v + \dots,$$

integrating the function (6.18) we get the following expression for $\mathcal{K}_2$ in (6.20)

$$\mathcal{K}_2 = \frac{3}{8}(1 + \lambda)\left(u^2 + v^2\right)^2.$$

Passing to action-angle coordinates (I, ϕ) given by $u = \sqrt{2I}\cos\phi,\ v = \sqrt{2I}\sin\phi$ we get (6.19) in the form

$$\mathcal{K} = I^2\Psi(\phi) + O\left(I^{5/2}\right), \tag{6.21}$$

where

$$\Psi(\phi) = \frac{3}{4}\epsilon^2(1 + \lambda).$$

Since $\Psi(\phi) \neq 0$ for all ϕ, the Stability Lemma 6.7 guarantees that the equilibrium $u = v = 0$ is Lyapunov stable. Consequently, the origin $x = 0$, $y = 0$ is a stable equilibrium for the Hamiltonian H.

By a slight modification of the proof of Lemma 6.7 we can extend the lemma to the case where the Hamiltonian also depends on a small parameter ϵ.

Lemma 6.8 *Let* $K(r, \phi, t, \epsilon) = \epsilon^k \Psi(\phi) r^n + \epsilon^{k+1} O\left(r^{n+\frac{1}{2}}\right)$ *be an analytic function of* $\sqrt{r}, \phi, t, \epsilon$, τ*-periodic in* ϕ *and* T*-periodic in* t*;* $n = \frac{m}{2}$ *and* $k \geq 0$ *and* $m \geq 3$ *are integers. Then, if* $\Psi(\phi) \neq 0$*, for all* ϕ*, and* ϵ *is sufficiently small, the equilibrium* $r = 0$ *of the Hamiltonian system defined by* K *is stable. If* Ψ *has a simple zero, that is a value* ϕ^* *such that* $\Psi(\phi^*) = 0$ *and* $\Psi'(\phi^*) \neq 0$*, and* ϵ *is sufficiently small, the equilibrium is unstable.*

Proof We prove the first part on the assumption that $\Psi(\phi) > 0$ for all ϕ. As before, we define the analytic function $I = I(k)$ for $k > 0$ by

$$I = \frac{1}{2\pi} \int_0^\tau r(k, \phi) d\phi, \qquad r(k, \phi) = \frac{k^{1/n}}{\Psi(\phi)^{1/n}}.$$

From here on, everything goes as before up to the point of getting from the Hamiltonian $K(r, \phi, t, \epsilon) = \epsilon^k \Psi(\phi) r^n + \epsilon^{k+1} O\left(r^{n+\frac{1}{2}}\right)$ the new Hamiltonian

$$K(I, W, t, \epsilon) = \epsilon^k \beta^n I^n + \epsilon^{k+1} O\left(I^{n+\frac{1}{2}}\right),$$

analytic in $\sqrt{I}, W, t, \epsilon$, 2π-periodic in W and T-periodic in t.

Now, exactly in the same way as before, we move from the variable I to a variable J in the interval $1 \leq J \leq 2$ by using a small parameter σ and get the Hamiltonian

$$\mathcal{K}(J, \psi, t, \epsilon) = \epsilon^k \frac{1}{n} \sigma^{n-1} J^n + \epsilon^{k+1} O\left(\sigma^{n-\frac{1}{2}}\right).$$

The corresponding system of Hamiltonian differential equations, with $\sigma = \epsilon^2$, is

$$\frac{dJ}{dt} = O\left(\epsilon^{2n+k}\right), \quad \frac{d\psi}{dt} = -\epsilon^{2n+k-2} J^{n-1} + O\left(\epsilon^{2n+k}\right),$$

where the right-hand sides are analytic in J, ψ, t, ϵ, 2π-periodic in ψ and T-periodic in t, with $1 \leq J \leq 2$. The flow $\Phi(J, \psi, t, \epsilon)$ of this system is a symplectic transformation, for each t, and so the period map $(J_1, \psi_1) = \Phi(J, \psi, T, \epsilon)$, which is given by

$$\begin{aligned} J_1 &= J + \epsilon^{2n+k} F_1(J, \psi, \epsilon), \\ \psi_1 &= \psi - \epsilon^{2n+k-2} T J^{n-1} + \epsilon^{2n+k} F_2(J, \psi, \epsilon) \end{aligned}$$

is an area-preserving transformation defined in the region $1 \leq J \leq 2$, $\psi \in \mathbf{R}$, $0 < \epsilon < \epsilon_0$ with the functions F_1, F_2 being 2π-periodic in ψ. But, in fact this period mapping has the intersection property because the period mapping of the flow defined by K is area-preserving in a punctured disk about the origin so that it has the intersection property. Therefore, we can apply Moser's invariant curve theorem and all the arguments used before apply to this case guaranteeing the stability of the origin $r = 0$.

The proof of the instability statement is exactly the same as before up to the point of defining the vector field $\mathbf{f}(r, \phi, t)$. In the present case we have

$$\mathbf{f}_\epsilon(r, \phi, t) = \big(K_\phi(r, \phi, t, \epsilon), -K_r(r, \phi, t, \epsilon)\big)$$

and we set $V_{\epsilon,t}(r, \phi) = \langle \nabla V(r, \phi), \mathbf{f}_\epsilon(r, \phi, t)\rangle$ with V given by (6.14). Since

$$K_r = \epsilon^k r^n \Psi'(\phi) + \epsilon^{k+1} O\big(r^{n+\frac{1}{2}}\big) \quad \text{and} \quad K_\phi = n\epsilon^k r^{n-1}\Psi(\phi) + \epsilon^{k+1} O\big(r^{n-\frac{1}{2}}\big)$$

we compute

$$V_{\epsilon,t}(r, \phi) = \epsilon^k n r^{2n-1}\left[\Psi'(\phi)\cos\Phi + \frac{\pi}{2\delta}\Psi(\phi)\sin\Phi\right] + \epsilon^{k+1} O\big(r^{2n-\frac{1}{2}}\big).$$

As the function inside the square brackets is positive for $|\phi - \phi^*| < \delta$, we conclude by the periodicity of $\mathbf{f}_\epsilon$ in t that $V_{\epsilon,t}$ is positive in $A \times [0, \infty)$ if ϵ is sufficiently small. So by Chetaev's theorem, the equilibrium is unstable. □

6.5 Arnold's Stability Theorem

Let us consider a two-degree of freedom Hamiltonian with an indefinite quadratic part and for which there exists a resonance of order m among the frequencies, $p\omega_1 = q\omega_2$, with $p + q = m$ and no other resonance relation of smaller order. Then, the Birkhoff normal form up to terms of order m can be found and as in (3.39) of Sect. 3.6 the normalized Hamiltonian is given by

$$H = H_2 + H_4 + \cdots + H_{2l-2} + H_m + O(\|I\|^{m+1}). \tag{6.22}$$

where $\|I\| = \sqrt{I_1^2 + I_2^2}$,

$$H_2 = \omega_1 I_1 - \omega_2 I_2, \tag{6.23}$$

H_{2j} is a homogeneous polynomial of degree j in the actions I_1, I_2 and $H_m(I_1, I_2, \phi_1, \phi_2)$ is a polynomial of degree m with respect to $\sqrt{I_1}, \sqrt{I_2}$ with the angles appearing only in the combination $p\phi_1 + q\phi_2$. Of course, H_{2j} is a polynomial of degree $2j$ in $\sqrt{I_1}, \sqrt{I_2}$.

We have the following version of Arnold's theorem.

Theorem 6.9 *If for some $j \geq 2$, $D_{2j} = H_{2j}(\omega_2, \omega_1) \neq 0$, then the equilibrium is stable.*

Proof See in §13.3 of [41] the proof and remarks on the theorem.

By using the arguments in [41] together with the Stability Lemmas 6.7 and 6.8 we can give a more extended statement. □

Theorem 6.10

(1) *If for some $j \geq 2$, $D_{2j} = H_{2j}(\omega_2, \omega_1) \neq 0$, then the equilibrium is stable.*
If $D_{2j} = 0$ for $j = 1, \dots l-1$, let $\Psi(\phi) = H_m(\omega_2, \omega_1, p\phi)$, where $\phi = \phi_1 + \frac{q}{p}\phi_2$. Then,
(2) *If $\Psi(\phi) \neq 0$ for all ϕ, the equilibrium is stable;*
(3) *If $\Psi(\phi)$ has a simple zero ϕ^*, that is $\Psi(\phi^*) = 0$ and $\Psi'(\phi^*) \neq 0$, the equilibrium is unstable.*

Proof We first prove the instability statement. Division of H_{2j} by H_2 leaves a constant remainder, $H_{2j}(I_1, I_2) = H_2(I_1, I_2)F_{2j-2}(I_1, I_2) + c$, so the condition $H_{2j}(\omega_2, \omega_1) = 0$ is equivalent to $H_{2j} = H_2F_{2j-2}$ and we can write (6.22) in the form

$$H = H_2F + H_m + O(\|I\|^{m+1}),$$

with $F = 1 + F_2 + \cdots + F_{2l-4}$. In a small neighborhood of the origin $I_1 = 0$, $I_2 = 0$, F is a positive function, so $1/F = 1 + \dots$, where the dots represent terms in I_1, I_2 of positive degree. Then, $\tilde{H} = (1/F)H$ is written in the form

$$\tilde{H} = H_2 + H_m + O(\|I\|^{m+1}). \tag{6.24}$$

Notice that to prove the instability of the equilibrium it suffices to prove its instability in the invariant level set $H = 0$.

Now $H = F\tilde{H}$, so $\nabla H = F\nabla\tilde{H} + \tilde{H}\nabla F$ and also in the considered neighborhood, $H = 0$ if, and only if, $\tilde{H} = 0$.

Therefore, inside this neighborhood the solutions of the Hamiltonian system $\dot{\mathbf{z}} = J\nabla H(\mathbf{z})$ on the level set $H = 0$ are the same as the solutions of the system $\dot{\mathbf{z}} = FJ\nabla\tilde{H}(\mathbf{z})$ on the level set $\tilde{H} = 0$. Since F is positive we can change time using $d\tau = F dt$ and the second system is written as the Hamiltonian system

$$\mathbf{z}' = J\nabla\tilde{H}(\mathbf{z}), \tag{6.25}$$

where the prime denotes derivative with respect to τ. Therefore, on the zero level set, the solutions $\mathbf{z}(t)$ of the Hamiltonian system for H are re-parameterizations of the solutions $\mathbf{z}(\tau)$ of the Hamiltonian system for $\tilde{H}$. So it suffices to prove the instability of the system (6.25) on the level set $\tilde{H} = 0$.

By (6.24), $\tilde{H} = 0$ is given by

$$0 = \omega_1 I_1 - I_2\omega_2 + H_m(I_1, I_2, p\phi_1 + q\phi_2) + O(\|I\|^{m+1}).$$

From this we get

$$\begin{aligned} I_2 &= \frac{\omega_1}{\omega_2} I_1 + \frac{1}{\omega_2} H_m(I_1, I_2, p\phi_1 + q\phi_2) + \dots \\ &= \frac{\omega_1}{\omega_2} I_1 + \frac{1}{\omega_2} H_m(I_1, \frac{\omega_1}{\omega_2} I_1, p\phi_1 + q\phi_2) + \dots, \end{aligned}$$

so,

$$I_2 = \frac{q}{p} I_1 + \frac{1}{\omega_2^{\frac{m+2}{2}}} H_m(\omega_2, \omega_1, p\phi_1 + q\phi_2) I_1^{\frac{m}{2}} + O\big(I_1^{\frac{m+1}{2}}\big).$$

Let $H^\dagger(I_1, \phi_1, \phi_2)$ be the negative of the right-hand side of the above equation, that is,

$$H^\dagger(I_1, \phi_1, \phi_2) = -\frac{q}{p} I_1 - \frac{1}{\omega_2^{\frac{m+2}{2}}} H_m(\omega_2, \omega_1, p\phi_1 + q\phi_2) I_1^{\frac{m}{2}} + O\big(I_1^{\frac{m+1}{2}}\big). \qquad (6.26)$$

For the Hamiltonian (6.24) we have $\phi_2' = -\tilde{H}_{I_2} = \omega_2 + \dots$ a positive function in a neighborhood of the origin, so ϕ_2 is an increasing function of τ and we can take it as the time. Then, $H^\dagger$ is an analytic function of $(\sqrt{I}_1, \phi_1, \phi_2)$, and defines a time-dependent Hamiltonian with one degree of freedom, 2π-periodic both in ϕ_1 and in the new time ϕ_2.

We now consider the time-dependent symplectic change of variables $r = I_1$, $\phi = \phi_1 + \frac{q}{p}\phi_2$, which is generated by the function

$$S(r, \phi_1, \phi_2) = \left(\phi_1 + \frac{q}{p}\phi_2\right) r.$$

Since $\dfrac{\partial S}{\partial \phi_2} = \dfrac{q}{p} r$, setting $\Psi(\phi) = -\dfrac{1}{\omega_2^{\frac{m+2}{2}}} H_m(\omega_2, \omega_1, p\phi)$ the Hamiltonian function (6.26) is transformed to the following

$$K(r, \phi, \phi_2) = \Psi(\phi) r^n + O\big(r^{n+\frac{1}{2}}\big), \quad n = \frac{m}{2}.$$

Notice that $K(r, \phi, \phi_2)$ is 2π-periodic in ϕ and $2p\pi$-periodic in ϕ_2. From Lemma 6.7 we conclude that if $\Psi(\phi)$ has a simple zero, then the equilibrium $r = 0$ is unstable, so the equilibrium for the Hamiltonian $\tilde{H}$ is unstable. Therefore, the equilibrium for H is unstable.

At this point we can also conclude by Lemma 6.7 that if $\Psi(\phi) \neq 0$ for all ϕ, the equilibrium is stable on the level set $H = 0$, because I_1 small gives I_2 small also. However, by using Lemma 6.8, we can prove the full stability statement. By the way, if some $D_{2j} \neq 0$ we take l to be the first j for which this occurs, then set $m = 2l$ and consider the constant function $\Psi(\phi) = H_m(\omega_2, \omega_1)$, so we prove (1) and (2) simultaneously. The argument goes as follows.

We first scale the action variables $I_i = \epsilon^2 J_i$, using a small $\epsilon \neq 0$. The angles being kept the same we get an ϵ^2-symplectic change of variables and the Hamiltonian (3.39) can be written as

$$H(J_1, J_2, \phi_1, \phi_2) = FH_2 + \epsilon^{m-2} H_m(J_1, J_2, p\phi_1 + q\phi_2) + \epsilon^{m-1} O(\|J\|^{m+1}), \tag{6.27}$$

where now $F = 1 + \epsilon^2 F_2 + \cdots + \epsilon^{2l-4} F_{2l-4}$. We fix a bounded neighborhood of the origin, say $|J_i| \leq 4$, so that the remainder term is uniformly $O(\epsilon^{m-1})$ in it and henceforth restrict our attention to this neighborhood.

Since for small ϵ, $1/F = 1 + O(\epsilon^2)$, we have

$$\begin{aligned} &\frac{1}{F}\left(\epsilon^{m-2} H_m(J_1, J_2, p\phi_1 + q\phi_2) + \epsilon^{m-1} O(\|J\|^{m+1})\right) \\ &\quad = \epsilon^{m-2} H_m(J_1, J_2, p\phi_1 + q\phi_2) + \epsilon^{m-1} O(\|J\|^{m+1}), \end{aligned}$$

we see from (6.27) that for any value of a parameter h in the interval $-1 \leq h \leq 1$, we have

$$H - \epsilon^{m-1} h = F\tilde{H},$$

where

$$\tilde{H} = \omega_1 J_1 - \omega_2 J_2 + \epsilon^{m-2} H_m(J_1, J_2, p\phi_1 + q\phi_2) + \epsilon^{m-1} O(\|J\|^{m+1}). \tag{6.28}$$

For ϵ sufficiently small the function F is positive in the considered neighborhood and so the level set $H = \epsilon^{m-1} h$ is the same as the level set $\tilde{H} = 0$. Let $\mathbf{z} = (J_1, J_2, \phi_1, \phi_2)$. Since $H - \epsilon^{m-1} h = F\tilde{H}$, the equation of motion is

$$\dot{\mathbf{z}} = J\nabla H = F(J\nabla\tilde{H}) + \tilde{H}(J\nabla F).$$

On the level set $\tilde{H} = 0$ the equation of motion is

$$\dot{\mathbf{z}} = F(J\nabla\tilde{H}).$$

As we noticed for small ϵ, F is positive, so the reparameterization $d\tau = F\,dt$ transforms this equation to the Hamiltonian system

$$\mathbf{z}' = J\nabla\tilde{H}(\mathbf{z}),$$

where the prime denotes derivative with respect to τ. Therefore, the solutions of the system $\mathbf{z}' = J\nabla\tilde{H}$ on the set $\tilde{H} = 0$ are reparameterizations of the solutions of the system $\dot{\mathbf{z}} = J\nabla H$ on the level set $H = \epsilon^{m-1}h$.

Now on the level set $\tilde{H} = 0$ we have

$$0 = \omega_1 J_1 - \omega_2 J_2 + \epsilon^{m-2} H_m(J_1, J_2, p\phi_1 + q\phi_2) + \epsilon^{m-1} O(\|J\|^{m+1}).$$

From this we get

$$\begin{aligned} J_2 &= \frac{\omega_1}{\omega_2} J_1 + \frac{\epsilon^{m-2}}{\omega_2} H_m(J_1, J_2, p\phi_1 + q\phi_2) + \epsilon^{m-1} O(\|J\|^{m+1}) \\ &= \frac{\omega_1}{\omega_2} J_1 + \frac{\epsilon^{m-2}}{\omega_2} H_m\left(J_1, \frac{\omega_1}{\omega_2} J_1, p\phi_1 + q\phi_2\right) + \epsilon^{m-1} O(\|J\|^{m+1}), \end{aligned}$$

so,

$$J_2 = \frac{q}{p} J_1 + \epsilon^{m-2} \frac{1}{\omega_2^{\frac{m+2}{2}}} H_m(\omega_2, \omega_1, p\phi_1 + q\phi_2) J_1^{\frac{m}{2}} + \epsilon^{m-1} O(\|J\|^{m+1}). \tag{6.29}$$

Let $H^\dagger(J_1, \phi_1, \phi_2)$ be the negative of the right-hand side of (6.29). For the Hamiltonian (6.24) we have $\phi_2' = \omega_2 + \ldots$ a positive function in a neighborhood of the origin, so ϕ_2 is an increasing function of τ and we can take it as the time. Then $H^\dagger$ is an analytic function of $(\sqrt{J}_1, \phi_1, \phi_2)$ and defines a time-dependent one degree of freedom Hamiltonian, 2π-periodic in ϕ_1 and ϕ_2.

We now consider the time-dependent symplectic change of variables $r = J_1$, $\phi = \phi_1 + \frac{q}{p}\phi_2$, which is generated by the function

$$S(r, \phi_1, \phi_2) = \left(\phi_1 + \frac{q}{p}\phi_2\right) r.$$

Since $\dfrac{\partial S}{\partial \phi_2} = \dfrac{q}{p} r$, setting

$$\Psi(\phi) = -\frac{1}{\omega_2^{\frac{m+2}{2}}} H_m(\omega_2, \omega_1, p\phi)$$

(or $\Psi(\phi) = -\dfrac{1}{\omega_2^{\frac{m+2}{2}}} H_m(\omega_2, \omega_1)$, as the case may be), the new Hamiltonian function is given by

$$K(r, \phi_1, \phi_2) = \epsilon^{m-2}\Psi(\phi) r^n + \epsilon^{m-1} O\left(r^{n+\frac{1}{2}}\right), \quad n = \frac{m}{2}.$$

Notice that $K(r, \phi, \phi_2)$ is 2π-periodic in ϕ and $2p\pi$-periodic in ϕ_2. From Lemma 6.8 we conclude that if $\Psi(\phi) \neq 0$ for all ϕ, then the equilibrium $r = 0$ is stable; hence, $J_1 = r$ remains small for all $\tau > 0$. Since $J_2 = \frac{\omega_1}{\omega_2} J_1 + O(\epsilon^2)$, J_2 also remains small for all $\tau > 0$. Therefore, $I_1 = \epsilon^2 J_1$ and $I_2 = \epsilon^2 J_2$ are small for all $t > 0$. For $I_1(0)$, $I_2(0)$ sufficiently small the solutions $(I_1(t), I_2(t), \phi_1(t), \phi_2(t))$ remain inside an open neighborhood contained in the union of the level sets $H = \epsilon^{m-1} h$. This proves that the equilibrium is stable. □

Example 6.11 (Stability of a System in Rotating Coordinates in the Plane) Let us consider a two-degree of freedom Hamiltonian system defined by a potential $V(\mathbf{x})$ in a rotating coordinate system in the plane with angular speed $\omega = 1$. It is described by the Hamiltonian

$$H(\mathbf{x}, \mathbf{y}) = \frac{1}{2}\|\mathbf{y}\|^2 - \mathbf{x}^T K \mathbf{y} - V(\mathbf{x}), \tag{6.30}$$

where $K = \begin{bmatrix} 0 & 1 \\ -1 & 0 \end{bmatrix}$. Its equilibria are defined by the equations

$$\mathbf{y} = -K\mathbf{x}, \qquad \nabla V(\mathbf{x}) = -\mathbf{x}.$$

Suppose that $P^* = (\mathbf{x}^*, \mathbf{y}^*)$ is an equilibrium point and let $\mathbf{V} = \begin{bmatrix} a & b \\ b & c \end{bmatrix}$ be the Hessian matrix of V at $\mathbf{x}^*$. The matrix of the linearized system at the point P^* is the Hamiltonian matrix

$$A = \begin{bmatrix} K & I \\ \mathbf{V} & K \end{bmatrix} \tag{6.31}$$

where I is the 2×2 identity matrix. In the computations below we profit from some information already obtained in Example 5.20.

An eigenvalue λ of A and corresponding eigenvectors $\mathbf{v} = (\mathbf{x}, \mathbf{y})$ are given by

$$\mathbf{y} = (\lambda I - K)\mathbf{x}, \qquad M_\lambda \mathbf{x} = 0,$$

where $M_\lambda = (\lambda I - K)^2 - \mathbf{V}$, that is,

$$M_\lambda = \begin{bmatrix} \lambda^2 - (a+1) & -(2\lambda + b) \\ 2\lambda - b & \lambda^2 - (c+1) \end{bmatrix}. \tag{6.32}$$

So $\mathbf{x}$ is a vector in the kernel of M_λ and λ is a root of the equation

$$\lambda^4 - (a + c - 2)\lambda^2 + (a+1)(c+1) - b^2 = 0. \tag{6.33}$$

The necessary and sufficient condition for the roots of (6.33) to be distinct purely imaginary numbers is that a, b, c satisfy the inequalities (5.83).

When a, b, c satisfy those inequalities the equilibrium P^* is linearly stable. If it happens that the Hessian matrix of H at P^* is positive or negative definite, then by the Dirichlet Theorem 2.22 the equilibrium is stable since H is a first integral of the system.

Suppose that a, b, c satisfy the inequalities (5.83) and let $\lambda = \pm i\beta_j$, with $\beta_j > 0$, $j = 1, 2$ be the four eigenvalues. Eigenvectors of $i\beta_j$ can be taken to be $\mathbf{v}_j = \mathbf{r}_j + i\mathbf{s}_j$, $j = 1, 2$, where

$$\begin{aligned}
\mathbf{r}_1 &= \left(b,\ -(\beta_1^2 + a + 1);\ -\beta_1^2 + a + 1,\ b\right),\\
\mathbf{s}_1 &= \left(2\beta,\ 0;\ \beta_1 b,\ -\beta_1(\beta_1^2 + a - 1)\right)\\
\mathbf{r}_2 &= \left(\beta_2^2 + c + 1,\ -b;\ b,\ -\beta_2^2 + c + 1\right),\\
\mathbf{s}_2 &= \left(0,\ 2\beta;\ \beta_2(\beta_2^2 + c - 1),\ -\beta_2 b\right).
\end{aligned} \tag{6.34}$$

If J is the 4×4 symplectic matrix, we compute from the expressions (5.85) the values of $(\mathbf{r}_1, J\mathbf{s}_1)$ and $(\mathbf{r}_2, J\mathbf{s}_2)$. If they have the same sign then the Hessian matrix of the Hamiltonian is positive or negative definite in a neighborhood of P^* (see Proposition 3.4), and the equilibrium is therefore stable by the Dirichlet Theorem. If they are of opposite signs the Hessian matrix is indefinite and the stability of the equilibrium has to be investigated using Arnold's theorem.

In the following example we consider again the restricted three-body problem described at the end of Example 2.2 and about which we already gave some information on Sect. 3.3.

Example 6.5 (Stability of the Lagrangian Equilibrium L_4) Consider the planar circular restricted three-body problem described in Sect. 2.2. We have seen in Example 3.3 that the characteristic equation at L_4 is given by

$$\lambda^4 + \lambda^2 + \frac{27}{4}\mu(1-\mu) = 0.$$

The region of the linear stability condition is defined by the inequality $\Delta = 1 - 27\mu(1-\mu) > 0$. This holds when $0 < \mu < \mu_1$, where μ_1 is the smaller root of the equation $\Delta = 0$, namely $\mu_1 = \frac{9-\sqrt{69}}{18} \cong 0.0385$. The value μ_1 is called the *Routh mass ratio*.

For the eigenvalues $\lambda_j = i\omega_j$, with $\omega_1 \geq \omega_2$, we have

$$\omega_1 = \sqrt{\frac{1}{2}\left(1 + \sqrt{1-\Delta}\right)} \quad \text{and} \quad \omega_2 = \sqrt{\frac{1}{2}\left(1 - \sqrt{1-\Delta}\right)}, \tag{6.35}$$

In Example 3.3 we exhibited the real and imaginary parts of the eigenvectors for the eigenvalues $i\omega_1, i\omega_2$ and computed in (3.37) the symplectic product $\{\mathbf{r}, J\mathbf{s}\}$, finding that $(\mathbf{r}_1, J\mathbf{s}_1) > 0$ and $(\mathbf{r}_2, J\mathbf{s}_2) < 0$ for $i\omega_1$ and $i\omega_2$ respectively.

So the quadratic part of the Hamiltonian is indefinite; hence, the stability of L_4 has to be decided using Arnold's theorem. Looking for resonances among the frequencies ω_1 and ω_2, $p\omega_1 = q\omega_2$, with p, q relatively prime positive integers, from (6.35) we get $\sqrt{1-\delta} = \dfrac{q^2 - p^2}{q^2 + p^2}$; hence, $\mu(1-\mu) = \dfrac{m}{n}$, where $\sqrt{m} = 2pq$ and $n = 27(q^2 + p^2)^2$. Since $\sqrt{m}$ is an even integer, the first four values of m are 4, 16, 36 and 64, which give the values $1, 2, 3$, and 4 respectively, for the product pq. The corresponding pairs of integers give the resonances $\omega_1 = \omega_2$, $\omega_1 = 2\omega_2$, $\omega_1 = 3\omega_2$, and $\omega_1 = 4\omega_2$ of orders 2, 3, 4 and 5 respectively. The resonance of order 2 corresponds to the value μ_1, whereas the values μ_2 and μ_3 for the resonances of order 3 and 4 can be computed from the equation $\mu(1-\mu) = \frac{m}{n}$. We find

$$\mu_1 = \frac{9 - \sqrt{69}}{18} > \mu_2 = \frac{45 - \sqrt{1833}}{90} > \frac{15 - \sqrt{213}}{30}.$$

If μ belongs to the interval $(0, \mu_1)$ of linear stability and $\mu \neq \mu_2, \mu_3$ then there is no resonance up to order 4 and we can write the Birkhoff normal form in terms of the action variables up to order four

$$H = \omega_1 I_1 - \omega_1 I_2 + \frac{1}{2}(AI_1^2 + 2BI_1I_2 + CI_2^2) + H_5.$$

The coefficients A, B, C were computed by Deprit and Bartholomé [21] and they found that the Arnold condition $D_4 = H(\omega_2, \omega_1) \neq 0$ was realized for all $\mu \neq \mu_1, \mu_2, \mu_3$ in the interval of linear stability but for an exceptional value μ_c that does not correspond to any resonance. So the equilibria L_4 and L_5 are stable for all μ in the interval of linear stability, except for these three values. Computing the normal form up to sixth order Meyer and Schmidt [42] also proved the stability for μ_c. The stability analysis for the resonant values μ_2 and μ_3 requires the use of the resonant stability theorem, Theorem 6.10. For this and more information on the stability of the restricted three-body problem, see [11, 39] and [41].

We finish this chapter with some consideration on the existence of zero eigenvalues for the linearized system at an equilibrium of a Hamiltonian system.

6.6 Elimination of a Zero Eigenvalue

On studying the stability of an equilibrium of a Hamiltonian system we frequently encounter the case where zero is an eigenvalue of the linearized system at the equilibrium. When this happens owing to the existence of a first integral in a neighborhood of the equilibrium we can eliminate the zero eigenvalue by passing to the quotient space defined by fixing the value of the integral. This section deals with this elimination process and an example is given to illustrate it.

We begin with the following proposition.

Proposition 6.11 *Let $\mathbf{x}_0$ be an equilibrium of the differential equation $\dot{\mathbf{x}} = f(\mathbf{x})$, $\mathbf{x} \in \mathbb{R}^m$ and let Ψ be a first integral for this system defined in a neighborhood of $\mathbf{x}_0$ and such that $\nabla\Psi(\mathbf{x}_0) \neq 0$. Then, $A = Df(\mathbf{x}_0)$ has a zero eigenvalue.*

Proof We can assume that $\mathbf{x}_0 = 0$, since we can achieve this by means of a translation to the origin. Then, the differential equation can be written in the form $\dot{\mathbf{x}} = A\mathbf{x} + O(\|\mathbf{x}\|^2)$.

Assume that A does not have a zero eigenvalue. Then, A is invertible; hence, there exists a vector $\mathbf{v}$ such that $A\mathbf{v} = \nabla\Psi(0)$. Now, for each $s > 0$ near 0, let $\mathbf{x}(t, s)$ be the solution of $\dot{\mathbf{x}} = f(\mathbf{x})$ such that $\mathbf{x}(0, s) = s\mathbf{v}$. Since $\Psi(\mathbf{x}(t, s)) \equiv \Psi(s\mathbf{v})$, differentiating with respect to t we obtain $\left\langle \nabla\Psi(\mathbf{x}(t, s)), \frac{\partial}{\partial t}\mathbf{x}(t, s)\right\rangle \equiv 0$. As $\frac{\partial}{\partial t}\mathbf{x}(t, s) = A\mathbf{x}(t, s) + O(\|\mathbf{x}(t, s\|^2)$, setting $t = 0$ we get $\langle \nabla\Psi(s\mathbf{v}), sA\mathbf{v} + O(\|s\mathbf{v}\|^2)\rangle \equiv 0$. Dividing by s and letting s approach zero we obtain, using the continuity of $\nabla\Psi$, $\langle\nabla\Psi(0), A\mathbf{v}\rangle = 0$; hence, $\|\nabla\Psi(0)\|^2 = 0$, contradicting the fact that $\nabla\Psi(0) \neq 0$. Therefore, A has a zero eigenvalue. □

The proof of the proposition below is adapted from that given in Siegel-Moser [55], §18.

Proposition 6.12 *Let $\mathbf{x}_0$ be an equilibrium of an autonomous Hamiltonian system and let Ψ be a first integral in a neighborhood of this equilibrium, which does not depend on time and is such that $\nabla\Psi(\mathbf{x}_0) \neq 0$. Then, the number of degrees of freedom of the system can be lowered by one.*

Proof Let $\Psi(\mathbf{x}, \mathbf{y})$ be the first integral of the n-degree of freedom Hamiltonian system defined by $H(\mathbf{x}, \mathbf{y})$. We search for a symplectic transformation $\phi : \mathbf{x} = \mathbf{x}(\mathbf{u}, \mathbf{v}),\ \mathbf{y} = \mathbf{y}(\mathbf{u}, \mathbf{v})$, generated by a function $W(\mathbf{u}, \mathbf{y})$, so that in the new variables $\mathbf{w} = (\mathbf{u}, \mathbf{v})$ we have $\psi(\mathbf{u}, \mathbf{v}) = v_n$, where $\psi(\mathbf{w}) = \Psi(\phi(\mathbf{w}))$. Since $\mathbf{x} = W_{\mathbf{y}}(\mathbf{u}, \mathbf{y})$ and $\psi(\mathbf{u}, \mathbf{v}) = \Psi(\mathbf{x}, \mathbf{y})$ we have the equation

$$v_n = \Psi(W_{\mathbf{y}}, \mathbf{y}). \tag{6.36}$$

Now assume that we have found a particular solution $W(\mathbf{u}, \mathbf{y})$ such that $\operatorname{Det} D^2W(\mathbf{u}, \mathbf{y}) \neq 0$. Then, W generates a symplectic transformation $\mathbf{z} = \phi(\mathbf{w})$ given by $\mathbf{x} = \mathbf{x}(\mathbf{u}, \mathbf{v}),\ \mathbf{y} = \mathbf{y}(\mathbf{u}, \mathbf{v})$.

Let $\mathcal{H}(\mathbf{w}) = H(\phi(\mathbf{w}))$ be the new Hamiltonian function. As $\nabla\psi(\mathbf{w}) = D\phi(\mathbf{w})^T \cdot \nabla\Psi(\mathbf{z})$ we have along the solution $\mathbf{z}(t) = \phi(\mathbf{w}(t))$ of the system $\dot{\mathbf{z}} = J\nabla H(\mathbf{z})$,

$$\begin{aligned}\frac{d}{dt}(\Psi(\mathbf{z}(t))) &= D\Psi(\phi(\mathbf{w}))\cdot\left(D\phi(\mathbf{w})\cdot\dot{\mathbf{w}}\right) = \langle\nabla\Psi(\mathbf{z}), D\phi(\mathbf{w})J\nabla\mathcal{H}(\mathbf{w})\rangle,\\ &= \langle D\phi(\mathbf{w})^T\nabla\Psi(\mathbf{z}), J\nabla\mathcal{H}(\mathbf{w})\rangle = \langle\nabla\psi(\mathbf{w}), J\nabla\mathcal{H}(\mathbf{w})\rangle.\end{aligned}$$

Then, as Ψ is a first integral of $\dot{\mathbf{z}} = J\nabla H(\mathbf{z})$ we have the equation

$$\langle \nabla\psi(\mathbf{w}), J\nabla\mathcal{H}(\mathbf{w})\rangle = 0.$$

Now, by the requirement $\psi(\mathbf{u}, \mathbf{v}) = v_n$ we have $\nabla\psi(\mathbf{w}) = (0, \dots, 0; 0, \dots, 0, 1)$, so this equation gives $\mathcal{H}_{u_n} = 0$; hence, $\mathcal{H}(\mathbf{u}, \mathbf{v})$ does not contain the variable u_n. As $\dot{v}_n = \mathcal{H}_{u_n} = 0$, the variable v_n is a constant of the motion. Fixing its value, v_n^0, the Hamiltonian $\mathcal{H}(u_1, \dots, u_{n-1}; v_1, \dots, v_{n-1}, v_n^0)$ has $n - 1$ degrees of freedom. □

Remark 6.13 Notice that in the Hessian $2n \times 2n$ matrix G of the Hamiltonian $\mathcal{H}(u, v)$, all the entries in the row corresponding to the variable u_n are zero. Since G is symmetric the entries in the nth column of G, and hence also in the nth column of $A = JG$, are zero too. Therefore, A has a zero eigenvalue and since this is a Hamiltonian matrix the zero eigenvalue appears once more. On reducing the dynamics to the manifold of a fixed level of the first integral one obtains a reduced Hamiltonian system in which the zero eigenvalue was eliminated, if it appeared only twice.

Example 6.6 (An Example of the Reduction Process) Consider the Hamiltonian H defined in a region of $\mathbb{R}^3 \times \mathbb{R}^3$ by

$$H(\mathbf{x}, \mathbf{y}) = \frac{1}{2}\|\mathbf{y}\|^2 - V(\mathbf{x}) - \langle \Sigma\mathbf{x}, \mathbf{y}\rangle \tag{6.37}$$

where $\Sigma = \begin{bmatrix} 0 & 1 & 0 \\ -1 & 0 & 0 \\ 0 & 0 & 0 \end{bmatrix}$ and the function $V(\mathbf{x})$ is invariant under rotations about the x_3-axis and has no critical point on this axis. The dynamics is given by the equations

$$\dot{\mathbf{x}} = H_{\mathbf{y}} = \mathbf{y} - \Sigma\mathbf{x}, \quad \dot{\mathbf{y}} = -H_{\mathbf{x}} = \nabla V(\mathbf{x}) - \Sigma\mathbf{y}, \tag{6.38}$$

so at an equilibrium $\mathbf{z}^* = (\mathbf{x}^*, \mathbf{y}^*)$ we have $\mathbf{y}^* = \Sigma\mathbf{x}^*$ and $\nabla V(\mathbf{x}^*) = \Sigma^2\mathbf{x}^*$.

Let us prove that the function $Q(\mathbf{x}, \mathbf{y}) = \langle \Sigma\mathbf{x}, \mathbf{y}\rangle$ is a first integral of the system. Using the equality $\Sigma^T = -\Sigma$, we see that along a solution of (6.38) we have

$$\dot{Q} = \langle \Sigma\dot{\mathbf{x}}, \mathbf{y}\rangle + \langle \Sigma\mathbf{x}, \dot{\mathbf{y}}\rangle = \langle \Sigma\mathbf{y} - \Sigma^2\mathbf{x}, \mathbf{y}\rangle + \langle \Sigma\mathbf{x}, \nabla V(\mathbf{x}) - \Sigma\mathbf{y}\rangle = \langle \Sigma\mathbf{x}, \nabla V(\mathbf{x})\rangle, \tag{6.39}$$

the last equality holding because

$$\langle \Sigma\mathbf{y} - \Sigma^2\mathbf{x}, \mathbf{y}\rangle + \langle \Sigma\mathbf{x}, -\Sigma\mathbf{y}\rangle = \langle \Sigma\mathbf{y}, \mathbf{y}\rangle + \langle -\Sigma^2\mathbf{x}, \mathbf{y}\rangle + \langle \Sigma^2\mathbf{x}, \mathbf{y}\rangle = 0.$$

Now we notice that $\Sigma = \Omega^{-1}\dot{\Omega}$, where Ω is the rotation matrix

$$\Omega = \begin{bmatrix} \cos t & \sin t & 0 \\ -\sin t & \cos t & 0 \\ 0 & 0 & 1 \end{bmatrix}.$$

Since V is invariant under the rotation $\Omega = \Omega(t)$, we have $V(\Omega\boldsymbol{\xi}) = V(\boldsymbol{\xi})$, for all $t, \boldsymbol{\xi}$ and differentiating this equality with respect to t we get $\langle \nabla V(\Omega\boldsymbol{\xi}), \dot{\Omega}\boldsymbol{\xi}\rangle =$ 0. The fact that Ω is an orthogonal matrix implies that $\nabla V(\Omega\boldsymbol{\xi}) = \Omega\nabla V(\boldsymbol{\xi})$. Therefore, $0 = \langle \nabla V(\boldsymbol{\xi}), \Omega^{-1}\dot{\Omega}\boldsymbol{\xi}\rangle = \langle \nabla V(\boldsymbol{\xi}), \Sigma\boldsymbol{\xi}\rangle$. We then see from (6.39) that $\dot{Q} = 0$, so Q is a first integral.

Checking for the nondegeneracy condition we compute $\nabla Q(\mathbf{z}^*)$ and we obtain $\nabla Q(\mathbf{z}^*) = (-\Sigma\mathbf{y}^*, \Sigma\mathbf{x}^*)$. If $\nabla Q(\mathbf{z}^*) = 0$ we have $\Sigma\mathbf{x}^* = 0$; hence, $\mathbf{x}^* = (0, 0, x_3^*)$, so $\nabla V(\mathbf{x}^*) = 0$, contrary to the hypothesis that V has no critical point on the x_3-axis. Therefore, $\nabla Q(\mathbf{z}^*) \neq 0$ and by Proposition 6.11 zero appears as a double eigenvalue of the linearized system at the equilibrium.

Let us make the reduction process described in the proof of Proposition 6.12 for the Hamiltonian (6.37) using the first integral

$$Q(\mathbf{x}, \mathbf{y}) = x_2 y_1 - x_1 y_2.$$

We search for a symplectic transformation $(u_1, u_2, u_3, v_1, v_2, v_3) \mapsto (x_1, x_2, x_3, y_1, y_2, y_3)$ such that in the variables $\mathbf{u}, \mathbf{v}$ we have $Q(\mathbf{u}, \mathbf{v}) = v_2$.

As Q is bilinear we take the generating function $W(\mathbf{u}, \mathbf{y})$ in the form of a bilinear function of $g_1, g_2, x_3, v_1, v_2, y_3$, namely

$$W(\mathbf{u}, \mathbf{y}) = g_1(u_1, u_2)y_1 + g_2(u_1, u_2)y_2 + u_3 y_3.$$

The Eq. (6.36) here means that

$$W_{u_2} = Q(W_{\mathbf{y}}, \mathbf{y}). \tag{6.40}$$

Now, $\mathbf{x} = W_{\mathbf{y}} = (g_1, g_2, u_3)$, so $Q(W_{\mathbf{y}}, \mathbf{y}) = g_2(u_1, u_2)y_1 - g_1(u_1, u_2)y_2$. On the other hand, $W_{u_2} = g_{1u_2} y_1 + g_{2u_2} y_2$, so comparison of both members in (6.40) gives the system of equations

$$g_{1u_2} = g_2(u_1, u_2), \quad g_{2u_2} = -g_1(u_1, u_2).$$

We readily check that

$$g_1(u_1, u_2) = u_1 \cos u_2, \quad g_2(u_1, u_2) = -u_1 \sin u_2$$

gives a particular solution of (6.40) for which Det $W_{\mathbf{u},\mathbf{y}} = -u_1$. Thus, in a region where $u_1 \neq 0$ the function $W(\mathbf{u}, \mathbf{y})$ generates the required symplectic transformation, which is given by

$$x_1 = u_1 \cos u_2, \quad x_2 = -u_1 \sin u_2, \quad x_3 = u_3$$

$$y_1 = v_1 \cos_2 - v_2 \frac{1}{u_1} \sin u_2, \quad y_2 = -v_1 \sin u_2 - v_2 \frac{1}{u_1} \cos u_2, \quad y_3 = v_3.$$

Then we have $x_1^2 + x_2^2 = u_1^2$ and $y_1^2 + y_2^2 = v_1^2 + \frac{1}{u_1^2} v_2^2$.

The function V has the form $V(x_1, x_2, x_3) = f(\sqrt{x_1^2 + x_2^2}, x_3)$, since it is invariant under rotations about the x_3-axis.

Now, as $Q(\mathbf{u}, \mathbf{v}) = v_2$ and v_2 is a constant of the motion, fixing its value $v_2 = \gamma$ we can ignore the term $\langle \Sigma \mathbf{x}, \mathbf{y} \rangle = Q(\mathbf{x}, \mathbf{y})$ in (6.37) and thus obtain the following expression for the reduced Hamiltonian

$$\mathcal{H}(u_1, u_3, v_1, v_3, \gamma) = \frac{1}{2}(v_1^2 + v_3^2) + \mathcal{V}(u_1, u_3, \gamma) \tag{6.41}$$

with $\mathcal{V}$ given by

$$\mathcal{V}(u_1, u_3, \gamma) = \frac{\gamma^2}{2u_1^2} - f(u_1, u_3).$$

Chapter 7
Stability of Linear Hamiltonian Systems

7.1 Introduction

The question of the nonlinear stability of an equilibrium point of an autonomous Hamiltonian system with more than two degrees of freedom is not trivial. There is no result such as Arnold's theorem if the autonomous system has more than two degrees of freedom. So in autonomous Hamiltonian systems with more than two degrees of freedom the problem of deciding the stability of an equilibrium is quite difficult. The same is true for non-autonomous Hamiltonian systems, even in the one-degree of freedom case.

However, the study of the linear stability of the equilibrium, that is, the stability of the linearized system about the equilibrium can be carried out for multi-dimensional autonomous systems and also for periodic Hamiltonian systems.

7.2 The Normed Linear Space of Square Matrices

In this chapter we need to introduce a norm in the space of matrices. Let V denote the real vector space $\mathbf{R}^m$ with the Euclidean inner product $\langle \mathbf{u}, \mathbf{v} \rangle = u_1 v_1 + \cdots + u_m v_m$ or the complex vector space $\mathbf{C}^m$ with the Hermitian inner product $(\mathbf{u}, \mathbf{v}) = u_1 \overline{v}_1 + \cdots + u_m \overline{v}_m$.

A real or complex $m \times m$ matrix A can be viewed as a linear operator $A : V \to V$ by taking $A(\mathbf{v}) = A\mathbf{v}$.

To talk about nearness of linear systems we introduce a norm in the space $\mathcal{M}_n(K)$ of real or complex square matrices, that is $K = \mathbf{R}$, or $K = \mathbf{C}$. Let $\| \ \|$ be the norm in V. If A is a constant matrix we will take the norm

$$\|A\| = \sup\{\|A\mathbf{x}\|; \ \|\mathbf{x}\| = 1\},$$

H. E. Cabral, L. Brandão Dias, *Normal Forms and Stability of Hamiltonian Systems*, Applied Mathematical Sciences 218,
https://doi.org/10.1007/978-3-031-33046-9_7

which has the nice properties

$$\|A\mathbf{x}\| \leq \|A\| \, \|\mathbf{x}\| \quad \text{and} \quad \|AB\| \leq \|A\| \, \|B\|.$$

Also, it has the following interesting geometric property: if A is an invertible symmetric real matrix the set of points $\mathbf{x} \in \mathbf{R}^n$ such that $\|A\mathbf{x}\| = 1$ is an ellipsoid and $\|A\|$ is the semi-major axis of this ellipsoid.

For a continuous τ-periodic matrix $A(t)$ we will take

$$||A|| = \int_0^\tau \|A(t)\| dt. \tag{7.1}$$

7.3 Logarithms of Square Matrices

We begin with the following result, which can be found in Hartman [28].

Proposition 7.1 *Consider an elementary Jordan block $D = \lambda I + N$, where $\lambda \neq 0$ and N is a nilpotent matrix as in* (5.25). *Then, D has a logarithm, that is there exists a complex matrix B, of the same order as D, such that $e^B = D$.*

Proof Write $D = \lambda(I + R)$ with $R = \frac{N}{\lambda}$ and consider the matrix $B = (\log \lambda)I + S$, where $S = \Sigma_{j=1}^{m-1}(-1)^{j+1}\frac{R^j}{j}$, m being the order of D and $\log \lambda$ one of the determinations of the logarithm of λ. Observe that $S = \log(I + R)$, that is S is obtained by substituting the variable z for R into the series $\log(1 + z) = \Sigma_{j=1}^{\infty}(-1)^{j+1}\frac{z^j}{j}$. We note that there is no problem as to convergence of the resulting series, since $R^j = 0$ for $j \geq m$. As $e^{\log(1+z)} = 1 + z$ and the verification of this property can be done in terms of formal power series, we conclude that $e^S = I + R$. As $(\log \lambda)I$ and S commute we have that $e^B = e^{(\log \lambda)I} e^S = \lambda(I + R) = D$. □

Remark 7.2 The matrix B is not unique because it depends on the choice of the determination of the logarithm of λ. If $\lambda = re^{i\theta}$, then $\log \lambda = \log r + i(\theta + 2k\pi)$, where $k \in \mathcal{Z}$, so for any other matrix $\tilde{B}$ such that $e^{\tilde{B}} = D$ we have $\tilde{B} = B + i2k\pi I$, for some integer k, where I is the $m \times m$ identity matrix. The matrix B is complex but when λ is a positive real number we can take $\log \lambda$ to be the real logarithm and get B real.

Let λ be an eigenvalue of a square matrix A. The Jordan λ-block of A is the block-diagonal matrix formed by all the elementary λ-blocks, in decreasing order of size. Now, every real or complex matrix is similar to the block-diagonal matrix formed by the Jordan λ_k-blocks, where $\lambda_1, \ldots, \lambda_p$ are the distinct eigenvalues of the given matrix. So the next proposition is an immediate consequence of the previous one.

Proposition 7.3 *Every real or complex invertible matrix A has a logarithm, that is, there exists a complex matrix B such that $e^B = A$.*

Proof Given A there is an invertible matrix P such that $P^{-1}AP = \text{diag}\,[D_1, \ldots, D_k]$, where D_j is an elementary Jordan block, that is, a matrix of the form $D = \lambda I + N$. By Proposition 7.1 there exist matrices $B_1, \ldots, B_k$ with $e^{B_j} = D_j$, so setting $\tilde{B} = \text{diag}\,[B_1 \ldots, B_k]$ we have $P^{-1}AP = e^{\tilde{B}}$; hence, $A = e^B$, where $B = P\tilde{B}P^{-1}$. □

Remark 7.4 Notice from the proof and Remark 7.2 that any two logarithms C_1, C_2 of A differ by a block diagonal matrix, $C_1 - C_2 = \text{diag}[K_1, \ldots, K_r]$ where $K_j = 2\pi i k_j I$, with k_j an integer and I the identity matrix of some order n_j.

We also notice that if a real matrix A has a real logarithm, then it is unique.

Existence of Real Logarithms

Let A be an $n \times n$ real square matrix. If λ is a complex eigenvalue of A so is $\bar{\lambda}$. Let $\mathbf{v}_1, \ldots, \mathbf{v}_k$ be a Jordan basis for λ, that is,

$$A\mathbf{v}_1 = \lambda\mathbf{v}_1 + \mathbf{v}_2, \quad \ldots, \quad A\mathbf{v}_{k-1} = \lambda\mathbf{v}_{k-1} + \mathbf{v_k}, \quad A\mathbf{v}_k = \lambda\mathbf{v}_k.$$

Then we also have

$$A\bar{\mathbf{v}}_1 = \bar{\lambda}\bar{\mathbf{v}}_1 + \bar{\mathbf{v}}_2, \quad \ldots, \quad A\bar{\mathbf{v}}_{k-1} = \bar{\lambda}\bar{\mathbf{v}}_{k-1} + \bar{\mathbf{v}}_k, \quad A\mathbf{v}_k = \bar{\lambda}\bar{\mathbf{v}}_k.$$

Let U be the subspace of $\mathbf{C}^n$ generated by the vectors $\mathbf{v}_1, \ldots, \mathbf{v}_k$. The subspace $U \oplus \overline{U} \subset \mathbf{C}^n$ is invariant under the linear operator $A : \mathbf{C}^n \to \mathbf{C}^n$ and the restriction of this operator to the subspace $U \oplus \overline{U}$ has on the basis $\mathbf{v}_1, \ldots, \mathbf{v}_k, \bar{\mathbf{v}}_1, \ldots \bar{\mathbf{v}}_k$ the matrix

$$[A] = \begin{bmatrix} \lambda I + N & O \\ O & \bar{\lambda} I + N \end{bmatrix}, \tag{7.2}$$

with N the $k \times k$ nilpotent matrix of maximal rank with 1 in the subdiagonal and zero elsewhere.

Since $U \oplus \overline{U}$ is clearly invariant under conjugation it is a complexification, so it has a real basis. In fact, if $\mathbf{v}_j = \mathbf{r}_j + i\mathbf{s}_j$ with $\mathbf{r}_j, \mathbf{s}_j$ real vectors, then the vectors $\mathbf{r}_1, \ldots, \mathbf{r}_k, \mathbf{s}_1, \ldots, \mathbf{s}_k$ form a real basis of $U \oplus \overline{U}$ and if $\lambda = a + ib$ we have for $j = 1, \ldots, k-1$,

$$\begin{aligned} A\mathbf{r}_j &= a\mathbf{r}_j - b\mathbf{s}_j + \mathbf{r}_{j+1}, \quad A\mathbf{s}_j = b\mathbf{r}_j + a\mathbf{s}_j + \mathbf{s}_{j+1} \quad \text{and} \\ A\mathbf{r}_k &= a\mathbf{r}_k - b\mathbf{s}_k, \quad A\mathbf{s}_k = b\mathbf{r}_k + a\mathbf{s}_k. \end{aligned}$$

Therefore, the matrix of the linear operator $A : U \oplus \overline{U} \to U \oplus \overline{U}$ in this real basis is

$$M = \left[\begin{array}{cccc|cccc} a & & & & b & & & \\ 1 & a & & & 0 & b & & \\ & \ddots & \ddots & & & \ddots & \ddots & \\ & & 1 & a & & & 0 & b \\ \hline -b & & & & a & & & \\ & -b & & & 1 & a & & \\ & & \ddots & & & \ddots & \ddots & \\ & & & -b & & & 1 & a \end{array}\right],$$

that is

$$M = aI + bJ + \mathcal{N}, \quad \text{where} \quad \mathcal{N} = \begin{bmatrix} N & O \\ O & N \end{bmatrix},$$

with N the $k \times k$ nilpotent matrix of maximal rank with 1 in the subdiagonal and zero elsewhere. We use I for the identity matrix leaving its order to be inferred from the context, for instance, in the above expression for M, the order of I is $2k$ whereas in $J = \begin{bmatrix} O & I \\ -I & O \end{bmatrix}$ the order of I is k.

Lemma 7.5 *The matrix $M = aI + bJ + \mathcal{N}$ has a real logarithm.*

Proof For the symplectic matrix J we easily compute

$$e^{\phi J} = \cos\phi I + \sin\phi J. \tag{7.3}$$

Using this we see that $e^{\alpha I + \beta J} = aI + bJ$ if and only if $e^{\alpha}\cos\beta = a$ and $e^{\alpha}\sin\beta = b$, that is, $Q = aI + bJ$ has the real logarithm $\alpha I + \beta J$, where $\alpha = \log\sqrt{a^2+b^2}$, $\cos\beta = a/\sqrt{a^2+b^2}$ and $\sin\beta = b/\sqrt{a^2+b^2}$.

From the identity $(xI + yJ)(aI + bJ) = (ax - by)I + (bx + ay)J$, we readily see that $Q^{-1} = \frac{a}{a^2+b^2}I - \frac{b}{a^2+b^2}J$.

Let $P = Q^{-1}M$. Since $M = Q + \mathcal{N}$ we have $P = I + Q^{-1}\mathcal{N}$. Now we readily check that Q^{-1} commutes with $\mathcal{N}$, so we have $\left(Q^{-1}\mathcal{N}\right)^l = (Q^{-1})^l\mathcal{N}^l = 0$, for any integer $l \geq k$ because $N^k = 0$. It follows that the series $\sum_{j=1}^{\infty}(-1)^{j+1}\frac{\left(Q^{-1}\mathcal{N}\right)^j}{j}$ reduces to a finite sum and thus defines a real matrix S. Since $S = \log(I + Q^{-1}\mathcal{N})$, see the Proof of Proposition 7.1, we have $P = e^S$. Therefore, $M = QP = e^{\alpha I + \beta J}e^S$. Now, J commutes with Q^{-1} and $\mathcal{N}$, so $S = \sum_{j=1}^{\infty}(-1)^{j+1}\frac{\left(Q^{-1}\mathcal{N}\right)^j}{j}$ commutes with $\alpha I + \beta J$. Therefore, $M = e^B$ with $B = \alpha I + \beta J + S$, a real matrix. □

Lemma 7.6 *Let $D = \lambda I + N$ be an elementary Jordan λ-block with λ a negative real number. Then, the matrix $M = \begin{bmatrix} D & O \\ O & D \end{bmatrix}$ has a real logarithm.*

Proof With $R = \lambda^{-1}N$ the matrix M can be written in the form

$$M = \lambda \begin{bmatrix} I+R & O \\ O & I+R \end{bmatrix}. \tag{7.4}$$

Since λ is negative, $\lambda = e^x$ with $x = \log|\lambda| + \pi i$. The identification of the matrix $\begin{bmatrix} 0 & 1 \\ -1 & 0 \end{bmatrix}$ with the complex number i suggests considering the real matrix $X = \log|\lambda| I + \pi J$. Then, using (7.3) we get

$$e^X = e^{\log|\lambda| I + \pi J} = e^{\log|\lambda| I} e^{\pi J} = |\lambda|\big(\cos\pi I + \sin\pi J\big) = \lambda I.$$

Let $\mathbf{S} = \begin{bmatrix} S & O \\ O & S \end{bmatrix}$ where $S = \Sigma_{j=1}^{m-1}(-1)^{j+1}\frac{R^j}{j}$. Notice that S is real because λ is real. In the Proof of Proposition 7.1 we have seen that $e^S = I + R$. Then from (7.4) we have

$$M = \lambda I \begin{bmatrix} e^S & O \\ O & e^S \end{bmatrix} = e^X e^{\mathbf{S}}.$$

We easily check that $\mathbf{S}$ commutes with J, hence also with X, and therefore we get $M = e^B$ with $B = X + \mathbf{S}$ a real matrix. Therefore, M has a real logarithm. □

Proposition 7.7 *Let A be a real invertible matrix. If for each negative eigenvalue λ of A the number of elementary Jordan λ-blocks of the same order is even, then A has a real logarithm.*

Proof The elementary Jordan λ-blocks with real positive eigenvalues λ have real logarithms (see Remark 7.2). By Lemma 7.5 the Jordan λ-blocks (7.2) coming from a pair of complex eigenvalues $\lambda, \overline{\lambda}$ give rise to real logarithms. By Lemma 7.6 a pair of elementary Jordan λ-blocks of same size coming from a real negative eigenvalue λ also has a real logarithm. Therefore, if the number of elementary Jordan λ-blocks of the same order with negative eigenvalue λ is even, then combining all the Jordan λ-blocks we get a real logarithm for A. □

Proposition 7.8 *Let M be an invertible matrix such that $M = X^2$, with X a real matrix. Then, M has a real logarithm.*

Proof The negative eigenvalues of M come from purely imaginary eigenvalues of X since for $\mu = -\omega^2$, $\omega > 0$ we have $M - \mu I = (X - i\omega I)(X + i\omega I)$.

Since X is real if $i\omega$ is an eigenvalue of X so is $-i\omega$ and these give a negative double eigenvalue for M. If $\lambda = i\omega$ is a simple eigenvalue of X the Jordan form of X contains the diagonal block $\begin{bmatrix} i\omega & 0 \\ 0 & -i\omega \end{bmatrix}$ and the Jordan form of M has two 1×1 blocks, $\begin{bmatrix} -\omega^2 & 0 \\ 0 & -\omega^2 \end{bmatrix}$. The same argument works if $\lambda = i\omega$ is an eigenvalue of multiplicity k and $X|_{\eta^\dagger(\lambda)}$ is diagonalizable since the Jordan form of X contains the diagonal block diag$[i\omega, \ldots, i\omega; -i\omega, \ldots, -i\omega]$; hence, the Jordan form of M contains $2k$ blocks, diag$[-\omega^2, \ldots, -\omega^2; -\omega^2, \ldots, -\omega^2]$, each of size one.

In the nondiagonalizable case, if $i\omega$ is a double eigenvalue of X, the Jordan form of X has two elementary Jordan 2-blocks

$$P^{-1}XP = \begin{bmatrix} i\omega & 0 & 0 & 0 \\ 1 & i\omega & 0 & 0 \\ 0 & 0 & -i\omega & 0 \\ 0 & 0 & 1 & -i\omega \end{bmatrix} \quad \text{and so} \quad P^{-1}MP = \begin{bmatrix} -\omega^2 & 0 & 0 & 0 \\ 2i\omega & -\omega^2 & 0 & 0 \\ 0 & 0 & -\omega^2 & 0 \\ 0 & 0 & -2i\omega & -\omega^2 \end{bmatrix},$$

which is a nondiagonalizable matrix. Therefore, the Jordan form of M contains two elementary Jordan 2-blocks

$$\begin{bmatrix} -\omega^2 & 0 & 0 & 0 \\ 1 & -\omega^2 & 0 & 0 \\ 0 & 0 & -\omega^2 & 0 \\ 0 & 0 & 1 & -\omega^2 \end{bmatrix}.$$

Proceeding in this way we conclude that if λ is a negative eigenvalue of M the number of elementary Jordan λ-blocks is even. Therefore, by Proposition 7.7, M has a real logarithm. □

Example 7.1 (Real Logarithms of Rotations) Let $R(\theta)$ be a rotation in the plane by the angle θ. Then, $R(\theta) = R(\theta/2)^2$; hence, by the above proposition $R(\theta)$ has a real logarithm

Example 7.2 (Real Logarithm of AA^T) For any nonsingular real matrix A, the matrix AA^T has a real logarithm.

Since the real matrix AA^T is symmetric, we know from Linear Algebra that there is a real orthogonal matrix P such that $P^{-1}AA^TP = \text{diag}[\lambda_1, \ldots, \lambda_n]$. As A is nonsingular, the matrix AA^T is positive definite so the eigenvalues $\lambda_1, \ldots, \lambda_n$ are all positive. Taking

$$B = \text{diag}[\sqrt{\lambda_1}, \ldots, \sqrt{\lambda_n}]$$

we have $AA^T = PB^2P^{-1} = (PBP^{-1})^2$. Therefore, by Proposition 7.8 the matrix AA^T has a real logarithm.

Proposition 7.9 *Let $\lambda_1, \ldots, \lambda_k$ be the distinct eigenvalues of the real or complex square matrix A with multiplicities $m_1, \ldots, m_k$ respectively. Then, the eigenvalues of e^A are $e^{\lambda_1}, \ldots, e^{\lambda_k}$ with the same multiplicities and the eigenvectors of A and e^A are the same.*

Proof In the case when the $n \times n$ matrix A is diagonalizable, this is clear because a basis of $\mathbf{C}^n$ formed by eigenvectors of A is also a basis of $\mathbf{C}^n$ formed by eigenvectors of e^A. Restricting A to an A-invariant subspace of $\mathbf{C}^n$, where A is diagonalizable, we have the same conclusion for this restriction of A. So we assume that all the Jordan blocks of A are nondiagonalizable.

Let $\lambda_1, \ldots, \lambda_k$ be the distinct eigenvalues of the real or complex $n \times n$ matrix A with multiplicities $m_1, \ldots, m_k$, respectively, so $m_1 + \cdots + m_k = n$. There exists a basis $\mathcal{B}$ of $\mathbf{C}^n$ such that

$$[A]_{\mathcal{B}} = \begin{bmatrix} A_1 & O & \ldots & O \\ O & A_2 & \ldots & O \\ O & O & \ldots & A_k \end{bmatrix},$$

where A_j is the Jordan matrix defined by λ_j, formed by elementary Jordan λ_j-blocks

$$A_j = \begin{bmatrix} B_1^{(j)} & \ldots & O \\ & \vdots & \\ O & \ldots & B_{l_j}^{(j)} \end{bmatrix},$$

with decreasing sizes $d_1^{(j)} \geq \cdots \geq d_{l_j}^{(j)}$, which add up to m_j.

For the elementary Jordan λ-block $B = \lambda I + N$, with $N = \begin{bmatrix} 0\,0\,0 \ldots \\ 1\,0\,0 \ldots \\ 0\,1\,0 \ldots \\ \vdots \end{bmatrix}$ the nilpotent matrix of order k, we have (elements above the diagonal are all zero)

$$e^B = e^\lambda \begin{bmatrix} 1 & & & \\ 1 & 1 & & \\ \frac{1}{2!} & 1 & 1 & \\ & & \cdots & \\ \frac{1}{(k-1)!} & \frac{1}{(k-2)!} & \cdots & 1\;1 \end{bmatrix}.$$

Therefore, e^B has the unique eigenvalue e^λ with algebraic multiplicity k and one-dimensional eigenspace generated by the k-dimensional vector $\mathbf{v} = (0, \ldots, 1)^T$, so e^λ has geometric multiplicity equal to one.

Now we have

$$e^A = \begin{bmatrix} e^{A_1} & \dots & O \\ & \vdots & \\ O & \dots & e^{A_k} \end{bmatrix}, \quad \text{with} \quad e^{A_j} = \begin{bmatrix} e^{B_1^{(j)}} & \dots & O \\ & \vdots & \\ O & \dots & e^{B_{l_j}^{(j)}} \end{bmatrix}.$$

The unique eigenvalue of e^{A_j} is e^{λ_j} with algebraic multiplicity m_j and geometric multiplicity l_j, the number of elementary Jordan λ_j-blocks in A_j.

Therefore, the eigenvalues of e^A are $e^{\lambda_1}, \dots, e^{\lambda_k}$ with algebraic multiplicities $m_1, \dots, m_k$ respectively and geometric multiplicities $l_1, \dots, l_k$. There are no other eigenvalues of e^A.

If $\mathbf{v}_s^{(j)} = (0, \dots, 1)^T \in \mathbf{R}^{d_s^{(j)}}$ is an eigenvector of $B_s^{(j)}$, then the vectors in $\mathbf{R}^{m_j}$

$$\mathbf{x}_1^{(j)} = (\mathbf{v}_1^{(j)}, \mathbf{0}, \dots, \mathbf{0}), \quad \mathbf{x}_2^{(j)} = (\mathbf{0}, \mathbf{v}_2^{(j)}, \dots, \mathbf{0}), \quad \dots, \quad \mathbf{x}_{l_j}^{(j)} = (\mathbf{0}, \mathbf{0}, \dots, \mathbf{v}_{l_j}^{(j)})$$

are generators for the eigenspaces of both A_j and e^{A_j}.

Now we consider the following $l = l_1 + l_2 + \dots + l_k$ vectors of $\mathbf{R}^n$

$$\begin{aligned} &\mathbf{z}_1^{(1)}=(\mathbf{x}_1^{(1)}, \mathbf{0}, \dots, \mathbf{0}), \quad \mathbf{z}_2^{(1)} = (\mathbf{x}_2^{(1)}, \mathbf{0}, \dots, \mathbf{0}), \quad \dots, \quad \mathbf{z}_{l_1}^{(1)} = (\mathbf{x}_{l_1}^{(1)}, \mathbf{0}, \dots, \mathbf{0}), \\ &\mathbf{z}_1^{(2)}=(\mathbf{0}, \mathbf{x}_1^{(2)}, \dots, \mathbf{0}), \quad \mathbf{z}_2^{(2)} = (\mathbf{0}, \mathbf{x}_2^{(2)}, \dots, \mathbf{0}), \quad \dots, \quad \mathbf{z}_{l_1}^{(2)} = (\mathbf{0}, \mathbf{x}_{l_2}^{(2)}, \dots, , \mathbf{0}), \\ &\vdots \\ &\mathbf{z}_1^{(k)}=(\mathbf{0}, \mathbf{0}, \dots, \mathbf{x}_1^{(k)}), \quad \mathbf{z}_2^{(k)} = (\mathbf{0}, \mathbf{0}, \dots, \mathbf{x}_2^{(k)}), \quad \dots, \quad \mathbf{z}_{l_k}^{(k)} = (\mathbf{0}, \mathbf{0}, \dots, \mathbf{x}_{l_k}^{(k)}). \end{aligned} \tag{7.5}$$

The vectors in the first row are generators of the eigenspace of e^A for the eigenvalue e^{λ_1}, those of the second row are the generators of the eigenspace of e^{λ_2} and the vectors of the last row are generators of the eigenspace of the eigenvalue e^{λ_k}.

Since the vectors (7.5) are also generators of the eigenspaces of A for the eigenvalues $\lambda_1, \lambda_2, \dots, \lambda_k$, the proposition is proved. □

7.4 Periodic Linear Systems

Consider a continuous τ-periodic real linear system

$$\dot{\mathbf{x}} = A(t)\mathbf{x}, \quad \mathbf{x} \in \mathbf{R}^m, \qquad A(t+\tau) = A(t) \tag{7.6}$$

and let $X(t)$ be the matrizant of this system, that is, the fundamental matrix solution with $X(0) = I$, where I is the $m \times m$ identity matrix; the columns of $X(t)$ are the solutions of (7.6) having the vectors of the canonical basis of $\mathbf{R}^m$ as initial conditions. The solution of (7.6) with the initial condition $\mathbf{x}_0$ is given by $\phi(t, \mathbf{x}_0) = X(t)\mathbf{x}_0$.

Theorem 7.10 (Floquet) *There exist $m \times m$ matrices, B and $Q(t)$, with B constant and $Q(t)$ τ-periodic such that*

$$X(t) = Q(t)e^{tB}. \tag{7.7}$$

Proof If $\mathbf{x}(t)$ is a solution of (7.6), then the translation by τ, $\mathbf{u}(t) = \mathbf{x}(t+\tau)$, is also a solution, so each column of the matrix $X(t+\tau)$ is a solution of (7.6); hence, it is a linear combination of the columns of $X(t)$. This means that there exists a constant matrix C such that $X(t+\tau) = X(t)C$. As $X(0) = I$ this equation can be written as

$$X(t+\tau) = X(t)X(\tau). \tag{7.8}$$

But $X(\tau)$ is invertible so by Proposition 7.3, $X(\tau) = e^{\tau B}$ for some complex matrix B.

Consider now the matrix $Q(t) = X(t)e^{-tB}$. Then, $X(t) = Q(t)e^{tB}$ and to conclude the proof of the theorem it suffices to prove that $Q(t)$ is τ-periodic, which is the case since, using (7.8)

$$Q(t+\tau) = X(t+\tau)e^{-(t+\tau)B} = X(t)X(\tau)e^{-\tau B}e^{-tB} = X(t)e^{-tB} = Q(t).$$

□

Remark 7.11

(1) We call the equality (7.7) a Floquet decomposition of the matrizant of the τ-periodic linear system $\dot{\mathbf{x}} = A(t)\mathbf{x}$.
(2) By (7.8) $X(2\tau) = X(\tau)^2$ and as $X(\tau)$ is a real matrix, $X(2\tau)$ has a real logarithm by Proposition 7.8. Writing it as $2\tau B$, we have $X(2\tau) = \mathrm{e}^{2\tau B}$, with B real. Since by (7.8) $X(t+2\tau) = X(t)X(2\tau)$ we easily check that the matrix $Q(t) = X(t)\mathrm{e}^{-tB}$ is 2τ-periodic.
    Therefore, in (7.7) we can take B real and $Q(t)$ a 2τ-periodic real matrix.
(3) Consider the Floquet decompositions $X(t) = Q(t)e^{tB}$ and $X(t) = \tilde{Q}(t)e^{t\tilde{B}}$ with $Q(t)$ 2τ-periodic, B real and $\tilde{Q}(t)$ τ-periodic. Then, $X(2\tau) = \mathrm{e}^{2\tau B}$ and $X(\tau) = \mathrm{e}^{\tau\tilde{B}}$. Since $X(2\tau) = X(\tau)^2$ we have $\mathrm{e}^{2\tau B} = \mathrm{e}^{2\tau\tilde{B}}$; hence, by Remark 7.4 we have

$$2\tau\tilde{B} = 2\tau B + \mathrm{diag}[K_1, \ldots, K_r],$$

where $K_j = 2\pi i k_j I$, k_j an integer and I the identity matrix of some order n_j.

Item (2) of the Remark deserves special mention, which we state as a proposition.

Proposition 7.12 *In its real form the Floquet decomposition of the matrizant is unique.*

Proof Indeed, let $X(t)$ be the matrizant of the τ-periodic real linear system $\dot{\mathbf{x}} = A(t)\mathbf{x}$ and let $X(t) = Q(t)\mathrm{e}^{tB} = \tilde{Q}(t)\mathrm{e}^{t\tilde{B}}$ be Floquet representations of $X(t)$ with B, $\tilde{B}$ real and $Q(t)$, $\tilde{Q}(t)$ periodic of period 2τ. Since $X(0) = I$ we have $Q(0) = \tilde{Q}(0) = I$, hence $Q(2\tau) = \tilde{Q}(2\tau) = I$. Therefore, $\mathrm{e}^{2\tau B} = \mathrm{e}^{2\tau\tilde{B}} = X(2\tau)$. As the real logarithm of a real matrix is unique (see Remark 7.4), we have $\tilde{B} = B$; hence, $\tilde{Q}(t) = Q(t)$. □

Definition 7.13 We call the Floquet representation $X(t) = Q(t)\mathrm{e}^{tB}$ with $Q(t)$ 2τ-periodic and B real the Floquet real form of the matrizant.

Proposition 7.14 *Given an $m \times m$ continuous τ-periodic matrix $A(t)$, there exists an invertible continuous τ-periodic matrix $Q(t)$ such that the change of coordinates $\mathbf{x} = Q(t)\mathbf{y}$ transforms the equation $\dot{\mathbf{x}} = A(t)\mathbf{x}$ into the equation $\dot{\mathbf{y}} = B\mathbf{y}$, with B constant.*

Proof Let $X(t) = Q(t)e^{tB}$ be a Floquet decomposition of the matrizant of $\dot{\mathbf{x}} = A(t)\mathbf{x}$. From $Q(t) = X(t)\mathrm{e}^{-tB}$ we get

$$\dot{Q} = \dot{X}\mathrm{e}^{-tB} - X\mathrm{e}^{-tB}B = AX\mathrm{e}^{-tB} - X\mathrm{e}^{-tB}B = AQ - QB.$$

Now differentiating $\mathbf{x} = Q(t)\mathbf{y}$ we get $\dot{\mathbf{x}} = \dot{Q}\mathbf{y} + Q\dot{\mathbf{y}} = AQ\mathbf{y} - QB\mathbf{y} + Q\dot{\mathbf{y}}$; therefore, $\dot{\mathbf{x}} = A\mathbf{x} + Q(\dot{\mathbf{y}} - B\mathbf{y})$.

Clearly, $\dot{\mathbf{x}} = A(t)\mathbf{x}$ if, and only if, $\dot{\mathbf{y}} = B\mathbf{y}$. □

The linear change of variables $\mathbf{x} = Q(t)\mathbf{y}$ and the system $\dot{\mathbf{y}} = B\mathbf{y}$ are real if we take $Q(t)$ as a 2τ-periodic matrix.

Definition 7.15 The matrizant of the τ periodic system $\dot{\mathbf{x}} = A(t)\mathbf{x}$ when evaluated at the chosen period, $X(\tau)$ or $X(2\tau)$, is called the monodromy matrix of the system. The eigenvalues of B are called the characteristic exponents of the system $\dot{\mathbf{x}} = A(t)\mathbf{x}$ and those of the monodromy matrix are called its multipliers.

The relation between the characteristic exponents λ and the multipliers ρ is $\rho = \mathrm{e}^{\tau\lambda}$ or $\rho = \mathrm{e}^{2\tau\lambda}$, depending on how we choose the period for $Q(t)$.

Let ρ be an eigenvalue of a symplectic matrix D. By the Lyapunov–Poincaré theorem, Proposition 4.24, the number ρ^{-1} is also an eigenvalue of D.

Consider the generalized eigenspaces $\eta^{\dagger}(\rho)$, $\eta^{\dagger}(\rho^{-1})$ defined in (4.18). Each of them is D-invariant, so the direct sum $U = \eta^{\dagger}(\rho) \oplus \eta^{\dagger}(\rho^{-1})$ is also D-invariant.

Proposition 7.16 *If the eigenvalue ρ has multiplicity k, then U is a symplectic subspace and:*

(1) *There is a symplectic basis of U in which $[D|_U] = \begin{bmatrix} J_\rho & O \\ O & J_\rho^{-T} \end{bmatrix}$, where J_ρ is the Jordan form of $D|_{\eta^{\dagger}(\rho)}$.*

(2) *If $D|_U$ is diagonalizable, then there is a symplectic basis of U in which the matrix has the $2k \times 2k$ diagonal form $[D|_U] = diag[\rho, \dots, \rho, \rho^{-1}, \dots, \rho^{-1}]$.*

Proof There is a D-invariant decomposition $\mathbf{C}^{2n} = U \oplus V$ such that all the eigenvalues of $D|_V$ are different from ρ and ρ^{-1}. By Lemma 4.37 the subspaces U and V are symplectically orthogonal; hence, by Proposition 4.7(a) U is a symplectic subspace (and V also).

The decomposition $U = \eta^{\dagger}(\rho) \oplus \eta^{\dagger}(\rho^{-1})$ is D-invariant and it is a Lagrangian splitting because if $\mathbf{u}, \mathbf{v} \in \eta^{\dagger}(\rho)$, then $\{\mathbf{u}, \mathbf{v}\} = 0$ by Lemma 4.37. Same argument for the subspace $\eta^{\dagger}(\rho^{-1})$, which, by the way, has the same dimension of the first.

Take a Jordan basis for $D|_{\eta^{\dagger}(\rho)}$ and let J_ρ be the corresponding Jordan form.

Complete this basis of $\eta^{\dagger}(\rho)$ with a basis of $\eta^{\dagger}(\rho^{-1})$ to form a symplectic basis of U, according to Proposition 4.21. Then, by Proposition 4.22, the matrix of D on this basis is $[D|_U] = \begin{bmatrix} J_\rho & O \\ O & J_\rho^{-T} \end{bmatrix}$. This proves (1). Item (2) is a consequence of (1). □

Now we state the symplectic version of Proposition 7.3.

Proposition 7.17 *If the constant nonsingular matrix D is symplectic there is a Hamiltonian matrix B such that $D = e^B$.*

Proof Since D is symplectic its distinct eigenvalues are $\rho_1, \dots, \rho_m; \rho_1^{-1}, \dots, \rho_m^{-1}$. By Proposition 3.10 and Proposition 7.16 there exists a symplectic basis such that the matrix of D on this basis is $\mathcal{D} = \begin{bmatrix} X & O \\ O & X^{-T} \end{bmatrix}$, where X is the Jordan form of D restricted to the invariant subspace corresponding to the first m eigenvalues. In block form $X = \mathrm{diag}[\mathcal{J}_1, \mathcal{J}_2, \dots, \mathcal{J}_m]$, where $\mathcal{J}_k = [\rho_k]$ is a diagonal matrix formed with the eigenvalue ρ_k if $D|_{\eta^{\dagger}(\rho_k)}$ is diagonalizable and $\mathcal{J}_k = \mathrm{diag}[J^{(1,k)}, \dots, J^{(p,k)}]$ if $D|_{\eta^{\dagger}(\rho_k)}$ is nondiagonalizable; each $J^{(l,k)}$ is an elementary Jordan block for the eigenvalue ρ_k.

By Proposition 7.1, we have $J^{(l,k)} = \mathrm{e}^{B_{l,k}}$, so $\mathcal{J}_k = \mathrm{e}^{B_k}$, where $B_k = \begin{bmatrix} B^{(1,k)} & & \\ & \ddots & \\ & & B^{(p,k)} \end{bmatrix}$. Also, $\mathcal{J}_k = \mathrm{e}^{B_k}$ for the diagonal matrix $\mathcal{J}_k = [\rho_k]$.

Therefore, $X = \mathrm{e}^B$ with $B = \mathrm{diag}[B_1, \dots, B_m]$. Consequently, $\mathcal{D} = \mathrm{e}^{\mathcal{B}}$, where $\mathcal{B} = \begin{bmatrix} B & O \\ O & -B^T \end{bmatrix}$, a Hamiltonian matrix. □

Example 7.3 (Real Hamiltonian Logarithm) If A is a real symplectic matrix, then the matrix AA^T has a real Hamiltonian logarithm.

Indeed, by Example 7.2, there is a real matrix C such that $AA^T = \mathrm{e}^C$. By Proposition 7.17 $AA^T = \mathrm{e}^B$ for some Hamiltonian matrix B. Since C is real, we have $C = B$ by Remark 7.4. Therefore, C is a real Hamiltonian matrix, that is, AA^T has a real Hamiltonian logarithm.

Theorem 7.18 (Floquet–Lyapunov) *For a linear Hamiltonian system* $\dot{\mathbf{x}} = A(t)\mathbf{x}$ *with a continuous* τ*-periodic matrix* $A(t)$*, we can have the Floquet decomposition of the matrizant* $X(t) = Q(t)e^{tB}$ *with* $Q(t)$ *symplectic and* B *Hamiltonian.*

Proof Since $X(t)$ is symplectic the equality $X(t+\tau) = X(t)C$ shows that the constant matrix C is symplectic. By the previous proposition, we can take B as a Hamiltonian matrix, so the matrix e^{tB} is symplectic for each t; hence, $Q(t) = X(t)\mathrm{e}^{-tB}$ is symplectic. □

Corollary 7.18.1 *The linear Hamiltonian system* $\dot{\mathbf{x}} = A(t)\mathbf{x}$ *with a continuous* τ*-periodic matrix* $A(t)$ *can be transformed into a constant linear Hamiltonian system* $\dot{\mathbf{y}} = B\mathbf{y}$ *by means of a linear symplectic change of coordinates* $\mathbf{x} = Q(t)\mathbf{y}$.

Proof The proof is that of Proposition 7.14 with the observation that we can take $Q(t)$ as a symplectic matrix, according to Theorem 7.18. □

Remark 7.19 In Theorem 7.18 and the corollary we can have B a real matrix if we take 2τ as the period of $Q(t)$.

7.5 Stable Linear Systems

In this section we consider the general properties of stable linear systems.

Definition 7.20 We say that the linear system (7.6) is stable if the origin is a stable equilibrium in both time directions, that is, given $\epsilon > 0$ there is a $\delta > 0$ such that if $\|\mathbf{x}\| < \delta$, the solution $\phi(t, \mathbf{x})$ remains inside the ball $B_\epsilon(\mathbf{x})$, for all time t.

Example 7.4 (Positively Stable, Nonstable System) Let A be a constant $n \times n$ real matrix with distinct eigenvalues $\lambda_1, \ldots, \lambda_n$, all having negative real parts $\alpha_1, \ldots, \alpha_n$. Then the conditions in this definition hold for $t \geq 0$, but the system $\dot{\mathbf{x}} = A\mathbf{x}$ is not stable.

Since the eigenvalues are distinct the matrix A is diagonalizable, so there is an invertible matrix P such that $P^{-1}AP = D$, where $D = \mathrm{diag}[\lambda_1, \ldots, \lambda_n]$. Setting $\mathbf{x} = P\mathbf{y}$ we have $\dot{\mathbf{y}} = D\mathbf{y}$; hence, $\mathbf{y}(t) = \sum_{j=1}^{n} c_j \mathrm{e}^{\lambda_j t}\mathbf{y}_j(0)$. Since $|\mathrm{e}^{\lambda_j t}| = \mathrm{e}^{\alpha_j t}$ we get $\|\mathbf{y}(t)\| \leq \sum_{j=1}^{n} |c_j| \mathrm{e}^{\alpha_j t}\|\mathbf{y}_j(0)\|$. Since $\alpha_j < 0$, for all j, given $\epsilon > 0$ we can take $\delta > 0$ sufficiently small to guarantee that if $\|\mathbf{y}(0)\| < \delta$ then, $\|\mathbf{y}(t)\| < \epsilon$, for all $t \geq 0$, hence the condition in Definition 7.20 holds for $t \geq 0$, but not for all t because $\alpha t \to +\infty$ when $t \to -\infty$, so $\mathbf{y}(t)$ can be unbounded.

We begin with the following result.

Proposition 7.21 *The continuous periodic linear system* (7.6) *is stable if, and only if, all its solutions are bounded.*

Proof Let $\phi(t, \mathbf{x})$ be the solution of (7.6) with an initial position $\mathbf{x}$ for $t = 0$. If the system is stable, given $\epsilon > 0$ there is a $\delta > 0$ such that if $\|\mathbf{x}\| < \delta$, then $\|\phi(t, \mathbf{x})\| < \epsilon$, for all t. Now let $\phi(t, \mathbf{x})$ be an arbitrary solution. Fix a positive real number b such that $b\|\mathbf{x}\| < \delta$. Then, $b\phi(t, \mathbf{x})$ is a solution of (7.6), which starts in the ball $B_\delta(\mathbf{0})$; hence, we have $\|\phi(t, \mathbf{x})\| < \frac{\epsilon}{b}$ for all t, so $\phi(t, \mathbf{x})$ is a bounded solution.

To prove the converse we first assume that A is a constant matrix. Let $\lambda = \alpha + i\beta$ be an eigenvalue of A and $\mathbf{v}$ a corresponding eigenvector. Then, $\mathbf{x}(t) = e^{\lambda t}\mathbf{v}$ is a solution of (7.6) and $\|\mathbf{x}(t)\| = e^{\alpha t}\|\mathbf{v}\|$. As the solutions are all bounded α must be zero. Therefore, if the solutions are bounded all the eigenvalues of A are purely imaginary numbers or zero. If the matrix A is not diagonalizable, there are solutions of the form t^r or $t^r e^{i\beta_k t}$, where r is some positive integer. But this implies the existence of unbounded solutions, contradicting our hypotheses. Therefore, A is diagonalizable. Let $\mathbf{v}_1, \ldots, \mathbf{v}_n$ be a basis of eigenvectors. The corresponding eigenvalues are purely imaginary $i\beta_1, \ldots, i\beta_n$, maybe with some β_j equal to zero and $\phi(t, \mathbf{v}_k) = e^{i\beta_k t}\mathbf{v}_k$. So, for any $\mathbf{x} = \sum c_k \mathbf{v}_k$ we have $\phi(t, \mathbf{x}) = \sum c_k e^{i\beta_k t}\mathbf{v}_k$; hence, $\|\phi(t, \mathbf{x})\| \le \sum |c_k|\|\mathbf{v}_k\|$. Now $\|\mathbf{x}\| \le \sum |c_k|\|\mathbf{v}_k\|$. Therefore, if for a given $\epsilon > 0$ we take $\delta = \epsilon$, then $\|\mathbf{x}\| < \delta$ implies $\|\phi(t, \mathbf{x})\| < \epsilon$, for all t, hence $\dot{\mathbf{x}} = A\mathbf{x}$ is stable.

The case of a periodic matrix $A(t)$ reduces to the constant case by using Theorem 7.10. Indeed, the change of coordinates $\mathbf{x} = Q(t)\mathbf{y}$ transforms the equation $\dot{\mathbf{x}} = A(t)\mathbf{x}$ into the equation $\dot{\mathbf{y}} = B\mathbf{y}$, with B constant. As $\|Q(t)^{-1}\|$ is continuous on the compact interval $0 \le t \le \tau$, it is bounded there and since it is τ-periodic it is bounded for all t. Hence, the boundedness of the solutions of (7.6) implies that the solutions $\mathbf{y}(t)$ of the constant system are bounded; hence, this system is stable and as $\|Q(t)\|$ is bounded for all t a simple argument shows that $\dot{\mathbf{x}} = A(t)\mathbf{x}$ is stable. □

From the arguments in the above proof, we immediately get the following result.

Proposition 7.22 *Let A be a constant square matrix. Then the system $\dot{\mathbf{x}} = A\mathbf{x}$ is stable if, and only if, all the eigenvalues of A are purely imaginary numbers, including 0, and A is diagonalizable over the complex numbers.*

We close this section with the following two results.

Proposition 7.23 *Let $X(t)$ be the matrizant of the τ-periodic linear system* (7.6). *Then, the system $\dot{\mathbf{x}} = A(t)\mathbf{x}$ is stable if, and only if, the sequence $X(\tau)^k$ is bounded.*

Proof Using (7.8) we conclude that $X(k\tau + t) = X(t)X(\tau)^k$. Since $\|X(t)\|$ is bounded on the compact interval $[0, \tau]$ we see that $\|X(t)\|$ is bounded for all t if, and only if, $\|X(\tau)^k\|$ is a bounded sequence. So the solutions $\mathbf{x}(t) = X(t)\boldsymbol{\xi}$ are bounded, that is, $\dot{\mathbf{x}} = A(t)\mathbf{x}$ is stable if and only if $\|X(\tau)^k\|$ is bounded. □

Proposition 7.24 *Let $X(t) = Q(t)e^{tB}$ be the Floquet decomposition for the continuous τ-periodic linear system $\dot{\mathbf{x}} = A(t)\mathbf{x}$. Then the system $\dot{\mathbf{x}} = A(t)\mathbf{x}$ is stable if, and only if, the system $\dot{\mathbf{y}} = B\mathbf{y}$ is stable.*

Proof Let $\mathbf{x}(t) = Q(t)\mathbf{y}(t)$. Since $\|Q(t)\|$ and $\|Q(t)^{-1}\|$ are bounded for all real t, we have that $\mathbf{x}(t)$ is bounded if, and only if, $\mathbf{y}(t)$ is bounded and the result follows from Proposition 7.21. □

Example 7.5 (A Nonstable System) Consider the linear system $\dot{\mathbf{x}} = A\mathbf{x}$, where A is the matrix

$$A = \begin{bmatrix} 0 & 1 & 1 & 0 \\ 0 & 0 & 0 & 1 \\ 1 & 0 & 0 & 0 \\ 0 & -1 & -1 & 0 \end{bmatrix}.$$

This system is not stable. Indeed, the matrix A has $\lambda = 0$ as an eigenvalue of multiplicity four. Looking for the eigenvectors of $\lambda = 0$ we see that they form a one-dimensional subspace, so this matrix is not diagonalizable. Therefore, by Proposition 7.22 the system is not stable.

7.6 Strongly Stable Linear Systems

In the study of stability of equilibria of Hamiltonian systems we lose strength when we pass from the study of stability to that of linear stability. But on the other hand we gain freedom as we can work with systems with a large number of variables. More than that we can strengthen the concept of stability by not only requiring a specific linear system to be stable but all nearby linear systems to also be stable. This is important because experiments and observations never give the system describing a real phenomenon with exactness. So if the linear system is stable in this stronger sense, this gives robustness to the consequences derived from the equations describing the phenomenon.

Definition 7.25 We say that a τ-periodic Hamiltonian linear system, $\dot{\mathbf{x}} = A(t)\mathbf{x}$, is strongly stable if it is stable and there exists an $\epsilon > 0$ such that any τ-periodic Hamiltonian linear system $\dot{\mathbf{x}} = \tilde{A}(t)\mathbf{x}$ with $||\tilde{A} - A|| < \epsilon$ is stable.

In this definition if the matrix of the system is constant we require the nearby matrix $\tilde{A}$ to also be constant.

We observe that the concept of strongly stability does not apply to general linear systems since in any neighborhood of a stable linear system there is an unstable linear system.

Indeed, if $\dot{\mathbf{x}} = A\mathbf{x}$ is a stable real constant system, for any $\epsilon > 0$, the matrix $A_\epsilon = A + \epsilon I$ is real, ϵ-close to A and it is unstable because there are eigenvalues with a real part ϵ. Therefore, A is not strongly stable. Now let $\dot{\mathbf{x}} = A(t)\mathbf{x}$ be a stable τ-periodic real linear system. Then, all of its solutions are bounded.

Let $X(t) = Q(t)e^{tB}$ be the 2τ-Floquet decomposition of $\dot{\mathbf{x}} = A(t)\mathbf{x}$, so B is real. As $A(t)$ is stable the matrix B is stable. Let B_ϵ be a real unstable matrix ϵ-close to B.

Let $X_\epsilon(t) = Q(t)e^{tB_\epsilon}$ and take $A_\epsilon(t) = \dot{X}_\epsilon(t)X_\epsilon(t)^{-1}$. Then, $X_\epsilon(t) = Q(t)e^{tB_\epsilon}$ is the Floquet decomposition of $\dot{\mathbf{x}} = A_\epsilon(t)\mathbf{x}$. Since B_ϵ is unstable, so is A_ϵ.

Since $\dot{X} = AX$ and $\dot{X}_\epsilon = A_\epsilon X_\epsilon$, we have

$$\dot{Q}e^{tB} + BQe^{tB} = AQe^{tB} \quad \text{and} \quad \dot{Q}e^{tB_\epsilon} + B_\epsilon Qe^{tB_\epsilon} = A_\epsilon Qe^{tB_\epsilon};$$

hence, $\dot{Q} + BQ = AQ$ and $\dot{Q} + B_\epsilon Q = A_\epsilon Q$. Therefore, $B - B_\epsilon = A - A_\epsilon$. This equality shows that A_ϵ is a real matrix ϵ-close to A and τ-periodic. Since A_ϵ is unstable, A is not strongly stable.

Proposition 7.26 *Let A be a constant Hamiltonian matrix. If the Hamiltonian H of the system $\dot{\mathbf{x}} = A\mathbf{x}$ is positive (or negative) definite, then this linear system is strongly stable.*

Proof Let $S = -JA$ be the $2n \times 2n$ symmetric matrix defining H. Then, S is positive (or negative) definite; hence, all symmetric matrices $\tilde{S}$ sufficiently near S are also positive (or negative) definite.[1] Therefore, the Hamiltonian $\tilde{H}(\mathbf{x}) = \frac{1}{2}\mathbf{x}^T\tilde{S}\mathbf{x}$ is a positive (or negative) definite first integral of the autonomous linear system $\dot{\mathbf{x}} = \tilde{A}\mathbf{x}$, $\tilde{A} = J\tilde{S}$, which is therefore stable by the Dirichlet theorem; hence, $\dot{\mathbf{x}} = A\mathbf{x}$ is strongly stable. □

Example 7.6 (A Strongly Stable Potential System) Consider the second-order differential equation $\ddot{\mathbf{x}} = -\nabla V(\mathbf{x})$, $\mathbf{x} \in \mathbf{R}^n$. Assume that the point $\mathbf{x}_*$ is a strict minimum of the potential V, so $\nabla V(\mathbf{x}_*) = 0$ and $D^2V(\mathbf{x}_*)$ is a positive definite matrix.

Setting $\mathbf{y} = \dot{\mathbf{x}}$, the second-order equation can be written as the first-order system $\dot{\mathbf{x}} = \mathbf{y}$, $\dot{\mathbf{y}} = -\nabla V(\mathbf{x})$, which is a Hamiltonian system of equation with Hamiltonian function given by

$$H(\mathbf{x}, \mathbf{y}) = \frac{1}{2}\|\mathbf{y}\|^2 + V(\mathbf{x}).$$

Let $\mathbf{z} = (\mathbf{x}, \mathbf{y})$. Then, $\mathbf{z}_* = (\mathbf{x}_*, 0)$ is an equilibrium of this Hamiltonian system and setting $\boldsymbol{\zeta} = \mathbf{z} - \mathbf{z}_*$ the linearized system about the equilibrium is given by

$$\dot{\zeta} = A\zeta, \quad \text{where} \quad A = JS,$$

with S the symmetric matrix $S = \begin{bmatrix} D^2V(\mathbf{x}_*) & O \\ O & I \end{bmatrix}$. Since the matrix $D^2V(\mathbf{x}_*)$ is positive definite so is the matrix S; hence, the Hamiltonian $H = \frac{1}{2}\zeta^T S\zeta$ of the linearized system is positive definite. Therefore, by Proposition 7.26, the linear system is strongly stable.

Proposition 7.27 *If the constant linear system $\dot{\mathbf{x}} = A\mathbf{x}$, $\mathbf{x} \in \mathbf{R}^m$, is strongly stable, then zero is not an eigenvalue of the matrix A.*

---

[1] We have $\langle x, \tilde{S}x\rangle = \langle x, Sx\rangle + \langle x, (\tilde{S} - S)x\rangle$ and $\langle x, Sx\rangle \geq \delta > 0$ for all $|x| = 1$. By the Cauchy–Schwarz inequality we have

$$-\,||\tilde{S} - S|||x|^2 \leq \langle x, (\tilde{S} - S)x\rangle \leq ||\tilde{S} - S|||x|^2,$$

so we get for all $|x| = 1$, $\langle x, \tilde{S}x\rangle \geq \delta - ||\tilde{S} - S|| > 0$ if $||\tilde{S} - S|| < \delta$; hence, $\tilde{S}$ is positive definite.

Proof Assume that 0 is an eigenvalue and complete a corresponding unit eigenvector $\mathbf{v}$ to form a basis of the space, $\mathbf{v}_1 = \mathbf{v}, \mathbf{v}_2, \ldots, \mathbf{v}_m$. For any $\epsilon > 0$ take $\tilde{A}$ such that $\tilde{A}\mathbf{v}_1 = \frac{1}{2}\epsilon\mathbf{v}_1$ and $\tilde{A}\mathbf{v}_j = A\mathbf{v}_j$, for $j \geq 2$. Then, $\|\tilde{A} - A\| = \frac{1}{2}\epsilon < \epsilon$. Since $\frac{1}{2}\epsilon$ is a real eigenvalue of $\tilde{A}$, $\dot{\mathbf{x}} = \tilde{A}\mathbf{x}$ is not stable by Proposition 7.21. Therefore, $\dot{\mathbf{x}} = A\mathbf{x}$ is not strongly stable, contradicting the hypothesis.

If the linear system is Hamiltonian we must take the nearby matrix $\tilde{A}$ also Hamiltonian and this requires additional care in the argument. Since A is Hamiltonian the eigenvalue zero has multiplicity of at least two. Since the system is stable A is diagonalizable. Take linearly independent unit vectors $\mathbf{v}_1$ and $\mathbf{v}_{n+1}$ of A corresponding to the zero eigenvalue and complete these vectors to form a symplectic basis of $\mathbf{R}^{2n}$, $\mathbf{v}_1 = \mathbf{v}, \mathbf{v}_2, \ldots, \mathbf{v}_{2n}$. Let $[A]$ be the matrix of the operator A on this basis. Since $A\mathbf{v}_1 = 0$ and $A\mathbf{v}_{n+1} = 0$, the first and $(n+1)$th columns of $[A]$ are zero. As $[A]$ is Hamiltonian its first and $(n+1)$th rows are also zero. Define $\tilde{A}$ by $\tilde{A}\mathbf{v}_1 = \frac{1}{2}\epsilon\mathbf{v}_1$, $\tilde{A}\mathbf{v}_{n+1} = -\frac{1}{2}\epsilon\mathbf{v}_{n+1}$, and $\tilde{A}\mathbf{v}_j = A\mathbf{v}_j$ if $j \neq 1, n+1$. Then $\tilde{A}$ is Hamiltonian and $\|\tilde{A} - A\| = \frac{1}{2}\epsilon < \epsilon$. Since $\tilde{A}$ has real eigenvalues the system $\dot{\mathbf{x}} = \tilde{A}\mathbf{x}$ is not stable, so the system $\dot{\mathbf{x}} = A\mathbf{x}$ is not strongly stable. □

Now, consider a stable linear Hamiltonian system $\dot{\mathbf{x}} = A\mathbf{x}$ with a real constant matrix A. Let $\pm i\beta_1, \ldots, \pm i\beta_m$ be the distinct, purely imaginary eigenvalues of A. For each $j = 1, \ldots, m$ consider the Hamiltonian H_j defined in the symplectic space (see Sect. 4.3), $Y_j = \eta^\dagger(i\beta_j) \oplus \eta^\dagger(-i\beta_j)$ by the restriction of (the complex extension of) A to it.

We now state and prove the Krein–Gel'fand–Lidskii theorem in the case of linear Hamiltonian systems $\dot{\mathbf{x}} = A\mathbf{x}$ with constant matrix A as it appears in [41].

Theorem 7.28 (Krein–Gel'fand–Lidskii) *Let $\dot{\mathbf{x}} = A\mathbf{x}$ be a stable linear Hamiltonian system with a real constant matrix A. Then, it is strongly stable if, and only if, 0 is not an eigenvalue of A and each H_j is either positive or negative definite.*

Proof Suppose the system is strongly stable. Then, 0 is not an eigenvalue of A, by Proposition 7.27. Now consider the Hamiltonian H_j defined in Y_j. By Corollary 4.39.1 there is a real symplectic basis of Y_j such that in the coordinates relative to this basis the Hamiltonian assumes the form

$$H_j = \frac{1}{2}\beta_j \sum_{s=1}^{m_j} \delta_s (x_s^2 + y_s^2). \tag{7.9}$$

If H_j is indefinite there is pair of consecutive $\delta's$ with opposite signs, say $\delta_1 = +1$ and $\delta_2 = -1$. Now for $\epsilon > 0$ consider the Hamiltonian $H_j^{(\epsilon)} = H_j + H_\epsilon$, where $H_\epsilon = \epsilon y_1 y_2$. Let $A_j^{(\epsilon)}$ be the matrix of the linear system defined by $H_j^{(\epsilon)}$. Then, the eigenvalues of $A_j^{(\epsilon)}$ are those of A_j corresponding to the coordinates x_s, y_s with $s \geq 3$ and the roots of the characteristic equation of the Hamiltonian

$$H_\epsilon = \frac{1}{2}\beta_j(x_1^2 + y_1^2) - \frac{1}{2}\beta_j(x_2^2 + y_2^2) + \epsilon y_1 y_2,$$

that is, the roots of the equation $\lambda^4 + 2\beta_j\lambda^2 + \beta_j^2(\beta_j^2 + \epsilon^2) = 0$. Since $\Delta = \beta_j^4 - \beta_j^2(\beta_j^2 + \epsilon^2) = -\beta_j^2\epsilon^2 < 0$, for any $\epsilon \neq 0$ these roots have a nonzero real part, so the linear system defined by $A_j^{(\epsilon)}$ is unstable. Since $A_j^{(\epsilon)}$ is ϵ-close to A_j this proves that $\dot{\mathbf{x}} = A\mathbf{x}$ is not strongly stable, a contradiction. So each H_j is either positive definite or negative definite.

We now prove the converse. Let $2n$ be the size of A. Since 0 is not an eigenvalue of A, $\mathbf{C}^{2n} = Y_1 \oplus \ldots \oplus Y_m$. Each H_j in (7.9) is either positive definite or negative definite by hypothesis. Let S be the symmetric matrix of the Hamiltonian H of the linear system $\dot{\mathbf{x}} = A\mathbf{x}$ and let S_j be the restriction of (the extension of) S to Y_j. Then, if $\epsilon > 0$ is sufficiently small, for any symmetric real matrix $\tilde{S}$ ϵ-close to S, the symmetric operators $\tilde{S}_j$ induced by the restrictions of (the extension of) S to Y_j are, together with S_j, either positive definite or negative definite on Y_j. Given a Hamiltonian matrix $\tilde{A}$ ϵ-close to A, the symmetric matrix $\tilde{S} = -J\tilde{A}$ is ϵ-close to S; hence, the Hamiltonians $\tilde{H}_j$ defined by $\tilde{S}_j$ on Y_j are either positive definite or negative definite, so each linear system $\dot{\mathbf{x}} = \tilde{A}_j\mathbf{x}$ is stable. From this it follows that $\dot{\mathbf{z}} = \tilde{A}\mathbf{z}$ is stable. Therefore, the system $\dot{\mathbf{x}} = A\mathbf{x}$ is strongly stable. □

Now we consider the periodic case.

Let $X_0(t) = Q_0(t)e^{tB_0}$ be the Floquet decomposition for the τ-periodic linear Hamiltonian system $\dot{\mathbf{x}} = A_0(t)\mathbf{x}$. For each constant Hamiltonian matrix B consider the symplectic matrix $X(t) = Q_0(t)e^{tB}$. Then, $A(t) = \dot{X}(t)X(t)^{-1}$ is a Hamiltonian matrix.[2] We have

$$A(t) = A^*(t) + A_B(t), \tag{7.10}$$

with $A^*(t) = \dot{Q}_0(t)Q_0(t)^{-1}$ and $A_B(t) = Q_0(t)BQ_0(t)^{-1}$. As $Q_0(t)$ is τ-periodic the matrices $A^*(t)$ and $A_B(t)$ are τ-periodic; hence, the matrix $A(t)$ is τ-periodic.

Since $\dot{X}_0 = A_0(t)X_0$ we have $A_0(t) = \dot{X}_0(t)X_0(t)^{-1} = \dot{Q}_0(t)Q_0^{-1}(t) + Q_0(t)B_0Q_0^{-1}(t)$, so we get $A(t) - A_0(t) = Q_0(t)(B - B_0)Q_0(t)^{-1}$; hence,

$$\|A(t) - A_0(t)\| \leq \|Q_0(t)\|\|Q_0(t)^{-1}\|\|B - B_0\|. \tag{7.11}$$

We can now prove the following result.

Proposition 7.29 *Let $X_0(t) = Q_0(t)e^{tB_0}$ be the Floquet decomposition of the real τ-periodic linear Hamiltonian system $\dot{\mathbf{x}} = A_0(t)\mathbf{x}$. If this system is strongly stable, so is the linear system $\dot{\mathbf{y}} = B_0\mathbf{y}$.*

Proof Since $\dot{\mathbf{x}} = A_0(t)\mathbf{x}$ is strongly stable there is a positive number $\epsilon > 0$ such that for any continuous τ-periodic real Hamiltonian matrix $A(t)$ with $||A - A_0|| < \epsilon$ the system $\dot{\mathbf{x}} = A(t)\mathbf{x}$ is stable. Let $\delta = \epsilon/M$, where M is the maximum of

[2] Since X is symplectic we have $XJX^T = J$, so differentiating we get $\dot{X}JX^T + XJ\dot{X}^T = 0$; hence, $(\dot{X}X^{-1})XJX^T + XJX^T(\dot{X}X^{-1})^T = 0$, that is, $AJ + JA^T = 0$, so A is Hamiltonian.

$\|Q_0(t)\|\|Q_0(t)^{-1}\|$ for $0 \le t \le \tau$. Given a Hamiltonian matrix B consider as above the symplectic matrix $X(t) = Q_0(t)\mathrm{e}^{tB}$ and the Hamiltonian matrix $A(t) = \dot{X}(t)X(t)^{-1}$. Then, if we take $\|B - B_0\| < \delta$, from (7.11) we have $||A - A_0|| < \epsilon$ so the system $\dot{\mathbf{x}} = A(t)\mathbf{x}$ is stable; hence, all its solutions $\mathbf{x}(t)$ are bounded.

We know that the change of variables $\mathbf{x} = Q_0(t)\mathbf{y}$ takes the equation $\dot{\mathbf{x}} = A_0(t)\mathbf{x}$ to the equation $\dot{\mathbf{y}} = B_0\mathbf{y}$. But it also takes the equation $\dot{\mathbf{x}} = A(t)\mathbf{x}$ to the equation $\dot{\mathbf{y}} = B\mathbf{y}$.

Indeed, since $Q_0 = X\mathrm{e}^{-tB}$ we get

$$\dot{Q}_0 = \dot{X}\mathrm{e}^{-tB} - X\mathrm{e}^{-tB}B = \dot{X}X^{-1}(X\mathrm{e}^{-tB}) - X\mathrm{e}^{-tB}B = AQ_0 - Q_0B.$$

Now, differentiating $\mathbf{x} = Q_0\mathbf{y}$ we get $\dot{\mathbf{x}} = Q_0\dot{\mathbf{y}} + \dot{Q}_0\mathbf{y}$ and using the above expression of $\dot{Q}_0$ we have

$$\dot{\mathbf{x}} = Q_0(\dot{\mathbf{y}} - B\mathbf{y}) + A\mathbf{x},$$

so $\dot{\mathbf{x}} = A\mathbf{x}$ if, and only if, $\dot{\mathbf{y}} = B\mathbf{y}$.

Since all the solutions $\mathbf{x}(t)$ of $\dot{\mathbf{x}} = A(t)\mathbf{x}$ are bounded, so are the solutions $\mathbf{y}(t) = Q_0(t)^{-1}\mathbf{x}(t)$ of $\dot{\mathbf{y}} = B\mathbf{y}$. Therefore, the system $\dot{\mathbf{y}} = B\mathbf{y}$ is stable by Proposition 7.21. This proves that $\dot{\mathbf{y}} = B_0\mathbf{y}$ is strongly stable. □

Now we aim to prove the converse statement.

Proposition 7.30 *If the system $\dot{\mathbf{y}} = B_0\mathbf{y}$ is strongly stable, then the periodic system $\dot{\mathbf{x}} = A_0(t)\mathbf{x}$ is strongly stable.*

The proof in this case is much harder and here we will go along the lines of ideas in [64].

We recall the Hermitian inner product in $\mathbf{C}^m$, $(\mathbf{z}, \boldsymbol{\zeta}) = z_1\overline{\zeta}_1 + \cdots + z_m\overline{\zeta}_m$ noticing that $(\mathbf{z}, \boldsymbol{\zeta})$ is linear in the first variable and that $(\mathbf{z}, \boldsymbol{\zeta}) = \overline{(\boldsymbol{\zeta}, \mathbf{z})}$; consequently, it is additive in the second variable and we have $(\alpha\mathbf{z}, \beta\boldsymbol{\zeta}) = \alpha\overline{\beta}(\mathbf{z}, \boldsymbol{\zeta})$. We also have the identity

$$(A\mathbf{z}, \boldsymbol{\zeta}) = (\mathbf{z}, A^*\boldsymbol{\zeta}), \tag{7.12}$$

where $A^* = \overline{A}^T$ is the adjoint of the $m \times m$ complex square matrix A.

For real vectors $\mathbf{z}$ and $\boldsymbol{\zeta}$ we have $(\mathbf{z}, \boldsymbol{\zeta}) = \langle \mathbf{z}, \boldsymbol{\zeta} \rangle$ the Euclidean inner product in $\mathbf{R}^m$, so if A is a real matrix the above identity becomes $\langle A\mathbf{z}, \boldsymbol{\zeta} \rangle = \langle \mathbf{z}, A^T\boldsymbol{\zeta} \rangle$.

Definition 7.31 The matrix A is self-adjoint if $A^* = A$. For a real matrix A self-adjoint means symmetric.

Let G be a self-adjoint non-singular matrix. We introduce the product in $\mathbf{C}^m$

$$(\mathbf{z}, \boldsymbol{\zeta})_G = (G\mathbf{z}, \boldsymbol{\zeta}) \tag{7.13}$$

and notice that it inherits the properties of $(\mathbf{z}, \boldsymbol{\zeta})$: $(\mathbf{z}, \boldsymbol{\zeta})_G$ is linear in the first variable and $(\mathbf{z}, \boldsymbol{\zeta})_G = \overline{(\boldsymbol{\zeta}, \mathbf{z})_G}$; consequently, it is additive in the second variable and it also satisfies the identity $(\alpha\mathbf{z}, \beta\boldsymbol{\zeta})_G = \alpha\overline{\beta}(\mathbf{z}, \boldsymbol{\zeta})_G$.

Since G is nonsingular, this product is nondegenerate, i.e., $(\mathbf{z}, \boldsymbol{\zeta})_G = 0$ for all $\boldsymbol{\zeta}$ implies $\mathbf{z} = 0$.

Definition 7.32 A complex $m \times m$ matrix X is said to be *G-unitary* if

$$(X\mathbf{z}, X\boldsymbol{\zeta})_G = (\mathbf{z}, \boldsymbol{\zeta})_G, \quad \text{for all} \quad \mathbf{z}, \boldsymbol{\zeta} \in \mathbf{C}^m.$$

It follows immediately from (7.12), (7.13) and the nondegeneracy condition that X is a G-unitary matrix if, and only if,

$$X^* G X = G. \tag{7.14}$$

Lemma 7.33 *Let X be a G-unitary matrix. Then:*

(1) *If $\rho \in S^1$ is a multiple eigenvalue of X and $X|_{\eta^\dagger(\rho)}$ is nondiagonalizable, there is an eigenvector $\mathbf{v}$ of ρ such that $(\mathbf{v}, \mathbf{v})_G = 0$.*
(2) *If $\rho \notin S^1$ is an eigenvalue of X with eigenvector $\mathbf{v}$ then $(\mathbf{v}, \mathbf{v})_G = 0$.*

Proof

(1) Since $X|_{\eta^\dagger(\rho)}$ is nondiagonalizable, there are vectors $\mathbf{v}$ and $\mathbf{w}$ in $\eta^\dagger(\rho)$ such that $X\mathbf{v} = \rho\mathbf{v}$ and $X\mathbf{w} = \rho\mathbf{w} + \mathbf{v}$. Since X is G-unitary we have $(X\mathbf{v}, X\mathbf{w})_G = (\mathbf{v}, \mathbf{w})_G$, from which we get using the properties of the product $(\ ,\ )_G$,

$$\rho\overline{\rho}(\mathbf{v}, \mathbf{w})_G + \rho(\mathbf{v}, \mathbf{v})_G = (\mathbf{v}, \mathbf{w})_G;$$

hence, $(|\rho|^2 - 1)(\mathbf{v}, \mathbf{w})_G + \rho(\mathbf{v}, \mathbf{v})_G = 0$. Since $\rho \in S^1$, we get $(\mathbf{v}, \mathbf{v})_G = 0$.
(2) From $(X\mathbf{v}, X\mathbf{v})_G = (\mathbf{v}, \mathbf{v})_G$ we get $(|\rho|^2 - 1)(\mathbf{v}, \mathbf{v})_G = 0$, so $(\mathbf{v}, \mathbf{v})_G = 0$.

□

We now prove the following theorem.

Theorem 7.34 (Krein) *Let $X^{(0)}$ be a G-unitary matrix and $\rho^{(0)}$ a multiple eigenvalue of $X^{(0)}$ such that $(\mathbf{v}, \mathbf{v})_G \neq 0$ for every eigenvector $\mathbf{v}$ of $\rho^{(0)}$. Then, there are positive constants γ and δ such that for any G-unitary matrix X with $\|X - X^{(0)}\| < \delta$, all the eigenvalues ρ of X with $|\rho - \rho^{(0)}| < \gamma$ lie in the unit circle S^1 and $X|_{\eta^\dagger(\rho)}$ is diagonalizable.*

Proof The negation of the conclusion of the theorem implies that there exists a sequence of G-unitary matrices X_n converging to $X^{(0)}$ with eigenvalues ρ_n converging to $\rho^{(0)}$ with the alternatives: either $|\rho_n| \neq 1$ or $|\rho_n| = 1$ and $X_n|_{\eta^\dagger}(\rho_n)$ nondiagonalizable. By Lemma 7.33 there are normalized eigenvectors $\mathbf{v}_n$, that is $X_n\mathbf{v}_n = \rho_n\mathbf{v}_n$, $\|\mathbf{v}_n\| = 1$ such that $(\mathbf{v}_n, \mathbf{v}_n)_G = 0$. As S^1 is compact the sequence of vectors $\{\mathbf{v}_n\}$ has a convergent subsequence $\mathbf{v}_{n_j} \to \mathbf{v}$. Letting $j \to \infty$ in the equalities

$$X_{n_j}\mathbf{v}_{n_j} = \rho_{n_j}\mathbf{v}_{n_j}, \quad \|\mathbf{v}_{n_j}\| = 1 \quad \text{and} \quad (\mathbf{v}_{n_j}, \mathbf{v}_{n_j})_G = 0,$$

we get $X^{(0)}\mathbf{v} = \rho^{(0)}\mathbf{v}$, $\|\mathbf{v}\| = 1$ and $(\mathbf{v}, \mathbf{v})_G = 0$, a contradiction to the hypothesis that $(\mathbf{v}, \mathbf{v})_G \neq 0$ for every eigenvector $\mathbf{v}$ of $\rho^{(0)}$. This contradiction shows that the conclusion of the theorem holds. □

Definition 7.35 Let $\rho \in S^1$ be an eigenvalue of the G-unitary matrix X. We say that ρ is an eigenvalue of the *first kind* if $(\mathbf{v}, \mathbf{v})_G > 0$, for every one of its eigenvectors. The eigenvalue is of the *second kind* if $(\mathbf{v}, \mathbf{v})_G < 0$, for every eigenvector $\mathbf{v}$.

Eigenvalues of the first and second kinds are said to be *definite*. The eigenvalue is said to be *indefinite* or of the *mixed kind* if $(\mathbf{v}, \mathbf{v})_G = 0$, for some eigenvector $\mathbf{v}$.

Now we consider a special self-adjoint matrix of particular interest for real Hamiltonian system theory, namely $G = -iJ$, where $J = \begin{bmatrix} O & I \\ -I & O \end{bmatrix}$ is the standard $2n \times 2n$ symplectic matrix. In this case a G-unitary matrix X satisfies the identity (just cancel the factor $-i$ in (7.14)):

$$X^* J X = J. \tag{7.15}$$

Definition 7.36 The complex matrix X is symplectic if it satisfies the Eq. (7.15).

Of course, this coincides with the notion of symplectic matrix if X is real.

Recalling that $(\mathbf{x}, \mathbf{y}) = \langle \mathbf{x}, \mathbf{y} \rangle$ for real vectors $\mathbf{x}$ and $\mathbf{y}$, we find for $\mathbf{z} = \mathbf{x} + i\mathbf{y}$ and $\boldsymbol{\zeta} = \boldsymbol{\xi} + i\boldsymbol{\eta}$,

$$(\mathbf{z}, \boldsymbol{\zeta})_G = \langle J\mathbf{y}, \boldsymbol{\xi} \rangle - \langle J\mathbf{x}, \boldsymbol{\eta} \rangle - i\big(\langle J\mathbf{x}, \boldsymbol{\xi} \rangle + \langle J\mathbf{y}, \boldsymbol{\eta} \rangle\big),$$

so in particular

$$(\mathbf{z}, \mathbf{z})_G = 2\langle \mathbf{x}, J\mathbf{y} \rangle = 2\{\mathbf{x}, \mathbf{y}\}. \tag{7.16}$$

Note that $\langle \mathbf{x}, J\mathbf{y} \rangle = \{\mathbf{x}, \mathbf{y}\}$, the standard symplectic product of $\mathbf{x}$ and $\mathbf{y}$.

Now consider the real continuous τ-periodic linear system in (7.6) with $\mathbf{x} \in \mathbf{R}^{2n}$ to be Hamiltonian so its matrizant $X(t)$ is a symplectic matrix.

Let $X(t) = Q(t)\mathrm{e}^{tB}$ be the Floquet real form of the matrizant (see Definition 7.13).

Then the monodromy matrix $X(2\tau)$ is related to B by $X(2\tau) = \mathrm{e}^{2\tau B}$ with B a real matrix.

Recall from Definition 7.15 that the eigenvalues of B are called the *characteristic exponents* and those of $X(2\tau)$ the *multipliers* of the linear system (7.6).

If (7.6) is a stable system, then all multipliers are on the unit circle S^1, by Propositions 7.22 and 7.24.

Let $\rho \in S^1$ be an eigenvalue of the monodromy matrix $X(2\tau)$ with eigenvector $\mathbf{v} = \mathbf{r} + i\mathbf{s}$. By (7.16) we have

$$(\mathbf{v}, \mathbf{v})_G = 2\langle \mathbf{r}, J\mathbf{s} \rangle = \{\mathbf{r}, \mathbf{s}\}. \tag{7.17}$$

Proposition 7.37 *Let X be a real symplectic matrix. If all the eigenvalues of X are definite, then neither $\rho = 1$ nor $\rho = -1$ can be an eigenvalue of X.*

Proof Suppose that $\rho = 1$ is an eigenvalue of X. Since X is symplectic, $\rho \in S^1$ is a multiple eigenvalue and as ρ is definite the matrix $X_{\eta^\dagger(\rho)}$ is diagonalizable, by Lemma 7.33. Since X and ρ are real there are two linearly independent real eigenvectors $\mathbf{u}$ and $\mathbf{v}$ of X for the eigenvalue $\rho = 1$. Then, $\mathbf{z} = \mathbf{u}+i\mathbf{v}$ and $\mathbf{z}' = \mathbf{u}-i\mathbf{v}$ are also eigenvectors of X belonging to ρ and by (7.17) we have

$$(\mathbf{z}, \mathbf{z})_G = 2\{\mathbf{u}, \mathbf{v}\} \quad \text{and} \quad (\mathbf{z}', \mathbf{z}')_G = 2\{\mathbf{v}, \mathbf{u}\}. \tag{7.18}$$

Since ρ is definite, assume it of first kind. Then, the product $(\boldsymbol{\zeta}, \boldsymbol{\zeta})_G > 0$ for all eigenvectors $\boldsymbol{\zeta}$ belonging to ρ. But (7.18) contradicts this condition. Therefore, $\rho = 1$ cannot be an eigenvalue of X. Same argument for $\rho = -1$. □

Let $\rho = e^{2\tau\lambda}$ be a multiplier of the τ-periodic linear Hamiltonian system $\dot{\mathbf{x}} = A(t)\mathbf{x}$. If $\mathbf{v} = \mathbf{r} + i\mathbf{s}$ is an eigenvector of $X(2\tau)$ belonging to ρ, then from (7.17) we see that in the terminology of Definition 7.35, the multiplier ρ is of the first kind if the symplectic product $\{\mathbf{r}, \mathbf{s}\} = \langle \mathbf{r}, J\mathbf{s}\rangle$ is positive for any $\mathbf{v}$ and it is of the second kind if $\{\mathbf{r}, \mathbf{s}\} < 0$, for any $\mathbf{v}$.

Proposition 7.38 *Suppose that ± 1 are not eigenvalues of $X(2\tau)$. Let $\rho \in S^1$ be a simple eigenvalue of $X(2\tau)$ and let $\mathbf{v} = \mathbf{r} + i\mathbf{s}$ be a corresponding eigenvector. Then, $\{\mathbf{r}, \mathbf{s}\} \neq 0$.*

Let ρ_j, ρ_j^{-1}, $j = 1, \ldots, s$ be the distinct eigenvalues of the symplectic matrix $X(2\tau)$. Then, $\mathbf{C}^{2n} = E_1 \oplus E_2 \oplus \ldots \oplus E_s$, where $E_j = \eta^\dagger(\rho_j) \oplus \eta^\dagger(\rho_j^{-1})$. As ± 1 are not eigenvalues of $X(2\tau)$, by Lemma 4.37 any two subspaces E_j, E_k are symplectically orthogonal, so by the argument right after (4.20), adapted from the case of Hamiltonian to symplectic matrices, the subspaces $E_1, \ldots, E_s$ are symplectic. If ρ_j is a simple eigenvalue, then E_j is two-dimensional; hence, $\{\mathbf{x}, \mathbf{y}\} \neq 0$ for any linearly independent vectors in E_j. Now $|\rho| = 1$, so $\rho^{-1} = \overline{\rho}$, and since $X(2\tau)$ is real corresponding to the eigenvector $\mathbf{u}$ of ρ we have the eigenvector $\overline{\mathbf{u}}$ of ρ^{-1}. Therefore, $\{\mathbf{u}, \overline{\mathbf{u}}\} \neq 0$, so if $\mathbf{u} = \mathbf{r} + i\mathbf{s}$ we get $\{\mathbf{r}, \mathbf{s}\} \neq 0$. □

In the Proof of Proposition 7.30 we use the fact that the solutions of an ordinary differential equation depend continuously on the data, which in our context is the following result.

Proposition 7.39 *Given $\gamma > 0$ there is $\eta > 0$ such that if $||A - A_0|| < \eta$ then $\|X(t) - X_0(t)\| < \gamma$, for all $t \in [0, 2\tau]$, where $X(t)$ and $X_0(t)$ are the matrizants of the τ-periodic real linear equations $\dot{\mathbf{x}} = A(t)\mathbf{x}$ and $\dot{\mathbf{x}} = A_0(t)\mathbf{x}$ respectively.*

We also use the fact that the roots of the characteristic equation of a matrix depend continuously on the entries of the matrix. The continuity of the roots of a polynomial is proved using Rouché's theorem of complex analysis (see [2]). We state it here for eigenvalues.

Proposition 7.40 *Let $\lambda_1, \ldots, \lambda_k$ be the distinct eigenvalues of the complex square matrix A with multiplicities $p_1, \ldots, p_k$ respectively. Let $r > 0$ be such that the open disks $D_r(\lambda_j)$, $j = 1, \ldots, k$ are pairwise disjoint. Then there is a number $\delta > 0$ such that any square matrix of the same order as A with $||B - A|| < \delta$ has p_j eigenvalues in $D_r(\lambda_j)$, $j = 1, \ldots, k$.*

Recalling from Theorem 4.39 and its corollary that δ_s in the partial Hamiltonian H_j in (7.9) is given by $\delta_s = \text{sign}\{\mathbf{r}_j, \mathbf{s}_j\}$, the hypothesis in Theorem 7.28, that is, the hypothesis that H_j is either positive or negative definite means that the eigenvalue ρ_j is definite.

Proof of Proposition 7.30 Since $\dot{\mathbf{y}} = B_0\mathbf{y}$ is strongly stable, Theorem 7.28 says that the multiple eigenvalues of the symplectic matrix $X_0(2\tau)$ lie on the unit circle S^1 and are all definite, so we can use Krein's Theorem 7.34 with $X^{(0)} = X_0(2\tau)$. Let $\rho_1^{(0)}, \ldots, \rho_k^{(0)}$ and their inverses be the multiple eigenvalues of $X^{(0)}$ with multiplicities $p_1, \ldots, p_k$ respectively.

By Theorem 7.34, for each $\rho_j^{(0)}$, $j = 1, \ldots, k$ there are positive constants γ_j and δ_j such that for any symplectic matrix X with $\|X - X^{(0)}\| < \delta_j$, if ρ_j is an eigenvalue of X with $|\rho_j - \rho_j^{(0)}| < \gamma_j$ we have $\rho_j \in S^1$ and $X|_{\eta^\dagger(\rho_j)}$ is diagonalizable. Choose such a pair γ_j, δ_j for each $j = 1, \ldots, k$.

Choose $r < \min\{\gamma_j\}$ and $\delta < \min\{\delta_j\}$ as in Proposition 7.40. By this proposition if $\|X - X^{(0)}\| < \delta$ then for each $\rho_j^{(0)}$, there are p_j eigenvalues ρ_j of X in the disk $|\rho_j - \rho_j^{(0)}| < r$. We have the same conclusion for the inverses of $\rho_1^{(0)}, \ldots, \rho_k^{(0)}$.

Consider the invariant direct sum decomposition for $X = X(2\tau)$, $E = \oplus\big(\eta^\dagger(\rho) \oplus \eta^\dagger(\rho^{-1})\big)$, given by all the eigenvalues of X coming from the multiple eigenvalues of X_0 as described above. Then, by the above discussion we see that X_E is diagonalizable and all the eigenvalues in the expression of E are on the unit circle S^1.

Now let $\rho_{01}, \ldots, \rho_{0l}$ and their inverses be the simple eigenvalues of $X^{(0)}$. Then, by Proposition 7.40, for each $j = 1, \ldots, l$, the disk $D_r(\rho_{0s})$ contains exactly one eigenvalue ρ'_s of $X(2\tau)$. The eigenvalues $\rho'_1, \ldots, \rho'_l$ are also on the unit circle S^1. Indeed, suppose that some ρ'_s were outside this circle. Since $X(2\tau)$ is real and symplectic the conjugate $\rho^*_s = \dfrac{\rho'_s}{|\rho'_s|^2}$ of the inverse of ρ'_s is also an eigenvalue of $X(2\tau)$. But ρ^*_s lies along the half straight line from the origin and passing through ρ'_s. The disk $D_r(\rho_{0s})$ cannot contain both ρ'_s and ρ^*_s, since it contains only one eigenvalue of X. Also these two eigenvalues of $X = X(2\tau)$ cannot lie in any of the disks with center at multiple eigenvalues of X_0. But this makes the number of eigenvalues of X be greater than $2n$, the order of the matrix X, a contradiction. Therefore, the eigenvalues $\rho'_1, \ldots, \rho'_l$ of X and their inverses are all on S^1. These eigenvectors are all simple so X restricted to the direct sum of their eigenspaces is diagonalizable.

Therefore, $X(2\tau)$ is a diagonalizable operator and all its eigenvalues are on the unit circle S^1. Consequently, the sequence $X(2\tau)^k$ is bounded, hence the sequence $X(\tau)^k$ is also bounded and by Proposition 7.23 the system $\dot{\mathbf{x}} = A(t)\mathbf{x}$ is stable. This proves that the system $\dot{\mathbf{x}} = A_0(t)\mathbf{x}$ is strongly stable. □

Propositions 7.29 and 7.30 together can be stated as the following theorem, our version of the Krein–Gel'fand–Lidskii Theorem.

Theorem 7.41 *Let $X(t) = Q(t)e^{tB}$ be the Floquet decomposition of the continuous τ-periodic linear Hamiltonian system $\dot{\mathbf{x}} = A(t)\mathbf{x}$ and let $\dot{\mathbf{y}} = B\mathbf{y}$ be the transformed constant Hamiltonian system through the symplectic coordinate change $\mathbf{x} = Q(t)\mathbf{y}$. Then, the system $\dot{\mathbf{x}} = A(t)\mathbf{x}$ is strongly stable if, and only if, $\dot{\mathbf{y}} = B\mathbf{y}$ is strongly stable.*

Compare the statement of Theorem 7.41 with that of Proposition 7.24.

This theorem can be re-stated in the terminology of Yakubovich and Starzhinskii [64]:

Theorem 7.42 (Krein–Gel'fand–Lidskii) *The real continuous τ-periodic linear Hamiltonian system $\dot{\mathbf{x}} = A(t)\mathbf{x}$ is strongly stable if, and only if, all of its multipliers are on the unit circle S^1 and all of them are definite.*

Proof Let $X(t) = Q(t)\mathrm{e}^{tB}$ be the Floquet real form of the matrizant of $\dot{\mathbf{x}} = A(t)\mathbf{x}$. The symplectic mapping $\mathbf{x} = Q(t)\mathbf{y}$ transforms the system $\dot{\mathbf{x}} = A(t)\mathbf{x}$ into $\dot{\mathbf{y}} = B\mathbf{y}$. Since $\dot{\mathbf{x}} = A(t)\mathbf{x}$ is strongly stable, so is the system $\dot{\mathbf{y}} = B\mathbf{y}$ by Theorem 7.41.

By Theorem 7.28 every eigenvalue λ_j of B is a nonzero purely imaginary number, $\lambda_j = i\beta_j$, with $\beta_j \neq 0$ and, moreover, the δ_s in (7.9) are all equal to $+1$ or all equal to -1. This implies that every multiplier $\rho_j = \mathrm{e}^{i2\tau\beta_j}$ is on the unit circle and are all definite.

We now prove the converse statement. The multipliers $\rho = \mathrm{e}^{2\tau\lambda}$ of the system $\dot{\mathbf{x}} = A(t)\mathbf{x}$ are the eigenvalues of the monodromy matrix $X(2\tau) = \mathrm{e}^{2\tau B}$. By hypothesis they are on the unit circle S^1 so the eigenvalues of B are purely imaginary numbers and as the multipliers are definite, Proposition 7.37 says that neither -1 nor $+1$ is among them, so 0 is not an eigenvalue of B. Let $\rho_1, \dots, \rho_k$ be the positive multipliers and $\rho_{k+1}, \dots, \rho_{2n}$ be the negative multipliers. Let $\mathbf{v}_j = \mathbf{r}_j + i\mathbf{s}_j$ be any eigenvector of B for the eigenvalue λ_j. Then, $\mathbf{v}_j$ is also an eigenvector of $X(2\tau)$ for the eigenvalue ρ_j. Therefore, $\delta_j = \{\mathbf{r}_j, \mathbf{s}_j\}$ is equal to $+1$ for $j = 1, \dots, k$ and equal to -1 for $j = k+1, \dots, 2n$. Then, the Hamiltonian H_j in (7.9) is positive definite for $j = 1, \dots, k$ and negative definite for $j = k+1, \dots, 2n$. Therefore, by Theorem 7.28 the system $\dot{\mathbf{y}} = B\mathbf{y}$ is strongly stable; hence, the system $\dot{\mathbf{x}} = A(t)\mathbf{x}$ is strongly stable, by Theorem 7.41. □

7.7 The Gelfand–Lidskii Theory

7.7.1 The Gelfand–Lidskii Theorem

In the article [24], Gelfand and Lidskii give a follow-up to the theory of Krein [34] on the strong stability of periodic linear Hamiltonian systems. They prove the sufficient part of a theorem of Krein obtaining what is nowadays known as the Krein–Gelfand–Lidskii theorem and then they make a study of the topological structure of the set of all strongly stable τ-periodic linear Hamiltonian systems.

Consider the strongly stable τ-periodic linear Hamiltonian system

$$\dot{\mathbf{x}} = A(t)\mathbf{x}, \tag{7.19}$$

together with the Floquet real form of its matrizant

$$Z(t) = Q(t)\mathrm{e}^{tB}, \tag{7.20}$$

that is, $Q(t)$, 2τ-periodic symplectic matrix and B, a constant real Hamiltonian matrix.

Let $\pm i\omega_1, \ldots, \pm i\omega_n$ be the characteristic exponents of the system (7.19), that is, the eigenvalues of the matrix B. We always take the frequencies $\omega_1, \ldots, \omega_n$ positive and assume the ordering $\omega_1 \geq \cdots \geq \omega_n$.

The eigenvalues of $Z(2\tau)$ are the multipliers of the system (7.19), so they are given by

$$\rho_j = \mathrm{e}^{i2\tau\omega_j}, \quad \rho_{-j} = \mathrm{e}^{-i2\tau\omega_j}, \qquad j = 1, \ldots, n.$$

Recall that the multiplier ρ_k is said to be of the first or second kind in accordance with the symplectic product $\{\mathbf{r}_k, \mathbf{s}_k\} = \mathbf{r}_k^T J\mathbf{s}_k$ being positive or negative respectively, for any eigenvector $\mathbf{v}_k = \mathbf{r}_k + i\mathbf{s}_k$ of $Z(2\tau)$ belonging to ρ_k. We say that a multiplier is definite if it is of the first or second kind; otherwise, we say it is indefinite.

The multipliers come in pairs ρ, ρ^{-1}. For a strongly stable linear system none of them is equal to 1 or -1 (see Proposition 7.37). Therefore, counting with multiplicities, half of the multipliers of a strongly stable system are situated on the upper semi-circle and half in the lower semi-circle.

Now, on perturbing the system, a group of multipliers of the same kind can permute their places among themselves on the upper semi-circle without violating the condition of strong stability during a collision between two of them, but multipliers of different kinds cannot interchange their places, because as they collapse at one point the symplectic product $\{\mathbf{r}, \mathbf{s}\}$ would be zero, violating the condition of strong stability.

Therefore, assigning the plus sign $\oplus$ to a multiplier of the first kind and the minus sign $\ominus$ to a multiplier of the second kind we consider the n multipliers $\rho_{k_1}, \ldots, \rho_{k_n}$

situated on the upper semi-circle with the corresponding sequence of signs, for instance, $\oplus, \ominus, \ominus, \oplus, \ldots, \oplus, \ominus$. Such an arrangement is called a *disposition* of the multipliers.

Consider two strongly stable τ-periodic linear Hamiltonian systems

$$\dot{\mathbf{x}} = J\nabla H_0(\mathbf{x}, t), \quad \dot{\mathbf{x}} = J\nabla H_1(\mathbf{x}, t). \tag{7.21}$$

Gelfand and Lidskii [24] define the *stability domain* as follows: the systems (7.21) are said to belong to the same stability domain if, and only if, there is a continuous quadratic Hamiltonian $H(\mathbf{x}, t, s)$, τ-periodic in t, $0 \le s \le 1$, such that

$$H(\mathbf{x}, t, 0) = H_0(\mathbf{x}, t), \quad H(\mathbf{x}, t, 1) = H_1(\mathbf{x}, t),$$

and for each s the linear system $\dot{\mathbf{x}} = J\nabla H(\mathbf{x}, t, s)\mathbf{x}$ is strongly stable.

Gelfand and Lidskii assign an integer to every strongly stable linear system, called the index of the system, and proving the following theorem (see [24]).

Theorem 7.43 (Gelfand-Lidskii) *The strongly stable linear systems* (7.21) *belong to the same stability domain if, and only if, their multipliers have the same disposition on the upper semi-circle and they have the same index.*

In the proof of this theorem they establish many properties of Hamiltonian systems, most of which appear scattered in several sections of this book. We prove this theorem using our approach to the study of linear Hamiltonian systems. Using our version of the Krein–Gelfand–Lidskii theorem, that is, Theorem 7.41, we formulate and prove the theorem in terms of characteristic exponents rather than of multipliers.

Let us identify the continuous τ-periodic real linear Hamiltonian system $\dot{\mathbf{x}} = A(t)\mathbf{x}$ with the matrix $A(t)$. We will consider therefore the space of continuous and τ-periodic linear Hamiltonian systems as the normed space $\mathcal{L}$ of continuous τ-periodic real Hamiltonian matrices $A(t)$. We will say that the continuous τ-periodic Hamiltonian matrix $A(t)$ is stable or strongly stable according to the linear system $\dot{\mathbf{x}} = A(t)\mathbf{x}$ being stable or strongly stable respectively.

We denote by $\mathcal{L}_s$ the set of stable matrices in the normed space $\mathcal{L}$ and by $\mathcal{L}_{ss} \subset \mathcal{L}$ the set of strongly stable matrices. Considering $\mathcal{L}$ with the topology induced by the norm we observe that the set $\mathcal{L}_{ss}$ is the interior of the set $\mathcal{L}_s$ and, therefore it is an open subset of $\mathcal{L}$.

Consider two strongly stable τ-periodic real linear Hamiltonian systems

$$\dot{\mathbf{x}} = A_0(t)\mathbf{x}, \quad \dot{\mathbf{x}} = A_1(t)\mathbf{x}. \tag{7.22}$$

As $A_0(t)$ and $A_1(t)$ are closed paths in the topological space $\mathcal{L}_{ss}$, to say that the systems (7.22) belong to the same stability domain is to say that there is a (free) homotopy $A(t, s)$ from $A_0(t)$ to $A_1(t)$, that is, there exists a continuous mapping

$$A : \mathbf{R} \times [0, 1] \to \mathcal{L}_{ss}, \tag{7.23}$$

τ−periodic in t such that $A(t, 0) = A_0(t), \quad A(t, 1) = A_1(t)$.

Free homotopy defines an equivalence relation $A_0(t) \sim A_1(t)$ in $\mathcal{L}_{ss}$. The equivalence classes $[A(t)]$ defined by this relation in $\mathcal{L}_{ss}$ are the *domains of stability* considered by Gelfand and Lidskii.

Let $\mathbf{v}_k = \mathbf{r}_k + i\mathbf{s}_k$ be an eigenvector for the eigenvalue $i\omega_k$ of the Hamiltonian matrix B in the decomposition (7.20). If $\{\mathbf{r}_k, \mathbf{s}_k\} > 0$, then the corresponding multiplier $\rho_k = e^{2i\omega_k\tau}$ is of the first kind. Indeed, $\mathbf{v}_k$ is also an eigenvector of $Z(2\tau) = e^{2\tau B}$ corresponding to ρ_k and since $A(t)$ is strongly stable the inequality $\{\mathbf{r}, \mathbf{s}\} > 0$ holds for any eigenvector $\mathbf{v} = \mathbf{r} + i\mathbf{s}$ of $Z(2\tau)$ corresponding to ρ_k. We say that the characteristic exponent $i\omega_k$, or the frequency ω_k, is also of the first kind. Analogously, we say that the characteristic exponent $i\omega_k$, or the frequency ω_k, is of the second kind if $\{\mathbf{r}_k, \mathbf{s}_k\} < 0$. Notice that $\overline{\mathbf{v}}_k = \mathbf{r}_k - i\mathbf{s}_k$ is an eigenvector of the real matrix B with eigenvalue $-i\omega_k$, so we have that $-i\omega_k$ is of the first (second) kind if $i\omega_k$ is of the second (first) kind.

Therefore, if $\delta_k = \text{sign}\{\mathbf{r}_k, \mathbf{s}_k\}$, then corresponding to the sequence of frequencies $\omega_1, \omega_2, \ldots, \omega_n$ we have the sequence of signs $\delta_1, \delta_2, \ldots, \delta_n$. We call the vector $\boldsymbol{\delta} = (\delta_1, \delta_2, \ldots, \delta_n)$ the *signature* of the frequency vector $\boldsymbol{\omega} = (\omega_1, \omega_2, \ldots, \omega_n)$.

We emphasize that the frequency vector $\boldsymbol{\omega}$ and its signature $\boldsymbol{\delta}$ are associated with the strongly stable linear system (7.19), so we write $\boldsymbol{\omega} = \boldsymbol{\omega}(A)$ and $\boldsymbol{\delta} = \boldsymbol{\delta}(A)$. Later on, in Proposition 7.46, we show that there is also an integer $n(A)$ associated with the strongly stable linear system (7.19), which we call the Gelfand–Lidskii index.

The main theorem of this section is the following version of the Gelfand–Lidskii theorem.

Theorem 7.44 (Gelfand–Lidskii) *The stability domain determined by* $A_0(t) \in \mathcal{L}_{ss}$ *is given by*

$$[A_0(t)] = \{A(t) \in \mathcal{L}_{ss};\ \boldsymbol{\delta}(A) = \boldsymbol{\delta}(A_0) \quad \text{and} \quad n(A) = n(A_0)\}.$$

An important point in the proof of this theorem is some knowledge on the topology of the real symplectic group $Sp(2n, \mathbf{R})$ so we make some considerations on that.

7.7.2 Topology of the Real Symplectic Group $Sp(2n, \mathbf{R})$

This group is a closed set of the normed linear space $M_{2n}(R)$ of real matrices of order $2n$. Let us prove that as a topological space $Sp(2n, \mathbf{R})$ *is homeomorphic to the product of a simply connected space* X *by the unit circle* S^1.

Let $Sp^+(2n)$ be the set of symplectic and positive definite symmetric matrices and $Sp^O(2n)$ the set of symplectic and orthogonal matrices.

By Proposition 2.31 we have a homeomorphism $Sp(2n) \sim Sp^{+}(2n) \times Sp^{O}(2n)$.

Denoting by $H_s(2n)$ the space of real symmetric Hamiltonian matrices the mapping $A \mapsto e^A$ establishes a homeomorphism from $H_s(2n)$ to $Sp^{+}(2n)$. Indeed, recall that the positive definite factor in the polar decomposition of e^A is the square root Y of the matrix $e^A\left(e^A\right)^T = e^{2A}$; hence, $Y = e^A$ showing that $e^A \in Sp^{+}(2n)$. The inverse of this homeomorphism is the mapping $Y \mapsto \log Y$.

Now $H_s(2n)$ is formed by the matrices A of the form

$$A = \begin{bmatrix} A_1 & A_2 \\ A_2 & -A_1 \end{bmatrix}, \quad \text{with real nxn symmetric matrices } A_1 \text{ and } A_2,$$

so the mapping $A \mapsto (A_1, A_2)$ establishes a homeomorphism from $H_s(2n)$ to $M_s(n) \times M_s(n)$, where $M_s(n)$ is the space of $n \times n$ real symmetric matrices. Consequently, the space $H_s(2n)$; hence, the space $Sp^{+}(2n)$ is also homeomorphic to the Euclidean space $\mathbf{R}^{n(n+1)}$.

Now $Sp^{O}(2n)$ is formed by the matrices A of the form $A = \begin{bmatrix} A_1 & A_2 \\ -A_2 & A_1 \end{bmatrix}$ with real nxn matrices A_1, A_2 such that

$$A_1^T A_1 + A_2^T A_2 = I \quad \text{and} \quad A_1^T A_2 \quad \text{is symmetric.} \tag{7.24}$$

Using Eqs. (7.24) we easily check that the mapping $A \mapsto A_1 + iA_2$ establishes a homeomorphism from $Sp^{O}(2n)$ to the space of unitary matrices $U(n)$, that is, the space of $n \times n$ complex matrices U such that $U^*U = I$, where $U^* = \overline{U}^T$.

Since $\det\overline{U} = \overline{\det U}$ from $U^*U = I$ we have $|\det U| = 1$, so $\det U = e^{i\varphi}$, for some real number $0 \leq \varphi < 2\pi$.

Given $U \in U(n)$, consider the complex diagonal matrix of order n $G(\varphi) = \mathrm{diag}[e^{i\varphi}, 1, \dots, 1]$. Then, $U = U^s G(\varphi)$ with $U^s \in U(n)$ having determinant equal to one.

Therefore, the mapping $U \mapsto (U^s, G(\varphi))$ establishes a homeomorphism from $U(n)$ to $SU(n) \times \{G(\varphi)\}$, where $SU(n)$ is the set of special unitary matrices, that is, those unitary matrices whose determinants are equal to one. It is clear that the set $\{G(\varphi)\}$ is homeomorphic to S^1.

Now we mention the fact that $SU(n)$ is a simply connected space (for a proof see [18]).

From the above considerations we see that $Sp(2n) \sim Sp^{+}(2n) \times Sp^{O}(2n) \sim X \times S^1$, where X is a simply connected space. Recall that this is a space where every closed path can be continuously deformed to a point.

Since the fundamental group of X is trivial, the fundamental group of $Sp(2n, \mathbf{R})$ is isomorphic to that of S^1, that is, the additive group of the integers, $\mathbf{Z}$.

We close this subsection by mentioning that $Sp(2n, \mathbf{R})$ is a path-connected space.

7.7.3 Some Preliminaries

To emphasize that the (free) homotopy $A(t,s)$ from $A_0(t)$ to $A_1(t)$ is a homotopy in the space $\mathcal{L}_{ss}$ we use the terminology strong homotopy. In the same vein, we say that B_0 is strongly connected to B_1 if there is a continuous path $B : [0,1] \to \mathcal{L}^0_{ss}$ such that $B(0) = B_0$ and $B(1) = B_1$. We use the superscript $\mathcal{L}^0$ to denote the space of constant real Hamiltonian matrices.

Proposition 7.45 *If the strongly stable systems* $\dot{\mathbf{x}} = A_0(t)\mathbf{x}$, $\dot{\mathbf{x}} = A_1(t)\mathbf{x}$ *are strongly homotopic, then the matrices* B_0 *and* B_1 *in the Floquet decompositions* (7.25) *are strongly connected. Moreover, the symplectic matrices* $Q_0(t)$ *and* $Q_1(t)$ *are homotopic.*

Proof Let $A(t,s) : A_0(t) \sim A_1(t)$ be a strong homotopy. Let $Z(t,s) = Q(t,s)\mathrm{e}^{tB(s)}$ be the Floquet real form of the matrizant of the τ-periodic linear Hamiltonian system $\dot{\mathbf{x}} = A(t,s)\mathbf{x}$. Since $A(t,0) = A_0(t)$ and $A(t,1) = A_1(t)$, we have that

$$Z(t,0) = Z_0(t) = Q_0(t)\mathrm{e}^{tB_0} \quad \text{and} \quad Z(t,1) = Z_1(t) = Q_1(t)\mathrm{e}^{tB_1}. \tag{7.25}$$

Since $Z(2\tau,s) = \mathrm{e}^{2\tau B(s)}$, the matrix $2\tau B(s)$ is the real logarithm of $Z(2\tau,s)$. Therefore, $B(s) = \frac{1}{2\tau}\log \circ Z(2\tau,s)$ is a continuous function and notice that from (7.25) we have $B(0) = \frac{1}{2\tau}\log \circ\, \mathrm{e}^{2\tau B_0} = B_0$ and similarly $B(1) = B_1$. Therefore, $B(s)$ is a continuous path in $\mathcal{L}$ from B_0 to B_1. Since the symplectic mapping $\mathbf{x} = Q(t,s)\mathbf{y}$ transforms the strongly stable linear system $\dot{\mathbf{x}} = A(t,s)\mathbf{x}$ into the linear system $\dot{\mathbf{y}} = B(s)\mathbf{y}$, this system is strongly stable by Theorem 7.41. Therefore, $B(s)$ is indeed a continuous path in $\mathcal{L}_{ss}$, so B_0 is strongly connected to B_1.

Now, the continuous matrix function

$$Q(t,s) = Z(t,s)\mathrm{e}^{-tB(s)}$$

is 2τ-periodic in t and is such that

$$Q(t,0) = Z_0(t)\mathrm{e}^{-tB_0} = Q_0(t) \quad \text{and} \quad Q(t,1) = Z_1(t)\mathrm{e}^{-tB_1} = Q_1(t),$$

so it gives a homotopy from $Q_0(t)$ to $Q_1(t)$. □

The following proposition is a key point in the proof of Theorem 7.44.

Proposition 7.46 *To each strongly stable* τ*-periodic linear Hamiltonian system* $\dot{\mathbf{x}} = A(t)\mathbf{x}$ *we can assign an integer* $n(A)$*, which is a homotopy invariant.*

Proof Given the strongly stable system (7.19) consider the Floquet decomposition (7.20) of its matrizant. Since $Q(t)$ is a closed path in $Sp(2n,\mathbf{R})$ and $\pi_1(Sp(2n,\mathbf{R})) = \mathbf{Z}$, we assign to $A(t)$ the homotopy class $m = [Q(t)]$ of $Q(t)$. From Sect. 7.7.2, $Q(t) = (\mathbf{q}(t), \mathrm{e}^{it}) \in X \times S^1$.

Now, let $\dot{\mathbf{x}} = A_0(t)\mathbf{x}$ and $\dot{\mathbf{x}} = A_1(t)\mathbf{x}$ be strongly stable systems and let $Q_0(t)$ and $Q_1(t)$ be the periodic matrices in the Floquet real form of their matrizants. By Proposition 7.45 if $A_0(t)$ is homotopic to $A_1(t)$, then $Q_0(t)$ is homotopic to $Q_1(t)$; hence, the integers m_0 and m_1 defined by them are equal. Therefore, the integer $m = n(A)$ is homotopy invariant. □

Definition 7.47 We call the integer $m = n(A)$ the *Gelfand–Lidskii index* of $A(t)$.

By Proposition 7.12 the index is well defined.

Consider the Floquet real form of the matrizants (7.25) of the linear systems (7.22), so $Q_i(t)$ are 2τ-periodic and B_i, constant real Hamiltonian matrices. We recall that the mappings $\mathbf{x} = Q_i(t)\mathbf{y}$ define canonical changes of variables that transform the time-dependent systems (7.22) into the autonomous linear Hamiltonian systems

$$\dot{\mathbf{y}} = B_0\mathbf{y}, \quad \dot{\mathbf{y}} = B_1\mathbf{y}. \tag{7.26}$$

We have the following relations between the matrices $A_i(t)$ and B_i for $i = 0, 1$,

$$\begin{aligned} A_i(t) &= \dot{Q}_i(t)Q_i(t)^{-1} + Q_i(t)B_iQ_i(t)^{-1}, \\ B_i &= Q_i(t)^{-1}A_i(t)Q_i(t) - Q_i(t)^{-1}\dot{Q}_i(t). \end{aligned} \tag{7.27}$$

which are readily seen to be valid by differentiating the matrizants given in (7.25).

For the converse of Proposition 7.45, we need an extra condition on the matrices $A_0(t)$ and $A_1(t)$.

Proposition 7.48 *If B_0 is strongly connected to B_1 and $n(A_0) = n(A_1)$, then $A_0(t)$ is strongly homotopic to $A_1(t)$.*

Proof Let $B(s)$ be a continuous path in $\mathcal{L}_{ss}$ from B_0 to B_1. Using the Floquet decompositions (7.25) consider the continuous paths $Q_0, Q_1 : \mathbf{R} \to X \times S^1$, defined by the 2τ-periodic symplectic matrices $Q_0(t)$ and $Q_1(t)$, where X is the simply connected space mentioned at the end of Sect. 7.7.2. Write $Q_j(t) = (\mathbf{q}_j(t), z_j(t)) \in X \times S^1$, $j = 0, 1$. Then, $\mathbf{q}_0(t), \mathbf{q}_1(t)$ are 2τ-periodic paths in X and since this space is simply connected there is a homotopy $\mathbf{q}(t, s)$ joining $\mathbf{q}_0(t)$ to $\mathbf{q}_1(t)$. Also, $z_0(t)$ and $z_1(t)$ are 2τ-periodic paths in S^1. Since $n(A_0) = n(A_1)$, there is a homotopy $\mathbf{z} : \mathbf{R} \times [0, 1]$ joining $z_0(t)$ to $z_1(t)$. Therefore, the 2τ-periodic mapping in t, $Q(t, s) = (\mathbf{q}(t, s), \mathbf{z}(t, s))$, gives a homotopy from $Q_0(t)$ to $Q_1(t)$. The continuous deformation $Q(t, s)$ can be smoothed out and so we assume that it has a continuous derivative with respect to t. We then set

$$A(t, s) = \dot{Q}(t, s)Q(t, s)^{-1} + Q(t, s)B(s)Q(t, s)^{-1}, \tag{7.28}$$

where the dot means derivative with respect to t. Then, using (7.27), we have

$$A(t,0) = \dot{Q}_0(t)Q_0(t)^{-1} + Q_0(t)B_0Q_0(t)^{-1} = A_0(t) \quad \text{and} \quad A(t,1) = A_1(t),$$

which gives a homotopy from $A_0(t)$ to $A_1(t)$, viewed as 2τ-periodic matrices, because $A(t, s)$ is 2τ-periodic in t.

To prove that it is a strong homotopy, let $Z(t, s) = Q(t, s)e^{tB(s)}$. Differentiate with respect to t, and use (7.28) to get

$$\dot{Z}(t,s) = \dot{Q}(t,s)e^{tB(s)} + Q(t,s)B(s)e^{tB(s)} = A(t,s)Z(t,s).$$

Since $Z(0, s) = I$, $Z(t, s)$ is the matrizant of $\dot{\mathbf{x}} = A(t, s)\mathbf{x}$ and $Z(t, s) = Q(t, s)e^{tB(s)}$ is the Floquet decomposition of this system. Therefore, the symplectic transformation $\mathbf{x} = Q(t, s)\mathbf{y}$ takes the 2τ-periodic linear system $\dot{\mathbf{x}} = A(t, s)\mathbf{x}$ to the constant system $\dot{\mathbf{y}} = B(s)\mathbf{y}$. Since this is strongly stable, Theorem 7.41 says that the periodic system $\dot{\mathbf{x}} = A(t, s)\mathbf{x}$ is strongly stable. Therefore, $A(t, s) : A_0(t) \sim A_1(t)$ is a strong homotopy.

Finally, to get a homotopy connecting the τ-periodic paths $A_0(t)$ and $A_1(t)$ take $\mathcal{A}(t, s)$ defined by

$$\mathcal{A}(t,s) = (1-s)A(2t,s), \quad \text{for } 0 \leq s \leq \frac{1}{2}$$

$$\mathcal{A}(t,s) = sA(2t,s), \quad \text{for } \frac{1}{2} \leq s \leq 1.$$

Clearly, for each s, $\mathcal{A}(t, s)$ is strongly stable. □

In the proof of Theorem 7.44 we use the Proposition 7.40. In the symplectic case this proposition can be refined to include in the disks the eigenvalues of the first and second kinds. In Section 2.10, Chapter 4 of the book [64] a theorem on the continuity of eigenvalues is concisely stated as follows.

Theorem 7.49 *The eigenvalues of the first kind of a G-unitary matrix X are continuous functions of X. The same holds for the eigenvalues of the second kind.*

In the proof of this theorem Yakubovich and Starzhinskii use arguments that prove a result, which we rephrase here as the following theorem.

Theorem 7.50 *Let X_0 be a G-unitary matrix and take $r > 0$ such that the disks $B_r(\alpha)$ are pairwise disjoint where α runs over the set of distinct eigenvalues of X_0. If r is sufficiently small there exists $\delta > 0$ such that when X is G-unitary and $||X - X_0|| < \delta$, then for any eigenvalue α of X_0, X possesses the same number of eigenvalues of the first kind inside the disk $B_r(\alpha)$ as does X_0. The same holds for eigenvalues of the second kind.*

Corollary 7.50.1 *Let $r > 0$ be such that the open disks $D_r(\rho)$ are pairwise disjoint as ρ runs over the distinct eigenvalues of the symplectic matrix M_0. Then, for r sufficiently small there is a number $\delta > 0$ such that for any symplectic matrix M with $||M - M_0|| < \delta$, if ρ_0 is an eigenvalue of the first kind and multiplicity k of*

M_0, then M has k eigenvalues of the first kind in the disk $D_r(\rho_0)$. The same holds for eigenvalues of the second kind.

Proof Indeed, taking r and δ as in Proposition 7.40 and Theorem 7.50, the Proposition 7.40 says that k eigenvalues of M are in the disk $D_r(\rho_0)$ and since ρ_0 is of the first kind and has multiplicity k Theorem 7.50 says that they are all of the first kind. The same holds if $\alpha = \rho_0$ is of the second kind. □

Now, let $Z(t,s) = Q(t,s)e^{tB(s)}$, $0 \le s \le 1$, be the Floquet decomposition of the real τ-periodic strongly stable linear Hamiltonian system $\dot{\mathbf{x}} = A(t,s)\mathbf{x}$, with $Q(t,s)$ 2τ-periodic in t and $B(s)$ real.

As $\dot{\mathbf{x}} = A(t,s)\mathbf{x}$ is strongly stable, neither 1 nor -1 are eigenvalues of $Z(s) = Z(2\tau, s)$ by Proposition 7.37. Also, it can easily be shown that distinct eigenvalues of $Z(s)$ are in an even number. Let them be $\rho_1, \ldots, \rho_k$, with $k = 2l$.

Proposition 7.51 *Let $0 \le s_0 \le 1$. Then there is a positive number γ such that for any $s \in (s_0 - \gamma, s_0 + \gamma) \cap [0, 1]$, the matrices $B(s)$ and $B(s_0)$ have the same signature, $\boldsymbol{\delta}(s) = \boldsymbol{\delta}(s_0)$.*

Proof Since $\dot{\mathbf{x}} = A(t,s)\mathbf{x}$ is strongly stable all the $k = 2l$ distinct eigenvalues $\rho_1(s), \ldots, \rho_k(s)$ of $Z(s) = Z(2\tau, s)$ are on the unit circle and they are either of the first kind or of the second kind. Take $0 \le s_0 \le 1$. Let $r > 0$ be sufficiently small and take a corresponding $\delta > 0$ as in the above corollary. Denote by $\rho_j^{(1)}(s_0)$ the eigenvalues of the first kind of $Z(s_0)$ and by $\rho_j^{(2)}(s_0)$ those of the second kind. By Corollary 7.50.1 if $M(s)$ is symplectic and $||M(s) - Z(s_0)|| < \delta$, then the eigenvalues $\rho(s)$ of $M(s)$ in the disk $B_r(\rho_j^{(1)}(s_0))$, which are equal in number to the multiplicity of $\rho_j^{(1)}(s_0)$, are of the first kind. Similarly, the eigenvalues of $M(s)$ in the disk in $B_r(\rho_j^{(2)}(s_0))$ are of the second kind.

Since $Z(s)$ is continuous, there is a positive number γ such that if $|s - s_0| < \gamma$, then $||Z(s) - Z(s_0)|| < \delta$. Consider the interval $I_\gamma = \{s;\ |s - s_0| < \gamma\} \cap [0, 1]$. Then, if s is in this interval, the eigenvalues $\rho(s)$ of $Z(s)$ in the disk $B_r(\rho_j^{(1)}(s_0))$ are of the first kind and those in the disk $B_r(\rho_j^{(2)}(s_0))$ are of the second kind.

Since the eigenvalues are on the unit circle and none of them is equal to 1 or -1, then half of them lie on the upper semi-circle and the other half on the lower semi-circle. By renumbering them we can assume that the first l eigenvalues $\rho_1(s), \ldots, \rho_l(s)$ lie on the upper semi-circle. We take them in increasing order of their principal arguments, that is, $\rho_j(s) = e^{2\tau i \omega_j(s)}$, then $0 < \omega_1(s) < \omega_2(s) < \cdots < \omega_l(s) < \pi$. We now arrange $\rho_1(s), \ldots, \rho_l(s)$ in groups of eigenvalues characterized by the condition that each group is contained in some disk $B_r(\rho_j^{(1)}(s_0))$ or $B_r(\rho_j^{(2)}(s_0))$. Therefore, each group is formed by eigenvalues of the same kind as the eigenvalue of $M(s_0)$ defining the center of the disk and in number equal to its multiplicity. Therefore, the eigenvalues $\rho_1(s), \ldots, \rho_l(s)$, for $s \in I_\gamma$ have the same disposition on the upper semi-circle as the centers $\rho_1(s_0), \ldots, \rho_l(s_0)$ and therefore we have the equality of the signatures $\boldsymbol{\delta}(s) = \boldsymbol{\delta}(s_0)$, for all $s \in I_\gamma$. □

7.7.4 Proof of Theorem 7.44

First, suppose that $A(t) \in [A_0(t)]$. Then, $A(t)$ is strongly homotopic to $A_0(t)$, so by Propositions 7.45 and 7.48 we have $n(A) = n(A_0)$. Now, let us prove that $\boldsymbol{\delta}(A) = \boldsymbol{\delta}(A_0)$.

By compactness, we can cover the interval [0, 1] with a finite number of intervals $I_{\gamma_i} = \{s;\ |s - s_i| < \gamma_i\} \cap [0, 1]$, $i = 1, \ldots, q$, given by Proposition 7.51 such that $0 \in I_{\gamma_1}$, $1 \in I_{\gamma_q}$, $s_i + \gamma_i \in I_{\gamma_{i+1}}$, $i = 1, \ldots, q-1$. Then, as seen at the end of Proposition 7.51, $\rho_1(s), \ldots, \rho_l(s)$ have the same disposition on the upper semicircle as the centers $\rho_1(s_i), \ldots, \rho_k(s_i)$ and signatures $\boldsymbol{\delta}(s) = \boldsymbol{\delta}(s_i)$, for all $s \in I_{\gamma_i}$. It then follows that $\boldsymbol{\delta}(0) = \boldsymbol{\delta}(1)$, which means that $\boldsymbol{\delta}(A) = \boldsymbol{\delta}(A_0)$. This proves the necessary part of the theorem.

We now prove the converse statement. Suppose that the pattern of signs for the frequency vector $\boldsymbol{\omega}^0 = (\omega_1^0, \omega_2^0, \ldots, \omega_n^0)$ of B_0 is the same as for the frequency vector $\boldsymbol{\omega}^1 = (\omega_1^1, \omega_2^1, \ldots, \omega_n^1)$ of B_1, that is, $A_0(t)$ and $A_1(t)$ have the same signature. Let us prove that B_0 is strongly connected to B_1.

We first assume the frequencies $\omega_1, \ldots, \omega_n$ to be distinct. In this case, we use the normalizing procedure given in Theorem 4.33. Consider the normal forms of B_0 and B_1, that is, take the real symplectic matrices P_0 and P_1 such that $P_j^{-1} B_j P_j = \begin{bmatrix} O & \Omega_j \\ -\Omega_j & O \end{bmatrix}$, $(j = 0, 1)$ where

$$\Omega_j = \begin{bmatrix} \delta_1 \omega_1^j & \ldots & 0 \\ \vdots & \ddots & \vdots \\ 0 & \ldots & \delta_n \omega_n^j \end{bmatrix}, \qquad j = 0, 1.$$

Let $\omega_j(s) = (1 - s)\omega_j^0 + s\omega_j^1$, $0 \le s \le 1$. As ω_j^0 and ω_j^1 are positive we have that $\omega_j(s) > 0$ for all $s \in [0, 1]$. Since the real symplectic group is connected there is a continuous path $P(s)$ in $Sp(2n, \mathbf{R})$ from P_0 to P_1. Consider the Hamiltonian matrix

$$B(s) = P(s) \begin{bmatrix} O & \Omega(s) \\ -\Omega(s) & O \end{bmatrix} P(s)^{-1}, \quad \text{where} \quad \Omega(s) = \begin{bmatrix} \delta_1 \omega_1(s) & \ldots & 0 \\ \vdots & \ddots & \vdots \\ 0 & \ldots & \delta_n \omega_n(s) \end{bmatrix} \tag{7.29}$$

and $\boldsymbol{\delta} = (\delta_1, \ldots, \delta_n)$ is the signature of the frequency vector $\boldsymbol{\omega}^0$. Let us prove that $B(s)$ is a strong path from B_0 to B_1.

The eigenvalues of $B(s)$ are $\pm i\omega_1(s), \ldots, \pm i\omega_n(s)$ and an eigenvector of $B(s)$ corresponding to $i\omega_j(s)$ is $\mathbf{v}_j = P(s)\mathbf{e}_j + i\delta_j P(s)\mathbf{e}_{n+j}$, with $\mathbf{e}_1, \ldots, \mathbf{e}_{2n}$ the canonical basis of $\mathbf{R}^{2n}$. Since $P(s)$ is symplectic and $\{\mathbf{e}_j, \mathbf{e}_{n+j}\} = 1$, we have $\{P(s)\mathbf{e}_j, P(s)\delta_j \mathbf{e}_{n+j}\} = \delta_j$, hence $i\omega_j$ is an eigenvalue of the first or second kind in accordance with $\delta_j = +1$ or $\delta_j = -1$. This shows that $B(s)$ is strongly stable. Therefore, $B(s)$ is a strong path from B_0 to B_1.

Since $A_0(t)$ and $A_1(t)$ have the same index, we conclude by Proposition 7.48 that they are strongly homotopic. Therefore, $A_1(t) \in [A_0(t)]$.

Now consider the case of multiple frequencies. The frequency vector $\boldsymbol{\omega}$ and its signature $\boldsymbol{\delta}$ are

$$\boldsymbol{\omega} = (\omega_1, \ldots, \omega_1; \omega_2, \ldots, \omega_2; \ldots; \omega_k, \ldots, \omega_k),$$
$$\boldsymbol{\delta} = (\delta_1, \ldots, \delta_1; \delta_2, \ldots, \delta_2; \ldots; \delta_k, \ldots, \delta_k).$$

Since B_0 is stable it is diagonalizable, so the restriction $B_0^{(l)}$ of B_0 to the subspace $\eta^\dagger(i\omega_l^0) \oplus \eta^\dagger(-i\omega_l^0)$, $l = 1 \ldots, k$ is diagonalizable. The same is true of B_1. So by Theorem 4.39, for each $l = 1, \ldots, k$ the normal forms of $B_0^{(l)}$ and $B_1^{(l)}$ are given by

$$P_{0,l}^{-1} B_0^{(l)} P_{0,l} = \begin{bmatrix} O & \Omega_l^{(0)} \\ -\Omega_l^{(0)} & O \end{bmatrix} \quad \text{and} \quad P_{1,l}^{-1} B_1^{(l)} P_{1,l} = \begin{bmatrix} O & \Omega_l^{(1)} \\ -\Omega_l^{(1)} & O \end{bmatrix}$$

where

$$\Omega_l^{(0)} = \begin{bmatrix} \delta_l \omega_l^0 & \ldots & 0 \\ & \ddots & \\ 0 & \ldots & \delta_l \omega_l^0 \end{bmatrix}, \quad \Omega_l^{(1)} = \begin{bmatrix} \delta_l \omega_l^1 & \ldots & 0 \\ & \ddots & \\ 0 & \ldots & \delta_l \omega_l^1 \end{bmatrix}.$$

Taking a symplectic path $P_l(s)$ connecting $P_{0,l}$ to $P_{1,l}$ we construct the continuous path $B_l(s)$ connecting $B_0^{(l)}$ to $B_1^{(l)}$ as we did in (7.29). Then, as before, it is shown that $B_l(s)$ is a strong path.

Finally, using these paths we construct the strong path given by $B(s) = \mathrm{diag}[B_1(s), \ldots, B_k(s)]$ connecting B_0 to B_1 and using the fact that $A_0(t)$ and $A_1(t)$ have the same index, we conclude that $A_1(t) \in [A_0(t)]$. □

The disposition in the case of Theorem 7.44 is made by arranging half of the characteristic exponents $i\omega_1, \ldots, i\omega_n$ in the positive imaginary axis according to the rule $\omega_1 > \cdots > \omega_n > 0$ with the signs determined by the sequence of δ's in the signature of the frequency vector $\boldsymbol{\omega} = (\omega_1, \ldots, \omega_n)$. This disposition induces an arrangement of all the characteristic exponents on the imaginary axis with the conjugate eigenvalues $-i\omega_1, \ldots, -i\omega_n$ displaced symmetrically with respect to the origin and accompanied by the corresponding symmetric values of the δ's.

Now, under the mapping $i\omega \mapsto e^{i2\tau\omega}$ the arrangement of all the characteristic exponents on the imaginary axis is carried to an arrangement of all the multipliers on the unit circle, with half of them lying on the upper semi-circle, half on the lower semi-circle, and none coinciding with the points $+1$ and -1. We assign to a multiplier the same plus sign, $\oplus$, or minus sign, $\ominus$, of the corresponding characteristic exponent. This gives an arrangement of half of the multipliers on the upper semi-circle with the associated sequence of signs, that is to say, a disposition of the multipliers on the upper semi-circle.

Conversely, a disposition of the multipliers on the upper semi-circle gives an arrangement of the characteristic exponents on the upper imaginary axis with the same sequence of signs.

The disposition of the eigenvalues on the positive imaginary axis with its sequence of signs is also clear for the frequency vector

$$\boldsymbol{\omega} = (\omega_1, \dots, \omega_1; \omega_2, \dots, \omega_2; \dots; \omega_k, \dots, \omega_k)$$

with signature $\boldsymbol{\delta} = (\delta_1, \dots, \delta_1; \delta_2, \dots, \delta_2; \dots; \delta_k, \dots, \delta_k)$.

This discussion shows the equivalence of Theorem 7.44 and the Gelfand–Lidskii Theorem 7.43.

7.7.5 A Formula for the Index

We now aim to find a formula to compute the index of the matrix $A(t)$.

In the article [24] Gelfand and Lidskii consider the upper-left $n \times n$ submatrix $z(t)$ of the matrix $M^* Z M$, where $Z(t)$ is the matrizant of the system $\dot{\mathbf{x}} = A(t)\mathbf{x}$, M is the $2n \times 2n$ unitary matrix $M = \frac{1}{\sqrt{2}} \begin{bmatrix} I & I \\ iI & -iI \end{bmatrix}$, I being the $n \times n$ identity matrix, and $M^* = \overline{M}^T$ is the adjoint of M. Then, they prove that the index m of $A(t)$ is given by the formula

$$m = \frac{1}{2\pi} \left(\text{Arg Det} z(t) \Big|_0^\tau - \sum_{k=1}^{n} \theta_k \right), \tag{7.30}$$

where θ_k is the principal argument of the complex number ρ_k.

Our approach to the computation of the index is as follows. We write the Floquet decomposition of the matrizant of the τ-periodic linear Hamiltonian system $\dot{\mathbf{x}} = A(t)\mathbf{x}$ in the form

$$Q(t) = Z(t) \mathrm{e}^{-tB}, \tag{7.31}$$

where $Q(t)$ is a 2τ-periodic real symplectic matrix and B is a constant real Hamiltonian matrix. Since $A(t)$ is strongly stable we have by Theorem 7.41 that B is strongly stable, so its normal form is obtained from the nondegenerate context of Chap. 4, Theorem 4.33,[3] which gives a real symplectic basis $\mathcal{B} = \{\mathbf{u}_1, \dots, \mathbf{u}_n, \mathbf{v}_1, \dots, \mathbf{v}_n\}$ of $\mathbf{C}^{2n}$ such that the matrix of B on this basis is

[3] In the case of multiple eigenvalues we use Theorem 4.39.

$$\Omega = \begin{bmatrix} O & \Omega_0 \\ -\Omega_0 & O \end{bmatrix}, \quad \text{with} \quad \Omega_0 = \operatorname{diag}[\delta_1\omega_1, \ldots, \delta_n\omega_n].$$

Recall that $\mathbf{v}_j = \mathbf{r}_j + i\mathbf{s}_j$ is an eigenvector of B corresponding to the eigenvalue $i\omega_j$ and that $\delta_j = \operatorname{sign}\{\mathbf{r}_j, \mathbf{s}_j\}$ where $\{\mathbf{r}_j, \mathbf{s}_j\} = \mathbf{r}_j^T J\mathbf{s}_j$. Since $B\mathbf{u}_j = -\delta_j\omega_j\mathbf{v}_j$ and $B\mathbf{v}_j = \delta_j\omega_j\mathbf{u}_j$ we easily compute

$$\mathrm{e}^{tB}\mathbf{u}_j = \cos(\delta_j\omega_j t)\mathbf{u}_j - \sin(\delta_j\omega_j t)\mathbf{v}_j \text{ and } \mathrm{e}^{tB}\mathbf{v}_j = \sin(\delta_j\omega_j t)\mathbf{u}_j + \cos(\delta_j\omega_j t)\mathbf{v}_j,$$

so the matrix of e^{tB} on the basis $\mathcal{B}$ is

$$\left[\mathrm{e}^{tB}\right]_{\mathcal{B}} = \begin{bmatrix} C(t) & S(t) \\ -S(t) & C(t) \end{bmatrix}, \tag{7.32}$$

where $C(t) = \operatorname{diag}[\cos\delta_1\omega_1 t, \ldots, \cos\delta_n\omega_n t]$ and the matrix $S(t)$ is given by $S(t) = \operatorname{diag}[\sin\delta_1\omega_1 t, \ldots, \sin\delta_n\omega_n t]$.

Now we consider the Euclidean inner product of $\mathbf{R}^{2n}$ associated with the basis $\mathcal{B}$, that is,

$$\langle \mathbf{u}, \mathbf{v}\rangle_{\mathcal{B}} = x_1\xi_1 + \cdots + x_n\xi_n + y_1\eta_1 + \cdots + y_n\eta_n, \tag{7.33}$$

where $x_1, \ldots, x_n, y_1, \ldots, y_n$ are the coordinates of $\mathbf{u}$ and $\xi_1, \ldots, \xi_n, \eta_1, \ldots, \eta_n$ the coordinates of $\mathbf{v}$ on the basis $\mathcal{B}$.

Look at the symplectic matrices $Q(t)$, $Z(t)$ and e^{-tB} as linear operators in the space $\mathbf{R}^{2n}$. The matrix of the third on the basis $\mathcal{B}$ is the transpose of the matrix in the right-hand side of (7.32). The matrices of the other two will be worked out below.

We first consider the polar decomposition of $[Q(t)]_{\mathcal{B}}$ and $[Z(t)]_{\mathcal{B}}$ relative to the inner product (7.33)

$$[Q(t)]_{\mathcal{B}} = Q_1Q_2, \quad [Z(t)]_{\mathcal{B}} = Z_1Z_2,$$

so Q_1, Z_1 are symplectic positive definite symmetric matrices and Q_2, Z_2 are symplectic matrices and orthogonal relative to $\langle\ ,\ \rangle_{\mathcal{B}}$, that is,

$$\langle Q_2\mathbf{u}, Q_2\mathbf{v}\rangle_{\mathcal{B}} = \langle \mathbf{u}, \mathbf{v}\rangle_{\mathcal{B}} \quad \text{and} \quad \langle Z_2\mathbf{u}, Z_2\mathbf{v}\rangle_{\mathcal{B}} = \langle \mathbf{u}, \mathbf{v}\rangle_{\mathcal{B}};$$

hence, we have the usual identities $Q_2Q_2^T = Q_2^TQ_2 = I$ and $Z_2Z_2^T = Z_2^TZ_2 = I$.

Therefore, see (7.24) in Sect. 7.7.2, Q_2 and Z_2 are matrices of the form

$$Q_2 = \begin{bmatrix} U & V \\ -V & U \end{bmatrix} \quad \text{and} \quad Z_2 = \begin{bmatrix} \mathcal{Z}_1 & \mathcal{Z}_2 \\ -\mathcal{Z}_2 & \mathcal{Z}_1 \end{bmatrix},$$

with U, V, Z_1, Z_2 real $n\times n$ matrices such that $U^TU + V^TV = I$, $\mathcal{Z}_1^T\mathcal{Z}_1 + \mathcal{Z}_2^T\mathcal{Z}_2 = I$, and U^TV, $\mathcal{Z}_1^T\mathcal{Z}_2$ symmetric. It follows from these equalities that $W = U + iV$

and $\mathcal{Z} = \mathcal{Z}_1 + i\mathcal{Z}_2$ are $n \times n$ unitary matrices. Let $e^{i\psi(t)} = \text{Det}W(t), 0 \le \psi < 2\pi$ and $e^{i\varphi(t)} = \text{Det}\mathcal{Z}(t), 0 \le \varphi < 2\pi$.

Let $M : \mathbf{C}^{2n} \to \mathbf{C}^{2n}$ be the linear operator whose matrix in the real basis $\mathcal{B}$ is the unitary matrix $M_{\mathcal{B}} = \frac{1}{\sqrt{2}}\begin{bmatrix} I & I \\ iI & -iI \end{bmatrix}$. Then, we have

$$M_{\mathcal{B}}^* Q_2 M_{\mathcal{B}} = \begin{bmatrix} W & O \\ O & \overline{W} \end{bmatrix}, \quad M_{\mathcal{B}}^* Z_2 M_{\mathcal{B}} = \begin{bmatrix} \mathcal{Z} & O \\ O & \overline{\mathcal{Z}} \end{bmatrix},$$

and $M_{\mathcal{B}}^* \left[e^{-tB}\right]_{\mathcal{B}} M_{\mathcal{B}} = \begin{bmatrix} C(t)-iS(t) & O \\ O & C(t)+iS(t) \end{bmatrix}$.

Using $[Q(t)]_{\mathcal{B}} = Q_1 Q_2$ we write

$$M_{\mathcal{B}}^*[Q(t)]_{\mathcal{B}} M_{\mathcal{B}} = (M_{\mathcal{B}}^* Q_1 M_{\mathcal{B}})(M_{\mathcal{B}}^* Q_2 M_{\mathcal{B}}) = \begin{bmatrix} Q_{11} & Q_{12} \\ Q_{21} & Q_{22} \end{bmatrix} \begin{bmatrix} W & O \\ O & \overline{W} \end{bmatrix}.$$

The top-left $n \times n$ block of $M_{\mathcal{B}}^*[Q(t)]_{\mathcal{B}} M_{\mathcal{B}}$ is the matrix $Q_{11}W$, which has a determinant equal to $(\text{Det}\, Q_{11})e^{i\psi(t)}$. Since $\left\langle \begin{bmatrix} Q_{11} & Q_{12} \\ Q_{21} & Q_{22} \end{bmatrix} \begin{bmatrix} \mathbf{x} \\ 0 \end{bmatrix}, \begin{bmatrix} \mathbf{x} \\ 0 \end{bmatrix} \right\rangle = \langle Q_{11}\mathbf{x}, \mathbf{x} \rangle$ and the matrix Q_1 is positive definite, so is the matrix Q_{11}, hence, Det Q_{11} is a positive real number. Therefore, the principal argument of the complex number $(\text{Det}Q_{11})e^{i\psi(t)}$ is $\text{Arg}\, e^{i\psi(t)} = \psi(t)$. As the matrix $[Q(t)]_{\mathcal{B}}$ is 2τ-periodic, so is the matrix $Q_{11}W$; hence, the variation in the argument of the complex number $(\text{Det}Q_{11})e^{i\psi(t)}$ is an integral multiple m of 2π, that is,

$$\psi(2\tau) - \psi(0) = 2\pi m. \tag{7.34}$$

Now we do the same for the matrix $[Z(t)]_{\mathcal{B}} \left[e^{-tB}\right]_{\mathcal{B}} = Z_1 Z_2 \left[e^{-tB}\right]_{\mathcal{B}}$, writing

$$M_{\mathcal{B}}^*[Z(t)]_{\mathcal{B}} \left[e^{-tB}\right]_{\mathcal{B}} M_{\mathcal{B}} = (M_{\mathcal{B}}^* Z_1 M_{\mathcal{B}})(M_{\mathcal{B}}^* Z_2 M_{\mathcal{B}})(M_{\mathcal{B}}^* \left[e^{-tB}\right]_{\mathcal{B}} M_{\mathcal{B}}).$$

Setting $Z_1 = \begin{bmatrix} Z_{11} & Z_{12} \\ Z_{21} & Z_{22} \end{bmatrix}$, the top-left $n \times n$ matrix block of the product $M_{\mathcal{B}}^*[Z(t)]_{\mathcal{B}} \left[e^{-tB}\right]_{\mathcal{B}} M_{\mathcal{B}}$ is the matrix $Z_{11}\mathcal{Z}\big(C(t) - iS(t)\big)$. Since $\text{Det}\mathcal{Z}(t) = e^{i\phi(t)}$ and

$$C(t)-iS(t) = \begin{bmatrix} e^{-i\delta_1\omega_1 t} & \dots & 0 \\ \vdots & \ddots & \vdots \\ 0 & \dots & e^{-i\delta_n\omega_n t} \end{bmatrix}$$

the determinant of this top-left matrix block is equal to

$$(\text{Det}Z_{11})e^{i\phi(t)}e^{-i\langle\boldsymbol{\delta},\boldsymbol{\omega}\rangle t},$$

where $\langle \boldsymbol{\delta}, \boldsymbol{\omega} \rangle = \delta_1\omega_1 + \cdots + \delta_n\omega_n$. Since Z_1 is a positive definite matrix so is Z_{11}; hence, $\mathrm{Det}Z_{11}$ is a positive real number. So the principal argument of the complex number $(\mathrm{Det}Z_{11})\mathrm{e}^{i\phi(t)}\mathrm{e}^{-i\langle \boldsymbol{\delta}, \boldsymbol{\omega}\rangle t}$ is equal to $\phi(t) - \langle \boldsymbol{\delta}, \boldsymbol{\omega}\rangle t$. Therefore, from (7.31) and (7.34) we have

$$m = \frac{1}{2\pi}\Big[\mathrm{Arg}\,\mathrm{Det}\mathcal{Z}(t)\Big|_0^{2\tau} - \langle \boldsymbol{\delta}, \boldsymbol{\omega}\rangle 2\tau\Big].$$

Let $m_j = \left[\frac{\tau\omega_j}{\pi}\right]$ be the largest integer less than or equal to $\frac{\tau\omega_j}{\pi}$. Then, $2\tau\omega_j = \theta_j + 2\pi m_j$, with $0 \le \theta_j < 2\pi$. Then, $\langle \boldsymbol{\delta}, \boldsymbol{\omega}\rangle 2\tau = \delta_1\theta_1 + \cdots + \delta_n\theta_n$ and the above formula for m becomes

$$m = \frac{1}{2\pi}\Big[\mathrm{Arg}\,\mathrm{Det}\mathcal{Z}(t)\Big|_0^{2\tau} - \sum_{k=1}^{n}\delta_j\theta_j\Big]. \tag{7.35}$$

Remark 7.52 If L is the transition matrix from the canonical basis $\mathbf{e}_1, \ldots, \mathbf{e}_{2n}$ of $\mathbf{R}^{2n}$ to the real symplectic basis $\mathcal{B}$, then L is a symplectic matrix and $[Q(t)]_{\mathcal{B}} = L^{-1}Q(t)L$. Let $L(s)$ be a continuous path in $Sp(2n, \mathbf{R})$ such that $L(0) = I$ and $L(1) = L$. Then, $Q(t, s) = L(s)^{-1}Q(t)L(s)$ is a homotopy from $Q(t)$ to $[Q(t)]_{\mathcal{B}}$; hence, the integer m in (7.35) is the index of $A(t)$.

We observe that the formula (7.35) depends on the basis $\mathcal{B}$, which is determined by the way in which we express the sequence of characteristic exponents $\pm i\omega_j$. For another way of writing down the sequence we get a matrix $\mathcal{B}_1$. But the matrices $[Q(t)]_{\mathcal{B}}$ and $[Q(t)]_{\mathcal{B}_1}$ are symplectically similar and as in the Remark 7.52, we conclude that the index determined by $[Q(t)]_{\mathcal{B}_1}$ is the same as the index determined by $[Q(t)]_{\mathcal{B}}$, so the expression (7.35) gives the same value for any choice of the ordering of the characteristic exponents.

Therefore, if instead of assuming $\omega_1 > \cdots > \omega_n > 0$ as we did, we take $i\omega_1, \ldots, i\omega_n$ as the eigenvalues of the first kind of B, then we have $\delta_1 = \cdots = \delta_n = +1$ and the index formula (7.35) is written as follows

$$m = \frac{1}{2\pi}\Big[\mathrm{Arg}\,\mathrm{Det}\mathcal{Z}(t)\Big|_0^{2\tau} - \sum_{k=1}^{n}\theta_j\Big], \tag{7.36}$$

which corresponds to the formula (7.30) given by Gelfand and Lidskii.

In (7.36) the matrizant is computed at 2τ whereas in (7.30) it is computed at $t = \tau$. Notice, however, that since

$$Z(2\tau) = Z(\tau)^2 \quad \text{and} \quad M^*\left[\mathrm{e}^{2\tau B}\right]_{\mathcal{B}} M = \left(M^*\left[\mathrm{e}^{\tau B}\right]_{\mathcal{B}} M\right)^2,$$

the principal arguments computed at 2τ have double the value when computed at τ.

Therefore, the index in Definition 7.47 is twice the index defined by Gelfand and Lidskii. We could take them to be the same by saying after the Proof of Proposition 7.46 that it will be proved later that the integer m is even and then define the index $n(A)$ as half this integer.

Remark 7.53 By taking the matrix $Q(t)$ in the Floquet decomposition with period 2τ we profited from the fact that B is real and then could use directly the topological structure of the real symplectic group to define the index of $A(t)$ as the index of the loop $Q(t)$ in $Sp(2n, \mathbf{R})$. By computing with the period τ the matrix B; hence, $Q(t)$ may no longer be real and since the fundamental group of the complex symplectic group $Sp(2n, \mathbf{C})$ is trivial Gelfand and Lidskii in [24] had to go through more involved arguments in order to get to the index formula.

Chapter 8
Parametric Resonance

8.1 Introduction

This chapter deals with periodic linear Hamiltonian systems that depend on parameters. Varying the parameters may set the system into resonances that are responsible for an increase in the amplitudes of the motion which provokes instabilities. In the parameter space the values of the parameters that correspond to stable and unstable motions are studied. The normal forms of Hamiltonian matrices studied in Chap. 5 will be very important in this study.

8.2 The Phenomenon of Resonance

As an illustrative example of the kind of problems and the methods developed in this chapter we consider the planar pendulum with a periodic external agent. In the Example 2.7 we saw that

$$\ddot{\theta} + \alpha \sin\theta = 0$$

defines the dynamics of the pendulum.

Releasing the pendulum from rest at the angle $\theta = a$, it accelerates until it reaches the lower position given by $\theta = 0$ and from this point on it decelerates until it comes to stop at the angle θ_0. The motion is reversed and continues forever in this oscillatory way. The angle θ_0 is the amplitude of this motion. The maximum speed $\dot{\theta}_{\max}$ is reached when $\theta = 0$, so the tension has the maximum value, $T_{\max} = mg + ml\dot{\theta}^2_{\max}$, at this point. If the rod is made of a material that does not support such an effort it breaks down. Now suppose that the pendulum is excited by a periodic external agent that gives an impulse on the bob every time that it passes through the lowest position and always in the same direction. Even with impulses

H. E. Cabral, L. Brandão Dias, *Normal Forms and Stability of Hamiltonian Systems*, Applied Mathematical Sciences 218,
https://doi.org/10.1007/978-3-031-33046-9_8

of small intensity the velocity after the first impulse is a little bigger than θ_0. The amplitude keeps increasing at successive impulses so at some time in the future it may reach a value that causes the rupture of the rod, since the maximal velocity increases with the amplitude. It is not necessary for the impulse to be applied at every moment that the bob of the pendulum is at the lowest position. The important thing is that the vibrating systems have a synchronization, for instance, during a time span in which the pendulum executes three oscillations of the same amplitude, the mechanism of applying the impulses is activated twice. If at an instant in time the impulse is applied when the pendulum is at the lowest position leading to a new value of the amplitude of oscillation, it needs six oscillations of the pendulum with the new amplitude for the impulse to be applied again on the bob of the pendulum at the same position. Nevertheless, at this instant of a new coincidence of lowest position and action of the impulse, the speed of the bob receives an increment that results in a higher value of the amplitude. As time goes on the oscillations get larger and larger amplitudes. This is the phenomenon of *resonance*. Depending on the fragility of the rod, the amplitude may eventually reach a value that provokes its rupture.

The phenomenon of resonance requires the commensurability of the frequencies of two interacting vibrating systems with the periodic application of a small driving force in synchronism with a natural frequency of the structure. Resonance may have made some contribution to the collapse of the Tacoma bridge in 1940. The long narrow bridge had a natural frequency of vertical oscillations and resonance provoked by gales may have played a role in that catastrophic event. Search the Web for the large oscillations performed by the bridge before its rupture.

The simple pendulum described by Eq. (2.29) has two equilibrium points, the lowest point P_0 given by $\theta = 0$ and the highest point P_1, given by $\theta = \pi$. The first equilibrium point is stable and the second, unstable.

Now consider a pendulum with a vertically oscillating suspension point that executes a harmonic motion of small amplitude a according to the law $\mathbf{r}_a = -a\cos t\,\mathbf{j}$ (see Fig. 8.1). The position of the bob of the pendulum is now given by $\mathbf{r} = \mathbf{r}_a + l\mathbf{e}_1$ and since $\ddot{\mathbf{r}} = a\cos t\,\mathbf{j} = a\cos t(\cos\theta\mathbf{e}_1 - \sin\theta\mathbf{e}_2)$ the motion is now governed by the equation

$$\ddot{\theta} + \alpha\sin\theta + \epsilon\cos t\sin\theta = 0,\ \ \alpha = \frac{g}{l},\ \ \epsilon = \frac{a}{l}. \tag{8.1}$$

Setting $x = \theta$ and $y = \dot{\theta}$ this equation can be written in the canonical form

$$\dot{x} = H_y, \quad \dot{y} = -H_x,$$

with the time-dependent Hamiltonian function

$$H(x, y, t, \alpha, \epsilon) = \frac{1}{2}y^2 - \alpha\cos x - \epsilon\cos t\cos x. \tag{8.2}$$

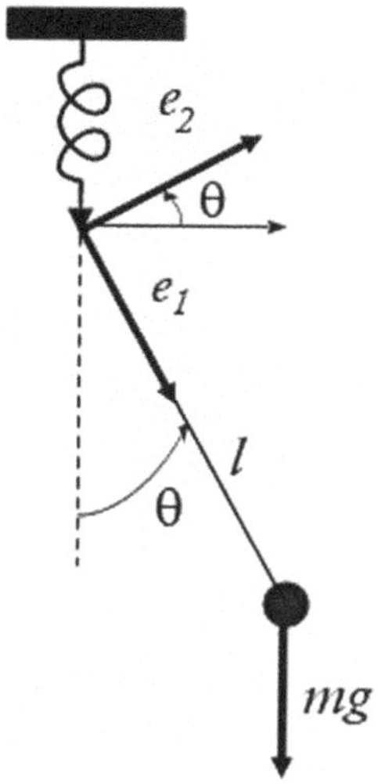

Fig. 8.1 Oscillating suspension point

This Hamiltonian system has the equilibria $P_0 = (0, 0)$ and $P_1 = (\pi, 0)$, for any values of the parameters α, ϵ. For $\epsilon = 0$ the Hamiltonian describes the dynamics of the simple pendulum (2.29) and we know that the equilibrium P_0 is stable. The question arises whether it is stable for the perturbed system (8.3). Because this Hamiltonian is time dependent this is a difficult question, even being a one-degree of freedom Hamiltonian.

So we restrict the analysis to the stability of P_0 for the linearized system around this equilibrium, which is governed by the quadratic Hamiltonian function

$$H(x, y, t, \alpha, \epsilon) = \frac{1}{2}y^2 + \frac{1}{2}\alpha x^2 + \frac{1}{2}\epsilon x^2 \cos t. \tag{8.3}$$

Making the symplectic change of coordinates

$$x = \frac{1}{\sqrt[4]{\alpha}}\xi, \quad y = \sqrt[4]{\alpha}\,\eta,$$

the Hamiltonian becomes

$$\mathcal{H}(\xi, \eta, t, \alpha, \epsilon) = \frac{1}{2}\omega(\xi^2 + \eta^2) + \frac{\epsilon}{2\omega}\xi^2 \cos t, \tag{8.4}$$

where $\omega = \sqrt{\alpha}$ is the frequency of the unperturbed system, which is now in normal form. The function (8.4) is the Hamiltonian for the Mathieu equation (see Sect. 8.5).

Let $A(t, \alpha, \epsilon) = JD^2\mathcal{H}(\xi, \eta, t, \alpha, \epsilon)$ and consider the linear Hamiltonian systems

$$\dot{\mathbf{x}} = A_0(\alpha)\mathbf{x} \quad \text{and} \quad \dot{\mathbf{x}} = A(t, \alpha, \epsilon)\mathbf{x},$$

where $A_0(\alpha) = A(t, \alpha, 0)$ is the Hamiltonian matrix corresponding to the unperturbed system, the simple pendulum.

Let $X(t, \alpha, \epsilon) = Q(t, \alpha, \epsilon)e^{tB(\alpha,\epsilon)}$ be the Floquet decomposition of the matrizant of the second system. Since the matrizant of the first system is given by the fundamental matrix $X_0(t, \alpha) = e^{tA_0(\alpha)}$ we can consider it as the Floquet decomposition of the first system. So the multipliers $\rho(\alpha, \epsilon) = e^{i\omega(\alpha,\epsilon)\tau}$ of the second system are related to the multipliers $\rho_0(\alpha) = e^{i\omega_0(\alpha)\tau}$ of the first system by $\rho(\alpha, \epsilon) = \rho_0(\alpha) + O(\epsilon)$. The Hamiltonian (8.4) is periodic with respect to time with period 2π.

Recall that $\dot{\mathbf{x}} = A(t, \alpha, \epsilon)\mathbf{x}$ is stable if, and only if, $\dot{\mathbf{y}} = B(\alpha, \epsilon)\mathbf{y}$ is stable, where $\mathbf{x} = Q(t, \alpha, \epsilon)\mathbf{y}$.

If the multipliers $\rho_0(\alpha_*)$ and $\rho_0(\alpha_*)^{-1}$ are distinct, then by continuity for (α, ϵ) near $(\alpha_*, 0)$ the multipliers $\rho(\alpha, \epsilon)$ and $\rho(\alpha, \epsilon)^{-1}$ are also distinct, so the eigenvalues $\pm i\omega(\alpha, \epsilon)$ of $B(\alpha, \epsilon)$ are distinct; hence, $\dot{\mathbf{y}} = B(\alpha, \epsilon)\mathbf{y}$ is stable and therefore the system $\dot{\mathbf{x}} = A(t, \alpha, \epsilon)\mathbf{x}$ is stable. Now, if the multipliers $\rho_0(\alpha_*)$ and $\rho_0(\alpha_*)^{-1}$ are not distinct, they collide at one point of the unit circumference. Under a small perturbation of the first system this double multiplier will split into a pair of distinct multipliers that are on the unit circle or on the real axis, the latter case being possible only if the collapsed multiplier is at one of the points $(1, 0)$ or $(-1, 0)$.[1] In the first case the perturbed system $\dot{\mathbf{x}} = A(t, \alpha, \epsilon)\mathbf{x}$ is stable and, in the second case, it is unstable.

A value α_* for which $\rho_0(\alpha_*)$ is a double multiplier of the system $\dot{\mathbf{x}} = A_0(\alpha_*)\mathbf{x}$ is called a *value of parametric resonance*.

The unperturbed system in (8.4) is stable for all $\alpha > 0$. By the above considerations the perturbed system is stable for (α, ϵ) near $(\alpha_*, 0)$, if $\rho(\alpha_*)$ is a simple multiplier of the system $\dot{\mathbf{x}} = A_0(\alpha_*)\mathbf{x}$.

Now, suppose that $\rho_0(\alpha_*) = e^{i\omega_0(\alpha_*)\tau}$ is a double multiplier of the unperturbed system $\dot{\mathbf{x}} = A_0(\alpha_*)\mathbf{x}$ different to ± 1. Then, $e^{i2\omega_0(\alpha_*)\tau} = 1$; hence, $2\omega_0(\alpha_*)\tau = 2\pi N$, for some positive integer N. Since $\tau = 2\pi$ and $\omega = \sqrt{\alpha}$ we get the value $\alpha_* = \frac{N^2}{4}$.

So the values of parametric resonance for the unperturbed system $\dot{\mathbf{x}} = A_0(\alpha_*)\mathbf{x}$ form the sequence of points

$$\alpha_N = \frac{N^2}{4}, \qquad N = 1, 2, 3, \ldots$$

in the half line $\alpha > 0$, which is the interval of stability of the unperturbed system.

For each $N = 1, 2, 3, \ldots$, as we enter the upper half of the parameter plane (α, ϵ) from the point $(\alpha_N, 0)$ we may encounter a stable or an unstable system in the family of perturbed linear Hamiltonian systems $\dot{\mathbf{x}} = A(t, \alpha, \epsilon)\mathbf{x}$.

---

[1] They cannot split into complex numbers outside the unit circle because, by the Lyapunov–Poincaré theorem, that would produce four roots of the characteristic equation, which has degree two.

This suggests searching for a continuous curve in the parameter plane (α, ϵ) emanating from the point $(\alpha_N, 0)$, which separates stable and unstable systems of the perturbed family of Hamiltonians (8.4). We look for such a curve in the form of a power series in ϵ,

$$\alpha = \alpha_N + \epsilon a_1 + \epsilon^2 a_2 + \dots, \tag{8.5}$$

and search to find the unknown coefficients $a_1, a_2, \dots$.

Replacing α from (8.5) into the Hamiltonian (8.4) and re-ordering the terms in powers of ϵ we get the new expansion of the quadratic Hamiltonian

$$H(\xi, \eta, t, \alpha, \epsilon) = \frac{1}{2}\omega_N(\xi^2 + \eta^2) + \epsilon H_1(\xi, \eta, t, a_1) + \epsilon^2 H_2(\xi, \eta, t, a_1, a_2) + \dots, \tag{8.6}$$

where for $j = 1, 2, 3, \dots$, $H_j(\xi, \eta, t, a_1, \dots, a_j)$ is a homogeneous quadratic polynomial in ξ, η which is τ-periodic in t and contains the coefficients $a_1, a_2, \dots$ up to a_j.

By means of a symplectic coordinate change, which is τ-periodic in the time t one searches to transform this periodic Hamiltonian to an autonomous Hamiltonian K. Writing the stability conditions in terms of the coefficients of K one determines successively the coefficients $a_1, a_2, a_3, \dots$ of the curve (8.5).

The process of obtaining K from H can be done through the method of Deprit–Hori seen in Chap. 3. In the next sections we use this method and the details for the one-degree case are seen Sect. 8.4, where we show that usually there are two curves in the parameter plane emanating from the point of resonance $(\alpha_N, 0)$. The parameters immediately to the right of the right curve and to the left of the left curve correspond to stable systems of the parametric family of Hamiltonians (8.4) and the parameters between the two curves correspond to unstable systems.

Since for an unstable system there are solution curves that begin near the equilibrium point and run away from it, resulting in increasing amplitudes of the motion, which is associated with the phenomenon of resonance, the unstable regions are called *regions of parametric resonance* and the problem of finding these regions is known as the *problem of parametric resonance*.

Finding the region of parametric resonance for each value of parametric resonance one decomposes a narrow band of the parameter plane bounded by the axis $\epsilon = 0$ into a succession of stability and instability regions. In Sect. 8.5 we give the boundary curves of the regions of parametric resonance for the Mathieu equation.

In Sect. 8.6 we begin the study of the two-degrees of freedom case. The starting point is to write the unperturbed quadratic Hamiltonian in normal form corresponding to the step done above when we passed from the Hamiltonian (8.3) to the Hamiltonian (8.4). The theory of normal forms developed in Chap. 5 is what we need here. After that we need the Deprit–Hori method studied in Chap. 3 in which the homological equation appears, which is the fundamental equation to be solved in order to get the autonomous Hamiltonian from the periodic Hamiltonian.

In Sect. 8.8 we work out some examples from satellite dynamics, getting the regions of parametric resonance for specific problems with two-degrees of freedom. In these examples we use the Hamiltonian, which describes the rotations of a satellite about its center of mass.

8.3 The Problem of Parametric Resonance

The Hamiltonian of a specific problem may depend on parameters, for instance, the masses, the eccentricity of an orbit in a celestial mechanics problem, the period of a periodic orbit, the length of a pendulum, the strength of a spring, an electric charge, the intensity of a vortex, etc.

We will define below the notion of μ-stability for a linear Hamiltonian system that contains parameters.

Let $\mathcal{L}$ be the normed linear space of τ-periodic Hamiltonian matrices or quadratic Hamiltonians $H(\mathbf{x}, t)$, with the norm (7.1). Definition 7.25 can be rephrased as: $H_0(\mathbf{x}, t) \in \mathcal{L}$ is *strongly stable* if it is stable and there exists $\epsilon > 0$ such that any $H(\mathbf{x}, t) \in B_\epsilon(H_0) \subset \mathcal{L}$ is stable.

Let $P \subset \mathbf{R}^k$ be a set of parameters and let $H : P \to \mathcal{L}$, $\mu \mapsto H(\mathbf{x}, t, \mu)$ be a family of τ-periodic quadratic Hamiltonians parametrized by $\mu \in P$.

Definition 8.1 We say that $H(\mathbf{x}, t, \mu_0) \in \mathcal{L}$ is *μ-stable* if it is stable and there exists $\upsilon > 0$ such that for any $\mu \in B_\upsilon(\mu_0) \subset \mathbf{R}^k$ the Hamiltonian $H(\mathbf{x}, t, \mu)$ is stable.

Instead of μ-stable, the natural name would be *parametrically stable*, but this is already used in the literature (see [41] and [44]), as a synonym for the term strongly stable, introduced by Krein, Gelfand, and Lidskii, around 1950 (see [24, 34]).

The notion of μ-stability is essentially distinct from the notion of strong stability as shown by the example below.

Example 8.1 (μ-Stability Versus Strong Stability) Consider the one-parameter family of Hamiltonian matrices $A(\mu) = \mu J$, where J is the standard symplectic matrix. The matrix $A(0)$ is μ-stable but in any neighborhood of it there is an unstable Hamiltonian matrix, namely $A_\epsilon = \begin{bmatrix} \epsilon I & O \\ O & -\epsilon I \end{bmatrix}$. So, $A(0)$ is not strongly stable.

Of course, if $H_0 = H(\mathbf{x}, t, \mu_0)$ is strongly stable, then it is μ-stable. The converse may not be true: the set $\{H(\mathbf{x}, t, \mu)\}$ has the dimension of the space of parameters, which may be much less than the dimension of the ball $B_\epsilon(H_0)$. This ball therefore may contain unstable Hamiltonians very close to $H(\mathbf{x}, t, \mu_0)$, which are not in the parametric family of Hamiltonians $H(\mathbf{x}, t, \mu)$.

Suppose that a real τ-periodic Hamiltonian $\tilde{H} = \tilde{H}(\mathbf{z}, t, \mu, \epsilon)$, $\mathbf{z} \in \mathbf{R}^{2n}$ depends on two real scalar parameters μ and ϵ, the latter being a small parameter. Let $\mathbf{z}^*$ be an equilibrium point of the Hamiltonian system, for all values of the parameters μ, ϵ. If the system is autonomous and the number of degrees of freedom is two, the stability of this equilibrium may be analyzed using Arnold's stability theorem by looking at the Arnold condition $\mathbf{H}_4(\omega_2, \omega_1, \mu, \epsilon) \neq 0$ (see Theorem 6.9), and analyzing the set of parameters (μ, ϵ) where it holds. If the Hamiltonian system has more than two-degrees of freedom, or even one-degree of freedom but is time-dependent, then the stability question is not trivial. Thus, we may have to analyze only the stability of the linearized system about the equilibrium.

Let $\mathbf{x} = \mathbf{z} - \mathbf{z}^*$. Then, the Hamiltonian system

$$\dot{\mathbf{z}} = JD^2\tilde{H}(\mathbf{z}^*, t, \mu, \epsilon) \cdot (\mathbf{z} - \mathbf{z}^*) + O(|\mathbf{z} - \mathbf{z}^*|^2)$$

linearized in a neighborhood of $\mathbf{z}^*$ gives the real τ-periodic linear Hamiltonian system

$$\dot{\mathbf{x}} = A(t, \mu, \epsilon)\mathbf{x}, \tag{8.7}$$

where $A(t, \mu, \epsilon) = JD^2\tilde{H}(\mathbf{z}^*, t, \mu, \epsilon)$.

Suppose that the unperturbed system $\dot{\mathbf{x}} = A(t, \mu, 0)\mathbf{x}$ is stable for μ in an interval I. The question is: if $\mu^* \in I$, is the perturbed system (8.7) stable for (μ, ϵ) near $(\mu^*, 0)$?

If the system $\dot{\mathbf{x}} = A(t, \mu^*, 0)\mathbf{x}$ is strongly stable, then as we have mentioned above, it is μ-stable, so $\dot{\mathbf{x}} = A(t, \mu, \epsilon)\mathbf{x}$ is stable for all (μ, ϵ) in a small neighborhood of $(\mu^*, 0)$. So the question is posed only if the system $\dot{\mathbf{x}} = A(t, \mu^*, 0)\mathbf{x}$ is not strongly stable.

We know that the unperturbed system is stable hence the multipliers $\rho_1(\mu^*), \ldots, \rho_{2n}(\mu^*)$ are all on the unit circumference. If they are all simple, that is, have multiplicity one, then by continuity of the eigenvalues, for μ near μ^* and ϵ small the system (8.7) has only simple multipliers. Moreover, all of them are on the unit circumference. Indeed, by Proposition 4.24 (Lyapunov-Poincaré theorem) they come in quadruples $\rho, \rho^{-1}, \overline{\rho}, \overline{\rho}^{-1}$, since the Hamiltonian system is real. So if a multiplier is off the unit circumference the number of multipliers would exceed $2n$, which is not possible. Consequently, if the multipliers are all simple, the system (8.7) is stable for (μ, ϵ) in a small neighborhood of $(\mu^*, 0)$.

Therefore, to continue the investigation we assume that there is at least one multiple multiplier $\rho_j^* = \rho_j(\mu^*)$ among $\rho_1(\mu^*), \ldots, \rho_{2n}(\mu^*)$.

Remark 8.2 Let $\lambda_k = i\omega_k$, $\omega_k \geq 0$ be a characteristic exponent of the linear system $\dot{\mathbf{x}} = A(t, \mu, 0)\mathbf{x}$ and let $\rho_k = e^{2\tau\lambda_k}$ be the corresponding multiplier. We have taken the period 2τ in order to work with a real constant matrix in the Floquet decomposition of the matrizant of $\dot{\mathbf{x}} = A(t, \mu, 0)\mathbf{x}$ (see Remark 7.11). As $\rho_k^{-1} = e^{-2\tau\lambda_k}$ is also a multiplier of this equation we see that if ρ_k is a multiple multiplier then either $\rho_k = \rho_l$ or $\rho_k = \rho_l^{-1}$, for some l.

Therefore, the existence of a multiple multiplier means that $\omega_k \pm \omega_l = \dfrac{\pi}{\tau}N$, for some integer N. By a change in time we can take the period of the system to be $\tau = \pi$, so this relation can be written as

$$\omega_k \pm \omega_l = N. \tag{8.8}$$

We will call *KGL resonance*[2] a resonance of the type (8.8). One such resonance is called a *basic resonance* if $2\omega_k = N$ where N is a natural number. If $\omega_k + \omega_l = N$ or $\omega_k - \omega_l = N$ with $2\omega_k$ or $2\omega_l$ an integer, we get a pair of basic resonances. A KGL resonance $\omega_k + \omega_l = N$ or $\omega_k - \omega_l = N$ with $2\omega_k$ or $2\omega_l$ not an integer is called a *combined resonance* .

A resonance is simple if there is only one of its type among the resonance relations (8.8); otherwise, we say that it is a multiple resonance.

The possible KGL resonances among the frequencies $\omega_1, \ldots, \omega_n$ for the first two values of n are the following;

(i) For $n = 1$, just basic resonances;
(ii) For $n = 2$, we have:

- two possible basic resonances, $2\omega_1 = N_1$ with $2\omega_2$ not an integer; $2\omega_2 = N_2$ with $2\omega_1$ not an integer;
- three possible combined resonances $\omega_1 \pm \omega_2 = N$ with $N \neq 0$ and $\omega_1 - \omega_2 = 0$;
- five double resonances $2\omega_1 = N_1, 2\omega_2 = N_2$, depending on the possibilities: (1) $N_1 \neq N_2$ both nonzero; (2) $N_1 = N_2 > 0$; (3) $N_1 > 0, N_2 = 0$; (4) $N_1 = 0, N_2 > 0$; (5) $N_1 = 0, N_2 = 0$.

For $n \geq 3$, there are more possibilities for a KGL resonance.

Example 8.2 (Examples of Concrete KGL Resonances) Specific cases of KGL resonances will appear in the examples on satellite dynamics considered at the end of this chapter.

Definition 8.3 A value $\mu^* \in I$ for which the system $\dot{\mathbf{x}} = A(t, \mu^*, 0)\mathbf{x}$ has a multiple multiplier is said to be a value of parametric resonance.

We assume that the values of parametric resonance are isolated in the interval I. Let μ^* be a value of parametric resonance. By the Krein–Gelfand–Lidskii Theorem 7.42 the unperturbed system $\dot{\mathbf{x}} = A(t, \mu^*, 0)\mathbf{x}$ may still be strongly stable because all the multipliers might be definite. In this case it is μ-stable, so for (μ, ϵ) near $(\mu^*, 0)$ the perturbed system $\dot{\mathbf{x}} = A(t, \mu, \epsilon)\mathbf{x}$ is stable.

Now assume that the unperturbed system, $\dot{\mathbf{x}} = A(t, \mu, 0)\mathbf{x}$, is not strongly stable. Then, in any neighborhood of it there are stable and unstable systems. Therefore, in the family (8.7) we may have parameters that give stable systems and parameters that give unstable systems. So maybe the parameters can be separated by continuous curves that bound, in the parameter plane, the regions of stable systems of the family and the regions of unstable systems.

The goal is then to look for such continuous curves in the parameter plane (μ, ϵ).

[2] The acronym KGL stands for Krein–Gelfand–Lidskii.

Let $S = D^2\tilde{H}(\mathbf{z}^*, t, \mu, \epsilon) = -JA(t, \mu, \epsilon)$. Then writing with $\mathbf{z}$ instead of $\mathbf{x}$, the quadratic form

$$H(\mathbf{z}, t, \mu, \epsilon) = \frac{1}{2}\mathbf{z}^T S(t, \mu, \epsilon)\mathbf{z}$$

is the Hamiltonian of the linear system (8.7).

We expand H in a power series of ϵ

$$H(\mathbf{z}, t, \mu, \epsilon) = H_0(\mathbf{z}, \mu) + \epsilon H_1(\mathbf{z}, t, \mu) + \epsilon^2 H_2(\mathbf{z}, t, \mu) + \dots, \tag{8.9}$$

where $H_0, H_1, H_2, \dots$ are homogeneous quadratic polynomials in $\mathbf{z}$ with $H_1, H_2, \dots$ having τ-periodic coefficients in t. We assume that the unperturbed Hamiltonian, $H_0(\mathbf{z}, \mu)$, does not depend on t, which is the case in many situations.[3]

We have assumed the Hamiltonian $H_0(\mathbf{z}, \mu)$ to be stable for μ in an open interval I. We will take $\lambda_j(\mu) = i\omega_j(\mu)$, $j = 1, \dots, n$ with $\omega_j = \omega_j(\mu) \geq 0$, for $\mu \in I$.

Let $\mu_0 \in I$ be a value of parametric resonance. Then, at least one of the KGL resonances (8.8) occurs when $\mu = \mu_0$. Since we have assumed that these values are isolated, the frequencies $\omega_1(\mu), \dots, \omega_n(\mu)$ are distinct for $\mu \neq \mu_0$ in a neighborhood of μ_0. Therefore, for each one of such values of μ we can find a symplectic linear mapping $q_k, p_k \mapsto x_k, y_k$, where $\mathbf{z} = (x_1, \dots, x_n, y_1, \dots, y_n)$ which depends continuously on μ and transforms $H_0(\mathbf{z}, \mu)$ to its normal form, an algebraic sum of harmonic oscillators

$$H_0(\mathbf{q}, \mathbf{p}, \mu) = \frac{1}{2}\sum_{k=1}^{n} \sigma_k(q_k^2 + p_k^2), \tag{8.10}$$

where $\sigma_k(\mu) = \delta_k\omega_k(\mu)$ with $\omega_k(\mu) > 0$ and $\delta_k = \pm 1$ well-determined in the normalization process (see Proposition 3.4).

However, at $\mu = \mu_0$ it may not be possible to write the normal form of $H_0(\mathbf{z}, \mu_0)$ in this manner, since the frequencies $\omega_1(\mu_0), \cdots, \omega_n(\mu_0)$ may not be distinct, for example. Furthermore, at least one of the KGL resonances occurs when $\mu = \mu_0$ so, using the linear normalization theory of Chaps. 4 and 5, in the Sect. 8.6 for the case of two-degrees of freedom, we will find the normal form of $H_0(\mathbf{z}, \mu_0)$ for each kind of KGL resonance and transform it subsequently to a convenient form suitable for the use of the Deprit–Hori method. Observe that the theory developed in these chapters allows finding the normal form, in a KGL resonance, for any number of degrees of freedom. The goal is to use this method, which is also valid in the restricted universe of quadratic Hamiltonians, to find a τ-periodic generating

[3] For instance in the elliptic restricted three-body problem in rotating coordinates, in periodic perturbations of the mathematical pendulum, in some problems of satellite dynamics, etc. In any case this can always be achieved by making a linear canonical transformation given by a Floquet decomposition of the matrizant of the unperturbed system $\dot{\mathbf{z}} = A(t, \mu, 0)\mathbf{z}$.

Fig. 8.2 Boundary curves

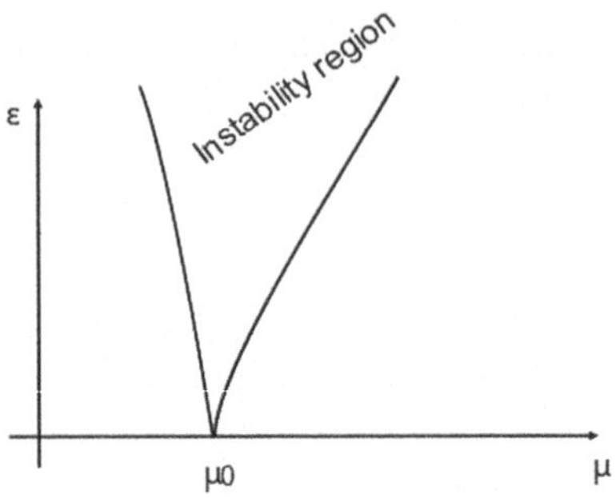

function giving a τ-periodic linear canonical change of coordinates that transforms the time-dependent Hamiltonian system into an autonomous one. The stability of the origin for the initial τ-periodic linear Hamiltonian system is equivalent to the stability of the origin for the autonomous system. Imposing the stability condition for the autonomous system, after developing the theory to find the mentioned curves in the parameter plane, will allow us to find the expressions for these curves. This program will be undertaken in the next sections. As $H_0(\mathbf{z}, \mu_0)$ lies in the boundary of the subsets of stable and unstable Hamiltonians in the linear space $\mathcal{L}$ of τ-periodic quadratic Hamiltonians $H(\mathbf{z}, t)$ with the norm (7.1), we will look for these curves as curves emanating from the point $(\mu_0, 0)$ expecting to find them in the parameter plane bounding the sets of stable systems and the sets of unstable systems (see Fig. 8.2). We will look for these curves in the form of a series expansion

$$\mu = \mu_0 + \mu_1\epsilon + \mu_2\epsilon^2 + \ldots, \tag{8.11}$$

and search to find the coefficients $\mu_1, \mu_2, \ldots$. The problem of determining the boundary curves of regions of stability and instability for the Hamiltonian (8.9) is the *problem of parametric resonance*.

8.4 Boundary Curves in the Parameter Plane

Replacing μ from (8.11) into the Hamiltonian (8.9) and rearranging it in powers of ϵ we get the new expansion in series

$$H(\mathbf{z}, t, \mu, \epsilon) = H_0 + \frac{\epsilon}{1!}H_1 + \frac{\epsilon^2}{2!}H_2 + \frac{\epsilon^3}{3!}H_3 + \ldots, \tag{8.12}$$

where $H_0 = H_0(\mathbf{z}, \mu_0)$ is the value of $H_0(\mathbf{z}, \mu)$ in (8.9) at $\mu = \mu_0$.

It is important to observe that we have adjusted the homogeneous polynomials in (8.12) to write them with the factorials in the denominators in conformity with the Deprit–Hori method.

The subsequent quadratic polynomials $H_1, H_2, H_3, \ldots$ can be expressed in terms of the derivatives of the quadratic Hamiltonian H. The first three of them are given by the expressions

$$
\begin{aligned}
H_1 &= \Big(\frac{\partial H}{\partial \mu}\Big)_0 \mu_1 + \Big(\frac{\partial H}{\partial \epsilon}\Big)_0,\\
H_2 &= 2\Big(\frac{\partial H}{\partial \mu}\Big)_0 \mu_2 + \Big(\frac{\partial^2 H}{\partial \mu^2}\Big)_0 \mu_1^2 + 2\Big(\frac{\partial^2 H}{\partial \mu \partial \epsilon}\Big)_0 \mu_1 + \Big(\frac{\partial^2 H}{\partial \epsilon^2}\Big)_0,\\
H_3 &= 6\Big(\frac{\partial H}{\partial \mu}\Big)_0 \mu_3 + 6\Big(\frac{\partial^2 H}{\partial \mu^2}\Big)_0 \mu_1\mu_2 + 6\Big(\frac{\partial^2 H}{\partial \mu \partial \epsilon}\Big)_0 \mu_2 + \Big(\frac{\partial^3 H}{\partial \mu^3}\Big)_0 \mu_1^3\\
&\quad + 3\Big(\frac{\partial^3 H}{\partial \mu^2 \partial \epsilon}\Big)_0 \mu_1^2 + 3\Big(\frac{\partial^3 H}{\partial \mu \partial \epsilon^2}\Big)_0 \mu_1 + \Big(\frac{\partial^3 H}{\partial \epsilon^3}\Big)_0,
\end{aligned}
\tag{8.13}
$$

where the subscript 0 means that the derivatives are computed at $(\mathbf{z}, t, \mu_0)$ and $\epsilon = 0$.

We notice that $H_j = H_j(\mathbf{z}, t, \mu_0, \mu_1, \ldots, \mu_j)$ it is linear in μ_j.

Recall that we are working with a value of parametric resonance μ_0, where therefore at least one resonance (8.8) holds. Now we submit the unperturbed Hamiltonian $H_0(\mathbf{x}, \mu_0)$ to a symplectic linear transformation to obtain its normal form. This transformation affects all the succeeding terms $H_1, H_2, \ldots$ in (8.12) and we get a Hamiltonian that has a new expansion in ϵ. We assume that this linear transformation $(\boldsymbol{\xi}, \boldsymbol{\eta}) \mapsto \mathbf{z} = (\mathbf{x}, \mathbf{y})$ has been done and in order not to overstuff notation, we assume that (8.12) is the series expansion of the new Hamiltonian, so $H_0 = H_0(\boldsymbol{\xi}, \boldsymbol{\eta})$ is already the normal form of $H_0(\mathbf{z}, \mu_0)$ and $H_j = H_j(\boldsymbol{\xi}, \boldsymbol{\eta}, t, \mu_1, \ldots, \mu_j)$.

If we go through the proof of Theorem 3.10 we check that everything works out in the degree two context, that is when the polynomials $H_i^0(\mathbf{z})$, $H_0^i(\mathbf{z})$ and $W_{i+1}(\mathbf{z})$ in the equalities (3.51), (3.52) and (3.54) are homogeneous polynomials of degree two. This means that formula (3.55) of the Deprit–Hori method and the corresponding Kamel formula (3.81) are also valid in the context of polynomials of degree two.

Now we apply the Deprit–Hori method to normalize the τ-periodic quadratic Hamiltonian (8.12) through an autonomous Hamiltonian K. We will use formula (3.55) of the method of Deprit–Hori in Kamel's formulation.

Recall from Chap. 3 that the new Hamiltonian K and the generating function W must be written in series expansions of the form

$$
K = K_0 + \frac{\epsilon}{1!}K_1 + \frac{\epsilon^2}{2!}K_2 + \frac{\epsilon^3}{3!}K_3 + \ldots, \qquad W = W_1 + \frac{\epsilon}{1!}W_2 + \frac{\epsilon^2}{2!}W_3 + \frac{\epsilon^3}{3!}W_4 + \ldots.
$$

We write Kamel's formula (3.81) in the form $K_m = H^{(m)} + \{H_0, W_m\} - \dfrac{\partial W_m}{\partial t}$, where

$$H^{(m)} = H_m + \sum_{j=1}^{m-1} \left[C_{m-1}^{j-1} L_j H_{m-j} + C_{m-1}^{j} K_{j,m-j} \right]. \tag{8.14}$$

The superscript m indicates that we are working at the mth step in the normalization process. So, at each stage of the method we are asked to solve an equation of the form

$$K = H + \{H_0, W\} - \frac{\partial W}{\partial t}, \tag{8.15}$$

for the quadratic forms K and W when the quadratic form H, τ-periodic in t, is given.

A homogeneous polynomial of degree two in $2n$ variables $x_1, \ldots, x_n, y_1, \ldots, y_n$ can be written in the form

$$\sum_{\nu,\mu} \gamma_{\nu_1 \ldots \nu_n; \mu_1 \ldots \mu_n} x_1^{\nu_1} \cdots x_n^{\nu_n} y_1^{\mu_1} \cdots y_n^{\mu_n}, \tag{8.16}$$

with the sum running over the indices ν, μ satisfying the conditions $\nu_1 + \cdots + \nu_n + \mu_1 + \ldots + \mu_n = 2, \quad \nu_i, \mu_i \geq 0$.

For $n = 1$ the polynomial has three coefficients $\gamma_{20}, \gamma_{11}, \gamma_{02}$ and for $n = 2$ it has ten coefficients

$$\gamma_{2000}, \gamma_{1010}, \gamma_{0020}, \gamma_{1100}, \gamma_{1001}, \gamma_{0110}, \gamma_{0011}, \gamma_{0200}, \gamma_{0101}, \gamma_{0002}. \tag{8.17}$$

In general it has $n(2n + 1)$ coefficients, so in particular for $n = 3$ the polynomial has 21 terms.

We end this section by working out the one-degree of freedom case. Let the autonomous part, $H_0 = H_0(\mathbf{z}, \mu)$, of the one-degree of freedom Hamiltonian (8.9) be given by

$$H_0 = \frac{1}{2}\left(ax_1^2 + 2bx_1y_1 + cy_1^2\right), \tag{8.18}$$

where $a = a(\mu)$, $b = b(\mu)$, $c = c(\mu)$ are functions of the parameter μ.

For $b^2 - ac < 0$, we have $H_0(\mathbf{z}, \mu)$ stable for μ in open interval I, in particular, $a \neq 0$.[4] After performing the symplectic transformation $x_1 = \xi - \dfrac{b}{a}\eta$, $y_1 = \eta$, follows that $H_0 = \dfrac{1}{2}a(\xi^2 + \gamma\eta^2)$, where $\gamma = \dfrac{ac - b^2}{a^2} > 0$. Applying the further symplectic transformation $\xi = \sqrt[4]{\gamma} q$, $\eta = \dfrac{1}{\sqrt[4]{\gamma}} p$ we obtain

[4] The boundary case $b^2 - ac = 0$ is discussed in the Remark 8.4.

$$H(q, p, t, \mu, \epsilon) = H_0(q, p, \mu) + \epsilon H_1(q, p, t, \mu) + \frac{\epsilon^2}{2!} H_2(q, p, t, \mu) + \cdots, \tag{8.19}$$

where

$$H_0 = \frac{1}{2}\delta\omega(\mu)(q^2 + p^2), \quad \omega(\mu) = \sqrt{ac - b^2}, \tag{8.20}$$

where $\delta = \pm 1$ is the sign of a. Let μ_0 a value where a KGL resonance occurs, for one-degree of freedom, the only possibility of a KGL resonance at μ_0 is $2\omega(\mu_0) = N$, N, a positive integer. We will take the expansion,

$$\mu = \mu_0 + \epsilon\mu_1 + \epsilon^2\mu_2 + \cdots$$

and, expanding $\omega(\mu)$ in (8.20) around $\epsilon = 0$, we obtain

$$H_0(q, p, \mu) = \frac{1}{2}\delta\omega(\mu_0)(q^2 + p^2) + \epsilon\frac{1}{2}\delta\omega'(\mu_0)(q^2 + p^2)\mu_1 + \mathcal{O}(\epsilon^2)$$

and so, the Hamiltonian in (8.19) becomes

$$\mathcal{H}(q, p, t, \mu, \epsilon) = \frac{1}{2}\delta\omega(\mu_0)(q^2 + p^2) + \epsilon\mathcal{H}_1(q, p, t, \mu_0, \mu_1) + \mathcal{O}(\epsilon^2) \tag{8.21}$$

where

$$\mathcal{H}_1(q, p, t, \mu_0, \mu_1) = \frac{1}{2}\delta\omega'(\mu_0)(q^2 + p^2)\mu_1 + H_1(q, p, t, \mu_0)$$

After this, since $2\omega(\mu_0) = N$, we make use of Proposition 2.15 with the rotation

$$q = x\cos\frac{\delta Nt}{2} + y\sin\frac{\delta Nt}{2}, \quad p = -x\sin\frac{\delta Nt}{2} + y\cos\frac{\delta Nt}{2}, \tag{8.22}$$

in order to get rid of the unperturbed term $\frac{1}{2}\delta\omega(\mu_0)(q^2 + p^2)$ in (8.12).

This rotation is 2π-periodic if N is even and 4π-periodic if N is odd. So, as $\mathcal{H}$ is π-periodic in t, the transformed Hamiltonian is 2π-periodic if N is even and 4π-periodic if N is odd. Let τ be the new period, 2π or 4π. The transformed τ-periodic Hamiltonian becomes

$$\begin{aligned}\mathcal{H}(x, y, t, \mu, \epsilon) &= \mathcal{H}_0(x, y, \mu_0) + \frac{\epsilon}{1!}\mathcal{H}_1(x, y, t, \mu_0, \mu_1) + \frac{\epsilon^2}{2!}\mathcal{H}_2(x, y, t, \mu_0, \mu_1, \mu_2) \\ &+ \frac{\epsilon^3}{3!}\mathcal{H}_3(x, y, t, \mu_0, \mu_1, \mu_2, \mu_3) + \ldots,\end{aligned} \tag{8.23}$$

where now $\mathcal{H}_0(x, y, \mu_0) = 0$.

For any $\mathcal{W}(x, y, t, \epsilon)$, we have $\{\mathcal{H}_0, \mathcal{W}\} = 0$ and the Eq. (8.15) reduces to $\mathcal{K} = \mathcal{H} - \dfrac{\partial \mathcal{W}}{\partial t}$. Looking for a τ-periodic generating function $\mathcal{W}(x, y, t, \epsilon)$ to get the near identity canonical transformation, $x = \xi + \Phi(\xi, \eta, t, \epsilon)$, $y = \eta + \Psi(\xi, \eta, t, \epsilon)$, integration from 0 to τ yields

$$\mathcal{K}(\xi, \eta, \epsilon) = \frac{1}{\tau}\int_0^\tau \mathcal{H}(x, y, t, \epsilon)dt, \ \mathcal{W}(x, y, t, \epsilon) = \int_0^t \mathcal{H}(x, y, s, \epsilon) - \mathcal{K}(\xi, \eta, \epsilon)ds.$$

For the normal form $\mathcal{K} = \sum_{m=0}^{\infty} \dfrac{\epsilon^m}{m!}\mathcal{K}_m$ write $\mathcal{K}_m = \sum_{\nu+\mu=2} k_{\nu\mu}^{(m)} \xi^\nu \eta^\mu$ with constant coefficients $k_{20}^{(m)}, k_{11}^{(m)}, k_{02}^{(m)}$. Then, the transformed autonomous Hamiltonian is

$$\begin{aligned}\mathcal{K}(\xi, \eta, \mu, \epsilon) &= \frac{\epsilon}{1!}\mathcal{K}_1(\xi, \eta, \mu_0, \mu_1) + \frac{\epsilon^2}{2!}\mathcal{K}_2(\xi, \eta, \mu_0, \mu_1, \mu_2) \\ &\quad + \frac{\epsilon^3}{3!}\mathcal{K}_3(\xi, \eta, \mu_0, \mu_1, \mu_2, \mu_3) + \ldots \\ &= k_{20}\xi^2 + k_{11}\xi\eta + k_{02}\eta^2,\end{aligned} \tag{8.24}$$

where

$$k_{20} = \sum_{l=1}^{\infty} \frac{\epsilon^l}{l!}k_{20}^{(l)}, \quad k_{11} = \sum_{l=1}^{\infty} \frac{\epsilon^l}{l!}k_{11}^{(l)}, \quad k_{02} = \sum_{l=1}^{\infty} \frac{\epsilon^l}{l!}k_{02}^{(l)}, \quad k_{ij}^{(l)} = \text{const.} \tag{8.25}$$

The characteristic equation for $\mathcal{K}$ is:

$$\lambda^2 - \left[k_{11}^2 - 4k_{20}k_{02}\right] = 0,$$

so the region of stability is given by $k_{11}^2 < 4k_{20}k_{02}$. The boundary of this region is defined by the equation

$$k_{11}^2 = 4k_{20}k_{02}. \tag{8.26}$$

Equating coefficients of same degree in ϵ in both members of Eq. (8.26), we find successively the coefficients $\mu_1, \mu_2, \ldots$ in the expansion (8.11).

Comparing coefficients of equal powers of ϵ in Eq. (8.26) we get

$$\begin{aligned} 4k_{20}^{(1)}k_{02}^{(1)} &= k_{11}^{(1)2} \\ 2k_{20}^{(1)}k_{02}^{(2)} + 2k_{02}^{(1)}k_{20}^{(2)} &= k_{11}^{(1)}k_{11}^{(2)} \\ \frac{2}{3}k_{20}^{(1)}k_{02}^{(3)} + k_{20}^{(2)}k_{02}^{(2)} + \frac{2}{3}k_{02}^{(1)}k_{20}^{(3)} &= \frac{1}{4}k_{11}^{(2)2} + \frac{1}{3}k_{11}^{(1)}k_{11}^{(3)} \\ \ldots\ldots\ldots\ldots\ldots\ldots &= \ldots\ldots\ldots\ldots\ldots\ldots \end{aligned} \tag{8.27}$$

As μ_l appears linearly in $\mathcal{K}_l(\xi, \eta, \mu_0, \mu_1, \dots, \mu_l)$ it follows that μ_l appears linearly in $k_{ij}^{(l)}(\mu_0, \mu_1, \dots, \mu_l)$, so the first Eq. (8.27) is quadratic in μ_1, the second is linear in μ_2, the third is linear in μ_3, and this linear dependence in the last parameter appearing in every succeeding equation continues for all of them.

We prove below that the first equation in (8.27), which is quadratic in μ_1, has only real roots. So the succeeding equations, each of which is linear in the last parameter μ_l, $l = 2, 3, 4, \dots$, has exactly one real root μ_l.

The conclusion is that there is one or two curves (8.11) emanating from the point $(\mu_0, 0)$ in the parameter plane (μ, ϵ). In the first case the systems corresponding to the parameters on this curve are unstable and the nearby points to the left or to the right of this curve correspond to stable systems. In the case where the first equation has two real roots we get two curves in the parameter plane emanating from the point $(\mu_0, 0)$ and they form the boundary of a region corresponding to unstable systems. This is called a *region of parametric resonance*. The points of the parameter plane near the left boundary of this region and to the left of it as well as the points near the right boundary and to the right of it correspond to stable systems.

Let us now prove that the first equation in (8.27) has only real roots. Consider the Hamiltonian function in (8.21). After a rotation $R(t)\boldsymbol{\zeta}$ such as (8.22) from $\boldsymbol{\zeta} = (\xi, \eta)$ to (q, p) in order to get the new unperturbed Hamiltonian $\mathcal{H}_0(\boldsymbol{\zeta}, \mu_0) = 0$, the Hamiltonian (8.21) becomes

$$\mathcal{H}(\boldsymbol{\zeta}, t, \mu, \epsilon) = \epsilon\Big[\mathcal{F}_0(\boldsymbol{\zeta}, \mu_0)\mu_1 + \mathcal{F}_1(\boldsymbol{\zeta}, t, \mu_0)\Big] + O(\epsilon^2),$$

where $\mathcal{F}_0, \mathcal{F}_1$ are given by

$$\mathcal{F}_0(\boldsymbol{\zeta}, \mu_0) = \frac{1}{2}\delta\omega'(\mu_0)(\xi^2 + \eta^2), \quad \mathcal{F}_1(\boldsymbol{\zeta}, t, \mu_0) = H_1(\boldsymbol{\zeta}, t, \mu_0).$$

Since $\mathcal{K}_1 = \dfrac{1}{\tau}\displaystyle\int_0^\tau \mathcal{H}_1(\xi, \eta, t, \mu_0, \mu_1)dt$ and $\mathcal{H}_1(\boldsymbol{\zeta}, t, \mu_0, \mu_1) = \mathcal{F}_0(\boldsymbol{\zeta}, \mu_0)\mu_1 + \mathcal{F}_1(\boldsymbol{\zeta}, t, \mu_0)$, then setting $\mathcal{F}_1 = f_{20}\xi^2 + f_{11}\xi\eta + f_{02}\eta^2$ and $\mathcal{K}_1 = k_{20}\xi^2 + k_{11}\xi\eta + k_{02}\eta^2$ we get

$$k_{20} = \frac{1}{2}\delta\omega'(\mu_0)\mu_1 + [f_{20}], \quad k_{02} = \frac{1}{2}\delta\omega'(\mu_0)\mu_1 + [f_{02}], \quad k_{11} = [f_{11}],$$

where $[f_{20}]$, $[f_{11}]$ and $[f_{02}]$ are the time averages of f_{20}, f_{11} and f_{02} respectively.

Therefore, we get

$$4k_{20}k_{02} - k_{11}^2 = \big(\omega'(\mu_0)\big)^2\mu_1^2 + 2\big([f_{20}] + [f_{02}]\big)\omega'(\mu_0)\mu_1 + 4\big([f_{20}][f_{02}] - [f_{11}]^2\big).$$

The discriminant Δ of the quadratic equation in μ_1, $4k_{20}k_{02} - k_{11}^2 = 0$, is

$$\begin{aligned}\Delta &= 4\big([f_{20}] + [f_{02}]\big)^2\big(\omega'(\mu_0)\big)^2 - 16\big(\omega'(\mu_0)\big)^2\big([f_{20}][f_{02}] - [f_{11}]^2\big)\\ &= 4\big(\omega'(\mu_0)\big)^2\big([f_{20}] - [f_{02}]\big)^2 + 16\big(\omega'(\mu_0)\big)^2[f_{11}]^2\big).\end{aligned}$$

We see that $\Delta \geq 0$, so the equation for μ_1, $4k_{20}k_{02} = k_{11}^2$, has one or two real roots.

Remark 8.4 If the coefficients a, b, c in $H_0 = \frac{1}{2}\big(ax_1^2 + 2bx_1y_1 + cy_1^2\big)$ are such that $b^2 - ac = 0$, then the corresponding matrix $A = J\,D^2H_0$ is nilpotent. In this case, there are two possibilities: A is zero matrix or there exists $\mathbf{v} = (x_1, y_1)$ such that $\{A\mathbf{v}, \mathbf{v}\} \neq 0$ and so, in according to Example 5.11, there exists a symplectic base that transforms H_0 into $\mathcal{H}_0 = \frac{1}{2}\delta y_1^2$. In the first case $\{H_0, W\} = 0$ since $H_0 = 0$ and in the last case, the term $\{\mathcal{H}_0, W\}$ is different than zero. Thus, in this case, the equation in (8.15) is a little more laborious than the previous case where $b^2 - ac < 0$ and the expression of autonomous Hamiltonian $\mathcal{K}$ in (8.24) is

$$\mathcal{K}(\xi, \eta, \mu, \epsilon) = k_{20}\xi^2 + k_{11}\xi\eta + \left(\frac{1}{2} + k_{02}\right)\eta^2.$$

The equations in (8.27), for these cases, are given by

$$\begin{aligned}2k_{20}^{(1)} &= 0\\ k_{11}^{(1)2} &= 4k_{20}^{(1)}k_{02}^{(1)} + k_{20}^{(2)}\\ k_{11}^{(1)}k_{11}^{(2)} &= 2k_{20}^{(1)}k_{02}^{(2)} + 2k_{02}^{(1)}k_{20}^{(2)} + \frac{1}{3}k_{20}^{(3)}\\ \cdots\cdots &= \cdots\cdots\end{aligned}$$

Observe that the first equation is linear in μ_1, the second equation is linear in μ_2, and so on.

Generically, in the one-degree of freedom case, we have two curves emanating from a KGL resonance. We can think of the case of a single curve as defining an empty region of parametric resonance, so the successive KGL resonances define regions of parametric resonance that together with the regions between the two of them, which correspond to stable systems, decompose a narrow band of the parameter plane, defined by ϵ small, into regions of stability and instability.

Remark 8.5 Let $\mu_0 < \mu_0^*$ be two consecutive values of parametric resonance for the unperturbed Hamiltonian $H_0(\mathbf{z}, \mu)$ in (8.9) and consider the stable region Ω in the space of parameters (μ, ϵ) for the system (8.12), for $\mu \in (\mu_0, \mu_0^*)$. This region is bounded by curves (8.11) that emanate from these resonant values of μ when $\epsilon = 0$. The boundary curves are constructed using the expansion (8.12).

For all values of μ in the interval $\mu_0 < \mu < \mu_0^*$, the multipliers of the linear system $\dot{\mathbf{z}} = A(t, \mu, 0)\mathbf{z}$ are simple so this system is strongly stable. As the set of strongly stable τ-periodic real linear Hamiltonian systems is open in the space $\mathcal{L}_{ss}$ the system $\dot{\mathbf{z}} = A(t, \mu, \epsilon)\mathbf{z}$ is strongly stable for (μ, ϵ) in Ω belonging to a disk centered at a point of the axis $\epsilon = 0$. The radius of this disk approaches zero as μ approaches one of the values μ_0 or μ_0^*.

The stable regions in the parameter space are constructed so that for a point (μ_1, ϵ_1) in Ω the system $\dot{\mathbf{z}} = A(t, \mu, \epsilon)\mathbf{z}$ is stable for (μ, ϵ) near (μ_1, ϵ_1), that is, the system $\dot{\mathbf{z}} = A(t, \mu_1, \epsilon_1)\mathbf{z}$ is μ-stable. By what we have seen above it is really strongly stable if the point (μ_1, ϵ_1) is near the axis $\epsilon = 0$. If ϵ_1 is big we can only guarantee the μ-stability of the system.

8.5 The Mathieu Equation

As a concrete example of the periodic one-degree of freedom case we consider the periodic Hamiltonian

$$H(x, y, t, \mu, \epsilon) = \frac{1}{2}y^2 + \frac{1}{2}(\mu + \epsilon \cos t)x^2,$$

associated to Mathieu's equation, a periodic differential equation of second order of the form

$$\frac{d^2q}{dt^2} + (\mu + \epsilon \cos t)q = 0.$$

This is an important equation that appears in many problems of dynamics; we have seen an example in Sect. 8.2. The regions of stability and instability in the parameter plane (μ, ϵ) have been found by many authors (see for instance [43, 57]).

We have computed the boundary curves of the regions of parametric resonance and found them to be given by the series expansions

$$\mu^{(0)} = -\frac{1}{2}\epsilon^2 + \frac{7}{32}\epsilon^4 - \frac{29}{144}\epsilon^6 + \frac{68687}{294912}\epsilon^8$$

$$\mu^{(1)} = \frac{1}{4} - \frac{1}{2}\epsilon - \frac{1}{8}\epsilon^2 + \frac{1}{32}\epsilon^3 - \frac{1}{384}\epsilon^4 - \frac{11}{4608}\epsilon^5 + \frac{49}{36864}\epsilon^6 - \frac{55}{294912}\epsilon^7 - \frac{83}{552960}\epsilon^8$$

$$\mu^{(2)} = \frac{1}{4} + \frac{1}{2}\epsilon - \frac{1}{8}\epsilon^2 - \frac{1}{32}\epsilon^3 - \frac{1}{384}\epsilon^4 + \frac{11}{4608}\epsilon^5 + \frac{49}{36864}\epsilon^6 + \frac{55}{294912}\epsilon^7 - \frac{83}{552960}\epsilon^8$$

$$\mu^{(3)} = 1 - \frac{1}{12}\epsilon^2 + \frac{5}{3456}\epsilon^4 - \frac{289}{4976640}\epsilon^6 + \frac{21391}{7166361600}\epsilon^8$$

$$\mu^{(4)} = 1 + \frac{5}{12}\epsilon^2 - \frac{763}{3456}\epsilon^4 + \frac{1002401}{4976640}\epsilon^6 - \frac{1669068401}{7166361600}\epsilon^8$$

$$\mu^{(5)} = \frac{9}{4} + \frac{1}{16}\epsilon^2 - \frac{1}{32}\epsilon^3 + \frac{13}{5120}\epsilon^4 + \frac{5}{2408}\epsilon^5 - \frac{1961}{1474560}\epsilon^6 + \frac{609}{3276800}\epsilon^7 + \frac{4957199}{33030144000}\epsilon^8$$

$$\mu^{(6)} = \frac{9}{4} + \frac{1}{16}\epsilon^2 + \frac{1}{32}\epsilon^3 + \frac{13}{5120}\epsilon^4 - \frac{5}{2408}\epsilon^5 - \frac{1961}{1474560}\epsilon^6 - \frac{609}{3276800}\epsilon^7 + \frac{4957199}{33030144000}\epsilon^8$$

$$\mu^{(7)} = 4 + \frac{1}{30}\epsilon^2 - \frac{317}{216000}\epsilon^4 + \frac{10049}{170100000}\epsilon^6 - \frac{93824197}{31352832000000}\epsilon^8$$

$$\mu^{(8)} = 4 + \frac{1}{30}\epsilon^2 + \frac{433}{216000}\epsilon^4 - \frac{5701}{170100000}\epsilon^6 - \frac{42214597}{31352832000000}\epsilon^8$$

$$\mu^{(9)} = \frac{25}{4} + \frac{1}{48}\epsilon^2 + \frac{11}{193536}\epsilon^4 - \frac{1}{18432}\epsilon^5 + \frac{37}{55738368}\epsilon^6 + \frac{7}{10616832}\epsilon^7$$

$$\mu^{(10)} = \frac{25}{4} + \frac{1}{48}\epsilon^2 + \frac{11}{193536}\epsilon^4 + \frac{1}{18432}\epsilon^5 + \frac{37}{55738368}\epsilon^6 - \frac{7}{10616832}\epsilon^7,$$

which agree with the values given in [40].

Other examples of one-degree of freedom can be found in [5, 9, 13, 14].

8.6 The Two-Degrees of Freedom Case

Here, we consider the Hamiltonian (8.9) in the two-degrees of freedom case, written in the canonical variables q_1, q_2, p_1, p_2. The frequencies are $\omega_1(\mu)$ and $\omega_2(\mu)$ and we take $\omega_1 \geq \omega_2 \geq 0$.

Let $\mathbf{q} = (q_1, q_2)$, $\mathbf{p} = (p_1, p_2)$ and recall that in the case of unequal frequencies ω_1, ω_2 the unperturbed Hamiltonian can be written in the normal form

$$H_0(\mathbf{q}, \mathbf{p}, \mu) = \frac{1}{2}\delta_1\omega_1(q_1^2 + p_1^2) + \frac{1}{2}\delta_2\omega_2(q_2^2 + p_2^2), \quad \delta_i = \pm 1, \quad \omega_i = \omega_i(\mu). \tag{8.28}$$

Let μ_0 be a value of parametric resonance. For each type of KGL resonance at μ_0 we will write the normal form $H_0(\mathbf{q}, \mathbf{p}, \mu_0)$ in a manner suitable for the solution of the homological equation

$$K_m = H^{(m)} + \{H_0, W_m\} - \frac{\partial W_m}{\partial t}. \tag{8.29}$$

8.6.1 Normal Forms at the Resonances

We enumerate the several cases of normal forms for a quadratic Hamiltonian. The normal form at one type of KGL resonance may depend on peculiarities of the matrix $A = JD^2H_0$, which makes the number of normal forms described below larger than the number of KGL resonances listed in (i)–(ii), page 268.

Remark 8.6 In the cases below, when we have a relation $2\omega_j = N$ and use it to make a rotation so as to eliminate a harmonic oscillator, the period of the original Hamiltonian (8.12), in the two-degrees of freedom case, becomes 2π if N is even and 4π if N is odd. If no such rotation is used to get the normal form the period continues to be $\tau = \pi$. In the enumeration we use the subscript R, for instance, $(1)_R$, $(4)_R$, to indicate that such a rotation has been performed.

First, we consider simple basic resonances $2\omega_1 = N$, $2\omega_2$ not an integer or $2\omega_2 = N$, $2\omega_1$ not an integer, $N > 0$. It suffices to consider one of them.

$(1)_R$ One basic resonance $2\omega_2 = N$, $N > 0$, $2\omega_1$ not an integer.
Recalling Proposition 2.15 we use the rotation

$$q_1 = x_1,\ p_1 = y_1,\ \ q_2 = x_2 \cos\omega t + y_2 \sin\omega t,\ \ p_2 = -x_2 \sin\omega t + y_2 \cos\omega t, \tag{8.30}$$

with $\omega = \omega_2 = \dfrac{\delta_2 N}{2}$ in order to get rid of the term $\frac{1}{2}\delta_2\omega_2(q_2^2 + p_2^2)$ in (8.28). As H is π-periodic in t, then if N is even we get the Hamiltonian in the variables $\mathbf{x}$, $\mathbf{y}$, periodic in t with period 2π. If N is odd the period is 4π. Denote by τ the new period, 2π or 4π.

After the rotation (8.30), the original Hamiltonian is written in the variables $(\mathbf{x}, \mathbf{y}) = (x_1, x_2, y_1, y_2)$ with the unperturbed term having the following normal form

$$\mathcal{H}_0(\mathbf{x}, \mathbf{y}, \mu_0) = \frac{1}{2}\delta_1\omega_1(x_1^2 + y_1^2), \qquad (\delta_1 = \pm 1). \tag{8.31}$$

Now we consider the case when $N = 0$. We have two possibilities, one in which $A\big|_{\eta^\dagger(0)} = 0$ and the other when $A\big|_{\eta^\dagger(0)} \neq 0$.

(2) Case when $\omega_2 = 0$, $2\omega_1$ not an integer and $A\big|_{\eta^\dagger(0)} = 0$.
We have seen in Example 5.12 that when $A\big|_{\eta^\dagger(0)} = 0$ the normal form is

$$\mathcal{H}_0(\mathbf{x}, \mathbf{y}, \mu_0) = \frac{1}{2}\delta_1\omega_1(x_1^2 + y_1^2), \qquad (\delta_1 = \pm 1). \tag{8.32}$$

(3) Case when $\omega_2 = 0$, $2\omega_1$ not an integer and $A\big|_{\eta^\dagger(0)} \neq 0$.
Example 5.12 also shows that in this case the normal form is

$$\mathcal{H}_0(\mathbf{x}, \mathbf{y}, \mu_0) = \frac{1}{2}\delta_1\omega_1(x_1^2 + y_1^2) + \frac{1}{2}\delta_2 y_2^2, \qquad (\delta_1 = \pm 1,\ \delta_2 = \pm 1). \tag{8.33}$$

Next, we consider combined resonances $\omega_1+\omega_2=N$ and $\omega_1-\omega_2=N$, with $2\omega_1$ not an integer.

(4)$_R$ The combined resonance $\omega_1+\omega_2=N$ with $2\omega_1$ not an integer.

Replacing $\omega_2=N-\omega_1$ into (8.28) and making the rotation (8.30) with $\omega=\delta_2 N$ to eliminate the term $\frac{1}{2}\delta_2 N(p_2^2+q_2^2)$ we get the Hamiltonian (8.23) having the unperturbed term now with normal form

$$\mathcal{H}_0(\mathbf{x},\mathbf{y},\mu_0)=\frac{1}{2}\delta_1\omega_1(x_1^2+y_1^2)-\frac{1}{2}\delta_2\omega_1(x_2^2+y_2^2), \qquad (\delta_1=\pm 1,\ \delta_2=\pm 1). \tag{8.34}$$

(5)$_R$ The combined resonance $\omega_1-\omega_2=N$, $N>0$, with $2\omega_1$ not an integer.

Replacing $\omega_2=-N+\omega_1$ into (8.28) and making the rotation (8.30) with $\omega=-\delta_2 N$ we get the normal form

$$\mathcal{H}_0(\mathbf{x},\mathbf{y},\mu_0)=\frac{1}{2}\delta_1\omega_1(x_1^2+y_1^2)+\frac{1}{2}\delta_2\omega_1(x_2^2+y_2^2), \qquad (\delta_1=\pm 1,\ \delta_2=\pm 1). \tag{8.35}$$

The combined resonance $\omega_1-\omega_2=0$, $2\omega_1$ not an integer is divided into two cases, one when the matrix $A-i\omega_1 I$ has rank 2, that is, $rk(A-i\omega_1 I)=2$, which means that it is diagonalizable, and the other when $rk(A-i\omega_1 I)=3$, which means that it is nondiagonalizable.

(6) Combined resonance $\omega_1=\omega_2=\omega\neq 0$, when $rk(A-i\omega I)=2$, 2ω not an integer.

We proved in Example 5.15 that in this case the normal form is given by

$$\mathcal{H}_0(\mathbf{x},\mathbf{y},\mu_0)=\frac{1}{2}\delta_1\omega(x_1^2+y_1^2)+\frac{1}{2}\delta_2\omega(x_2^2+y_2^2), \qquad (\delta_1=\pm 1,\ \delta_2=\pm 1) \tag{8.36}$$

which is the Hamiltonian given by equation (4.26) in Chapter 4 of [40].

(7) Combined resonance $\omega_1=\omega_2=\omega\neq 0$, when $rk(A-i\omega I)=3$, 2ω not an integer.

Example 5.16 treats the nondiagonalizable case and there we proved that the Hamiltonian is given by

$$H_0(\mathbf{u},\mathbf{v},\mu_0)=\frac{1}{2}\delta(u_1^2+u_2^2)+\omega(u_1v_2-u_2v_1).$$

Using the linear change of canonical coordinates $u_1=y_1, u_2=y_2, v_1=-x_1, v_2=-x_2$ we can write it as equation (4.23) in Chapter 4 of [40], namely

$$H_0(\mathbf{x},\mathbf{y},\mu_0)=\frac{1}{2}\delta(y_1^2+y_2^2)+\omega(x_1y_2-x_2y_1), \quad (\delta=\pm 1). \tag{8.37}$$

Now we consider the case of two basic resonances $2\omega_1=N_1, 2\omega_2=N_2$.

(8)$_R$ Two basic resonances $2\omega_1 = N_1, 2\omega_2 = N_2$ with $N_1 \neq N_2$, both nonzero.

The symplectic mapping $x_1, x_2, y_1, y_2 \mapsto q_1, q_2, p_1, p_2$ given by the double rotation

$$q_1 = x_1 \cos \frac{\delta_1 N_1 t}{2} + y_1 \sin \frac{\delta_1 N_1 t}{2}, \quad p_1 = -x_1 \sin \frac{\delta_1 N_1 t}{2} + y_1 \cos \frac{\delta_1 N_1 t}{2},$$

$$q_2 = x_2 \cos \frac{\delta_2 N_2 t}{2} + y_2 \sin \frac{\delta_2 N_2 t}{2}, \quad p_2 = -x_2 \sin \frac{\delta_2 N_2 t}{2} + y_2 \cos \frac{\delta_2 N_2 t}{2},$$

reduces the Hamiltonian (8.28) to the normal form

$$\mathcal{H}_0 = 0. \tag{8.38}$$

Now consider two basic resonances with equal frequencies $\omega_1 = \omega_2 = \omega$, $2\omega = N$.

Just as in the cases of the combined resonances (6) and (7) we here also distinguish the cases when $A - i\omega I$ has rank 2 and rank 3.

(9)$_R$ Case $\omega_1 = \omega_2 = \omega, 2\omega = N$ and $rk(A - i\omega I) = 2$.

In this case Example 5.15 gives the Hamiltonian (8.35), which we write with capital letters

$$H_0(\mathbf{X}, \mathbf{Y}, \mu_0) = \frac{1}{2}\delta\omega(X_1^2 + Y_1^2) + \frac{1}{2}\delta\omega(X_2^2 + Y_2^2), \qquad (\delta = \pm 1).$$

With the symplectic mapping $x_1, x_2, y_1, y_2 \mapsto X_1, X_2, Y_1, Y_2$ given by the double rotation with angles $\frac{\delta N_1 t}{2}$ and $\frac{\delta N_2 t}{2}$ we eliminate both harmonic oscillators and the normal form is $\mathcal{H}_0 = 0$, that is (8.38).

(10)$_R$ Case $\omega_1 = \omega_2 = \omega, 2\omega = N$ and $rk(A - i\omega I) = 3$.

The normal form as in (7) is given by

$$H_0(\mathbf{u}, \mathbf{v}, \mu_0) = \frac{1}{2}\delta(u_1^2 + u_2^2) + \omega(u_1 v_2 - u_2 v_1)$$

but now $2\omega = N$ and we can use this fact to reduce the Hamiltonian to a sum of squares. Indeed, submit the Hamiltonian to the symplectic transformation $u_1, u_2, v_1, v_2 \mapsto \xi_1, \xi_2, \eta_1, \eta_2$ given by the double rotation, the first in the (ξ_1, ξ_2)-plane, the second in the (η_1, η_2)-plane

$$u_1 = \xi_1 \cos \omega t + \xi_2 \sin \omega t, \qquad u_2 = -\xi_1 \sin \omega t + \xi_2 \cos \omega t,$$

$$v_1 = \eta_1 \cos \omega t + \eta_2 \sin \omega t, \qquad v_2 = -\eta_1 \sin \omega t + \eta_2 \cos \omega t.$$

This canonical transformation is generated by the time-dependent function

$$W(\xi_1, \xi_2, v_1, v_2, t) = \xi_1(v_1 \cos \omega t - v_2 \sin \omega t) + \xi_2(v_1 \sin \omega t + v_2 \cos \omega t).$$

We compute the partial derivative and find $W_t = -\omega(u_1 v_2 - u_2 v_1)$. So by Proposition 2.15 the new Hamiltonian becomes

$$\mathcal{H}_0 = \frac{1}{2}\delta(\xi_1^2 + \xi_2^2), \quad (\delta = \pm 1).$$

The further symplectic transformation $\xi_1 = y_1, \xi_2 = y_2, \eta_1 = -x_1, \eta_2 = -x_2$ leads to the normal form

$$\mathcal{H}_0 = \frac{1}{2}\delta(y_1^2 + y_2^2), \quad (\delta = \pm 1). \tag{8.39}$$

Notice that the period becomes $\tau = 2\pi$ if N is even and $\tau = 4\pi$ if N is odd.

Next we consider the two basic resonances when $N_2 = 0$ and here we distinguish the cases $A\big|_{\eta^\dagger(0)} = 0$ and $A\big|_{\eta^\dagger(0)} \neq 0$.

(11)$_R$ Case $2\omega_1 = N_1, \omega_2 = 0$, with $A\big|_{\eta^\dagger(0)} = 0$.

For this case the Hamiltonian is given in Example 5.12 and the normal form is the first one in (5.74), namely

$$\mathcal{H}_0 = \frac{1}{2}\delta_1\omega_1(x_1^2 + y_1^2), \qquad (\delta_1 = \pm 1).$$

The rotation $x_1 = \xi_1 \cos\delta_1\omega_1 t + \eta_1 \sin\delta_1\omega_1 t, \quad y_1 = -\xi_1 \sin\delta_1\omega_1 t + \eta_1 \cos\delta_1\omega_1 t$, by Proposition 2.15 leads this Hamiltonian to the normal form $\mathcal{H}_0 = 0$, that is (8.38).

(12)$_R$ Case $2\omega_1 = N_1, \omega_2 = 0$, with $A\big|_{\eta^\dagger(0)} \neq 0$.

In this case the Hamiltonian is also given in Example 5.12 and the normal form is the second one in (5.74), namely in terms of capital letters

$$H_0 = \frac{1}{2}\delta_1\omega_1(X_1^2 + Y_1^2) + \frac{1}{2}\delta_2 Y_2^2, \qquad (\delta_1 = \pm 1, \ \delta_2 = \pm 1).$$

With a rotation given by the angle $\frac{\delta_1 N_1 t}{2}$ to eliminate the harmonic oscillator we get the normal form

$$\mathcal{H}_0(\mathbf{x}, \mathbf{y}, \mu_0) = \frac{1}{2}\delta_2 y_2^2, \qquad (\delta_2 = \pm 1). \tag{8.40}$$

Finally, we consider the case when $\omega_1 = \omega_2 = 0$.

This condition means that the 4×4 Hamiltonian matrix of the linear system defined by the unperturbed Hamiltonian $H_0(\mathbf{q}, \mathbf{p}, \mu_0)$ in (8.28) is nilpotent, so we make use here of Example 5.17 where the Hamiltonians determined by the symmetric matrices $S = -JA$ associated with the normal forms of A were found according to $r = \text{rank}(A)$.

(13) Case $\omega_1 = \omega_2 = 0$ and $r = 1$.

For $r = 1$, Example 5.17 gives two Hamiltonians for possible normal forms

$$H_0^{(6)} = -\frac{1}{2}X_1^2 \quad \text{and} \quad H_0^{(7)} = \frac{1}{2}Y_1^2.$$

The linear mappings

$$\begin{aligned} &X_1 = y_2, \quad X_2 = y_1, \quad Y_1 = -x_2, \quad Y_2 = -x_1 \quad \text{and} \\ &X_1 = x_2, \quad X_2 = x_1, \quad Y_1 = y_2, \quad Y_2 = y_1 \end{aligned}$$

are symplectic and they transform the Hamiltonians $H_0^{(6)}$ and $H_0^{(7)}$ respectively into the Hamiltonians

$$\mathcal{H}_0^{(6)} = -\frac{1}{2}y_2^2 \quad \text{and} \quad \mathcal{H}_0^{(7)} = \frac{1}{2}y_2^2,$$

which can be encompassed in the single formula

$$\mathcal{H}_0(\mathbf{x}, \mathbf{y}, \mu_0) = \frac{1}{2}\delta y_2^2, \qquad (\delta = \pm 1). \tag{8.41}$$

(14) Case $\omega_1 = \omega_2 = 0$ and $r = 2$.

For $r = 2$, Example 5.17 gives five Hamiltonians for possible normal forms

$$\begin{aligned} &H_0^{(1)} = X_1 Y_2, \quad H_0^{(2)} = -\frac{1}{2}(X_1^2 + X_2^2), \quad H_0^{(3)} = \frac{1}{2}(Y_1^2 + Y_2^2), \\ &H_0^{(4)} = \frac{1}{2}(-X_1^2 + Y_2^2), \quad H_0^{(5)} = \frac{1}{2}(-X_2^2 + Y_1^2). \end{aligned}$$

The following linear transformations

$$\begin{aligned} &X_1 = y_1, \quad X_2 = y_2, \quad Y_1 = -x_1, \quad Y_2 = -x_2; \\ &X_1 = x_1, \quad X_2 = x_2, \quad Y_1 = y_1, \quad Y_2 = y_2; \\ &X_1 = y_1, \quad X_2 = x_2, \quad Y_1 = -x_1, \quad Y_2 = y_2; \\ &X_1 = x_1, \quad X_2 = y_2, \quad Y_1 = y_1, \quad Y_2 = -x_2; \end{aligned}$$

are symplectic and take $H_0^{(2)}$, $H_0^{(3)}$, $H_0^{(4)}$ and $H_0^{(5)}$ respectively to the Hamiltonians

$$\mathcal{H}_0^{(2)} = -\frac{1}{2}(y_1^2 + y_2^2), \quad \mathcal{H}_0^{(3)} = \frac{1}{2}(y_1^2 + y_2^2), \quad \mathcal{H}_0^{(4)} = \frac{1}{2}(-y_1^2 + y_2^2), \quad \mathcal{H}_0^{(5)} = \frac{1}{2}(y_1^2 - y_2^2).$$

Finally, for any real numbers a, b, c, d such that $a + d = b - c$, the matrix

$$P = \begin{bmatrix} 0 & 0 & \gamma & -\gamma \\ \gamma & \gamma & a & b \\ -\gamma & \gamma & c & d \\ 0 & 0 & \gamma & \gamma \end{bmatrix}, \quad \gamma = 1/\sqrt{2}$$

is symplectic and defines a symplectic change of coordinates in which $X_1 = \frac{1}{\sqrt{2}}(y_1 - y_2)$ and $Y_2 = \frac{1}{\sqrt{2}}(y_1 + y_2)$, so this mapping takes $H_0^{(1)}$ to the Hamiltonian

$$\mathcal{H}_0^{(1)} = \frac{1}{2}(y_1^2 - y_2^2).$$

Therefore, the five Hamiltonians $H_0^{(1)}, \ldots, H_0^{(5)}$ can be encompassed in a single formula, namely

$$\mathcal{H}_0(\mathbf{x}, \mathbf{y}, \mu_0) = \frac{1}{2}(\delta_1 y_1^2 + \delta_2 y_2^2), \qquad (\delta_1 = \pm 1,\ \delta_2 = \pm 1). \tag{8.42}$$

(15) Case $\omega_1 = \omega_2 = 0$ and $r = 3$.
For $r = 3$, the normal form of the Hamiltonian is given by

$$H_0 = \frac{1}{2}\delta X_2^2 + X_1 Y_2, \qquad (\delta = \pm 1).$$

Making the linear symplectic change of coordinates $X_1 = x_2, X_2 = -y_1, Y_1 = y_2, Y_2 = x_1$, then H_0 is transformed into

$$\mathcal{H}_0(\mathbf{x}, \mathbf{y}, \mu_0) = \frac{1}{2}\delta y_1^2 + x_1 x_2, \quad (\delta = \pm 1). \tag{8.43}$$

In the Table 8.1 we summarize the above discussion to get the normal form in all cases of KGL resonances.

Remark 8.7 Details about each item in the Table 8.1, for instance $\tau = 2\pi$ or $\tau = 4\pi$, can be seen in the corresponding item of the above considerations. In those examples one can find the canonical linear transformation that will affect all the polynomials in the expansion of the Hamiltonian as a power series in ϵ. This linear transformation must be composed of linear canonical transformations that may have been used in the above considerations.

Table 8.1 Normal forms in all cases of KGL resonances

Case	Nature of frequencies	Peculiarities	Normal form $\mathcal{H}_0$	Period
(1)$_R$	$2\omega_2 = N \neq 0,\ 2\omega_1 \notin \mathbf{Z}$		(8.31)	τ
(2)	$\omega_2 = 0,\ 2\omega_1 \notin \mathbf{Z}$	$A_{\eta^\dagger(0)} = 0$	(8.32)	π
(3)	$\omega_2 = 0,\ 2\omega_1 \notin \mathbf{Z}$	$A_{\eta^\dagger(0)} \neq 0$	(8.33)	π
(4)$_R$	$\omega_1 + \omega_2 = N,\ 2\omega_1 \notin \mathbf{Z}$		(8.34)	τ
(5)$_R$	$\omega_1 - \omega_2 = N,\ 2\omega_1 \notin \mathbf{Z}$		(8.35)	τ
(6)	$\omega_1 = \omega_2 = \omega,\ 2\omega \notin \mathbf{Z}$	$r = 2$	(8.36)	π
(7)	$\omega_1 = \omega_2 = \omega,\ 2\omega \notin \mathbf{Z}$	$r = 3$	(8.37)	π
(8)$_R$	$2\omega_1 = N_1,\ 2\omega_2 = N_2,\ N_1, N_2 \neq 0$		$\mathcal{H}_0 = 0$	τ
(9)$_R$	$\omega_1 = \omega_2 = \omega,\ 2\omega = N$	$r = 2$	$\mathcal{H}_0 = 0$	τ
(10)$_R$	$\omega_1 = \omega_2 = \omega,\ 2\omega = N$	$r = 3$	(8.39)	τ
(11)$_R$	$2\omega_1 = N_1,\ \omega_2 = 0$	$A_{\eta^\dagger(0)} = 0$	$\mathcal{H}_0 = 0$	τ
(12)$_R$	$2\omega_1 = N_1,\ \omega_2 = 0$	$A_{\eta^\dagger(0)} \neq 0$	(8.40)	τ
(13)	$\omega_1 = 0,\ \omega_2 = 0$	$r = 1$	(8.41)	π
(14)	$\omega_1 = 0,\ \omega_2 = 0$	$r = 2$	(8.42)	π
(15)	$\omega_1 = 0,\ \omega_2 = 0$	$r = 3$	(8.43)	π

8.6.2 *Matrix Form of the Homological Equation*

Having found the normal form of H_0 for each KGL resonance we now write the homological equation (8.29) in matrix form. We consider the monomials of the quadratic homogeneous polynomials taken in the order

$$x_1^2,\ x_1y_1,\ y_1^2,\ x_1x_2,\ x_1y_2,\ x_2y_1,\ y_1y_2,\ x_2^2,\ x_2y_2,\ y_2^2, \tag{8.44}$$

and group the coefficients of $\mathcal{W}$ and $\mathcal{B}$ accordingly using the vectors

$$\begin{aligned} X &= \big(w_{2000}, w_{1010}, w_{0020}, w_{1100}, w_{1001}, w_{0110}, w_{0011}, w_{0200}, w_{0101}, w_{0002}\big) \\ \mathcal{B} &= \big(b_{2000}, b_{1010}, b_{0020}, b_{1100}, b_{1001}, b_{0110}, b_{0011}, b_{0200}, b_{0101}, b_{0002}\big). \end{aligned} \tag{8.45}$$

Then the homological equation (8.29) can be written in the matrix form

$$\dot{X}^{(m)} = MX^{(m)} + \mathcal{B}^{(m)}, \text{ where } MX^{(m)} = \{\mathcal{H}_0, \mathcal{W}_m\} \text{ and } \mathcal{B}^{(m)} = \mathcal{H}^{(m)} - \mathcal{K}_m, \tag{8.46}$$

where superscript m indicates that we are working at the mth step in the normalization process. We now observe that except for the cases **(7)** and **(15)** all the other cases run under the general formula

$$\mathcal{H}_0 = \frac{1}{2}a_1x_1^2 + \frac{1}{2}b_1y_1^2 + \frac{1}{2}a_2x_2^2 + \frac{1}{2}b_2y_2^2. \tag{8.47}$$

Computing the Poisson brackets in (8.46) for the general case (8.47) we get

$$
\begin{aligned}
&\{\mathcal{H}_0, \mathcal{W}\}_{2000} = a_1 w_{1010}, && \{\mathcal{H}_0, \mathcal{W}\}_{1010} = 2a_1 w_{0020} - 2b_1 w_{2000},\\
&\{\mathcal{H}_0, \mathcal{W}\}_{0020} = -b_1 w_{1010}, && \\
&\{\mathcal{H}_0, \mathcal{W}\}_{1100} = a_1 w_{0110} + a_2 w_{1001}, && \{\mathcal{H}_0, \mathcal{W}\}_{1001} = a_1 w_{0011} - b_2 w_{1100},\\
&\{\mathcal{H}_0, \mathcal{W}\}_{0110} = -b_1 w_{1100} + a_2 w_{0011}, && \{\mathcal{H}_0, \mathcal{W}\}_{0011} = -b_1 w_{1001} - b_2 w_{0110},\\
&\{\mathcal{H}_0, \mathcal{W}\}_{0200} = a_2 w_{0101}, && \{\mathcal{H}_0, \mathcal{W}\}_{0101} = 2a_2 w_{0002} - 2b_2 w_{0200},\\
&\{\mathcal{H}_0, \mathcal{W}\}_{0002} = -b_2 w_{0101} && .
\end{aligned}
$$

So for the coefficient matrix M of the linear system (8.46) we have the following 10×10 matrix

$$
M = \begin{bmatrix} M_1 & O & O \\ O & M_2 & O \\ O & O & M_3 \end{bmatrix} \tag{8.48}
$$

where M_1 and M_3 are 3×3 matrices, M_2 is a 4×4 matrix and they are given by

$$
M_1 = \begin{bmatrix} 0 & a_1 & 0 \\ -2b_1 & 0 & 2a_1 \\ 0 & -b_1 & 0 \end{bmatrix}, \quad M_3 = \begin{bmatrix} 0 & a_2 & 0 \\ -2b_2 & 0 & 2a_2 \\ 0 & -b_2 & 0 \end{bmatrix}, \tag{8.49}
$$

$$
M_2 = \begin{bmatrix} 0 & a_2 & a_1 & 0 \\ -b_2 & 0 & 0 & a_1 \\ -b_1 & 0 & 0 & a_2 \\ 0 & -b_1 & -b_2 & 0 \end{bmatrix}. \tag{8.50}
$$

The system (8.46) (we will omit the superscript m) obviously decouples into three nonhomogeneous linear systems. The first is the three-dimensional system

$$
\dot{X}_1 = M_1 X_1 + \mathcal{B}_1 \tag{8.51}
$$

where $X_1 = \left(w_{2000}, w_{1010}, w_{0020}\right)$, $\mathcal{B}_1 = \left(b_{2000}, b_{1010}, b_{0020}\right)$.

The second system is four dimensional and is defined by the matrix M_2,

$$
\dot{X}_2 = M_2 X_2 + \mathcal{B}_2, \tag{8.52}
$$

where $X_2 = \left(w_{1100}, w_{1001}, w_{0110}, w_{0011}\right)$, $\mathcal{B}_2 = \left(b_{1100}, b_{1001}, b_{0110}, b_{0011}\right)$ and the third is the three-dimensional system

$$
\dot{X}_3 = M_3 X_3 + \mathcal{B}_3, \tag{8.53}
$$

where $X_3 = \left(w_{0200}, w_{0101}, w_{0002}\right)$, $\mathcal{B}_3 = \left(b_{0200}, b_{0101}, b_{0002}\right)$.

8.6.3 The Coefficient Matrix for Each Resonance

To solve the above linear systems we will find the matrix M for each case (1)–(15) listed in Table 8.1. For the meaning of the subscript R see Remark 8.6.

(1)$_R$ The normal form is (8.31), hence in (8.47) $a_1 = b_1 = \delta_1\omega_1$, $a_2 = b_2 = 0$, so we have

$$M_1 = \delta_1\omega_1 C, \quad M_2 = \delta_1\omega_1 J_4, \quad M_3 = O, \tag{8.54}$$

where C is the matrix of Example 1.4 and J_4 is the standard symplectic matrix of order 4.

(2) The normal form is (8.32) hence the matrices M_1, M_2 and M_3 are also those in (8.54).

(3) The normal form is (8.33) hence $a_1 = b_1 = \delta_1\omega_1$, $a_2 = 0$ and $b_2 = \delta_2$, so the matrices are

$$M_1 = \delta_1\omega_1 C, \quad M_2 = \delta_1\omega_1 J_4 - \delta_2 F_2, \quad M_3 = -\delta_2 E_1, \tag{8.55}$$

where E_1 is the matrix of Example 1.6 with $a = 2$, $b = 1$, and F_2 the matrix of Example 1.7.

(4)$_R$ Here the normal form is (8.34), hence $a_1 = b_1 = \delta_1\omega_1$, $a_2 = b_2 = -\delta_2\omega_1$, so

$$M_1 = \delta_1\omega_1 C, \quad M_2 = \delta_1\omega_1 J_4 - \delta_2\omega_1 B, \quad M_3 = -\delta_2\omega_1 C, \tag{8.56}$$

where B is the matrix of Example 1.3.

(5) The normal form is (8.35), hence $a_1 = b_1 = \delta_1\omega_1$, $a_2 = b_2 = \delta_2\omega_1$, so

$$M_1 = \delta_1\omega_1 C, \quad M_2 = \delta_1\omega_1 J_4 + \delta_2\omega_1 B, \quad M_3 = \delta_2\omega_1 C. \tag{8.57}$$

where B, the matrix of Example 1.3 and C the matrix of Example 1.4.

(6) The normal form is (8.36) so in (8.47) $a_1 = b_1 = \delta_1\omega$, $a_2 = b_2 = \delta_2\omega$, hence

$$M_1 = \delta_1\omega C, \quad M_2 = \delta_1\omega J_4 + \delta_2\omega B, \quad M_3 = \delta_2\omega C. \tag{8.58}$$

(7) We consider this case at the end together with case (**15**).

(8)$_R$ Normal form is $\mathcal{H}_0 = 0$, so we have $M_1 = 0$, $M_2 = 0$, $M_3 = 0$.

(9)$_R$ Here, the normal form is also $\mathcal{H}_0 = 0$; hence, $M_1 = 0$, $M_2 = 0$ and $M_3 = 0$.

(10)$_R$ The matrices here are those of the case (**14**) below with $\delta_1 = \delta_2 = \delta$.

(11)$_R$ The normal form is $\mathcal{H}_0 = 0$, so $M_1 = 0$, $M_2 = 0$, $M_3 = 0$.

(12)$_R$ The normal form is (8.40), so $a_1 = 0$, $b_1 = 0$, $a_2 = 0$ and $b_2 = \delta_2 = \pm 1$; hence,

$$M_1 = O, \quad M_2 = -\delta_2 F_2, \quad M_3 = -\delta_2 E_1 \tag{8.59}$$

E_1 of Example 1.6 with $a = 2$ and $b = 1$ and F_2 of Example 1.7.

(13) The normal form is (8.41), so the matrices are as in case **(12)**$_R$ with $\delta_2 = \delta = \pm 1$ and here the period is $\tau = \pi$.

(14) The normal form is (8.42) so $a_1 = 0$, $b_1 = \delta_1$, $a_2 = 0$, $b_2 = \delta_2$, and therefore

$$M_1 = -\delta_1 E_1, \quad M_2 = -\delta_1 F_1 - \delta_2 F_2, \quad M_3 = -\delta_2 E_1, \tag{8.60}$$

where E_1 is the matrix in **(12)** and F_1, F_2 are the matrices of Example (1.18).

(15) The normal form is given by (8.43) and the corresponding 10×10 matrix M will be found at the end of the next section.

In the next section we solve each one of the above linear systems.

8.6.4 The Autonomous Hamiltonian $\mathcal{K}$

We solve the homological equation considering that the generating function $\mathcal{W}$ is τ-periodic. We recall that originally we have taken $\tau = \pi$ but if a rotation involving an integer N owing to a resonance was needed to get the normal form in one of the items (1)–(15) then the period becomes $\tau = 2\pi$ if N is even and $\tau = 4\pi$ if N is odd.

(1)$_R$ (a) Solving system (8.51).

By the variation of parameters formula (1.23) we have

$$X_1^{(m)}(t) = \mathrm{e}^{tM_1} X_1^{(m)}(0) + \mathrm{e}^{tM_1} \int_0^t \mathrm{e}^{-sM_1} \mathcal{B}_1^{(m)}(s) ds. \tag{8.61}$$

The τ-periodicity of $\mathcal{W}$ leads to the equation

$$(I_4 - \mathrm{e}^{\tau M_1}) X_1^{(m)}(0) = \mathrm{e}^{\tau M_1} \int_0^\tau \mathrm{e}^{-sM_1} \big(\mathcal{H}_1^{(m)}(s) - \mathcal{K}_1^{(m)}(s)\big) ds. \tag{8.62}$$

Remark 8.8 The superscript m in $\mathcal{H}_1^{(m)}$ and $\mathcal{K}_1^{(m)}$ in (8.62) concerns the mth step in the normalization process and the subscript 1 concerns the X_1. Observe that $m \geq 1$ since step zero is already defined because $\mathcal{K}_0 = \mathcal{H}_0$.

Since $M_1 = aC$, where $a = \delta_1 \omega_1$, we have by Example 1.4

$$e^{tM_1} = \frac{1}{2}\begin{bmatrix} 1+\cos 2at & \sin 2at & 1-\cos 2at \\ -2\sin 2at & 2\cos 2at & 2\sin 2at \\ 1-\cos 2at & -\sin 2at & 1+\cos 2at \end{bmatrix}. \tag{8.63}$$

The first and third rows of the matrix $I - e^{tM_1}$ differ only in sign so this matrix is not invertible whatever be the value of t.

The kernel, χ_τ, of the matrix $I - e^{\tau M_1}$ is formed by the vectors (ξ_1, ξ_2, ξ_3) such that

$$\begin{aligned} &(1-\cos 2a\tau)(\xi_1-\xi_3) - (\sin 2a\tau)\xi_2 = 0, \\ &(\sin 2a\tau)(\xi_1-\xi_3) + (1-\cos 2a\tau)\xi_2 = 0. \end{aligned} \tag{8.64}$$

The determinant of the coefficient matrix of this 2×2 linear system is equal to $4\sin^2\delta_1\omega_1\tau$. If N is even the period is $\tau = 2\pi$ and $\sin^2\delta_1\omega_1\tau = 0$ cannot happen since $2\omega_1$ is not an integer. If N is odd the period is $\tau = 4\pi$ and the equality $\sin^2\delta_1\omega_1\tau = 0$ could happen, but only in the case when $2\omega_1$ is a half integer.

Suppose that $\sin^2\delta_1\omega_1\tau \neq 0$. Then the kernel χ_τ of $I - e^{\tau M_1}$ is generated by the vector $(1, 0, 1)$, so $X_1^{(m)} = (w_{2000}^{(m)}, w_{1010}^{(m)}, w_{0020}^{(m)})$ lies in χ_τ if, and only if, $w_{2000}^{(m)} = w_{0020}^{(m)}$ and $w_{1010}^{(m)} = 0$. This suggests considering the linear transformation

$$x_1 = w_{2000}^{(m)}, \quad x_2 = w_{1010}^{(m)}, \quad x_3 = w_{2000}^{(m)} - w_{0020}^{(m)},$$

which carries the kernel χ_τ to the x_1-axis and the differential equation (8.51) becomes the system

$$\dot{x}_1 = ax_2 + \beta_1, \quad \dot{x}_2 = -2ax_3 + \beta_2, \quad \dot{x}_3 = 2ax_2 + \beta_3, \tag{8.65}$$

where

$$\beta_1 = h_{2000}^{(m)} - k_{2000}^{(m)}, \quad \beta_2 = h_{1010}^{(m)} - k_{1010}^{(m)}, \quad \beta_3 = (h_{2000}^{(m)} - h_{0020}^{(m)}) - (k_{2000}^{(m)} - k_{0020}^{(m)}).$$

The solution of the subsystem $x_2\, x_3$ in (8.65) is given by

$$X^{(m)}(t) = e^{tM}X^{(m)}(0) + e^{tM}\int_0^t e^{-sM}\mathcal{B}(s)ds,$$

where $X = (x_2, x_3)$, $\mathcal{B}(s) = (\beta_2, \beta_3)$ and

$$e^{Mt} = \begin{pmatrix} \cos 2at & -\sin 2at \\ \sin 2at & \cos 2at \end{pmatrix}.$$

The τ-periodicity of $\mathcal{W}$ leads to the equation

$$(I_2 - e^{\tau M})X^{(m)}(0) = e^{\tau M}\int_0^\tau e^{-sM}\mathcal{B}^{(m)}(s)ds,$$

which has a solution for any set of constants $k_{2000}^{(m)}, k_{1010}^{(m)}, k_{0200}^{(m)}$ since $(I_2 - e^{\tau M})$ is invertible. We take these constants satisfying the conditions

$$k_{1010}^{(m)} = 0, \quad k_{2000}^{(m)} = k_{0020}^{(m)}. \tag{8.66}$$

The first equation in (8.65) with $\beta_1 = h_{2000}^{(m)}(t) - k_{2000}^{(m)}$ can then be integrated by inserting in it the function $x_2(t)$ already found. This solution gives the components $w_{2000}^{(m)}(t)$, $w_{1010}^{(m)}(t)$, $w_{0020}^{(m)}(t)$ of the vector $X_1(t) = X_1^{(m)}(t)$. The τ-periodicity of $x_1(t)$ allows one to find the value

$$k_{2000}^{(m)} = k_{0200}^{(m)} = \frac{1}{\tau}\int_0^\tau h_{2000}^{(m)}(s)ds + \frac{\delta_1\omega_1}{\tau}\int_0^\tau x_2(s)ds.$$

Here, we take the initial condition $x_1(0)$ equal to zero.

Another way to solve this case is explained below. Observe that for a constant vector $\mathcal{K}_1$, the Eq. (8.62) can be written as

$$EX_1^{(m)}(0) = \int_0^\tau e^{-sM_1}(\mathcal{K}_1^{(m)}(s) - \mathcal{H}_1^{(m)}(s))ds, \tag{8.67}$$

where

$$E = I_4 - \mathrm{e}^{-\tau M_1} = \sin a\tau \begin{pmatrix} \sin a\tau & \cos a\tau & -\sin a\tau \\ -2\cos a\tau & 2\sin a\tau & 2\cos a\tau \\ -\sin a\tau & -\cos a\tau & \sin a\tau \end{pmatrix}. \tag{8.68}$$

The vector in the right-hand side of (8.67) lies in the image of the operator E, which is the subspace of points (y_1, y_2, y_3) satisfying the equation $y_1 + y_3 = 0$. Let

$$\int_0^\tau e^{-sM_1}\mathcal{H}_1^{(m)}(s)ds = (h_1, h_2, h_3)^T.$$

Now, computing the matrix

$$P(\tau) = \int_0^\tau e^{-sM_1}ds = \frac{1}{2}\begin{bmatrix} \tau + \frac{1}{2a}\sin 2a\tau & \frac{1}{2a}(\cos 2a\tau - 1) & \tau - \frac{1}{2a}\sin 2a\tau \\ \frac{1}{a}(1 - \cos 2a\tau) & \frac{1}{a}\sin 2a\tau & \frac{1}{a}(\cos 2a\tau - 1) \\ \tau - \frac{1}{2a}\sin 2a\tau & \frac{1}{2a}(1 - \cos 2a\tau) & \tau + \frac{1}{2a}\sin 2a\tau \end{bmatrix},$$

we readily see that

$$P(\tau)(1, 0, 1)^T = (\tau, 0, \tau)^T.$$

Considering the choice in (8.66) of constant vector $\mathcal{K}_1^{(m)}$ of the form $\mathcal{K}_1^{(m)} = (k_1, 0, k_1)^T$ we have

$$\int_0^\tau e^{-sM_1}\mathcal{K}_1^{(m)}(s)ds = (k_1\tau, 0, k_1\tau)^T.$$

We see that the right-hand side of (8.67) belongs to the image of E if, and only if,

$$k_1 = \frac{1}{2\tau}(h_1 + h_3). \tag{8.69}$$

Therefore, if $X_1^{(m)}(0) = (\xi, \eta, \zeta)$, using (8.68) we get

$$\begin{aligned}
\sin a\tau\xi + \cos a\tau\eta - \sin a\tau\zeta &= \frac{1}{\sin a\tau}(k_1\tau - h_1)\\
-\cos a\tau\xi + \sin a\tau\eta + \cos a\tau\zeta &= -\frac{1}{2\sin a\tau}h_2\\
-\sin a\tau\xi - \cos a\tau\eta + \sin a\tau\zeta &= \frac{1}{\sin a\tau}(k_1\tau - h_3);
\end{aligned}$$

hence, from (8.69), we obtain $k_1\tau - h_1 = \frac{1}{2}(h_3 - h_1)$ and we find

$$\xi = \frac{1}{2}[h_3 - h_1 + h_2\cot a\tau] + \zeta, \quad \eta = \frac{1}{2}[-h_2 + (h_3 - h_1)\cot a\tau],$$

with ζ arbitrary.

(b) Solving system $\dot{X}_2 = M_2X_2 + \mathcal{B}_2$.

By the variation of parameters formula (1.23) we have

$$X_2(t) = e^{tM_2}X_2(0) + e^{tM_2}\int_0^t e^{-sM_2}\mathcal{B}_2(s). \tag{8.70}$$

The τ-periodicity of $\mathcal{W}$ leads to the equation

$$(I_4 - e^{\tau M_2})X_2^{(m)}(0) = e^{\tau M_2}\int_0^\tau e^{-sM_2}\big(\mathcal{H}_2^{(m)}(s) - \mathcal{K}_2^{(m)}(s)\big)ds. \tag{8.71}$$

Using Example 1.2 with $\alpha = \delta_1\omega_1$ and $J = J_4$, we compute the matrix

$$I_4 - e^{\tau M_2} = \begin{bmatrix} (1-\cos\delta_1\omega_1\tau)I_2 & -\sin\delta_1\omega_1\tau I_2 \\ \sin\delta_1\omega_1\tau I_2 & (1-\cos\delta_1\omega_1\tau)I_2 \end{bmatrix},$$

where I_2 is the 2×2 identity matrix. The determinant of the matrix $I_4 - e^{\tau M_2}$ is

$$\left[(1-\cos\delta_1\omega_1\tau)^2+\sin^2\delta_1\omega_1\tau\right]^2=16\sin^4\frac{\omega_1\tau}{2}.$$

This determinant is zero if, and only if, $\omega_1\tau = 2k\pi$, for some integer k. But this cannot happen since the period τ is either 2π, when N is even or 4π, when N is odd and by hypothesis $2\omega_1$ is not an integer. Therefore, the matrix $I_4-e^{\tau M_2}$ is invertible; hence, we can take $\mathcal{K}_2^{(m)}=0$, that is,

$$k_{1100}=0,\quad k_{1001}=0,\quad k_{0110}=0,\quad k_{0011}=0.$$

The initial condition $X_2^{(m)}(0)$ is determined by (8.71) and the components $w_{1100}^{(m)}(t)$, $w_{1001}^{(m)}(t)$, $w_{0110}^{(m)}(t)$, $w_{0011}^{(m)}(t)$ of $X_2(t) = X_2^{(m)}(t)$ are then found from (8.70) with $\mathcal{B}_2(s)=\mathcal{H}_2^{(m)}(s)$.

(c) Solving system $\dot{X}_3 = M_3X_3+\mathcal{B}_3$.

Since $M_3=0$, the solution of the third system is given by

$$X_3(t)=X_3(0)+\int_0^t(\mathcal{H}_3^{(m)}(s)-\mathcal{K}_3^{(m)}(s))ds, \tag{8.72}$$

so we get the constant vector $\mathcal{K}_3^{(m)}=(k_{0200}^{(m)},k_{0101}^{(m)},k_{0002}^{(m)})$ by quadrature

$$\mathcal{K}_3^{(m)}=\frac{1}{\tau}\int_0^\tau\mathcal{H}_3^{(m)}(t)dt,$$

and for arbitrary values $w_{0200}^{(m)}(0)$, $w_{0101}^{(m)}(0)$, $w_{0002}^{(m)}(0)$ we get the components of the vector $X_3(t)=\mathcal{W}_3^{(m)}(t)$ from (8.72). Here we can choose the simplest initial conditions, namely $w_{0200}^{(m)}(0)=0$, $w_{0101}^{(m)}(0)=0$, $w_{0002}^{(m)}(0)=0$.

Therefore, if $\sin\delta_1\omega_1\tau\neq 0$, we have computed in (a)-(b)-(c), at the mth stage, all the components of the constant vector $\mathcal{K}^{(m)}$ five of which are zero, namely

$$k_{1010}^{(m)}=0,\quad k_{1100}^{(m)}=0,\quad k_{1001}^{(m)}=0,\quad k_{0110}^{(m)}=0,\quad k_{0011}^{(m)}=0 \tag{8.73}$$

and we also have the equality $k_{2000}^{(m)}=k_{0020}^{(m)}$.

We have also found the initial conditions to solve the equations that give the components of the generating function $\mathcal{W}^{(m)}(t)$.

Remark 8.9 To determine effectively the components of $\mathcal{W}$ and $\mathcal{K}$ at the mth stage we need to know $\mathcal{H}^{(m)}$. This is given by formula (8.14) (see Theorem 3.22 for the explanation of the terms in this formula). We start with $\mathcal{H}^{(1)}=\mathcal{H}_1$.

In the next cases we omit the superscript (m) but we are reminded that we are working with the Hamiltonian in the mth stage of the Deprit–Hori method.

(2) Solving systems (8.51), (8.52), (8.53) is exactly as in **(1)** and since here the period is $\tau = \pi$ and $2\omega_1$ not integer, we always have $\sin^2 \omega_1 \tau \neq 0$, so we get the equalities (8.73) and also $k_{2000} = k_{0020}$.
(3) (a) Solving system (8.51) is exactly as in **(1)** and here also the inequality $\sin^2 \omega_1 \tau \neq 0$ holds so we have

$$k_{1010} = 0, \quad k_{2000} = k_{0020}.$$

(b) Let us solve system (8.52). The matrix is

$$M_2 = \begin{bmatrix} 0 & 0 & \delta_1\omega_1 & 0 \\ -\delta_2 & 0 & 0 & \delta_1\omega_1 \\ -\delta_1\omega_1 & 0 & 0 & 0 \\ 0 & -\delta_1\omega_1 & -\delta_2 & 0 \end{bmatrix}.$$

The period is $\tau = \pi$ and the system of equations to be solved is:

$$\begin{aligned} \dot{x}_1 &= \delta_1\omega_1 x_3 + b_1, \\ \dot{x}_2 &= -\delta_2 x_1 + \delta_1\omega_1 x_4 + b_2, \\ \dot{x}_3 &= -\delta_1\omega_1 x_1 + b_3 \\ \dot{x}_4 &= -\delta_1\omega_1 x_2 - \delta_2 x_3 + b_4, \end{aligned}$$

where

$$x_1 = w_{1100},\ x_2 = w_{1001},\ x_3 = w_{0110},\ x_4 = w_{0011}, \quad (b_1, b_2, b_3, b_4) = \mathcal{H}_2 - \mathcal{K}_2.$$

Since $2\omega_1$ is not an integer the system $\dot{x}_1 = \delta_1\omega_1 x_3 + b_1,\ \dot{x}_3 = -\delta_1\omega_1 x_1 + b_3$ can be solved for any value of the constants, so we take the simplest values

$$k_{1100} = k_{0110} = 0$$

and get the τ-periodic functions $x_1(t)$ and $x_3(t)$. Then we insert these functions into the system for x_2 and x_4, which can also be solved for any values of the constants that we also take as the simplest values

$$k_{1001} = k_{0011} = 0$$

and then solve to get the τ-periodic functions $x_2(t)$ and $x_4(t)$.

In the solutions of the systems for x_1, x_3 and x_2, x_4 the coefficient matrix is $M = \delta_1\omega_1 J_2$ and the matrix $I_2 - e^{\tau M}$ is invertible because $2\omega_1$ is not an integer. Therefore, the initial conditions $w_{1100}(0), w_{0110}(0)$ for the first system and $w_{1001}(0), w_{0011}(0)$ for the second, are completely determined in the process.

(c) Solving system (8.53).
Let $x_1 = w_{0200}$, $x_2 = w_{0101}$ and $x_3 = w_{0002}$. The system to be solved here is

$$\dot{x}_1 = b_1, \quad \dot{x}_2 = -2\delta_2 x_1 + b_2, \quad \dot{x}_3 = -\delta_2 x_2 + b_3.$$

We readily get the solution, which is given by

$$\begin{aligned} x_1(t) &= x_1(0) + \int_0^t h_{0200}(s)ds - k_{0200}t, \\ x_2(t) &= x_2(0) + f(t) - k_{0101}t, \\ x_3(t) &= x_3(0) + g(t) - k_{0002}t, \end{aligned}$$

where

$$\begin{aligned} f(t) &= -\left[2\delta_2 x_1(0)t + 2\delta_2 \int_0^t \int_0^s h_{0200}(u)du\,ds - \delta_2 k_{0200}t^2\right] + \int_0^t h_{0101}(s)ds, \\ g(t) &= -\left[\delta_2 x_2(0)t + \delta_2 \int_0^t f(s)ds - \frac{1}{2}\delta_2 k_{0101}t^2\right] + \int_0^t h_{0002}(s)ds, \end{aligned}$$

Choose the initial condition $x_1(0)$ so that $f(\tau) = 0$. Then the τ-periodicity of $x_2(t)$ implies that $k_{0101} = 0$. Now take $x_2(0) = 0$ and the expression of $g(t)$ is completely determined. Using it the τ-periodicity of $x_3(t)$ gives the value $k_{0002} = \frac{1}{\tau}g(\tau)$.

The initial conditions $w^{(m)}_{0200}(0) = x_1(0)$, $w^{(m)}_{0101}(0) = x_2(0) = 0$ have already been chosen, and $w^{(m)}_{0002}(0) = x_3(0)$ remains arbitrary. So in this case we have

$$k_{1100} = 0, \quad k_{1010} = 0, \quad k_{1001} = 0, \quad k_{0110} = 0, \quad k_{0101} = 0, \quad k_{0011} = 0,$$

and moreover, $k_{2000} = k_{0020}$.

(4)$_R$ (a) The solution of the system (8.51) is exactly as in (1), so we get the values (8.66).

(b) Solving the system (8.52).

Let us find the components of $w_{1100}, w_{1001}, w_{0110}, w_{0011}$ of $X_2(t) = X_2^{(m)}(t)$. Here we use Eq. (8.70) with $M_2 = \delta_1\omega_1 J_4 - \delta_2\omega_1 B$, $2\omega_1$ not an integer.

Let $d = -\delta_2/\delta_1$ so the matrix M_2 is given by

$$M_2 = \delta_1\omega_1 \begin{bmatrix} 0 & d & 1 & 0 \\ -d & 0 & 0 & 1 \\ -1 & 0 & 0 & d \\ 0 & -1 & -d & 0 \end{bmatrix}.$$

The kernel of M_2 is generated by the vectors $(1, 0, 0, d)$ and $(0, 1, -d, 0)$ which are also, for any $t \neq 0$, the generators of the kernel of the matrix $I - e^{tM_2}$.

By Eq. (8.70), the requirement that $X_2(t)$ is τ-periodic and $\mathcal{K}_2$ is constant leads to the equation

$$(I - e^{-\tau M_2})X_2(0) = P(\tau)\mathcal{K}_2 - \mathbf{h}_2, \tag{8.74}$$

where $P(t) = \int_0^t e^{-sM_2}\,ds$ and $\mathbf{h}_2 = \int_0^\tau e^{-sM_2}\mathcal{H}_2(s)\,ds$.

From Examples 1.3 and 1.2 with $\alpha = \delta_1\omega_1$ and $\beta = d\delta_1\omega_1 = d\alpha$, we find

$$I - e^{tM_2} = \begin{bmatrix} I_2 - \cos\alpha t R_{\beta t} & -\sin\alpha t R_{\beta t} \\ \sin\alpha t R_{\beta t} & I_2 - \cos\alpha t R_{\beta t} \end{bmatrix}. \tag{8.75}$$

For the points $\begin{bmatrix} \mathbf{u} \\ \mathbf{v} \end{bmatrix}$ in the image of the matrix $I - e^{tM_2}$, we have

$$(I_2 - \cos\alpha t R_{\beta t})\mathbf{x} - \sin\alpha t R_{\beta t}\mathbf{y} = \mathbf{u}, \quad \sin\alpha t R_{\beta t}\mathbf{x} + (I_2 - \cos\alpha t R_{\beta t})\mathbf{y} = \mathbf{v};$$

hence, multiplying the first equation by $-\sin\alpha t R_{\beta t}$, the second by $(I_2 - \cos\alpha t R_{\beta t})$ and adding the results we find

$$\big(I_2 - 2\cos\alpha t R_{\beta t} + R_{\beta t}^2\big)\mathbf{y} = -\sin\alpha t R_{\beta t}\mathbf{u} + (I_2 - \cos\alpha t R_{\beta t})\mathbf{v}.$$

Since $d = \pm 1$, we have $\cos\alpha t = \cos\beta t$ and a computation shows that $I_2 - 2\cos\alpha t R_{\beta t} + R_{\beta t}^2 = 0$. Therefore, the points in the image of $I - e^{tM_2}$ are defined by the equation

$$-\sin\alpha t R_{\beta t}\mathbf{u} + (I_2 - \cos\alpha t R_{\beta t})\mathbf{v} = 0.$$

Using the fact that $\sin\alpha t = d\sin\beta t$ and taking $t = -\tau$, we obtain from the above equation that the points $\begin{bmatrix} \mathbf{u} \\ \mathbf{v} \end{bmatrix}$ in the image of the operator $I - e^{-\tau M_2}$ satisfies

$$\sin\beta\tau\Big(-d\begin{bmatrix} \cos\beta\tau & -\sin\beta\tau \\ \sin\beta\tau & \cos\beta\tau \end{bmatrix}\mathbf{u} + \begin{bmatrix} -\sin\beta\tau & -\cos\beta\tau \\ \cos\beta\tau & -\sin\beta\tau \end{bmatrix}\mathbf{v}\Big) = 0.$$

Now, $\sin\beta\tau = 0$ means that $\omega_1\tau = k\pi$, for some integer k. If N is even, $\tau = 2\pi$ and this cannot happen because by hypothesis $2\omega_1$ in not an integer. It could happen if N is odd, unless $4\omega_1$ is not an integer.

If $\sin\beta\tau \neq 0$ we find from the above equation[5] that $\mathbf{u}$ and $\mathbf{v}$ are related by the equation

$$d\begin{bmatrix} 1 & 0 \\ 0 & 1 \end{bmatrix}\mathbf{u} + \begin{bmatrix} 0 & 1 \\ -1 & 0 \end{bmatrix}\mathbf{v} = 0, \quad \text{that is} \quad du_1 + v_2 = 0,\ du_2 - v_1 = 0,$$

[5] Multiplying the above equation by the inverse matrix of the first matrix.

so the image of the operator $E = I - e^{-\tau M_2}$ is generated by the vectors

$$(1, 0, 0, -d) \quad \text{and} \quad (0, 1, d, 0). \tag{8.76}$$

By (8.74) we have to take $\mathcal{K}_2$ so that the vector $P\mathcal{K}_2 - \mathbf{h}_2$ belongs to $Im\,E$. We will take the point $\mathcal{K}_2$ such that $P\mathcal{K}_2 = \mathbf{h}_2 + \text{Proj}_{(Im\,E)}\mathbf{h}_2$,[6] so we get

$$\begin{aligned} EX_2(0) = Proj_{(Im\,E)}\mathbf{h}_2 &= \tfrac{1}{2}(\langle \mathbf{h}_2, (1,0,0,-d)\rangle\,(1,0,0,-d) \\ &\quad + \langle \mathbf{h}_2, (0,1,d,0)\rangle\,(0,1,d,0)) \\ &= \tfrac{1}{2}(h_1 - dh_4, h_2 + dh_3, dh_2 + h_3, -dh_1 + h_4), \end{aligned}$$

where $\mathbf{h}_2 = (h_1, h_2, h_3, h_4)$. Therefore, we get the vector $\mathcal{K}_2$ given by

$$\mathcal{K}_2 = \frac{1}{2}P(\tau)^{-1}\big(3h_1 - dh_4, 3h_2 + dh_3, dh_2 + 3h_3, -dh_1 + 3h_4\big).$$

To get the initial conditions $X_2(0)$ we use the expression of $E = I - e^{-\tau M_2}$ obtained from (8.75) by setting $t = -\tau$ and we solve the system $E(\mathbf{x}, \mathbf{y}) = Proj_{(Im\,E)}\mathbf{h}_2$ with $(\mathbf{x}, \mathbf{y}) \notin Ker(E)$, that is, since $\mathbf{R}^4 = ker\,E \oplus Im E$, we solve this system taking $(\mathbf{x}, \mathbf{y}) \in Im E$. In this way, we get

$$\mathbf{x} = \frac{1}{2\sin\alpha\tau}\begin{bmatrix} \sin\alpha\tau & -d\cos\alpha\tau \\ d\cos\alpha\tau & \sin\alpha\tau \end{bmatrix}\mathbf{h}, \quad \mathbf{y} = d\begin{bmatrix} 0 & 1 \\ -1 & 0 \end{bmatrix}\mathbf{x},$$

with $\mathbf{h} = \frac{1}{2}(h_1 - dh_4, h_2 + dh_3)$.

(c) The solution of the system (8.53) here is just as in **(1)** (a) with $a = -\delta_2\omega_1$ and we get the equalities $k_{0101} = 0$ and $k_{0200} = k_{0002}$.

Therefore, in this case we get two zero constants

$$k_{1010} = 0, \; k_{0101} = 0 \quad \text{and} \quad k_{2000} = k_{0020}, \; k_{0200} = k_{0002}. \tag{8.77}$$

(5) Solving systems (8.51), (8.52), (8.53) here is just as in the case **(4)**, with the only difference that $d = \delta_2/\delta_1$ and $a = \delta_2\omega_1$. So we get the equalities (8.77)
(6) Exactly as the two previous cases and therefore we have the equalities (8.77).
(7) The solution of this case will be undertaken at the end, after case **(15)**.
(8) Solving the three systems in this case is immediate and we have

$$X_j(t) = X_j(0) + \int_0^t \mathcal{H}_j(s)ds - t\mathcal{K}_j, \qquad j = 1, 2, 3$$

[6] An outstanding point in $Im E$ is that which minimizes the distance to the point $\mathbf{h}_2$ is the orthogonal projection $Proj_{(Im\,E)}\mathbf{h}_2$.

the required τ-periodicity of $X_j(t)$ giving the ten constant values for the components of the vector $\mathcal{K}$:

$$k_{ijkl} = \frac{1}{\tau}\int_0^\tau h_{ijkl}(t)dt.$$

(9) Exactly as the case **(8)**.
(10) This is a particular case of (**14**) and later on will simply set $\delta_1 = \delta_2 = \delta$ in the solution.
(11) Exactly as the case (**8**).
(12) (a) For system (8.51) we have

$$X_1(t) = X_1(0) + \int_0^t \mathcal{B}_1(s)ds, \quad \text{so} \quad \mathcal{K}_1 = \frac{1}{\tau}\int_0^\tau \mathcal{H}_1(t)dt$$

and this determines completely the values k_{2000}, k_{1010} and k_{0020}.
(b), (c) Solving the systems (8.52) and (8.53).
For $x_4 = w_{1100}$, $x_5 = w_{1001}$, $x_6 = w_{0110}$, $x_7 = w_{0011}$ the system (8.52) is

$$\dot{x}_4 = b_4, \quad \dot{x}_5 = -\delta_2 x_4 + b_5, \quad \dot{x}_6 = b_6, \quad x_7 = -\delta_2 x_6 + b_7$$

and with $x_8 = w_{0200}$, $x_9 = w_{0101}$, $x_{10} = w_{0002}$ the system (8.53) is

$$\dot{x}_8 = b_8, \quad \dot{x}_9 = -2\delta_2 x_8 + b_9, \quad \dot{x}_{10} = -\delta_2 x_9 + b_{10}.$$

For constant values of $\mathcal{K}_{ijkl}$ these systems are readily solved and we get

$$\begin{aligned}
x_4(t) &= x_4(0) + \int_0^t h_{1100}(s)ds - -tk_{1100} \\
x_5(t) &= x_5(0) + f_5(t) - tk_{1001}, \\
x_6(t) &= x_6(0) + \int_0^t h_{0110}(s)ds - tk_{0110} \\
x_7(t) &= x_7(0) + f_7(t) - tk_{0011},
\end{aligned} \tag{8.78}$$

where

$$f_5(t) = -\delta_2 x_4(0)t - \delta_2 \int_0^t \int_0^s h_{1100}(u)duds + \frac{1}{2}\delta_2 t^2 k_{1100} + \int_0^t h_{1001}(s)ds,$$

$$f_7(t) = -\delta_2 x_6(0)t - \delta_2 \int_0^t \int_0^s h_{0110}(u)duds + \frac{1}{2}\delta_2 t^2 k_{0011} + \int_0^t h_{0110}(s)ds$$

and

$$x_8(t) = x_8(0) + \int_0^t h_{0200}(s)ds - tk_{0200}$$

$$x_9(t) = x_9(0) + f_9(t) - tk_{0101}$$

$$x_{10}(t) = x_{10}(0) + f_{10}(t) - tk_{0002}$$

where

$$f_9(t) = -2\delta_2 x_8(0)t - 2\delta_2 \int_0^t \int_0^s h_{0200}(u)du\,ds + \delta_2 t^2 k_{0200} + \int_0^t h_{0101}(s)ds,$$

$$f_{10}(t) = -\delta_2 x_9(0)t - \delta_2 \int_0^t f_9(s)ds + \frac{1}{2}\delta_2 t^2 k_{0101} + \int_0^t h_{0002}(s)ds.$$

The required τ-periodicity of $x_4(t)$, $x_6(t)$ and $x_8(t)$ gives the values

$$k_{1100} = \frac{1}{\tau}\int_0^\tau h_{1100}(t)dt, \quad k_{0110} = \frac{1}{\tau}\int_0^\tau h_{0110}(t)dt, \quad k_{0200} = \frac{1}{\tau}\int_0^\tau h_{0200}(t)dt.$$

These values, together with the initial conditions $x_4(0)$, $x_6(0)$, and $x_8(0)$ determine the functions $f_5(t)$, $f_7(t)$ and $f_9(t)$ and the periodicity of $x_5(t)$, $x_7(t)$ and $x_9(t)$ leads to the equalities

$$k_{1001} = \frac{1}{\tau} f_5(\tau), \quad k_{0011} = \frac{1}{\tau} f_7(\tau), \quad k_{0101} = \frac{1}{\tau} f_9(\tau).$$

The values k_{0101} with the initial condition $x_9(0)$ determine the function $f_{10}(t)$ and the periodicity of $x_{10}(t)$ gives the equality $k_{0002} = \frac{1}{\tau} f_{10}(\tau)$.

Therefore, by a choice of the initial conditions we can have the four equalities

$$k_{1001} = 0, \quad k_{0011} = 0, \quad k_{0101} = 0, \quad k_{0002} = 0, \tag{8.79}$$

or any combination of p of them, for $p = 1, 2, 3$.

(13) The solution is just as the case (12), except that the period is now equal to π. Therefore, we get the four zero components (8.79) of $\mathcal{K}^{(m)}$ also in this case.

(14) Let $Z = (x_1, \ldots, x_{10})$ and $B = (b_1, \ldots, b_{10})$ have the components expressed in the order given in (8.45). The differential equations decouple into the three following systems

$$\dot{x}_1 = b_1, \quad \dot{x}_2 = -2\delta_1 x_1 + b_2, \quad \dot{x}_3 = -\delta_1 x_2 + b_3, \tag{8.80}$$

$$\dot{x}_4 = b_4, \quad \dot{x}_5 = -\delta_2 x_4 + b_5, \quad \dot{x}_6 = -\delta_1 x_4 + b_6, \; \dot{x}_7 = -\delta_1 x_5 - \delta_2 x_6 + b_7, \tag{8.81}$$

$$\dot{x}_8 = b_8, \quad \dot{x}_9 = -2\delta_2 x_8 + b_9, \quad \dot{x}_{10} = -\delta_2 x_9 + b_{10}. \tag{8.82}$$

The solutions of system (8.80) and (8.82) are given by

$$x_1(t) = x_1(0) + \int_0^t h_{2000}(s)ds - tk_{2000}, \qquad x_8(t) = x_8(0) + \int_0^t h_{0200}(s)ds - tk_{0200},$$

$$x_2(t) = x_2(0) + f_2(t) - tk_{1010}, \qquad x_9(t) = x_9(0) + f_9(t) - tk_{0101}$$

$$x_3(t) = x_3(0) + f_3(t) - tk_{0020}, \qquad x_{10}(t) = x_{10}(0) + f_{10}(t) - tk_{0002},$$

where

$$f_2(t) = -2\delta_1 x_1(0)t - 2\delta_1 \int_0^t \int_0^s h_{2000}(u)duds - \delta_1 t^2 k_{2000} + \int_0^t h_{1010}(s)ds,$$

$$f_3(t) = -\delta_1 x_2(0)t - \delta_1 \int_0^t f_2(s)ds - \frac{1}{2}\delta_1 t^2 k_{1010} + \int_0^t h_{0020}(s)ds,$$

$$f_9(t) = -2\delta_2 x_8(0)t - 2\delta_2 \int_0^t \int_0^s h_{0200}(u)duds - \delta_2 t^2 k_{0200} + \int_0^t h_{0101}(s)ds,$$

$$f_{10}(t) = -\delta_2 x_9(0)t - \delta_2 \int_0^t f_9(s)ds - \frac{1}{2}\delta_2 t^2 k_{0101} + \int_0^t h_{0002}(s)ds.$$

The solution of the system (8.81) is given by

$$x_4(t) = x_4(0) + \int_0^t h_{1100}(s)ds - tk_{1100},$$
$$x_5(t) = x_5(0) + f_5(t) - tk_{1001},$$
$$x_6(t) = x_6(0) + f_6(t) - tk_{0110},$$
$$x_7(t) = x_7(0) + f_7(t) - tk_{0011},$$

where

$$f_5(t) = -\delta_2 x_4(0)t - \delta_2 \int_0^t \int_0^s h_{1100}(u)duds - \frac{1}{2}\delta_2 t^2 k_{1100} + \int_0^t h_{1001}(s)ds,$$

$$f_6(t) = -\delta_1 x_4(0)t - \delta_1 \int_0^t \int_0^s h_{1100}(u)duds - \frac{1}{2}\delta_1 t^2 k_{1100} + \int_0^t h_{0110}(s)ds,$$

$$f_7(t) = -\delta_1 x_5(0)t - \delta_2 x_6(0)t - \delta_1 \int_0^t f_5(s)ds - \delta_2 \int_0^t f_6(s)ds$$
$$-\frac{1}{2}\delta_1 t^2 k_{1001} - \frac{1}{2}\delta_2 t^2 k_{0110} + \int_0^t h_{0011}(s)ds.$$

The τ-periodicity of $x_1(t)$, $x_4(t)$ and $x_8(t)$ leads to the values

$$k_{2000} = \frac{1}{\tau}\int_0^\tau h_{2000}(s)ds, \quad k_{1100} = \frac{1}{\tau}\int_0^\tau h_{1010}(s)ds, \quad k_{0200} = \frac{1}{\tau}\int_0^\tau h_{0200}(s)ds.$$

These values together with the initial conditions $x_1(0)$, $x_4(0)$ and $x_8(0)$ determine the functions $f_2(t)$, $f_5(t)$ and $f_9(t)$. So, the τ-periodicity of $x_2(t)$, $x_5(t)$ and $x_9(t)$ leads to the values

$$k_{1010} = \frac{1}{\tau} f_2(\tau), \quad k_{1001} = \frac{1}{\tau} f_5(\tau), \quad k_{0101} = \frac{1}{\tau} f_9(\tau).$$

The value k_{1100} and the initial condition $x_4(0)$ also determine the function $f_6(t)$ but generically $f_6(\tau) \neq f_5(\tau)$.

The initial conditions $x_2(0)$ and $x_9(0)$ determine the functions $f_3(t)$ and $f_{10}(t)$ if the values k_{1010} and k_{0101} are known. So, the τ-periodicity of $x_3(t)$ and $x_{10}(t)$ gives the values

$$k_{0020} = \frac{1}{\tau} f_3(\tau), \quad k_{0002} = \frac{1}{\tau} f_{10}(\tau).$$

Once the functions $f_5(t)$ and $f_6(t)$ are known the initial conditions $x_5(0)$ and $x_6(0)$ determine the function $f_7(t)$ and the τ-periodicity of $x_7(t)$ gives the value $k_{0011} = \frac{1}{\tau} f_7(\tau)$.

Choosing $x_4(0)$ so that $f_5(\tau) = 0$, we get $k_{1001} = 0$. The choice of $x_4(0)$ determines the function $f_6(t)$ so we get $k_{0110} = \frac{1}{\tau} f_6(\tau)$. Choosing $x_5(0)$ and $x_6(0)$ so that $f_7(\tau) = 0$, we get $k_{0011} = 0$. Therefore, we can choose the initial conditions in order to have all the following six equalities

$$k_{1010} = 0, \quad k_{1001} = 0, \quad k_{0101} = 0, \quad k_{0011} = 0, \quad k_{0020} = 0, \quad k_{0002} = 0,$$

or any p combinations of these, for $p = 1, \ldots, 5$.

(15) We now consider the first system that does not fall under the general pattern (8.47) the normal form of which is

$$\mathcal{H}_0 = \frac{1}{2} \delta y_1^2 + x_1 x_2.$$

Computing the Poisson brackets $\{H_0, W\}$ we find that the matrix M of the system (8.46) is

$$M = \begin{bmatrix} 0 & 0 & 0 & 0 & 1 & 0 & 0 & 0 & 0 & 0 \\ -2\delta & 0 & 0 & 0 & 0 & 0 & 1 & 0 & 0 & 0 \\ 0 & -\delta & 0 & 0 & 0 & 0 & 0 & 0 & 0 & 0 \\ 0 & 1 & 0 & 0 & 0 & 0 & 0 & 0 & 1 & 0 \\ 0 & 0 & 0 & 0 & 0 & 0 & 0 & 0 & 0 & 2 \\ 0 & 0 & 2 & -\delta & 0 & 0 & 0 & 0 & 0 & 0 \\ 0 & 0 & 0 & 0 & -\delta & 0 & 0 & 0 & 0 & 0 \\ 0 & 0 & 0 & 0 & 0 & 1 & 0 & 0 & 0 & 0 \\ 0 & 0 & 0 & 0 & 0 & 0 & 1 & 0 & 0 & 0 \\ 0 & 0 & 0 & 0 & 0 & 0 & 0 & 0 & 0 & 0 \end{bmatrix}.$$

Let $Z = (x_1, \ldots, x_{10})$ and $B = (b_1, \ldots, b_{10})$ have the components expressed in the order given in (8.45). Then the system of equations to be solved is

$$\begin{aligned} \dot{x}_1 &= x_5 + b_1, & \dot{x}_6 &= 2x_3 - \delta x_4 + b_6 \\ \dot{x}_2 &= -2\delta x_1 + x_7 + b_2, & \dot{x}_7 &= -\delta x_5 + b_7 \\ \dot{x}_3 &= -\delta x_2 + b_3, & \dot{x}_8 &= x_6 + b_8 \\ \dot{x}_4 &= x_2 + x_9 + b_4, & \dot{x}_9 &= x_7 + b_9 \\ \dot{x}_5 &= 2x_{10} + b_5, & \dot{x}_{10} &= b_{10}. \end{aligned}$$

Recall that with $\mathcal{B} = \mathcal{H} - \mathcal{K}$, we have

$$X = (w_{2000}, w_{1010}, w_{0020}, w_{1100}, w_{1001}, w_{0110}, w_{0011}, w_{0200}, w_{0101}, w_{0002}),$$

$$\mathcal{B} = (b_{2000}, b_{1010}, b_{0020}, b_{1100}, b_{1001}, b_{0110}, b_{0011}, b_{0200}, b_{0101}, b_{0002}).$$

The system is easily solved, beginning with the last equation and working with pairs of equations.

$$x_{10}(t) = x_{10}(0) + \int_0^t h_{0002}(s)ds - tk_{0002}$$

$$x_5(t) = x_5(0) + f_5(t) - tk_{1001},$$

where

$$f_5(t) = 2x_{10}(0)t + 2\int_0^t \int_0^s h_{0002}(u)du\,ds - t^2 k_{0002} + \int_0^t h_{1001}(s)ds.$$

Next we work with the first and seventh equations to get

$$x_1(t) = x_1(0) + f_1(t) - tk_{2000}$$

$$x_7(t) = x_7(0) + f_7(t) - tk_{0011},$$

where

$$f_1(t) = x_5(0)t + \int_0^t f_5(s)ds - \frac{1}{2}t^2 k_{1001} + \int_0^t h_{2000}(s)ds$$

$$f_7(t) = -\delta x_5(0)t - \delta\int_0^t f_5(s)ds - \frac{1}{2}\delta t^2 k_{1001} + \int_0^t h_{0011}(s)ds.$$

In the sequence we work with the second and ninth equations and get

$$x_2(t) = x_2(0) + f_2(t) - tk_{1010}$$

$$x_9(t) = x_9(0) + f_9(t) - tk_{0101},$$

where

$$f_2(t) = -\big(2\delta x_1(0) - x_7(0)\big)t - \int_0^t \big(2\delta f_1(s) - f_7(s)\big)ds + \\ + \frac{1}{2}t^2\big(2\delta k_{2000} - k_{0011}\big) + \int_0^t h_{1010}(s)ds,$$

$$f_9(t) = x_7(0)t + \int_0^t f_5(s)ds - \frac{1}{2}t^2 k_{0011} + \int_0^t h_{0101}(s)ds.$$

This allows the third and fourth equations to be integrated, which give

$$x_3(t) = x_3(0) + f_3(t) - tk_{0020}$$

$$x_4(t) = x_4(0) + f_4(t) - tk_{1100},$$

where

$$f_3(t) = -\delta x_2(0)t - \delta \int_0^t f_2(s)ds + \frac{1}{2}\delta t^2 k_{1010} + \int_0^t h_{0020}(s)ds$$

$$f_4(t) = \big(x_2(0) + x_9(0)\big)t + \int_0^t \big(f_2(s) + f_9(s)\big)ds \\ -\frac{1}{2}t^2(k_{1010} + k_{0101}) + \int_0^t h_{1100}(s)ds.$$

Finally, we can integrate the sixth and the eighth equations:

$$x_6(t) = x_6(0) + f_6(t) - tk_{0110}$$

$$x_8(t) = x_8(0) + f_8(t) - tk_{0200},$$

where

$$f_6(t) = \big(2x_3(0) - \delta x_4(0)\big)t + \int_0^t \big(2f_3(s) - \delta f_4(s)\big)ds \\ -\frac{1}{2}t^2\big(2k_{2000} - \delta k_{1100}\big) + \int_0^t h_{0110}(s)ds.$$

$$f_8(t) = x_6(0)t + \int_0^t f_6(s)ds - \frac{1}{2}t^2 k_{0110} + \int_0^t h_{0200}(s)ds.$$

Now we discuss these equations to get the values of the constants k_{ijkl}. First, take $x_{10}(0)$ so that $f_5(\tau) = 0$. This gives

$$k_{1001} = 0 \quad \text{and} \quad k_{0002} = \frac{1}{\tau}\int_0^\tau h_{0002}(s)\,ds.$$

Next we choose $x_5(0)$ so as to have $f_1(\tau) = 0$, which gives

$$k_{2000} = 0 \quad \text{and} \quad k_{0011} = \frac{1}{\tau} f_7(\tau) = \frac{1}{\tau}\int_0^\tau \big(\delta h_{2000}(s) + h_{0011}(s)\big)\,ds.$$

Take $x_1(0)$ and $x_7(0)$ such that $f_2(\tau) = 0$. Then we get

$$k_{1010} = 0 \quad \text{and} \quad k_{0101} = \frac{1}{\tau} f_9(\tau).$$

Choosing $x_2(0)$ such that $f_3(\tau) = 0$ gives

$$k_{0020} = 0 \quad \text{and} \quad k_{1100} = \frac{1}{\tau} f_4(\tau).$$

Choosing $x_3(0)$ and $x_4(0)$ such that $f_6(\tau) = 0$ and finally choosing $x_6(0)$ such that $f_8(\tau) = 0$ we get $k_{0110} = 0$ and $k_{0200} = 0$.

Therefore, in this case we get the six vanishing constant components of $\mathcal{K}$:

$$k_{2000} = 0, \quad k_{1010} = 0, \quad k_{1001} = 0, \quad k_{0020} = 0, \quad k_{0110} = 0, \quad k_{0200} = 0. \tag{8.83}$$

(7) We now consider the last system, the second one that does not fall under the general pattern (8.47), and whose normal form is

$$\mathcal{H}_0 = \frac{1}{2}\delta(y_1^2 + y_2^2) + \omega(x_1 y_2 - x_2 y_1).$$

Computing the Poisson brackets $\{H_0, W\}$ we find that the matrix M of the system (8.46) is

$$M = \begin{bmatrix} 0 & 0 & 0 & -\omega & 0 & 0 & 0 & 0 & 0 & 0 \\ -2\delta & 0 & 0 & 0 & -\omega & -\omega & 0 & 0 & 0 & 0 \\ 0 & -\delta & 0 & 0 & 0 & 0 & -\omega & 0 & 0 & 0 \\ 2\omega & 0 & 0 & 0 & 0 & 0 & 0 & -2\omega & 0 & 0 \\ 0 & \omega & 0 & -\delta & 0 & 0 & 0 & 0 & -\omega & 0 \\ 0 & \omega & 0 & -\delta & 0 & 0 & 0 & 0 & -\omega & 0 \\ 0 & 0 & 2\omega & 0 & -\delta & -\delta & 0 & 0 & 0 & -2\omega \\ 0 & 0 & 0 & \omega & 0 & 0 & 0 & 0 & 0 & 0 \\ 0 & 0 & 0 & 0 & \omega & \omega & 0 & -2\delta & 0 & 0 \\ 0 & 0 & 0 & 0 & 0 & 0 & \omega & 0 & -\delta & 0 \end{bmatrix}.$$

As always with $\mathcal{B} = \mathcal{H} - \mathcal{K}$, we take

$$Z = (w_{2000}, w_{1010}, w_{0020}, w_{1100}, w_{1001}, w_{0110}, w_{0011}, w_{0200}, w_{0101}, w_{0002}),$$
$$B = (b_{2000}, b_{1010}, b_{0020}, b_{1100}, b_{1001}, b_{0110}, b_{0011}, b_{0200}, b_{0101}, b_{0002}).$$

Let $Z = (x_1, \ldots, x_{10})$ and $B = (b_1, \ldots, b_{10})$. We will split the system of equations $\dot{Z} = MZ + B$ into three subsystems. The first for the variables x_1, x_4, x_8, the second for the variables x_2, x_5, x_6, x_9, and the third for x_3, x_7, x_{10}. The first system is

$$\dot{x}_1 = -\omega x_4 + b_1, \quad \dot{x}_4 = 2\omega x_1 - 2\omega x_4 + b_4, \quad \dot{x}_8 = \omega x_4 + b_8. \tag{8.84}$$

Setting

$$\tilde{b}_2 = -2\delta x_1 + b_2, \quad \tilde{b}_5 = -\delta x_4 + b_5, \quad \tilde{b}_6 = -\delta x_4 + b_6, \quad \tilde{b}_9 = -2\delta x_8 + b_9$$

we write the second system in the form

$$\begin{aligned}
\dot{x}_2 &= \quad - \omega x_5 - \omega x_6 \quad + \tilde{b}_2, \\
\dot{x}_5 &= \omega x_2 \quad - \omega x_9 + \tilde{b}_5, \\
\dot{x}_6 &= \omega x_2 \quad - \omega x_9 + \tilde{b}_6, \\
\dot{x}_9 &= \quad \omega x_5 + \omega x_6 \quad + \tilde{b}_9.
\end{aligned} \tag{8.85}$$

The third system we write as

$$\dot{x}_3 = -\omega x_7 + \tilde{b}_3, \quad \dot{x}_7 = 2\omega x_3 - 2\omega x_{10} + \tilde{b}_7, \quad \dot{x}_{10} = \omega x_7 + \tilde{b}_{10}, \tag{8.86}$$

where

$$\tilde{b}_3 = -2\delta x_2 + b_3, \quad \tilde{b}_7 = -\delta x_5 - \delta x_6 + b_7, \quad \tilde{b}_{10} = -\delta x_9 + b_{10}.$$

We first solve the system (8.84) to get the functions $\tilde{b}_2(t), \tilde{b}_5(t), \tilde{b}_6(t)$ and $\tilde{b}_9(t)$. Then, inserting these functions into the system (8.85) we can solve it to get the functions $\tilde{b}_3(t), \tilde{b}_7(t), \tilde{b}_{10}(t)$. Finally, we solve the system (8.86).

Let M_1,M_2 and M_3 be the coefficient matrices of the above systems. Then, with $Z_1 = (x_1, x_4, x_8)$, $B_1 = (b_1, b_4, b_8)$, $Z_2 = (x_2, x_5, x_6, x_9)$, $\tilde{B}_2 = (\tilde{b}_2, \tilde{b}_5, \tilde{b}_6, \tilde{b}_9)$ and $Z_3 = (x_3, x_7, x_{10})$, $\tilde{B}_3 = (\tilde{b}_3, \tilde{b}_7, \tilde{b}_{10})$ these systems can be written as

$$\dot{Z}_1 = M_1 Z_1 + B_1, \quad \dot{Z}_2 = M_2 Z_2 + \tilde{B}_2, \quad \dot{Z}_3 = M_3 Z_3 + \tilde{B}_3. \tag{8.87}$$

We have $M_1 = M_3 = -\omega C$, where C is the matrix of Example 1.4 and $M_2 = -\omega B - \omega J_4$, where B is the matrix of Example 1.3 and J_4 is the standard symplectic matrix of order four.

Table 8.2 Normalized constant Hamiltonian function

Case	$k_{ijkl}^{(m)} = 0 \; m \geq 1$			Additional
(1), (2)	$k_{1100}^{(m)} = 0,$	$k_{1010}^{(m)} = 0,$	$k_{1001}^{(m)} = 0,$	$k_{2000}^{(m)} = k_{0020}^{(m)}$
	$k_{0110}^{(m)} = 0,$	$k_{0011}^{(m)} = 0,$		
(3)	$k_{1100}^{(m)} = 0,$	$k_{1010}^{(m)} = 0,$	$k_{1001}^{(m)} = 0,$	$k_{2000}^{(m)} = k_{0020}^{(m)}$
	$k_{0110}^{(m)} = 0,$	$k_{0101}^{(m)} = 0,$	$k_{0011}^{(m)} = 0$	
(4), (5), (6)	$k_{1010}^{(m)} = 0,$	$k_{0101}^{(m)} = 0,$		$k_{2000}^{(m)} = k_{0020}^{(m)}$
				$k_{0200}^{(m)} = k_{0002}^{(m)}$
(7)	$k_{1100}^{(m)} = 0,$	$k_{0011}^{(m)} = 0,$		$k_{2000}^{(m)} = k_{0200}^{(m)}$
				$k_{0020}^{(m)} = k_{0002}^{(m)}$
(8), (9)	–	–	–	–
				–
(10), (14)	$k_{1010}^{(m)} = 0,$	$k_{1001}^{(m)} = 0,$	$k_{0101}^{(m)} = 0,$	Or any $0 \leq p \leq 5$
	$k_{0011}^{(m)} = 0,$	$k_{0020}^{(m)} = 0,$	$k_{0002}^{(m)} = 0$	of these equalities
(11)	–	–	–	–
				–
(12), (13)	$k_{1001}^{(m)} = 0,$	$k_{0101}^{(m)} = 0,$		Or any $0 \leq p \leq 3$
	$k_{0011}^{(m)} = 0,$	$k_{0002}^{(m)} = 0,$		of these equalities
(15)	$k_{1010}^{(m)} = 0,$	$k_{1001}^{(m)} = 0,$	$k_{0110}^{(m)} = 0,$	–
	$k_{2000}^{(m)} = 0,$	$k_{0200}^{(m)} = 0,$	$k_{0020}^{(m)} = 0$	–

The first and third systems in (8.87) can be solved as in item (a) of (**1**) and the second system can be solved as in item b of item (**4**). As a result we get the equalities

$$k_{1100} = 0, \quad k_{2000} = k_{0200} \quad \text{and} \quad k_{0011} = 0, \quad k_{0020} = k_{0002}.$$

The Table 8.2 shows the results obtained for the constant coefficients $k_{ijkl}^{(m)}$, $m \geq 1$.

8.7 Characteristic Equation of the Autonomous Hamiltonian at the Resonances

Applying the Deprit–Hori method to a quadratic τ-periodic Hamiltonian depending on a small parameter ϵ,

$$H(\mathbf{x}, \mathbf{y}, t, \epsilon) = H_0(\mathbf{x}, \mathbf{y}) + \frac{\epsilon}{1!} H_1(\mathbf{x}, \mathbf{y}, t) + \frac{\epsilon^2}{2!} H_2(\mathbf{x}, \mathbf{y}, t) + \dots, \tag{8.88}$$

in which the unperturbed term is time independent, then under the presence of a KGL resonance for $H_0(\mathbf{x}, \mathbf{y})$ we can find an autonomous quadratic Hamiltonian

$$\mathcal{K}(\boldsymbol{X}, \boldsymbol{Y}, \epsilon) = \mathcal{K}_0(\boldsymbol{X}, \boldsymbol{Y}) + \frac{\epsilon}{1!}\mathcal{K}_1(\boldsymbol{X}, \boldsymbol{Y}) + \frac{\epsilon^2}{2!}\mathcal{K}_2(\boldsymbol{X}, \boldsymbol{Y}) + \ldots \tag{8.89}$$

as described in items **(1)**–**(15)** of the previous section. The first term $\mathcal{K}_0(\boldsymbol{X}, \boldsymbol{Y})$ is the normal form of $H_0(\mathbf{x}, \mathbf{y})$ at the given resonance. We start with the unperturbed term in (8.88) already in normal form, so $H_0(\mathbf{x}, \mathbf{y}) = \mathcal{K}_0(\mathbf{X}, \mathbf{Y})$. Now, for each $m \geq 1$, let

$$H_m = \sum_{\nu,\mu} h^{(m)}_{\nu_1\nu_2\mu_1\mu_2}(t) x_1^{\nu_1} x_2^{\nu_2} y_1^{\mu_1} y_2^{\mu_2},$$

and

$$\mathcal{K}_m = \sum_{\nu,\mu} k^{(m)}_{\nu_1\nu_2\mu_1\mu_2} X_1^{\nu_1} X_2^{\nu_2} Y_1^{\mu_1} Y_2^{\mu_2},$$

where the sums run over the non-negative integers $\nu_1, \nu_2, \mu_1, \mu_2$ such that $\nu_1 + \nu_2 + \mu_1 + \mu_2 = 2$. Then we have

$$\mathcal{K} = \sum_{\nu,\mu} k_{\nu_1\nu_2\mu_1\mu_2} X_1^{\nu_1} X_2^{\nu_2} Y_1^{\mu_1} Y_2^{\mu_2},$$

where

$$k_{\nu_1\nu_2\mu_1\mu_2} = k^{(0)}_{\nu_1\nu_2\mu_1\mu_2} + k^{(\infty)}_{\nu_1\nu_2\mu_1\mu_2}, \quad \text{where} \quad k^{(\infty)}_{\nu_1\nu_2\mu_1\mu_2} = \sum_{m=1}^{\infty} \frac{\epsilon^m}{m!} k^{(m)}_{\nu_1\nu_2\mu_1\mu_2}.$$

For each m we have computed the coefficients $k^{(m)}_{\nu_1\nu_2\mu_1\mu_2}$ of K_m in terms of the coefficients $h^{(s)}_{\nu_1\nu_2\mu_1\mu_2}$ of some H_s. The Hamiltonian $\mathcal{K}$ is formed, in general, by ten monomials,

$$\begin{aligned}\mathcal{K}(X_1, X_2, Y_1, Y_2) = k_{2000}X_1^2 + k_{1010}X_1Y_1 + k_{0200}X_2^2 + k_{1100}X_1X_2 + k_{1001}X_1Y_2 + \\ + k_{0110}X_2Y_1 + k_{0011}Y_1Y_2 + k_{0200}X_2^2 + k_{0101}X_2Y_2 + k_{0002}Y_2^2.\end{aligned} \tag{8.90}$$

But, independently of the terms $H_1, H_2, \ldots,$ of the Hamiltonian H, we know that at certain KGL resonances the Hamiltonian $\mathcal{K}$ contains less than ten monomials, because some of the coefficients are zero (see Table 8.2). The Hessian matrix G of $\mathcal{K}$ is

$$G = \begin{bmatrix} 2k_{2000} & k_{1100} & k_{1010} & k_{1001} \\ k_{1100} & 2k_{0200} & k_{0110} & k_{0101} \\ k_{1010} & k_{0110} & 2k_{0020} & k_{0011} \\ k_{1001} & k_{0101} & k_{0011} & 2k_{0002} \end{bmatrix}.$$

The characteristic equation of the matrix $A = JG$ has the form

$$\lambda^4 + a\lambda^2 + b = 0. \tag{8.91}$$

For the general case the expressions of a and b are the following:

$$a = 4k_{2000}k_{0020}+4k_{0200}k_{0002}-k_{1010}^2-k_{0101}^2+2k_{1100}k_{0011}-2k_{1001}k_{0110}, \tag{8.92}$$

$$\begin{aligned} b = {} & k_{1010}^2k_{0101}^2 + k_{1001}^2k_{0110}^2 + k_{1100}^2k_{0011}^2 \\ & -2k_{1010}k_{1100}k_{0101}k_{0011} - 2k_{1010}k_{1001}v_{0101}k_{0110} - 2k_{1001}k_{1100}k_{0110}k_{0011} \\ & +4k_{2000}k_{0110}k_{0101}k_{0011} + 4k_{0200}k_{1001}k_{1010}k_{0011} \\ & +4k_{0020}k_{1001}k_{1100}k_{0101} + 4k_{0002}k_{1100}k_{1010}k_{0110} \\ & -4k_{2000}k_{0200}k_{0011}^2 - 4k_{2000}k_{0020}k_{0101}^2 - 4k_{2000}k_{0002}k_{0110}^2 \\ & -4k_{0200}k_{0020}k_{1001}^2 - 4k_{0200}k_{0002}k_{1010}^2 - 4k_{0020}k_{0002}k_{1100}^2 \\ & +16k_{2000}k_{0200}k_{0020}k_{0002}. \end{aligned} \tag{8.93}$$

In each case, on computing the above expressions of a and b we have to consider if the subscript of some coefficient of the unpertubed terms is present in the equations of the second column. For instance, this happens in cases 10 and 12 to 15.

For each of the cases in Table 8.2, we have to select the maximal number of equations from the second column so that the resulting constant Hamiltonian $\mathcal{K}$ have coefficients that make possible the conditions required for stability, that is in terms of the coefficients a and b of the characteristic equation (8.91) the inequalities must be satisfied

$$a > 0, \quad b > 0, \quad \Delta = a^2 - 4b > 0. \tag{8.94}$$

A smaller number of equalities may also give a convenient constant Hamiltonian but its expression gets more complex as the number of chosen equalities is taken smaller.

(1) and **(2)** We readily find from (8.92) and (8.93) that a and b are given by

$$\begin{aligned} a &= 4k_{2000}k_{0020} + 4k_{0200}k_{0002} - k_{0101}^2, \\ b &= -4k_{2000}k_{0020}k_{0101}^2 + 16k_{2000}k_{0200}k_{0020}k_{0002}. \end{aligned}$$

The characteristic equation can be easily factored out and we find

$$[\lambda^2 + 4k_{2000}k_{0020}][\lambda^2 - (k_{0101}^2 - 4k_{0200}k_{0002})] = 0.$$

(3) The characteristic equation is

$$(\lambda^2 + 4k_{2000}k_{0020})(\lambda^2 + 4k_{0200}k_{0002}) = 0.$$

(4), **(5)**, **(6)** In these cases we find from (8.92) and (8.93)

$$\begin{aligned}
a = &4k_{2000}k_{0020} + 4k_{0200}k_{0002} + 2k_{1100}k_{0011} - 2k_{1001}k_{0110},\\
b = &k_{1001}^2k_{0110}^2 + k_{1100}^2k_{0011}^2 - 2k_{1001}k_{1100}k_{0110}k_{0011}\\
&- 4k_{2000}k_{0200}k_{0011}^2 - 4k_{2000}k_{0002}k_{0110}^2\\
&- 4k_{0200}k_{0020}k_{1001}^2 - 4k_{0020}k_{0002}k_{1100}^2\\
&+ 16k_{2000}k_{0200}k_{0020}k_{0002}
\end{aligned}$$

(7) The values of a and b are:

$$\begin{aligned}
a = &4k_{2000}k_{0020} + 4k_{0200}k_{0002} - k_{1010}^2 - k_{0101}^2 - 2k_{1001}k_{0110},\\
b = &k_{1010}^2k_{0101}^2 + k_{1001}^2k_{0110}^2 - 2k_{1010}k_{1001}k_{0101}k_{0110}\\
&- 4k_{2000}k_{0020}k_{0101}^2 - 4k_{2000}k_{0002}k_{0110}^2\\
&- 4k_{0200}k_{0020}k_{1001}^2 - 4k_{0200}k_{0002}k_{1010}^2\\
&+ 16k_{2000}k_{0200}k_{0020}k_{0002}.
\end{aligned}$$

(8), **(9)** and **(11)** Coefficients a and b are given by the general expressions (8.92) and (8.93).
(10), **(14)** In these cases if we consider the six equations of the second colunm of Table 8.2 , since $k_{0002}^{(0)} \neq 0$ and $k_{0020}^{(0)} \neq 0$, we get for a and b the expressions

$$a = 2(k_{2000}^{(\infty)}\delta_1 + k_{0200}^{(\infty)}\delta_2)$$

$$b = -((2(k_{0110}^{(\infty)})^2k_{2000}^{(\infty)} + ((k_{1100}^{(\infty)})^2 - 4k_{0200}^{(\infty)}k_{2000}^{(\infty)})\delta_1)\delta_2). \tag{8.95}$$

For case 10, make $\delta_1 = \delta$ and $\delta_2 = \delta$ in the expressions a and b given above.
(12), **(13)** In these cases if we consider the four equations of the second colunm of Table 8.2 and knowing that, for theses cases, $k_{0002}^{(0)} \neq 0$, we get for a and b the expressions

$$a = -(k_{1010}^{(\infty)})^2 + 4k_{0020}^{(\infty)}k_{2000}^{(\infty)} + 2k_{0200}^{(\infty)}\delta_2$$

$$b = -2(-k_{0110}^{(\infty)}k_{1010}^{(\infty)}k_{1100}^{(\infty)} + k_{0020}(k_{1100}^{(\infty)})^2 + (k_{0110}^{(\infty)})^2k_{2000}^{(\infty)} + k_{0200}^{(\infty)}((k_{1010}^{(\infty)})^2$$
$$-4k_{0020}^{(\infty)}k_{2000}^{(\infty)}))\delta_2.$$

For case 13, make $\delta_2 = \delta$ in the expressions a and b given above.

If we omit the equation $k_{0002} = 0$ the expression of b has seven terms so this Hamiltonian is more complicated than any one of the above three possibilities.

(15) The values of the coefficients a and b are

$$a = -(k_{0101}^{(\infty)})^2 + 2k_{0011}^{(\infty)}(1 + k_{1100}^{(\infty)})$$

$$b = (1 + k_{1100}^{(\infty)})^2((k_{0011}^{(\infty)})^2 - 2k_{0002}^{(\infty)}\delta).$$

We have thus finished the consideration of all cases for the two-degrees of freedom.

8.8 Construction of Boundary Curves

Now we consider a τ-periodic Hamiltonian, which, besides the small parameter ϵ contains another parameter μ,

$$H(\mathbf{x}, t, \mu, \epsilon) = H_0(\mathbf{x}, \mu) + \frac{1}{1!}\epsilon H_1(\mathbf{x}, t, \mu) + \frac{1}{2!}\epsilon^2 H_2(\mathbf{x}, t, \mu), \ldots$$

To proceed with the task of obtaining the boundary curves of the stability domains we must have explicitly the polynomials $H_0(\mathbf{x}, \mu)$, $H_1(\mathbf{x}, t, \mu)$, $H_2(\mathbf{x}, t, \mu), \ldots$ in order to be able to use the Deprit–Hori method to find the homogeneous polynomials $\mathcal{H}_j$ to be used in getting the constant polynomials $\mathcal{K}_j$ of the constant Hamiltonian $\mathcal{K}$. Sections 8.8.1 and 8.8.2 are devoted to some examples where we use these techniques to construct the boundary curves in two degrees of freedom. These examples are described in [40] and come from satellite dynamics where the motion of a rigid body about its center of mass, which revolves around a Newtonian center of attraction, a satellite, is studied. We add more detail around these examples by applying the Deprit–Hori method to obtain the boundary curves of the stability/instability regions.

We know from Example 2.15 that if the angles ψ, θ, ϕ are taken as the position variables and dimensionless momenta p_ψ, p_θ, p_ϕ are introduced, then taking the true anomaly ν as the new time the motion of the satellite about its center of mass is governed by the system of canonical differential equations

$$\begin{aligned}\frac{d\psi}{d\nu} &= \frac{\partial H}{\partial p_\psi}, \quad \frac{d\theta}{d\nu} = \frac{\partial H}{\partial p_\theta}, \quad \frac{d\phi}{d\nu} = \frac{\partial H}{\partial p_\phi}, \\ \frac{dp_\psi}{d\nu} &= -\frac{\partial H}{\partial \psi}, \quad \frac{dp_\theta}{d\nu} = -\frac{\partial H}{\partial \theta}, \quad \frac{dp_\phi}{d\nu} = -\frac{\partial H}{\partial \phi},\end{aligned} \tag{8.96}$$

where the Hamiltonian function $H = H(\psi, \theta, \phi, p_\psi, p_\theta, p_\phi, \nu)$ is given by

$$\begin{aligned}H = &\frac{\sin^2\phi + \theta_A \cos^2\phi}{2(1+e\cos\nu)^2 \sin^2\theta}(p_\psi - p_\phi\cos\theta)^2 + \frac{\cos^2\phi + \theta_A \sin^2\phi}{2(1+e\cos\nu)^2} p_\theta^2 + \\ &+ \frac{\frac{\theta_A}{\theta_C}}{2(1+e\cos\nu)^2} p_\phi^2 + \frac{(1-\theta_A)\sin 2\phi}{2(1+e\cos\nu)^2\sin\theta} p_\theta(p_\psi - p_\phi\cos\theta) - \\ &- \cos\psi\cot\theta p_\psi - \sin\psi p_\theta + \frac{\cos\psi}{\sin\theta} p_\phi + \\ &+ \frac{3}{2}(1+e\cos\nu)\left[\left(\frac{1}{\theta_A} - 1\right)\sin^2\theta\cos^2\phi + \left(\frac{\theta_C}{\theta_A} - 1\right)\cos^2\theta\right],\end{aligned} \tag{8.97}$$

where $\theta_A = \dfrac{A}{B}$ and $\theta_C = \dfrac{C}{B}$. This is a time-dependent Hamiltonian that contains the small parameter e and is 2π-periodic in the time ν.

We know that the principal moments of inertia A, B, C satisfy the equations

$$A + B = C + \int_S z^2 dm, \quad A + C = B + \int_S y^2 dm, \quad B + C = A + \int_S x^2 dm, \tag{8.98}$$

so the admissible values of the parameters θ_A and θ_C are given by the inequalities

$$\theta_A + 1 \geq \theta_C, \quad \theta_A + \theta_C \geq 1, \quad 1 + \theta_C \geq \theta_A. \tag{8.99}$$

In the plane θ_C, θ_A this region is represented, in the Fig. 8.3, as an unbounded band lying between the parallel lines $\theta_A = \theta_C - 1$ and $\theta_A = \theta_C + 1$ and extending to

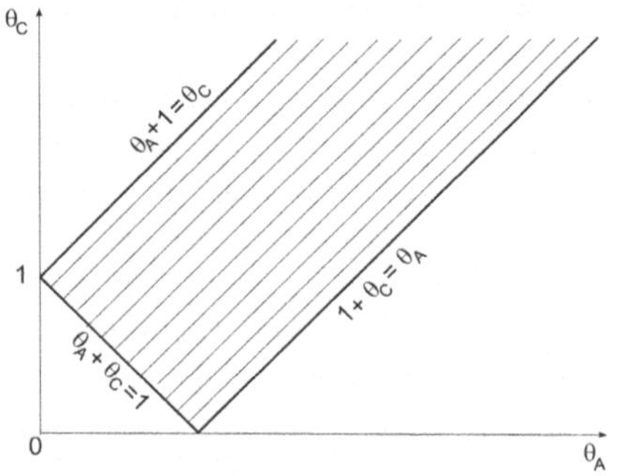

Fig. 8.3 Admissible region

the right and above the line $\theta_A = -\theta_C + 1$. We see from (8.98) that the portions of the boundary of the region (8.99) given by the equations $\theta_A = \theta_C - 1, \theta_A = -\theta_C + 1$ and $\theta_A = \theta_C + 1$ correspond to the satellites named (in the dynamical sense) plates; they are situated respectively in the planes O_{xy}, O_{xz}, O_{yz}. The points $(1, 0)$ and $(0, 1)$ represent pivots and lie along the axes Ox and Oz. The infinitely distant point of the region corresponds to the pivot located at the axis Oy.

In Sect. 8.6.4 we computed the coefficients of the autonomous Hamiltonian $\mathcal{K}$ for each one of the KGL resonances in the two-degrees of freedom case. We now use this information to construct the boundary curves of the regions of stability and instability for some specific problems.

When $A = B$ (dynamic symmetry) the Hamiltonian function corresponding to the motion of a satellite relative to the center of mass (8.97) is given by

$$H = \frac{(p_\psi - p_\phi \cos\theta)^2}{2(1 + e\cos\nu)^2 \sin^2\theta} + \frac{p_\theta^2}{2(1 + e\cos\nu)^2} + \frac{Ap_\phi^2}{2C(1 + e\cos\nu)^2} - \\ - \cos\psi\, cotg\,\theta\, p_\psi - \sin\psi\, p_\theta + \frac{\cos\psi}{\sin\theta} p_\phi + \tfrac{3}{2}(1 + e\cos\nu)\left((\tfrac{C}{A} - 1)\cos^2\theta\right), \tag{8.100}$$

so $\dot{p}_\phi = 0$ and since from (2.69) and (2.71), it follows that $p_\phi = \dfrac{Cr(1 - e^2)^{3/2}}{A\omega_0}$. This way, r is a constant, which we denote by r_*. As a result, the Hamiltonian function (8.100) reduces to

$$H = \frac{p_\psi^2}{2(1 + e\cos\nu)^2 \sin^2\theta} + \frac{p_\theta^2}{2(1 + e\cos\nu)^2} - \\ - \left[\frac{\alpha\beta(1 - e^2)^{3/2}\cos\theta}{(1 + e\cos\nu)^2 \sin^2\theta} + \cos\psi\cot\theta\right] p_\psi - \sin\psi\, p_\theta + \frac{\alpha^2\beta^2(1 - e^2)^3}{2(1 + e\cos\nu)^2}\cot^2\theta + \\ + \alpha\beta(1 - e^2)^{3/2}\frac{\cos\psi}{\sin\theta} + \frac{3}{2}(\alpha - 1)(1 + e\cos\nu)\cos^2\theta + \frac{\alpha\beta^2(1 - e^2)^3}{2(1 - e\cos\nu)^2}, \tag{8.101}$$

where $\alpha = \dfrac{C}{A}, \beta = \dfrac{r_*}{\omega_0}$ $(0 < \alpha \leq 2, -\infty < \beta < \infty)$ and from (2.74)

$$r_* = \dot{\psi}\cos\theta + \dot{\phi} - \dot{\nu}\cos\psi\sin\theta. \tag{8.102}$$

The last term in (8.101) does not depend on the variables $\psi, \theta, \phi, p_\psi, p_\theta, p_\phi$ so it can be omitted from the Hamiltonian.

Remark 8.10 We know that $\theta_A = \frac{A}{B}, \theta_C = \frac{C}{B}$ and the admissible region is given by (8.99). In this way, it follows that when $A = B$ we have $\theta_C = \alpha$ and

$$0 \leq \alpha \leq 2. \tag{8.103}$$

The inertia ellipsoid of the satellite is given by (x, y, z) such that $Ax^2 + By^2 + Cz^2 = 1$. When $A = B$, from (8.103) follows that $0 \leq C \leq 2A$ and so

- If $C = 0$, we have a cylinder.
- If $0 < C < A$, we have a prolate ellipsoid.
- If $C = A$, we have a sphere.
- If $A < C \leq 2A$, we have a oblate ellipsoid. When $C = 2A$ the oblate ellipsoid is called a plate.

The differential equations corresponding to (8.97) are

$$\frac{d\psi}{dv} = \frac{\partial H}{\partial p_\psi} \quad \frac{d\theta}{dv} = \frac{\partial H}{\partial p_\theta}$$
$$\frac{dp_\psi}{dv} = -\frac{\partial H}{\partial \psi} \quad \frac{dp_\theta}{dv} = -\frac{\partial H}{\partial \theta}. \tag{8.104}$$

After solving the system (8.104) we get the rotation ϕ of the satellite about the symmetry axis from (8.102) by quadrature.

We can check that the system (8.104) has the equilibrium solution

$$\theta_0 = \frac{\pi}{2}, \;\; \psi_0 = \pi, \;\; p_{\theta_0} = 0, \;\; p_{\psi_0} = 0 \tag{8.105}$$

which corresponds to the rotation of the satellite about the axis of symmetry perpendicular to the plane of the orbit (cylindrical precession). This solution exists for every values of the parameters α, β and of the eccentricity e. Making the translation

$$\theta = \frac{\pi}{2} + Q_1, \;\; \psi = \pi + Q_2, \;\; p_\theta = P_1, \;\; p_\psi = P_2,$$

we obtain from (8.97) the Hamiltonian function corresponding to the linearized system

$$\begin{aligned} H =& \frac{1}{2(1+e\cos v)^2}(P_1^2 + P_2^2) + \\ &+ \frac{1}{2}\left[\frac{\alpha^2\beta^2(1-e^2)^3}{(1+e\cos v)^2} - \alpha\beta(1-\epsilon^2)^{3/2} + 3(\alpha - 1)(1+e\cos v)\right] Q_1^2 + \\ &+ \frac{1}{2}\alpha\beta(1-e^2)^{3/2}Q_2^2 + \left[\frac{\alpha\beta(1-e^2)^{3/2}}{(1+e\cos v)^2} - 1\right] Q_1P_2 + Q_2P_1. \end{aligned} \tag{8.106}$$

For e small, we write (8.106) in the following way

$$H = H_0 + eH_1 + e^2H_2 + \cdots,$$

where

$$H_0 = \frac{1}{2}(P_1^2 + P_2^2) + P_1 Q_2 + (\alpha\beta - 1)P_2 Q_1 + \frac{1}{2}(\alpha^2\beta^2 - \alpha\beta + 3\alpha - 3)Q_1^2 + \frac{1}{2}\alpha\beta Q_2^2$$

$$H_1 = \left(-(P_1^2 + P_2^2) - 2\alpha\beta P_2 Q_1 + (-\alpha^2\beta^2 + \frac{3}{2}\alpha - \frac{3}{2})Q_1^2\right)\cos(\nu)$$

$$H_2 = \frac{3}{4}\left((1 + \cos 2\nu)(P_1^2 + P_2^2) - \alpha\beta[(\alpha\beta - 1) - \alpha\beta\cos 2\nu]Q_1^2\right) +$$

$$+\frac{3}{2}\alpha\beta P_2 Q_1 \cos(2\nu) - \frac{3}{4}\alpha\beta Q_2^2$$

In the next sections we consider the problem of the linear stability of the satellite's cylindrical precession for small values of the eccentricity of the orbit.

8.8.1 Satellites with Mass Geometry of the Plate

When $A = B$, the case of mass geometry given by $2A = C$ or $\alpha = 2$ is called the plate. So, for $\alpha = 2$ we look for the equation of the boundary curves of the regions of stability and instability in the parameter plane (β, e) in the form

$$\beta = \beta_0 + e\beta_1 + e^2\beta_2 + e^3\beta^3 \cdots \tag{8.107}$$

So, making $\alpha = 2$, replacing (8.107) in Hamiltonian (8.106) and making its expansion as a series in power of e we get

$$H = H_0 + eH_1 + e^2 H_2 + \cdots , \tag{8.108}$$

where

$$H_0 = \frac{1}{2}(P_1^2 + P_2^2) + P_1 Q_2 + (2\beta_0 - 1)P_2 Q_1 + \frac{1}{2}(4\beta_0^2 - 2\beta_0 + 3)Q_1^2 + \beta_0 Q_2^2$$

$$H_1 = Q_2^2\beta_1 - (P_1^2 + P_2^2)\cos\nu + 2P_2 Q_1(\beta_1 - 2\beta_0\cos\nu) + \frac{1}{2}Q_1^2(2\beta_1(-1 + 4\beta_0)$$

$$+(3 - 8\beta_0^2)\cos\nu)$$

$$H_2 = Q_2^2(\beta_2 - \frac{3}{2}\beta_0) + \frac{3}{2}(P_1^2 + P_2^2)\cos^2\nu + P_2 Q_1(2\beta_2 - 4\beta_1\cos\nu + 3\beta_0\cos 2\nu) +$$

$$+\frac{1}{2}Q_1^2(3\beta_0 - 6\beta_0^2 + 4\beta_1^2 - 2\beta_2 + 8\beta_0\beta_2 - 16\beta_0\beta_1\cos\nu + 6\beta_0^2\cos 2\nu). \tag{8.109}$$

The characteristic equation corresponding to the unperturbed system is

$$\lambda^4 + (4\beta_0^2 - 4\beta_0 + 5)\lambda^2 + 4\beta_0^2 + 2\beta_0 - 2 = 0$$

and making $\pm i\omega_1, \pm i\omega_2$ the characteristic exponents for $e = 0$, we obtain

$$\omega_1 = \frac{\sqrt{5 - 4\beta_0 + 4\beta_0^2 + \sqrt{33 - 48\beta_0 + 40\beta_0^2 - 32\beta_0^3 + 16\beta_0^4}}}{\sqrt{2}}$$

$$\omega_2 = \frac{\sqrt{5 - 4\beta_0 + 4\beta_0^2 - \sqrt{33 - 48\beta_0 + 40\beta_0^2 - 32\beta_0^3 + 16\beta_0^4}}}{\sqrt{2}}$$

For $-1 < \beta_0 < \frac{1}{2}$ we have instability of cylindrical precession of the plate-satellite in a circular orbit (see [40]) and, parametric resonance occurs for β_0 equal to

$$1)\ \beta_0 = \frac{1}{2}, \quad 2)\ \beta_0 = 1, \quad 3)\ \beta_0 = -1.$$

The respective frequencies for these three values are

$$1)\ \omega_1 = 2,\ \ \omega_2 = 0, \quad 2)\ \omega_1 = 2,\ \ \omega_2 = 1, \quad 3)\ \omega_1 = \sqrt{13},\ \ \omega_2 = 0.$$

Example 8.3 For the value $\beta_0 = \dfrac{1}{2}$, we have resonance $2\omega_1 = 4\ \ \omega_2 = 0$.

For $\beta_0 = \dfrac{1}{2}$, the unperturbed Hamiltonian in (8.108) reduces to

$$H_0(Q, P) = \frac{1}{2}\left((P_1^2 + P_2^2) + 3Q_1^2 + 2P_1Q_2 + Q_2^2\right)$$

and we obtain the frequencies $\omega_1 = 2$ and $\omega_2 = 0$. So, according to the steps described in Examples 5.11 and 5.12, we obtain a normalizing base $\mathbf{u}_1, \mathbf{u}_2, \mathbf{v}_1, \mathbf{v}_2$, where

$$\mathbf{u}_1 = -\kappa_1 s_1 = -\tfrac{1}{\sqrt{8}}(0, -1, -3, 0), \ \mathbf{v}_1 = \delta_1\kappa_1 r_1 = \tfrac{1}{\sqrt{8}}(-2, 0, 0, 2)$$

$$\mathbf{u}_2 = \kappa_2 T\mathbf{v} = \tfrac{1}{\sqrt{12}}(0, 3, -3, 0), \qquad \mathbf{v}_2 = \delta_2\kappa_2\mathbf{v} = \tfrac{1}{\sqrt{12}}(1, 0, 0, 3)$$

and so the linear normalizing transformation is given by

$$Q_1 = -\frac{\sqrt{2}}{2}p_1 + \frac{\sqrt{3}}{6}p_2 \ \ P_1 = \frac{3\sqrt{2}}{4}q_1 - \frac{\sqrt{3}}{2}q_2$$

$$Q_2 = \frac{\sqrt{2}}{4}q_1 + \frac{\sqrt{3}}{2}q_2 \quad P_2 = \frac{\sqrt{2}}{2}p_1 + \frac{\sqrt{3}}{2}p_2$$

and, in these new variables, for $\beta_0 = \dfrac{1}{2}$, the Hamiltonian (8.108) is such that H_0 becomes

$$F_0(q, p) = \frac{1}{2}\omega_1(q_1^2 + p_1^2) + \frac{1}{2}p_2^2. \tag{8.110}$$

So, since ω_1 is an integer number, we take the rotation by angle $\delta_1\omega_1 t$,

$$\begin{aligned} q_1 &= x_1 \cos\delta_1\omega_1 t + y_1 \sin\delta_1\omega_1 t \quad & q_2 = x_2 \\ p_1 &= -x_1 \sin\delta_1\omega_1 t + y_1 \cos\delta_1\omega_1 t \quad & p_2 = y_2, \end{aligned} \tag{8.111}$$

in order toeliminate the term $\frac{1}{2}\omega_1(q_1^2 + p_1^2)$ in (8.110). In this way, the Hamiltonian function obtained after the rotation is given by

$$F(x_1, x_2, y_1, y_2, \nu, e, \mu) = F_0 + eF_1 + \cdots \tag{8.112}$$

where

$$\begin{aligned} F_0 &= \tfrac{1}{2}y_2^2 \\ F_1 &= \tfrac{1}{48}{x_1}^2(-9\beta_1 + 15\beta_1\cos(4t) - 9\cos(t) - 45\cos(t)\cos(4t)) + \\ &+ \tfrac{1}{48}x_1x_2\left(12\sqrt{6}\beta_1\cos(2t) + 36\sqrt{6}\cos(t)\cos(2t)\right) + \\ &+ \tfrac{1}{48}x_1y_1(30\beta 1\sin(4t) - 90\sin(4t)\cos(t)) \\ &+ \tfrac{x_1y_2}{48}\left(24\sqrt{6}\beta_1\sin(2t) + 12\sqrt{6}\sin(2t)\cos(t)\right) + \\ &+ \tfrac{1}{48}{x_2}^2(36\beta_1 - 36\cos(t)) + \tfrac{1}{48}x_2y_1\left(12\sqrt{6}\beta_1\sin(2t) + 36\sqrt{6}\sin(2t)\cos(t)\right) + \\ &+ \tfrac{1}{48}{y_1}^2(-9\beta_1 - 15\beta_1\cos(4t) - 9\cos(t) + 45\cos(t)\cos(4t)) + \\ &+ \tfrac{1}{48}y_1y_2\left(-24\sqrt{6}\beta_1\cos(2t) - 12\sqrt{6}\cos(t)\cos(2t)\right) + \tfrac{1}{48}{y_2}^2(28\beta_1 - 58\cos(t)) \end{aligned} \tag{8.113}$$

In this case, the matrices M_1, M_2 and M_3 in (8.48) are such that $M_1 = O$, $M_2 = -\delta_2 F_2$ and $M_3 = -\delta_2 E_1$, with $a = 2, b = 1$, where E_1 and F_2 are given in Examples 1.6 and 1.7 in Chap. 1. So, according to the case (**12**), applying the Deprit–Hori method to the fourth approximation, we obtain[7]

$$K(X_1, X_2, Y_1, Y_2, \nu, e, \mu) = K_0 + eK_1 + \frac{e^2}{2!}K_2 + \cdots + \frac{e^4}{4!}K_4 \tag{8.114}$$

[7] On applying the Deprit–Hori method, we choose all initial conditions $w_{\nu_1\nu_2\iota_1\iota_2}(0) = 0$.

where

$$
\begin{aligned}
K_0 &= \frac{Y_2^2}{2}\\
K_1 &= -\tfrac{3}{16}\beta_1 X_1{}^2 + \tfrac{3\beta_1 X_2{}^2}{4} - \tfrac{3\beta_1 Y_1{}^2}{16} - \tfrac{\sqrt{6}}{8}(\beta_1+4)Y_1Y_2 + \tfrac{1}{12}(7\beta_1 - 18)Y_2{}^2\\
K_2 &= \tfrac{3}{256}X_1{}^2\left(39\beta_1{}^2 + 32\beta_1 - 32\beta_2 + 64\right) + \tfrac{3\sqrt{6}}{32}(\beta_1(5\beta_1 - 12) - 6)X_1X_2-\\
&\quad -\tfrac{3}{16}X_2{}^2((\beta_1 - 24)\beta_1 - 8\beta_2) + \tfrac{3}{256}Y_1{}^2(\beta_1(67\beta_1 + 32) - 32(\beta_2 - 6))+\\
&\quad +\tfrac{\sqrt{6}}{192}Y_1Y_2(\beta_1(31\beta_1 - 404) - 48\beta_2 + 483)\\
&\quad +\tfrac{1}{192}Y_2{}^2\left(-17\beta_1{}^2 - 3624\beta_1 + 224\beta_2 + 1020\right)
\end{aligned}
\tag{8.115}
$$

So, the Hamiltonian function in (8.114) assumes the form

$$
K = k_{2000}X_1^2 + k_{1100}X_1X_2 + k_{0020}Y_1^2 + k_{0011}Y_1Y_2 + k_{0200}X_2^2 + \left(\frac{1}{2} + k_{0002}\right)Y_2^2,
$$

where the coefficients $k_{\nu_1\nu_2\iota_1\iota_2}$ are functions of $e, \beta_0, \beta_1, \cdots$ The characteristic polynomial is given by $\lambda^4 + a\lambda^2 + b$ and, in this case, the coefficients a and b are

$$
a = \tfrac{3\beta_1}{2}e + \mathcal{O}(e^2) \qquad b = \tfrac{27\beta_1{}^3}{128}e^3 + \mathcal{O}(e^4)
$$

and $d = a^2 - 4b$ is given by

$$
d = \tfrac{9\beta_1{}^2}{4}e^2 + \mathcal{O}(e^3)
$$

We know that the boundary curves are given by condition $b = 0$ and $a > 0$ or $d = 0$ and $a > 0$. So, note that from the expressions of b and d above, we obtain $\beta_1 = 0$ on the boundary curves. In this way, taking $\beta_1 = 0$, the expressions of a, b and d reduce to

$$
\begin{aligned}
a &= \tfrac{3\beta_2}{2}e^2 + \tfrac{3\beta_3}{2}e^3 + \mathcal{O}(e^4)\\
b &= \left(\tfrac{27\beta_2{}^3}{128} - \tfrac{27\beta_2{}^2}{32} + \tfrac{4185\beta_2}{4096} - \tfrac{729}{2048}\right)e^6 + \left(\tfrac{27(96\beta_2{}^2 - 256\beta_2 + 155)\beta_3}{4096}\right)e^7\\
&\quad \left(\tfrac{-873\beta_2{}^4}{2048} - \tfrac{2637\beta_2{}^3}{1024} + \tfrac{27\beta_2{}^2(3072\beta_4 + 38185)}{131072} + \tfrac{27\beta_2(1536\beta_3{}^2 - 4096\beta_4 - 9495)}{65536} -\right.\\
&\quad \left. -\tfrac{9(786432\beta_3{}^2 - 952320\beta_4 - 18603)}{8388608}\right)e^8 + \mathcal{O}(e^9)\\
d &= \tfrac{9\beta_2{}^2}{4}e^4 + \tfrac{9\beta_2\beta_3}{2}e^5 + \left(\tfrac{3(2912\beta_2{}^3 + 6528\beta_2{}^2 + 3\beta_2(1024\beta_4 - 663) + 1536\beta_3{}^2 + 972)}{2048}\right)e^6 + \mathcal{O}(e^7)
\end{aligned}
\tag{8.116}
$$

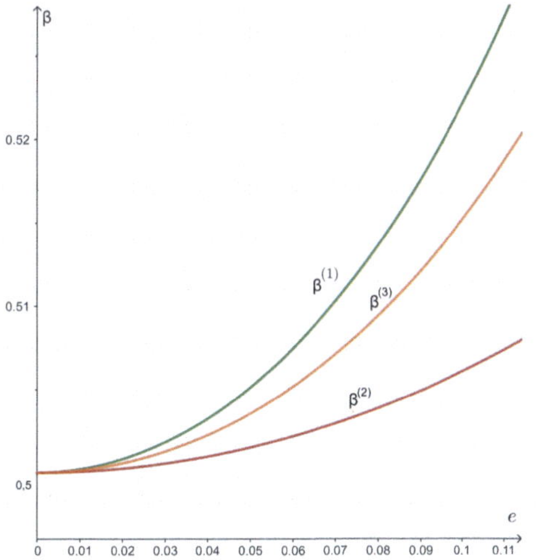

Fig. 8.4 Boundary curves for $\beta_0 = 1/2$. Region of instability is below $\beta^{(2)}$ and between $\beta^{(1)}$ and $\beta^{(3)}$

Considering the condition $b = 0$, we obtain the curves

$$\begin{aligned}
\beta^{(1)} &= \tfrac{1}{2} + 2e^2 + \tfrac{129871}{6144}e^4 \\
\beta^{(2)} &= \tfrac{1}{2} + \tfrac{1}{8}\left(8 - \sqrt{10}\right)e^2 + \frac{\left(79567-23564\sqrt{10}\right)e^4}{12288} \\
\beta^{(3)} &= \tfrac{1}{2} + \tfrac{1}{8}\left(8 + \sqrt{10}\right)e^2 + \frac{\left(79567+23564\sqrt{10}\right)e^4}{12288}.
\end{aligned} \tag{8.117}$$

According to (8.116), the coefficient of the second order in e in the expression of a will be a positive number since $\beta_2 > 0$ and the coefficient of fourth order in e in the expression of d is a positive number. So, for e small, we have $d > 0$ and, in this way, we will not obtain boundary curves from the condition $d = 0$ and $a > 0$. Observe that for ϵ small, $a > 0$ on the curves $\beta^{(j)}$, $j = 1, 2, 3$ and, so the curves in (8.117) are the boundary curves whose graphics are described in Fig. 8.4.

Note that from expression b in (8.116), for e sufficiently small, we have $b < 0$ below of $\beta^{(2)}$ and between $\beta^{(1)}$ and $\beta^{(3)}$. So, the region of instability is shaded in the figure above.

Example 8.4 For the value $\beta_0 = 1$, we obtain resonance $2\omega_1 = 4 \quad 2\omega_2 = 2$.

In this case, H_0 in (8.108) reduces to

$$H_0 = \frac{1}{2}(P_1^2 + P_2^2) + P_1Q_2 + P_2Q_1 + \frac{5}{2}Q_1^2 + Q_2^2$$

and, for this case, we obtain frequencies $\omega_1 = 2$ and $\omega_2 = 1$. So, according to the steps described in Theorem 4.33 we obtain $\delta_1 = 1$, $\delta_2 = 1$ and the linear normalizing transformation is given by

$$Q_1 = -\frac{\sqrt{2}}{2}p_1,\; Q_2 = q_2,\; P_1 = \sqrt{2}q_1 - q_2,\; P_2 = \frac{\sqrt{2}}{2}p_1 + p_2,$$

which carries (8.108), for $\beta_0 = 1$, to the Hamiltonian function

$$F = F_0 + \epsilon F_1 + \epsilon^2 F_2 +, \cdots$$

where

$$F_0(q_1, q_2, p_1, p_2) = \frac{1}{2}\omega_1(q_1^2 + q_1^2) + \frac{1}{2}\omega_2(q_2^2 + p_2^2).$$

And after the double rotation by angle $\delta_1\omega_1 t$ and $\delta_2\omega_2 t$

$$q_1 = x_1 \cos\delta_1\omega_1 t + y_1 \sin\delta_1\omega_1 t \qquad q_2 = x_2 \cos\delta_2\omega_2 t + y_2 \sin\delta_2\omega_2 t$$

$$p_1 = -x_1 \sin\delta_1\omega_1 t + y_1 \cos\delta_1\omega_1 t \quad p_2 = -x_2 \sin\delta_2\omega_2 t + y_2 \cos\delta_2\omega_2 t.$$

the function F_0 is equal to zero. So, the Hamiltonian function (8.108), for $\beta_0 = 1$, after these canonical transformations becomes

$$F(x_1, x_2, y_1, y_2, \nu, e, \mu) = F_0 + eF_1 + \cdots \tag{8.118}$$

where

$$F_0 = 0$$

$$\begin{aligned}
F_1 = {} & \frac{1}{2}\beta_1(y_1\cos 2t - x_1\sin 2t)\left(-x_1\sin 2t - 2\sqrt{2}(y_2\cos t - x_2\sin t) + y_1\cos 2t\right) + \\
& +2\sqrt{2}(x_1\cos 2t + y_1\sin 2t)(x_2\cos t + y_2\sin t)\cos t + \\
& +\sqrt{2}(y_1\cos 2t - x_1\sin 2t)(y_2\cos t - x_2\sin t)\cos t - \\
& -2(x_1\cos 2t + y_1\sin 2t)^2\cos t + \frac{1}{4}(y_1\cos 2t - x_1\sin 2t)^2\cos t - \\
& -(x_2\cos t + y_2\sin t)^2\cos t - (y_2\cos t - x_2\sin t)^2\cos t + \beta_1(x_2\cos t + y_2\sin t)^2.
\end{aligned} \tag{8.119}$$

Since F_0 in (8.119) is equal to zero, the matrices M_1, M_2 and M_3 in (8.48) are null, and, according to the case (**8**), the Hamiltonian function in the Deprit–Hori variables is given by[8]

$$K(X_1, X_2, Y_1, Y_2, \nu, e, \mu) = K_0 + eK_1 + \frac{e^2}{2}K_2 + \cdots + \frac{e^4}{4!}K_4, \quad (8.120)$$

where

$$K_0 = 0$$

$$K_1 = \frac{\beta_1 X_1{}^2}{4} + \frac{3X_1X_2}{2\sqrt{2}} + \frac{\beta_1 X_2{}^2}{2} + \frac{\beta_1 Y_1{}^2}{4} + \frac{3Y_1Y_2}{2\sqrt{2}} + \frac{\beta_1 Y_2^2}{2}$$

$$K_2 = X_1{}^2\left(\frac{55\beta_1{}^2}{48} + \frac{7\beta_1}{5} + \frac{\beta_2}{2} + \frac{57}{160}\right) + \frac{80\beta_1{}^2 + 165\beta_1 - 72}{40\sqrt{2}}X_1X_2 +$$

$$+\left(-\frac{\beta_1{}^2}{6} - \beta_1 + \beta_2 + \frac{21}{32}\right)X_2{}^2 + \frac{(670\beta_1{}^2 + 768\beta_1 + 240\beta_2 - 369}{480}Y_1{}^2 +$$

$$+\frac{3(24 - 5\beta_1)}{40\sqrt{2}}Y_1Y_2 + \left(-\frac{13\beta_1{}^2}{6} - 2\beta_1 + \beta_2 + \frac{9}{32}\right)Y_2{}^2 \quad (8.121)$$

So, the Hamiltonian function in (8.120) assumes the form

$$K = k_{2000}X_1^2 + k_{1100}X_1X_2 + k_{0200}X_2^2 + k_{0020}Y_1^2 + k_{0011}Y_1Y_2 + k_{0002}Y_2^2,$$

where the coefficients $k_{\nu_1\nu_2\iota_1\iota_2}$ are functions of $e, \beta_0, \beta_1, \cdots$ In this case, the polynomial characteristic has coefficients

$$a = 4k_{0002}k_{0200} + 2k_{0011}k_{1100} + 4k_{0020}k_{2000}$$

$$b = 16k_{0002}k_{0020}k_{0200}k_{2000} - 4k_{0002}k_{0020}k_{1100}^2 + k_{0011}^2\left(k_{1100}^2 - 4k_{0200}k_{2000}\right)$$

and $d = a^2 - 4b$.

The expressions of a, b and d represented in power series of e are

[8] We compute the Hamiltonian $\mathcal{K}$ to the fourth order, as well as all the other autonomous Hamiltonians $\mathcal{K}$ shown in this section. We chose to display the expression of the $\mathcal{K}_j$ up to order two because as the order j increases, their expressions become bigger.

$$
\begin{aligned}
a &= \tfrac{1}{4}(5\beta_1{}^2+9)e^2+\tfrac{1}{160}\beta_1(-170\beta_1{}^2+400\beta_2+567)e^3+\cdots\\
b &= \tfrac{1}{64}(9-4\beta_1{}^2)^2e^4+\tfrac{\beta_1(4\beta_1{}^2-9)(220\beta_1{}^2+320\beta_2-441)}{1280}e^5+\cdots\\
d &= \tfrac{9}{16}\beta_1{}^2(\beta_1{}^2+18)e^4+\tfrac{1}{320}\beta_1(-1730\beta_1{}^4+720(\beta_1{}^2+9)\beta_2\\
&\quad +5049\beta_1{}^2+1134)e^5+\cdots
\end{aligned}
\tag{8.122}
$$

Observe that $a > 0$, for e small. So, the boundary of the regions of stability and instability are defined by $b = 0$ or $d = 0$.

Let us consider the condition $b = 0$. Setting equal to zero the coefficients of b in (8.122), we see that the coefficient of the fourth order in e becomes zero if $\beta_1 = \pm\frac{3}{2}$ and, so the coefficients of order five, six, and seven in e of b reduce to

$$
\begin{aligned}
b_5 &= 0\\
b_6 &= \tfrac{9(160\beta_2-3)(160\beta_2+57)}{102400}\\
b_7 &= \tfrac{\pm3\beta_2{}^3}{2}\pm\tfrac{3213\beta_2{}^2}{320}+\beta_2\left(\tfrac{9\beta_3}{2}\mp\tfrac{88209}{25600}\right)+\tfrac{243\beta_3}{320}\mp\tfrac{3420657}{4096000}
\end{aligned}
\tag{8.123}
$$

and, in this way, we obtain four solutions for $b = 0$

$$
\beta^{(1)}(e) = 1-\frac{3}{2}e-\frac{57}{160}e^2-\frac{48511}{25600}e^3-\frac{1876167}{128000}e^4,\ \beta^{(3)}(e)=\beta^{(1)}(-e),
$$

$$
\beta^{(2)}(e) = 1-\frac{3}{2}e+\frac{3}{160}e^2-\frac{27191}{25600}e^3-\frac{1291167}{128000}e^4,\ \beta^{(4)}(e)=\beta^{(2)}(-e).
$$

With respect to equality $d = 0$, we have that $\beta_1 = 0$ and, in this way, the coefficients of order six and seven with respect to e in d respectively reduce to

$$
\begin{aligned}
d_6 &= \tfrac{81(40\beta_2-3)(40\beta_2+17)}{12800}\\
d_7 &= \tfrac{81}{160}(40\beta_2+7)\beta_3\\
d_8 &= \tfrac{9\beta_2{}^4}{16}+\tfrac{5049\beta_2{}^3}{320}+\tfrac{761121\beta_2{}^2}{25600}+\beta_2\left(\tfrac{81\beta_4}{4}-\tfrac{4033953}{512000}\right)+\tfrac{81\beta_3{}^2}{8}+\tfrac{567\beta_4}{160}-\tfrac{162415233}{20480000}
\end{aligned}
\tag{8.124}
$$

So, from $d = 0$ we obtain two curves

$$
\beta^{(5)}(e) = 1-\frac{17}{40}e^2-\frac{46033}{576000}e^4,
$$

$$
\beta^{(6)}(e) = 1+\frac{3}{40}e^2+\frac{13191}{8000}e^4.
$$

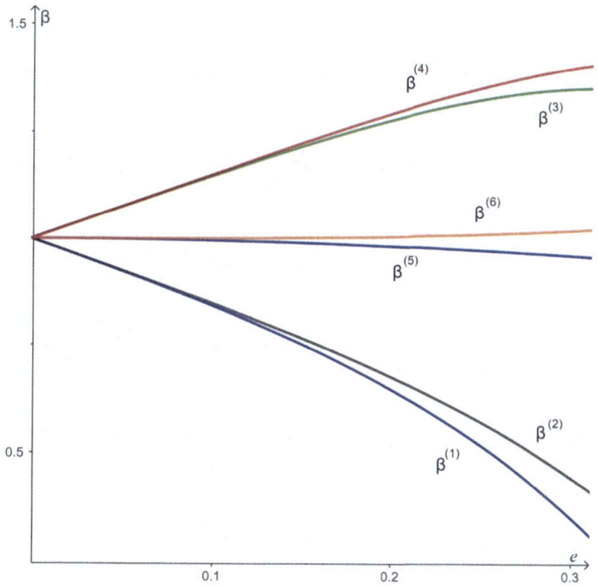

Fig. 8.5 Six boundary curves for $\beta_0 = 1$. Regions of instability are between the curves $\beta^{(1)}$ and $\beta^{(2)}$, $\beta^{(3)}$ and $\beta^{(4)}$ and $\beta^{(5)}$ and $\beta^{(6)}$

The boundaries curves obtained above are shown in Fig. 8.5. Observe that, in this case, the boundaries curves intersect, but the regions of stability and instability remain well defined. For e sufficiently small, from expressions of b and d in (8.123) and (8.124), we obtain the regions of instability described in the Fig. 8.5. The next case, when $\beta_0 = -1$, we obtain one boundary curve.

Example 8.5 For the value $\beta_0 = -1$, we obtain $2\omega_1 \notin \mathbb{Z}$ $\omega_2 = 0$.

In this case, H_0 in (8.108) reduces to

$$H_0 = \frac{1}{2}(P_1^2 + P_2^2) + P_1 Q_2 - 3P_2 Q_1 + \frac{9}{2}{Q_1}^2 - {Q_2}^2.$$

and, for this case, we obtain frequencies $\omega_1 = \sqrt{13}$ and $\omega_2 = 0$. In according to Examples 5.11 and 5.12, the linear normalizing transformation is given

$$Q_1 = \frac{\sqrt{39}}{13} q_2 - \frac{4}{\sqrt[4]{13^3}} p_1 \;\; P_1 = \frac{3}{\sqrt[4]{13}} q_1 + \frac{\sqrt{39}}{39} p_2$$

$$Q_2 = \sqrt[4]{\frac{1}{13}} q_1 - \frac{4}{\sqrt{39}} p_2 \;\; P_2 = \frac{3\sqrt{39}}{13} q_2 + \sqrt[4]{\frac{1}{13^3}} p_1$$

In this way, the Hamiltonian function H is such that the unperturbed Hamiltonian function corresponding is equal to

$$H_0 = \frac{1}{2}\sqrt{13}(q_1^2 + p_1^2) - \frac{1}{2}p_2^2,$$

and H_1 is given by

$$H_1 = \beta_1 \left(\frac{q_1}{\sqrt[4]{13}} - \frac{4p_2}{\sqrt{39}}\right)^2 - \cos(t)\left(\left(\frac{3q_1}{\sqrt[4]{13}} + \frac{p_2}{\sqrt{39}}\right)^2 + \left(3\sqrt{\frac{3}{13}}q_2 + \frac{p_1}{13^{3/4}}\right)^2\right) +$$
$$\frac{-5}{338}\left(\sqrt{39}q_2 - 4\sqrt[4]{13}p_1\right)^2 (2\beta_1 + \cos(t)) +$$
$$\frac{2}{169}\left(\sqrt{39}q_2 - 4\sqrt[4]{13}p_1\right)\left(3\sqrt{39}q_2 + \sqrt[4]{13}p_1\right)(\beta_1 + 2\cos(t))$$

In this case, the rotation is not applied and so, according to the case (**3**), the matrices M_1, M_2 and M_3 are given as in (8.55). The Hamiltonian function up to the fourth order in the Deprit–Hori variables is given by

$$K(X_1, X_2, Y_1, Y_2, \nu, e, \mu) = K_0 + eK_1 + \frac{e^2}{2}K_2 + \cdots + \frac{e^4}{4!}K_4, \tag{8.125}$$

where

$$\begin{aligned}
K_0 &= \tfrac{1}{2}\sqrt{13}(X_1^2 + Y_1^2) - \tfrac{1}{2}Y_2^2\\
K_1 &= -\frac{75\beta_1 X_1{}^2}{26\sqrt{13}} + \frac{3\beta_1 X_2{}^2}{13} - \frac{75\beta_1 Y_1{}^2}{26\sqrt{13}} + \tfrac{1}{39}(16\beta_1 + 9)Y_2{}^2\\
K_2 &= \frac{3X_1{}^2(76772\beta_1{}^2+169(817-3400\beta_2))}{298792\sqrt{13}} + \frac{3X_2{}^2(704\beta_1{}^2+312\beta_1+676\beta_2+507)}{4394} +\\
&\quad + \frac{3Y_1{}^2(76772\beta_1{}^2+169(817-3400\beta_2))}{298792\sqrt{13}} + \frac{Y_2{}^2(-3200\beta_1{}^2-136116\beta_1+43264\beta_2+48633)}{52728}.
\end{aligned} \tag{8.126}$$

In this case, the coefficients of the characteristic polynomial are given by

$$\begin{aligned}
a &= 13 - 12\beta_1 e + (\tfrac{165}{68} + 4\beta_1^2 - 12\beta_2)e^2 + (\beta_1(8\beta_2 - \tfrac{5467}{2312}) - 12\beta_3)e^3 + \cdots\\
b &= -6\beta_1 e + (4\beta_1^2 - 6\beta_2 - \tfrac{9}{2})e^2 + (\beta_1(\tfrac{375}{68} + 8\beta_2) - 6\beta_3)e^3 +\\
&\quad (\tfrac{5663\beta_1{}^2}{2312} + 8\beta_1\beta_3 + 4\beta_2{}^2 + \tfrac{375\beta_2}{68} - 6\beta_4 - \tfrac{1371}{544})e^4 + \cdots
\end{aligned} \tag{8.127}$$

Analyzing the equation $b = 0$, we obtain $\beta_1 = 0, \beta_2 = -3/4, \beta_3 = 0, \beta_4 = -47/64$. So, since $\beta_1 = 0$ and $\beta_2 < 0$ follows that $a > 0$, for e sufficiently small. And, in this way, we obtain the boundary curve up to fourth order (Fig. 8.6)

$$\beta = -1 - \frac{3}{4}e^2 - \frac{47}{64}e^4.$$

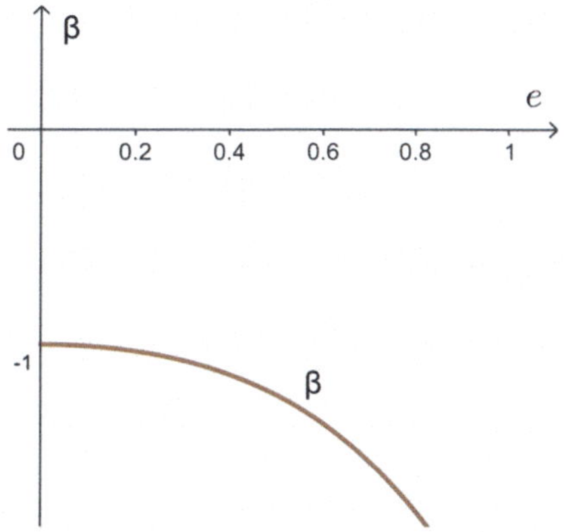

Fig. 8.6 Boundary curve for $\beta_0 = -1$. Region of instability is above the curve β

The condition $a > 0, d = 0$ has no solution because

$$d = 169 - 288\beta_1 e + \left(\frac{2757}{34} + 232\beta_1^2 - 288\beta_2\right) e^2 + \cdots$$

8.8.2 Satellite with Translational Motion

In this subsection, we study the case of the cylindric precession with projection of the angular velocity of the satellite on its symmetry axis equal to zero ($r_0 = 0$), that is, the motion of the satellite is translational.

Setting $\beta = 0$ in (8.106) we obtain the Hamiltonian function

$$H = \frac{1}{2(1 + e\cos\nu)^2}(P_1^2 + P_2^2) + Q_2 P_1 - Q_1 P_2 + \frac{3}{2}(\alpha - 1)(1 + e\cos\nu)Q_1^2. \tag{8.128}$$

We construct the boundary curves of the regions of stability and instability, in the parameter plane (α, e), of cylindrical precession of a dynamical symmetric satellite with the projection of the angular velocity equal to zero (translational motion, $\beta = 0$). In this way, replacing

$$\alpha = \alpha_0 + \alpha_1 e + \alpha_2 e^2 + \alpha_3 e^3 + \cdots$$

in (8.128) and expanding in power series around circular orbit, $e = 0$, we have

$$H = H_0 + eH_1 + e^2 H_2 + \cdots, \tag{8.129}$$

where

$$H_0 = \tfrac{1}{2}(P_1^2 + P_2^2) - P_2 Q_1 + P_1 Q_2 + \tfrac{3}{2} Q_1^2(\alpha_0 - 1)$$

$$H_1 = -(P_1^2 + P_2^2)\cos\nu + \tfrac{3}{2} Q_1^2(\alpha_1 + (\alpha_0 - 1)\cos\nu)$$

According to [40], if the orbit of the center mass of the satellite is circular, $e = 0$, and the value of the dimensionless inertial parameter $\alpha_0 = \frac{C}{A}$ lies outside the interval

$$1 \leq \alpha_0 \leq \frac{4}{3} \tag{8.130}$$

then the considered translational motion of the satellite is unstable. If, however, the value of α_0 lies inside this interval, then we have stability.

The frequencies ω_1, ω_2 corresponding to Hamiltonian function H_0 in (8.129) are the roots of the equation

$$\omega^4 - (3\alpha_0 - 1)\omega^2 + 4 - 3\alpha_0 = 0. \tag{8.131}$$

For $0 < e < 1$, instability is possible if parametric resonance occurs. In this problem, we know that, for α_0 inside of the interval (8.130), parametric resonances exist. Analyzing (8.131) we verify that the relation $k_1\omega_1 + k_2\omega_2 = n$, where $|k_1| + |k_2| = 2$ with k_1, k_2 integers numbers is satisfied for $n = 1$ obtaining $\omega_1 - \omega_2 = 1$ for $\alpha_0 = \frac{2\sqrt{3}}{3}$, $2\omega_2 = 1$ for $\alpha_0 = 23/20$ and $\omega_1 + \omega_2 = 1$ for $\alpha_0 = -\frac{2\sqrt{3}}{3}$. The last relation is discarded since α_0 does not belong to the admissibility interval (8.103). For $n = 3$ and $\alpha_0 = \frac{181}{156}$ the relation $2\omega_1 = 3$ is obtained. Here, we give more information about the case $\alpha_0 = 23/20$.

Example 8.6 For $\alpha_0 = \frac{23}{20}$, we obtain $2\omega_1 \notin \mathbb{Z}$ and $2\omega_2 = 1$.

For $\alpha_0 = \frac{23}{20}$, the Hamiltonian function H_0 in (8.129) reduces to

$$H_0 = \frac{1}{2}(P_1^2 + P_2^2) - P_2 Q_1 + \frac{9}{40} Q_1^2 + P_1 Q_2, \tag{8.132}$$

and the frequencies associated with the linear Hamiltonian system corresponding to H_0 are $\omega_1 = \sqrt{\frac{11}{5}}$ and $\omega_2 = \frac{1}{2}$. According to the steps described in Theorem 4.33, we have $\delta_1 = 1, \delta_2 = -1$ and the linear normalizing transformation is given by

$$Q_1 = -\tfrac{8m}{n\sqrt{39}} p_1 - 5\sqrt{\tfrac{2}{39}} p_2 \quad P_1 = \sqrt{\tfrac{3}{13}\tfrac{n}{m}} q_1 + \sqrt{\tfrac{3}{26}} q_2$$

$$Q_2 = \tfrac{m^3 n}{\sqrt{39}} q_1 - 4\sqrt{\tfrac{2}{39}} q_2 \quad P_2 = \sqrt{\tfrac{3}{13}\tfrac{m}{n}} p_1 - \sqrt{\tfrac{6}{13}} p_2,$$

where $m = \sqrt[4]{5}$ and $n = \sqrt[4]{11}$. So, in these new variables, the Hamiltonian function in (8.129) is such that unperturbed Hamiltonian (8.132) becomes

$$H_0 = \frac{1}{2}\omega_1(q_1^2 + p_1^2) - \frac{1}{2}\omega_2(q_2^2 + p_2^2).$$

After the rotation described in (8.30) we obtain the Hamiltonian function H_0 equal to (8.31) and, following the steps described in the Sect. 8.6.4, case $(\mathbf{1})_R$, we obtain the autonomous Hamiltonian up to 4 degree in e given by

$$K = K_0 + eK_1 + \frac{e^2}{2!}K_2 + \frac{e^3}{3!}K_3 + \frac{e^4}{4!}K_4, \tag{8.133}$$

where

$$K_0 = \tfrac{1}{2}\sqrt{\tfrac{11}{5}}\left({X_1}^2 + {Y_1}^2\right)$$

$$K_1 = \tfrac{16}{13}\sqrt{\tfrac{5}{11}}\alpha_1 {X_1}^2 + \tfrac{1}{208}(200\alpha_1 + 3){X_2}^2 + \tfrac{16}{13}\sqrt{\tfrac{5}{11}}\alpha_1 {Y_1}^2 + \tfrac{1}{208}(200\alpha_1 - 3){Y_2}^2$$

The coefficients a, b and the quantity $d = a^2 - 4b$ up to third order are given by

$$a = \frac{11}{5} + \frac{64\alpha_1}{13}e + \left(-\frac{7625{\alpha_1}^2}{6591} + \frac{64\alpha_2}{13} + \frac{17431863}{703040}\right)e^2$$

$$+ \left(\frac{2693840000{\alpha_1}^3 - 494832000\alpha_2\alpha_1 + 1143983161989\alpha_1}{213864768} + \frac{64\alpha_3}{13}\right)e^3,$$

$$b = \frac{11(40000{\alpha_1}^2 - 9)}{54080}e^2 + \frac{\alpha_1(46160000{\alpha_1}^2 + 89232000\alpha_2 + 388098621)}{5483712}e^3,$$

$$d = \frac{121}{25} + \frac{1408\alpha_1}{65}e + \left(-\frac{88306{\alpha_1}^2}{6591} + \frac{1408\alpha_2}{65} + \frac{191763363}{1757600}\right)e^2 +$$

$$\left(\frac{5539600000{\alpha_1}^3 - 14326765440\alpha_2\alpha_1 + 12562986109833\alpha_1}{534661920} + \frac{1408\alpha_3}{65}\right)e^3.$$

In this way, the boundary condition $a > 0, d = 0$ is not possible and from boundary condition $a > 0, b = 0$ we have the following curves (see Fig. 8.7)

$$\alpha^{(1)} = \tfrac{23}{20} - \tfrac{3}{200}e - \tfrac{69591}{16000}e^2 + \tfrac{9088323}{1280000}e^3 + \tfrac{190618300131}{102400000}e^4$$

$$\alpha^{(2)} = \tfrac{23}{20} + \tfrac{3}{200}e - \tfrac{69591}{16000}e^2 - \tfrac{9088323}{1280000}e^3 + \tfrac{190618300131}{102400000}e^4$$

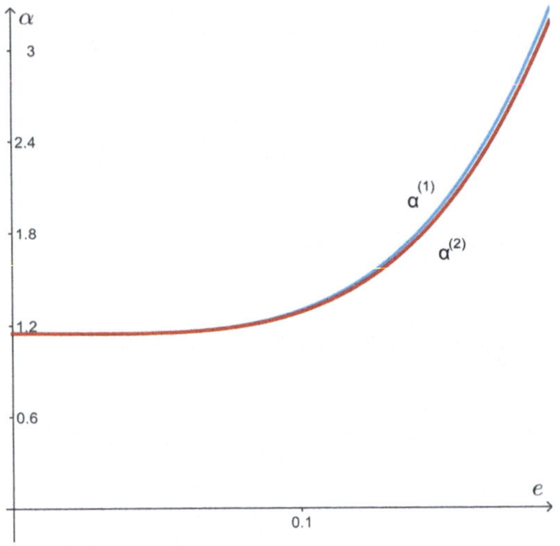

Fig. 8.7 Region of instability is between the curves $\alpha^{(1)}$ and $\alpha^{(2)}$

8.8.3 *Other Cases on the Stability of Cylindrical Precession*

In this subsection, we analyse two more cases. Fixing $\alpha = \frac{3}{4}$ we analyze boundary curves when $\beta_0 = \frac{8}{3}$ and $\beta_0 = \frac{4}{3}$.

Taking $\alpha = \frac{3}{4}$ and replacing β by serie (8.107) in the Hamiltonian (8.106), we obtain

$$H = H_0 + eH_1 + e^2 H_2 + \cdots, \tag{8.134}$$

where

$$\begin{aligned} H_0 &= \tfrac{1}{2}({P_1}^2 + {P_2}^2) + P_1 Q_2 + (\tfrac{3\beta_0}{4} - 1) P_2 Q_1 + \tfrac{3}{32}(3\beta_0^2 - 4\beta_0 - 4){Q_1}^2 + \tfrac{3\beta_0 {Q_2}^2}{8} \\ H_1 &= -({P_1}^2 + {P_2}^2)\cos t + \tfrac{3}{16}{Q_1}^2(3\beta_0\beta_1 - 2\beta_1 - 3\beta_0^2 \cos t - 2\cos t) + \\ &\quad + \tfrac{3}{4} P_2 Q_1 (\beta_1 - 2\beta_0 \cos t) + \tfrac{3\beta_1 {Q_2}^2}{8} \end{aligned} \tag{8.135}$$

Example 8.7 Parametric resonance value $\beta_0 = 8/3$.

For $\beta_0 = \frac{8}{3}$, the Hamiltonian function H_0 in (8.135) will be

$$H_0 = \frac{1}{2}({P_1}^2 + {P_2}^2) + P_1 Q_2 + P_2 Q_1 + \frac{5{Q_1}^2}{8} + {Q_2}^2$$

and the frequencies corresponding to the unperturbed system are

$$\omega_1 = 1, \quad \omega_2 = \frac{1}{2}.$$

with signs $\delta_1 = 1$ and $\delta_2 = 1$. Applying the double rotation (8.30) and following the way corresponding to the case $(\mathbf{8})_R$ described in the Sect. 8.6.4, we obtain the autonomous Hamiltonian K, up to order four in e,

$$K = K_0 + eK_1 + \frac{e^2}{2!}K_2 + \frac{e^3}{3!}K_3 + \frac{e^4}{4!}K_4,$$

where

$$K_0 = 0$$

$$K_1 = \tfrac{3\beta_1 X_1{}^2}{16} + \tfrac{3}{16}(2\beta_1 + 3)X_2{}^2 + \tfrac{3\beta_1 Y_1{}^2}{16} + \tfrac{3}{16}(2\beta_1 - 3)Y_2{}^2$$

$$\begin{aligned} K_2 = {} & \tfrac{3(15\beta_1{}^2+22\beta_1+12)X_1X_2}{20\sqrt{2}} + \tfrac{3(13\beta_1-48)Y_1Y_2}{20\sqrt{2}} + \tfrac{3}{640}X_1{}^2\left(95\beta_1{}^2 + 80\beta_2 - 64\right) \\ & + \tfrac{3}{640}Y_1{}^2\left(35\beta_1{}^2 + 80\beta_2 - 704\right) + \tfrac{3}{640}X_2{}^2\left(-400\beta_1{}^2 - 700\beta_1 + 160\beta_2 + 151\right) \\ & + \tfrac{3}{640}Y_2{}^2\left(80\beta_1{}^2 - 20\beta_1 + 160\beta_2 + 151\right) \end{aligned}$$

The characteristic equation corresponding to K, $\lambda^4 + a\lambda^2 + b = 0$ is such that

$$\begin{aligned} a &= \tfrac{9}{64}(5\beta_1{}^2 - 9)e^2 - \tfrac{9\beta_1(255\beta_1{}^2-400\beta_2-938)}{2560}e^3 + \cdots \\ b &= \tfrac{81\beta_1{}^2(4\beta_1{}^2-9)}{4096}e^4 - \tfrac{81\beta_1(60\beta_1{}^4+80(9-8\beta_1{}^2)\beta_2+799\beta_1{}^2-3456)}{163840}e^5 + \cdots \end{aligned} \tag{8.136}$$

And the quantity $d = a^2 - 4b$ given up to the fifth order is

$$d = \frac{729(\beta_1{}^2 - 3)^2}{4096}e^4 - \frac{243\beta_1(\beta_1{}^2 - 3)(385\beta_1{}^2 - 240\beta_2 - 1706)}{81920}e^5.$$

From equality $b = 0$ we obtain

$$\begin{aligned} \beta^{(1)} &= \tfrac{8}{3} - \tfrac{3}{2}e - \tfrac{301}{160}e^2 + \tfrac{117047}{12800}e^3 - \tfrac{17349077}{1433600}e^4 \\ \beta^{(2)} &= \tfrac{8}{3} + \tfrac{4}{5}e^2 - \tfrac{5443}{2625}e^4 \\ \beta^{(3)} &= \tfrac{8}{3} + \tfrac{44}{5}e^2 - \tfrac{506783}{2625}e^4 \\ \beta^{(4)} &= \beta^{(1)}(-e) \end{aligned}$$

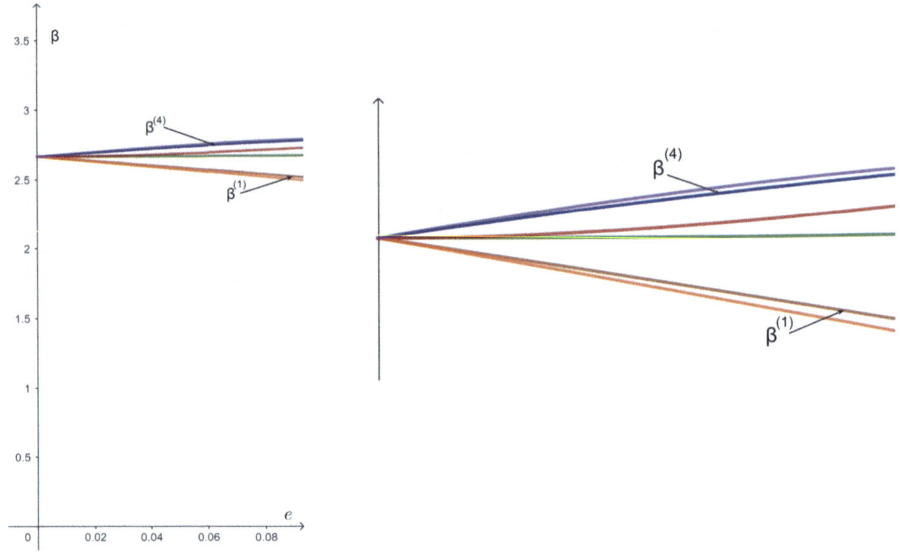

Fig. 8.8 Region of instability is between $\beta^{(1)}$ and $\beta^{(4)}$

And from $d = 0$ we obtain

$$\beta^{(5)} = \tfrac{8}{3} - \sqrt{3}e - \tfrac{551}{240}e^2 + \frac{3256649}{115200\sqrt{3}}e^3 - \frac{4535585381}{96768000}e^4$$

$$\beta^{(6)} = \tfrac{8}{3} + \sqrt{3}e - \tfrac{551}{240}e^2 - \frac{3256649}{115200\sqrt{3}}e^3 - \frac{4535585381}{96768000}e^4$$

The boundary curves are obtained of the conditions $b = 0, a > 0$ and $d = 0, a > 0$, so, from the expressions of a, b in (8.136), we have that the curves $\beta^{(2)}$ and $\beta^{(3)}$ do not satisfy the condition $a > 0$, for e sufficiently small. The regions of stability, for e sufficiently small, are above the curve $\beta^{(4)}$ and below the curve $\beta^{(1)}$ ($b > 0$ and $a > 0$ in these regions, for e small). So, since the curves $\beta^{(5)}$ and $\beta^{(6)}$, obtained from $d = 0$, belong to this stability region, it follows that the region of instability is between the curves $\beta^{(1)}$ and $\beta^{(4)}$, as shown in the Fig. 8.8.

Example 8.8 Parametric resonance value $\beta_0 = 4/3$.
For $\beta_0 = \frac{4}{3}$, the Hamiltonian function H_0 in (8.135) will be

$$H_0 = \frac{1}{2}({P_1}^2 + {P_2}^2) - \frac{3}{8}{Q_1}^2 + P_1 Q_2 + \frac{{Q_2}^2}{2}$$

and the frequencies corresponding to the unperturbed system are

$$\omega_1 = \frac{1}{2}, \quad \omega_2 = 0$$

with signs $\delta_1 = 1$ and $\delta_2 = -1$, obtained according to the steps described in Examples 5.11 and 5.12. In this case, the linear normalizing transformation obtained is given by

$$Q_1 = -\sqrt{2}p_1 - \tfrac{4\sqrt{3}}{3}p_2 \quad Q_2 = 2\sqrt{2}q_1 - \sqrt{3}q_2$$

$$P_1 = -\tfrac{3\sqrt{2}}{2}q_1 + \sqrt{3}q_2 \quad P_2 = \sqrt{2}p_1 + \sqrt{3}p_2$$

Applying the rotation by the angle $\delta_1\omega_1 t$ and following the way corresponding to the case $(\mathbf{12})_R$ described in the Sect. 8.6.4, the new Hamiltonian function K, up to the fourth order in e, is given by

$$K = K_0 + eK_1 + \frac{e^2}{2!}K_2 + \frac{e^3}{3!}K_3 + \frac{e^4}{4!}K_4,$$

where

$$K_0 = -\tfrac{1}{2}Y_2^2$$

$$K_1 = \tfrac{3}{16}(6\beta_1 - 5){X_1}^2 + \tfrac{9\beta_1 {X_2}^2}{8} + \tfrac{3}{16}(6\beta_1 + 5){Y_1}^2 - \sqrt{6}(3\beta_1 + 2)Y_1Y_2 + (-\beta_1 - 6){Y_2}^2$$

$$\begin{aligned} K_2 = &-\tfrac{1}{640}15{X_1}^2\left(432{\beta_1}^2 - 1404\beta_1 - 96\beta_2 + 997\right) - \tfrac{480}{640}\sqrt{6}\left(45{\beta_1}^2 - 39\beta_1 - 10\right)X_1X_2 \\ &-\tfrac{160}{640}\left(9{X_2}^2\left(12{\beta_1}^2 + 12\beta_1 - \beta_2\right) + {Y_2}^2\left(231{\beta_1}^2 - 588\beta_1 + 8\beta_2 - 35\right)\right) \\ &-\tfrac{15}{640}{Y_1}^2\left(4176{\beta_1}^2 + 5676\beta_1 - 96\beta_2 + 2021\right) \\ &+\tfrac{16}{640}\sqrt{6}Y_1Y_2\left(4830{\beta_1}^2 + 955\beta_1 - 240\beta_2 - 1716\right) \end{aligned}$$

and the coefficients of the characteristic equation corresponding to K are

$$\begin{aligned} a \; &= -\tfrac{9}{4}\beta_1 e + \tfrac{9}{64}\left(196{\beta_1}^2 - 16\beta_2 - 25\right)e^2 \\ &+\left(-\tfrac{133488}{256}{\beta_1}^3 + \tfrac{441}{256}\beta_1(32\beta_2 + 83) - \tfrac{576}{256}\beta_3\right)e^3 + \\ &\tfrac{9(111490560{\beta_1}^4 - 240{\beta_1}^2(59328\beta_2 + 178261) + 501760\beta_1\beta_3 + 15680\beta_2(16\beta_2 + 83) - 20480\beta_4 + 394803)}{81920}e^4 + \cdots \\ b &= -\tfrac{81}{256}\beta_1(36{\beta_1}^2 - 25)e^3 + \tfrac{81(4896{\beta_1}^4 - {\beta_1}^2(432\beta_2 + 3019) + 100\beta_2)}{1024}e^4 + \cdots, \end{aligned} \tag{8.137}$$

The expression $d = a^2 - 4b$ is given by

$$\begin{aligned} d = &\tfrac{81}{16}\beta_1^2 e^2 - \tfrac{81}{128}\beta_1(25 + 124\beta_1^2 - 16\beta_2)e^3 + \\ &+\tfrac{81}{4096}(78736\beta_1^4 + 16\beta_1^2(373 - 744\beta_2) + (25 - 16\beta_2)^2 + 512\beta_1\beta_3)e^4 + \cdots \end{aligned} \tag{8.138}$$

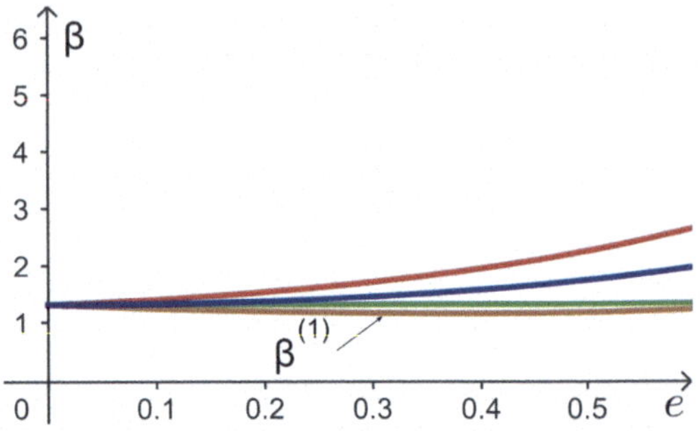

Fig. 8.9 Solution curves of the equations $b = 0$, $d = 0$. Region of instability is above the curve $\beta^{(1)}$

Of the equations $b = 0$ and $d = 0$ we obtain, respectively,

$$\beta^{(1)} = \tfrac{4}{3} - \tfrac{5}{6}e + \tfrac{127}{96}e^2 - \tfrac{4597}{4608}e^3 + \tfrac{2272399}{1843200}e^4$$

$$\beta^{(2)} = \tfrac{4}{3} + \tfrac{9}{40}e^4$$

$$\beta^{(3)} = \beta^{(1)}(-e)$$

and

$$\beta^{(4)} = \frac{4}{3} + \frac{25}{16}e^2 + \frac{16413}{20480}e^4$$

We know that the boundary curves are obtained of the conditions $b = 0$ and $a > 0$ or $d = 0$ and $a > 0$. In this case, analyzing the signs of a, b and d from the expressions in (8.137) and (8.138) it is possible to conclude that $\beta^{(2)}$, $\beta^{(3)}$ and $\beta^{(4)}$, for e sufficiently small, do not satisfy these conditions. The region of instability is above the curve $\beta^{(1)}$, as shown in the Fig. 8.9.

Other examples of construction of stability curves in the parameter plane can be seen in [4, 5, 9, 13, 14, 58].

Reference

1. Albouy, A.: The symmetric central configurations of four equal masses. Contemp. Math. **198**, 131–135 (1996)
2. Alexanderian, A.: On Continuous Dependence of Roots of Polynomials on Coefficients. North Carolina State University, USA (2013)
3. Anjos, A.: EU, 29ª Edição. Livraria São José, Rio de Janeiro (1963)
4. Araújo, G.C., Cabral, H.E.: Parametric stability in a $P + 2$-body problem. J. Dynam. Differ. Equ. **30**, 719–742 (2018)
5. Araújo, G.C., Cabral, H.E.: Parametric stability of a charged pendulum with an oscillating suspension point. Regul. Chaotic Dyn. **26**(1), 39–60 (2021)
6. Arnold, V.I.: Mathematical Methods of Classical Mechanics. Springer, New York (1989)
7. Barrow-Green, J.: Poincaré and the Three-Body Problem. AMS Publication, Providence, RI (1997)
8. Birkhoff, G.D.: Dynamical Systems. AMS Publication, Providence, RI (1927)
9. Brandão Dias, L., Cabral, H.E.: Parametric stability in a Sitnikov-like restricted P-body problem. J. Dyn. Differ. Equ. **30**, 81–82 (2018)
10. Cabral, H.E.: Constant inclination solutions in the three-body problem. J. Differ. Equ. **84**, 215–227 (1990)
11. Cabral, H.E.: Normal Forms of Hamiltonian Systems and Stability of Equilibria. In: Classical and Celestial Mechanics, The Recife Lectures. Princeton University Press, Princeton (2002)
12. Cabral, H.E., Meyer, K.R.: Stability of equilibria and fixed points of conservative systems. Nonlinearity **12**, 1351–1362 (1999)
13. Cabral, H.E., Carvalho, A.C.: Parametric stability of a charged pendulum with oscillating suspension point. J. Differ. Equ. **284**, 23–38 (2021)
14. Cabral, H.E., Menezes Neto, J.L.: Parametric stability of a pendulum with variable length in an elliptic orbit. Regul. Chaotic Dyn. **25**(4), 323–329 (2020)
15. Cartan, É.: Léçons sur les invariants integraux, Paris, 1971 (First edition, 1922)
16. Chazy, J.: Sur l'allure du mouvement dans le problème des trois corps. Annales Scientifiques de l'École Normale Superieure **39**(Series 3), 29–130 (1922)
17. Chenciner, A., Montgomery, R.: A remarkable periodic solution of the three body problem in the case of equal masses. Ann. Math. **152**, 881–901 (2000)
18. Chevalley, C.: Theory of Lie Groups. Princeton University Press, Princeton, NJ (1946)
19. Cherry, T.M.: On periodic solutions of Hamiltonian systems of differential equations. Phil. Trans. Roy. Soc. **A 227**, 137–221 (1928)

H. E. Cabral, L. Brandão Dias, *Normal Forms and Stability of Hamiltonian Systems*, Applied Mathematical Sciences 218,
https://doi.org/10.1007/978-3-031-33046-9

20. Deprit, A.: Canonical transformations depending on a small parameter. Celest. Mech. **1**, 12–30 (1969)
21. Deprit, A., Deprit-Bartholomé, A.: Stability of the triangular Lagrangian points. Astron. J. **72**, 173–179 (1967)
22. Diacu, F., Fugiwara, T., Pérez-Chavela, E., Santoprete, M.: Saari's homographic conjecture of the three-body problem. Trans. Am. Math. Soc. **360**, 6447–6473 (2008)
23. Donsub, R.: An elementary proof that symplectic matrices have determinant one. Adv. Dyn. Syst. Appl. **12**(1), 15–20 (2017). arXiv:1505.04240v4
24. Gelfand, I.M., Lidskii, V.B.: On the structure of the regions of stability of linear canonical systems of differential equations with periodic coefficients. Uspekhi Mat. Nauk **10**, 3–40 (1955). In Russian
25. Goffman, C.: Calculus of Several Variables. A Harper International Student Reprint. Harper and Row, New York (1965)
26. Goldstein, H.: Classical Mechanics. Addison-Wesley, Boston (1959)
27. Gustavson, F.: On constructing formal integrals of a Hamiltonian system near an equilibrium point. Astron. J. **71**, 670–686 (1966)
28. Hartman, P.: Ordinary Differential Equations, 2nd edn. Birkhäuser, Basel (1982)
29. Herman, M.: Sur les courbes invariantes par les difféomorphismes de l'anneau. Astérisque, vol. 144. Société Mathématique de France, CNRS (1986)
30. Hirsch, M., Smale, S.: Differential Equations, Dynamical Systems, and Linear Algebra. Academic Press, Cambridge (1974)
31. Hori, I.: Theory of general perturbations with unspecified canonical variables. Publ. Astron. Soc. Jpn. **18**, 287–296 (1966)
32. Kamel, A.: Expansion formulae in canonical transformations depending on a small parameter. Celest. Mech. **1**, 190–199 (1969)
33. Kamel, A.: Perturbation method in the theory of nonlinear oscillations. Celest. Mech. **3**, 90–106 (1970)
34. Krein, M.G.: Generalization of some investigations of A. M. Lyapunov on linear differential equations with periodic coefficients. Dokl. Akad. Nauk **73**, 445–448 (1950)
35. Laub, A.L., Meyer, K.: Canonical forms for symplectic and Hamiltonian matrices. Celest. Mech. **9**, 213–238 (1974)
36. Lie, S.: Theorie der Transformationgruppen, I,§12. Teubner, Leipzig (1888)
37. Loomis, L., Sternberg, S.: Advanced Calculus. Jones and Barrtlett Publishers, Burlington (1990)
38. Marcolongo, R.: Il Problema dei tre corpi da Newton al nostri giorni. Hoepli, Milan (1919)
39. Markeev, A.P.: Libration Points in Celestial Mechanics and Astrodynamics. Moskva, Nauka (1978). In Russian
40. Markeev, A.P.: Linear Hamiltonian Systems and Some Problems of Stability of Motion of Satellites. R&C Dynamics, Moscow-Izhevsk (2009). In Russian
41. Meyer, K., Offin, D.: Introduction to Hamiltonian Dynamical Systems and the N-Body Problem, 3rd edn. Springer (Applied Mathematical Sciences vol. 90), New York (2017)
42. Meyer, K., Schmidt, D.: The stability of the Lagrange triangular point and a theorem of Arnold. J. Differ. Equ. **62**, 222–236 (1986)
43. McLachlan, N.W.: Theory and Application of Mathieu Functions. The Clarendon Press, Oxford (1947)
44. Moser, J.K.: New aspects in the theory of stability of Hamiltonian systems. Commun. Pure Appl. Math. **XI**, 81–114 (1958)
45. Moser, J.K.: On invariant curves of area-preserving mappings of an annulus. Nachr. Akad. Wiss. Göttingen Math. Phys., Kl. 2, 1–20 (1962)
46. Moser, J.K.: Lectures on Hamiltonian Systems, vol. 81. Memoirs of the American Mathematical Society, Providence, RI (1968)
47. Moulton, F.R.: The straight line solutions of the problem of N bodies. Ann. Math. **12**, 1–17 (1910)

48. Newton, I.: Sir Isaac Newton's Mathematical Principles of Natural Philosophy and His System of the World. In: Mote, A. (Translator); Cajori, F. (Translator & Editor). University of California Press, Berkeley, CA (1947)
49. Pérez-Marco, R.: Convergence or generic divergence of the Birkhoff normal form. Ann. Math. **157**, 557–574 (2003)
50. Poincaré, H.: Les Méthodes Nouvelles da la Mécanique Céleste, vols. I-III. Gautthier-Villars, Paris (1892, 1893, 1899)
51. Roberts, G.E.: A continuum of relative equilibria in the five-body problem. Phys. D Nonlinear Phenomena **127**, 141–145 (1999)
52. Rudin, W.: Principles of Mathematical Analysis. McGraw-Hill, New York (1964)
53. Saari, D.: On bounded solutions of the n-body problem. In: Giacaglia, G.E.O. (ed.) Periodic Orbits, Stability and Resonances. Progr. Math. Birkhäuser, Basel (2001)
54. Siegel, C.L.: Über die Existenz einer Normalform analytischer Hamiltonischer Differrential-gleichungen in der Nähe einer Gleichgewichtlösung. Math. Ann. **128**, 144–170 (1954)
55. Siegel, C.L., Moser, J.K.: Lectures on Celestial Mechanics. Springer, New York (1971)
56. Smale, S.: Mathematical problems for the next century. Math. Intell. **20**, 7–15 (1998)
57. Stoker, J.J.: Nonlinear Vibrations in Mechanical and Electrical Systems. Interscience Publishers, New York (1950)
58. Valeriano, R.L.: Parametric stability in Robe's problem. Regul. Chaotic Dyn. **21**(1), 126–135 (2016)
59. Williamson, J.: On the algebraic problem concerning the normal forms of linear dynamical systems. Am. J. Math. **58**, 141–163 (1936)
60. Williamson, J.: On the normal forms of linear canonical transformations in dynamics. Am. J. Math. **59**, 599–617 (1937)
61. Williamson, J.: The exponential representation of canonical matrices. Am. J. Math. **61**, 897–911 (1939)
62. Wintner, A.: The Analytical Foundations of Celestial Mechanics. Princeton University Press, Princeton (1941)
63. Xia, Z.: The existence of noncollision singularities in Newtonian systems. Ann. Math. **135**, 411–468 (1992)
64. Yakubovich, V.A., Starzhinskii, V.M.: Linear Differential Equations with Periodic Coefficients. Wiley, New York (1975)

Index

H. E. Cabral, L. Brandão Dias, *Normal Forms and Stability of Hamiltonian Systems*, Applied Mathematical Sciences 218,
https://doi.org/10.1007/978-3-031-33046-9

SPRINGER NATURE

GPSR Compliance

The European Union's (EU) General Product Safety Regulation (GPSR) is a set of rules that requires consumer products to be safe and our obligations to ensure this.

If you have any concerns about our products, you can contact us on ProductSafety@springernature.com

In case Publisher is established outside the EU, the EU authorized representative is:

Springer Nature Customer Service Center GmbH
Europaplatz 3
69115 Heidelberg, Germany

The manufacturer's authorised representative in the EU is Springer Nature Customer Service Centre GmbH, Europaplatz 3, 69115 Heidelberg, Germany. If you have any concerns regarding our products, please contact ProductSafety@springernature.com

Printed and bound by CPI Group (UK) Ltd, Croydon, CR0 4YY
26/03/2026
02078992-0002